Mottistone

December 1995

THE DEVELOPMENT OF RADAR EQUIPMENTS
FOR THE ROYAL NAVY, 1935–45

photograph by Seaman Photographers, Sheffield

Cecil Everard Horton, CBE, MA

father of British Naval radar

(by kind permission of Desmond Horton, Esq.)

The Development of Radar Equipments for the Royal Navy, 1935–45

Edited by F. A. Kingsley
on behalf of the Naval Radar Trust

© Naval Radar Trust 1995

All rights reserved. No reproduction, copy or transmission of this publication may be made without written permission.

No paragraph of this publication may be reproduced, copied or transmitted save with written permission or in accordance with the provisions of the Copyright, Designs and Patents Act 1988, or under the terms of any licence permitting limited copying issued by the Copyright Licensing Agency, 90 Tottenham Court Road, London W1P 9HE.

Any person who does any unauthorised act in relation to this publication may be liable to criminal prosecution and civil claims for damages.

First published 1995 by
MACMILLAN PRESS LTD
Houndmills, Basingstoke, Hampshire RG21 2XS
and London
Companies and representatives
throughout the world

ISBN 0-333-61210-8

A catalogue record for this book is available from the British Library.

10 9 8 7 6 5 4 3 2 1
04 03 02 01 00 99 98 97 96 95

Printed and bound in Great Britain by
Antony Rowe Ltd
Chippenham, Wiltshire

Contents

List of Illustrations — xiii
List of Tables — xvii
Preface — xix
Tribute – Cecil Horton: Father of British Naval Radar — xxv
Development and Installation of British Naval Radar – Some Significant Milestones — xxxiii
Notes on the Contributors — xxxvii

PART I   RADAR EQUIPMENT DEVELOPMENTS, 1935–45

Editorial Note — 3

1   The Origins and Development of Radar in the Royal Navy, 1935–1945, with Particular Reference to Decimetric Gunnery Equipments — 5
J. F. Coales

Summary — 5
Introduction — 6
The research background, 1915–35 — 6
The formative years, 1935–7 — 11
The 1937 reorganisation of Naval radar research — 16
Preparing for war, 1938–9 — 21
Early developments in 50-cm radar equipment — 23
Wartime developments — 29
The need for a surface detection capability — 31
The small-ship radar requirement, 1940 — 33
The requirement for gunnery and fire-control radar (50-cm) — 34
Development of Naval gunnery and fire-control radar systems — 36
Trials of main armament and high-angle director radars — 43
The birth of Naval 10-cm radar — 44
The application of 10-cm radar in the U-boat war — 46
Parallel developments — 48
Gunnery radar developments and improvements, 1942 onwards — 49

2 Basic Science and Research for Naval Radar, 1935–1945   67
B. W. Lythall

Summary   67
Early history – and two missed opportunities   67
    The invention of radar   68
    The multiple-cavity anode magnetron   68
The environment for research   69
    1938–42   69
    1942–5   70
The patterns of research   71
    Research themes   71
    The need to predict performance   71
    Early methods of height estimation   74
Research on the radar environment   75
    Radio propagation   75
    Target reflection characteristics   79
    Noise   80
    Sea clutter   82
Interference with the environment: anti-jamming research   83
    Electronic jamming   83
    Non-electronic jamming: Window   84
    Radar camouflage and echo enhancement   85
Interaction between theory and experiment   86
    Antenna developments   87
    Transmission lines and waveguides   89

3 Valve Developments for Naval Radar Applications, 1935–45   95
F. M. Foley

Summary   95
Introduction   95
Early history of the valve section, HM Signal School   96
Silica as a valve-envelope material   98
Silica-valve technology   98
    The electrode system   99
    The silica parts   100
    The lead-out seals   101
    Valve assembly   103
    Processing   103
    Valve repair   104
Silica valves for radio transmitters   104
Silica valves for radar   106
    The start   106
    Special developments   107

Contents vii

The end of the era	109
Output power valves	109
Metric-wave valves	109
Decimetre-wave valves	110
Microwave valves	111
Pulse modulators	114
Hard valves	114
Thyratrons	114
Spark gaps	115
Low-power RF valves	116
Duplexer valves	121
Metric valves	121
Microwave valves	121
Mixers	122
General-purpose valves	123
Cathode-ray valves	123
Appendix 1 HM Signal School staff working on valve development up to 1945	128
Appendix 2 Silica valve types	129

4 Royal Navy Metric Warning Radar, 1935–45   133
J. S. Shayler

Summary	133
Introduction	134
Types 79X and 79Y (the first successful metric wavelength radar for shipborne installation)	135
Type 79Z (the improved equipment)	140
Type 279 (introduction of ranging panel, becoming Type 279)	144
Types 79B/279B (development of common transmit/receive switch, for single antenna operation)	147
Type 281 (development of a 90MHz equipment for increased ranges on aircraft, plus a surface detection capability)	151
Type 281B (development of a T/R switch for Type 281)	161
Type 281BQ (start of experimental work on a continuously rotating antenna system, for use with a PPI display)	162
Dual installations of Types 79B and 281BQ (development of dual installation for gapless cover on aircraft targets)	165
Type 286 (adaption of RAF ASV set as small-ship radar)	169
Type 290 (attempt to develop an improved Type 286)	172
Type 291 (successful development of replacement for Type 286)	173
Type 960 (initiation of development of a replacement for the combined Type 79 and 281 equipments)	175

viii Contents

Postscript		177
Appendix 1	Chronology of development of metric radars	181
Appendix 2	Summary of main characteristics of metric radars	182

5 Development of Naval Warning and Tactical Radar Operating in the 10-cm Band, 1940–45   185
C. A. Cochrane

Summary	185
Preface	185
Introduction	188
Origins of the first Naval cm-radar	189
The first operational centimetric radar	195
Technical constraints and characteristics of Type 271X	197
In the wake of HMS *Orchis*	203
Coast and harbour defence applications	206
Types 271, 272, 273 in service, 1941-2	208
Anti-submarine range performance of 5 KW centimetric radar	211
The high-power magnetrons	215
The Mark 4 development: Types 271Q and 273Q	217
The Mark 5 development	224
The Mark 5 experimental shore trials – Type 277T	226
The Mark 5 experimental shipborne radar – Type 277X	228
Sea reflection and target indication – Types 276 and 293	231
WS tactical radar – low air warning – heightfinding – Type 277	239
Further development of Type 277 – Types 277P and 277Q	245
Centimetric fighter direction (FD) radar – initial proposals	246
Revised proposals for FD radar – Types 980 and 981	249
The unknown factor in Type 980 (982) performance	255
Conclusions	258
Appendix 1 The complimentary role of X-band for WS radar	259
Appendix 2 Chronology of developments of Naval radar for air and surface warning	267
Appendix 3 Characteristic data for Naval S-band warning radar	271

6 The Royal Navy and IFF – Identification Friend or Foe, 1935–45   277
J. S. Shayler

Summary	277
Introduction	277
The Beginning	278

IFF Mark I	278
IFF Mark II	279
IFF Mark III	281
IFF Mark IV	289
IFF Mark V	289
Postscript	290

PART II RADAR SHIP-FITTING AND MAINTENANCE, 1939–45

Editorial Note   293

7 Radar Ship-Fitting and Maintenance in the Royal Navy, 1939–45: Experiences at Scapa Flow, May 1940 to April 1942   295
B. G. H. Rowley

Summary	295
Appointment to HM Signal School	295
Appointment to the staff of C-in-C, Home Fleet	296
Scapa Flow	297
Expanding activities	299
Summer 1940	299
Autumn 1940	302
Early 1941	302
Summer 1941	303
Early 1942	304

8 Radar Maintenance at Sea: A Personal Story, 1940–5   305
R. A. Laws

Summary	305
Introduction	305
Author's radar experience	306
Personnel	306
Fitting out and commissioning	307
Acceptance of a new ship	310
At sea	311
Handbooks and test gear	312
Spares	313
Calibration and setting up	314
Metric sets	314
Range scales	315
Index errors	316
Angular alignment	316
Anti-jamming	317

Receivers	317
Transponders	317
Preventive maintenance	317
State of readiness	317
Logs	318
Open-wire feeders	318
Tuning the transmitter	318
Common T/R	319
Malfunctions	319
Types 282–5	319
Antenna control	320
Breakdowns	320
Modifications	322
Ancillary services	322
Cooling	322
Power supplies	323
Data transmission	323
Tribute to the Service	323
Appendix 1 Main radar sets encountered by the author	324
Appendix 2 Author's collection of useful texts	324

9 Naval Radar, Fitting Policy, Matériel Procurement, Installation, Sea-Trials and Shore-Based Maintenance   325
A.M. Patrick

Summary	325
Introduction	325
Formulation of Admiralty radar-fitting policy and programmes	329
Procurement of radar equipment	330
From Naval Stores to dockyards/shipyards	331
The ship fitting-out task, 1939–45: an overview	334
Pre-fitting work on-board ship	335
Electrical work	337
Radar installation, testing and trials	338
Equipment trials	339
Shore maintenance	340
Replenishment	341
The Port Radar Officer Organisation	342
General	342
Sherbrooke House, Glasgow	343
The radar officer organisation, 1939–45	344
Conclusion	347
Postscript	348

Appendix 1	Type Numbers of Naval Radar Sets, Operational or Designed, 1935–45	353
Appendix 2	Tabulations of Radar System Data, 1935–45	359
Appendix 3	Ship-Fitting Tables, 1938–45, Arranged by Class of Ship	387
Appendix 4	Table of Manufacturers Employed on Naval Radar Developments	401
Glossary		405
Select Bibliography		415
Index		421

List of Illustrations

1.1	Organisation of Experimental Department, HM Signal School, 1919	8
1.2	Organisation of the Experimental Department of Signal School in September 1937	18
1.3	The November 1937 reorganisation of research (R Department)	20
1.4	Close-range director with experimental 600 MHz Yagi antennae at the AA range, Eastney	35
1.5	(a) Horizontal polar diagram of the Yagi antenna installation; (b) vertical polar diagram of the Yagi antenna installation	35
1.6	Experimental 24-dipole tapered array for Type 284 undergoing field trials	39
1.7	(a) Horizontal polar diagram of 24-dipole 'pig-trough' array; (b) vertical polar diagram of 24-dipole 'pig-trough' array	40
1.8	Office equipment of Types 282, 284 and 285 (600 MHz); modulator and receiver on the right, transmitter in resonant cylinder centre, display with rangefinder on the left	40
1.9	Diagram of 24-dipole array, showing how power is partitioned	42
1.10	Type 284 Yagi array on the LA director in *Suffolk*, with Type 285 Yagi array on the HA director, further aft	42
1.11	Organisation of Radar department under C. E. Horton in October, 1942	50
1.12	Diagram of 6-Yagi antenna array with rotary switch	52
1.13	(a) Horizontal polar diagram of switched 24-dipole array; (b) horizontal polar diagram of switched 6-Yagi array	52
1.14	Vertical polar diagrams of antennae of Type 279 (43 MHz) and Type 281 (90 MHz), with that of Type 273 (3000 MHz) added	55
1.15	Organisation of Radar Department under Dr S. E. A. Landale in October 1943	58
1.16	Organisation of the Communications Department, ASE in October, 1944	63
3.1	Corrugated envelope for NT22B	100
3.2	Typical bulb with end-cap fitted	101
3.3	Lead seal	102
3.4	Graded-glass seal	103
3.5	H. G. Hughes and T. E. Goldup sealing-off a silica valve, 1922	105

List of Illustrations

3.6	The NT57T silica valve	108
3.7	A 'Micropup' triode, NT99	111
3.8	An early magnetron, NT98	112
3.9	Magnetron cavity strapping	113
3.10	High-power thyratron modulator, CV22	115
3.11	Outline of 'acorn' valve	117
3.12	600 MHz amplifier triode, CV53	118
3.13	Local-oscillator klystron, NR89 (CV35)	119
3.14	Cross-section of local-oscillator klystron, CV67	120
3.15	Duplexer arrangement	122
3.16	The microwave crystal mixer, CV101	122
3.17	Receiving valve outlines	124
3.18	Electrostatic focus and deflection CRT	125
3.19	12-in electromagnetic focus and deflection CRT	126
3.20	The Skiatron tube for optical projection, NC17	127
4.1	Type 79/279 antenna array, with Type 243 IFF antenna mounted above	137
4.2	Rear of Type 79 equipment in the Transmitter Office	143
4.3	Type 79 equipment in the Receiver Office	144
4.4	Schematic of Type 279B T/R switch	149
4.5	Vertical lobe structure, Type 79/279 and Type 281/281B	150
4.6	Type 281 equipment in the Transmitter Office	155
4.7	Type 281 transmitter array	156
4.8	Type 281 receiver array	157
4.9	Effect of switching on Type 281 receiver array polar diagram	158
4.10	Equipment in a Type 281 Receiver Office	160
4.11	Type 281BQ RF sliprings	164
4.12	Equipment in Type 281BQ Receiver Office	166
4.13	Early Radar Display Room	168
4.14	214 MHz array for later Type 286s and Types 290 and 291	171
5.1	Rear view of the Type 271 antenna	197
5.2	The 'office'/lantern installation on HMS *Periwinkle* constructed by the Dockyard, shown before the installation of Type 271	198
5.3	Front view of the Type 271 antenna	199
5.4	Diagram of the feed dipole and rod reflector of the Type 271 'cheese' antenna	199
5.5	Schematic diagram of crystal holder and crystal for the mixer unit: (a) original full-wave line form for TRE laboratory crystalr; (b) adaptor to US capsule-type production crystal; (c) quarter-wave inductance line mixe	201
5.6	Type 271X panels in the original vertical format manufactured in the experimental workshops and by Allen West Ltd: (1) high-voltage rectifier; (2) modulator; (3) tetrode	

List of Illustrations

	series-modulator valve; (4) power-supply panel; (5) A-scan display; (6) receiver panel	202
5.7	Rear view of Type 273 antenna	205
5.8	Type 271P panels: the early Type 271 panels redesigned for main production. The modulator with power supply is below; the receiver, display, and the associated power supplies are immediately above; the G82 tuning test set is on top	209
5.9	Effect of the lantern on the Type 271 radiation pattern (a) Teak lantern with plane perspex windows; (b) cylindrical perspex lantern	211
5.10	HMS *Suffolk* with Type 273 raised on stalk for clear view over high-angle director to left and low-angle director to right	212
5.11	Type 271Q panels designed to fit the same space as Type 271P panels, for ease in conversion: (1) discharge-line modulator unit; (2) receiver and display unit	218
5.12	Type 271Q transmitter mounted behind the antenna: (1) magnet; (2) magnetron; (3) flexible bellows coupling of magnetron to output line; (4) Current transformer to monitor magnetron pulse; (5) cooling fan; (6) matching adjustments on output line; (7) voltage step-up transformer for modulator pulse; (8) waveguide to antenna	219
5.13	Type 273Q antenna with waveguide transmitter feed	221
5.14	Type 273Q antenna – rear view showing (1) waveguide feed to horn; (2) transmitter unit; (3) stabiliser to hold beam horizontal; (4) gas switch to protect receiver crystal; (5) crystal mixer and IF head amplifier	221
5.15	Type 277T trailer installation with large 'cheese' antenna in a fixed position on the roof: (1a) horn feed; (1b) 'cheese'-type reflector; (2) interrogater antenna (a) dipole feed, (b) reflector; (3) rotatable cabin having transmitter, receiver and display	227
5.16	The Type 277X antenna – note the sheet-metal paraboloid	229
5.17	HMS *Janus* with the experimental Outfit AUR antenna of Type 293X on top of a lattice mast – note the HF DF pole-mast installation behind	233
5.18	Predicted vertical coverage diagram for Type 293X	234
5.19	Bar-chart presentation of PPI 'signal/no signal' versus range for 14 radical flights at constant aircraft height	235
5.20	Vertical coverage of Type 293X as determined by trials against a Boston aircraft target: (a) height versus range – aircraft opening; (b) height versus range – aircraft closing	236
5.21	Type 276 (Outfit AUJ) antenna identical to the Type 271Q transmitting antenna with aperture sealed to permit use in exposed sites without the need for a radome	237

5.22	Vertical coverage of Type 276 as determined by trials	238
5.23	Type 293M antenna replacing the Outfit AUR used in the Type 293X trials	239
5.24	Type 277 antenna Outfit AUK: (1) wire-mesh parabolodial reflector; (2) waveguide rotating-joint on elevation axis; (3) gyro vertical-stabiliser	240
5.25	Type 277 in HMS *Campania*: height versus range diagram for a single Fulmar aircraft	241
5.26	HMS *Campania*: analysis relating probability of 'paint' to target-acquisition probability	243
5.27	Comparison of vertical coverage diagrams of Types 281 and 277	244
5.28	Types 982, 983 and 960 at the Royal Naval Air Station, Kete	250
5.29	Vertical coverage diagrams of the tilted 'cheese' array for different phase conditions of the feed	251
5.30	Type 980: trials against a Meteor aircraft at 20 000-ft: analysis of 10 flights, opening and closing range	253
5.31	HMS *Eagle*, circa 1951	254
5.32	Air-defence picket *Llandaff*	257
5.33	Type 261 installed in *Saltburn* for trials	261
5.34	The Type 268 radar antenna	265
5.35	The Solent as seen on the PPI display of a 3-cm radar	266
6.1	Type 243 antenna and feeder system	283
6.2	Type 243 antenna mounted above Type 281 array	284
6.3	Type 941 antenna mounted above Type 281BQ antenna	286
7.1	Scapa flow: the Home Fleet anchorage	297
8.1	The modified mainmast of *Anson*	309
9.1	Formulation of Admiralty radar fitting-out programme	327
9.2	Implementation of radar ship-fitting programme	328
9.3	Plan-packing scheme for a hypothetical radar	332
9.4	Relevant sections of plan-packing notes covering complete delivery in three consignments	333
9.5	Sherbrooke House, Glasgow	343

List of Tables

1.1	Characteristics of typical Naval gunnery radar sets	61
4.1	Detection ranges on aircraft and ships in the *Saltburn* trials	175
5.1	Radar detection ranges on U-boats, 1942–5	214
5.2	Type 261 trials, November 1942	262

Preface

This book contains a series of technical monographs dealing with various aspects of British Naval radar from its inception in 1935 until the end of World War 2. It stems from several years of collective historical research by a group of scientists, Naval officers and certain representatives of the electronics industry, all personally involved some forty or fifty years before. It is one of two such volumes, both of which are complementary to Derek Howse's book *Radar at Sea – the Royal Navy in World War 2*, published in 1993, which is addressed more to the general reader. The background research, preparation and publication of both books has been sponsored by the Naval Radar Trust.

Whereas *Radar at Sea* is a carefully researched historical treatise by a single author, this book is a collection of accounts by people who actually worked at HM Signal School (later the Admiralty Signal Establishment) – or were associated closely with it – during the period in question. The subjects are treated in considerably more technical detail than was possible in *Radar at Sea*. With few exceptions they are based on the individual authors' own contemporary experiences, supplemented by extensive archival research and discussions with surviving colleagues in order to safeguard against the fallibility of human memory.

THE NAVAL RADAR TRUST

The idea that sparked off this venture was the brain-child of Professor J. F. Coales, who had been intimately involved with Naval radar both before and throughout World War 2. In June 1985, half a century after the first historic experiments for the Air Ministry, the Institution of Electrical Engineers organised a seminar on 'Fifty Years of Radar', to which Coales, in collaboration with the late J. D. S. Rawlinson, contributed a paper dealing with the early stages of Naval radar in Britain. The realization that so little else had been included about the Navy's contribution, as opposed to the other two Services, led him to put forward the idea of assembling a comprehensive collection of archives on British Naval radar, not only for the historical record, but also in the hope that one day it would lead to a published account.

A start was made by contacting those civilian and Naval officers involved whose whereabouts were known, and by gathering archival material – personal notebooks, recollections, photographs and so on. In

December 1985 a working reunion of more than 40 wartime colleagues was held at Churchill College, Cambridge, at which it was agreed to proceed with Coales' idea. Since all concerned were at least in their sixties, and many in their seventies and eighties, it seemed important to get on with the collection and digestion of data as soon as practicable.

From these beginnings the project steadily gained momentum. An Administrative Committee was elected to manage the enterprise; this was subsequently formed into the Naval Radar Trust, with charitable status and with the following membership:

- Sir Hermann Bondi, KCB, FRS, then Master of Churchill College, Cambridge; formerly Chief Scientific Adviser, Ministry of Defence.
- Professor J. F. Coales, CBE, ScD, FEng, FRS, Emeritus Professor of Engineering, University of Cambridge.
- Basil Lythall, CB, formerly Chief Scientist, Royal Navy, Member of the Admiralty Board, and Deputy Controller of the Navy for Research and Development.
- D. Stewart Watson, CB, OBE, formerly Director of the Admiralty Surface Weapons Establishment; Deputy Chief Scientist, Navy; and Director General of Establishments, Ministry of Defence.

By December 1986 Coales had contacted some 150 wartime colleagues. At a second reunion it was resolved to continue with archival research and to aim towards the preparation of a book, for the general reader, which would tell the story of the early development of British Naval radar and its operational use at sea. In the hope that adequate financial support would eventually be forthcoming Derek Howse was appointed to be the author designate – a major act of faith which eventually proved justified when *Radar at Sea* was published at the beginning of 1993.

This book could only tell the technical story in very general terms, so it was also decided to prepare a series of more definitive technical papers, both as authoritative technical background for the general book and to supplement the growing archival collection. Working groups were set up, each with a convener who volunteered to start the preparation of a monograph on a selected topic, such as an individual family of radars, a specialised set of techniques, or a particular aspect of the use of radar at sea. The next few years saw several more reunions; the majority of the monographs reached completion, each in its turn being added to the archives, and the original list of topics was extended to make the collection more comprehensive.

It has now become possible to publish all these monographs, together with additional reference data. In view of the large amount of material the collection has had to be split into two separate books, each with an integrated bibliography and index. The present volume is concerned with radar development; it gives an overview of work in the Experimental

Department of HM Signal School (later the Admiralty Signal Establishment), and describes each of the main programmes of radar equipment development, the underlying research, and some of the problems of installation, operation and maintenance at sea. A companion volume[1] describes the application of radars in systems – for target indication, weapon direction, command-and-control, and fighter direction. It also includes the story of British Naval radar countermeasures, a technical history of HF DF (which, in conjunction with radar, made a most important contribution to the Battle of the Atlantic) and an essay on parallel developments in German naval radar over the same period.

Although all the monographs were initiated as part of a common venture, each one was originally prepared as an independent contribution dealing with one major topic, and not necessarily depending on other monographs to provide background or to set the general scene. Not surprisingly there were considerable areas of overlap. There were also the expected differences of style, balance and depth of technical detail, and a few apparent inconsistencies. It has been possible to address some of these aspects in editing the present volume, but inevitably examples of overlap must remain.

SOURCES

A primary source of information has been the surviving records of the Experimental Department of HM Signal School, and its successor the Admiralty Signal Establishment. Some of these are held at the Public Record Office; others remain in the Defence Research Establishment, Portsdown, now part of the Defence Research Agency, and are not yet available to the general public. Other important sources of information exist at the Defence Research Establishment, Malvern (in the wartime archives of the Telecommunications Research Establishment); at HMS *Collingwood*; at HMS *Dryad*; and at the Ministry of Defence's Naval Historical Branch in London. A certain amount of material is also to be found, rather widely scattered, in other files at the Public Record Office.

There is one major published source of technical information, concerned with the whole range of Service radar developments during the war. This is the 'Proceedings of the Radiolocation Convention' held by the Institution of Electrical Engineers in London in 1946. A few more papers were subsequently published in the Institution's Journal, and in other scientific and mathematical journals.

Supplementing this is a wealth of collateral information received from private individuals. Many scientists and serving officers attending the reunions have written their own recollections, lent or given personal papers, and provided other information. Tape recordings have been made of recollections of the few people now available who worked on radar well before the war. To this has been added the very extensive collection of historical detail and personal reminiscences assembled by Derek Howse during the preparation of *Radar at Sea*. The whole now forms a most valuable archive, which is to be deposited in the Archive Centre at Churchill College, Cambridge, where it will be cared for professionally, in company with many other Naval papers of World War 2.

Sources are provided in more detail in the Bibliography.

ACKNOWLEDGEMENTS

Our first thanks must go to all the authors of individual monographs, particularly to John Coales, without whom the project would never have been started. The source material is now very diffuse, and a great deal of painstaking work has been necessary for each author to piece together the various elements of the story as accurately as possible. For the appendices Alan Laws prepared the collection of technical data sheets, and Derek Howse provided the guide to the complex ramifications of radar type numbers and the tables of ships fitted with radar – most valuable reference material compiled during the preparation of his own book. Alex Rae prepared indexes for both volumes – as well as compiling a series of staff lists to be deposited in the Churchill Archive Centre. Thanks are also due to other members of the original working groups, and to many other colleagues, for helpful contributions and discussion. Some are acknowledged in specific monographs; it is impossible to mention all the many others who have contributed so enthusiastically in one way or another.

We are grateful to many Defence authorities for allowing access to their archival collections, particularly Janet Dudley, formerly Senior Librarian at the Defence Research Establishment, Malvern; John Briggs, Librarian at the Defence Research Establishment, Portsdown, Lieutenant-Commander Bill Legg of HMS *Collingwood*, and Lieutenant-Commander Peter Lee of HMD *Dryad* for their willing assistance. John Briggs was exceptionally patient and helpful in responding to numerous requests for access to the many old technical reports, memoranda and miscellaneous uncatalogued papers and photographs that remain at Portsdown, as well as providing copies for use as working material. Here a special word of thanks is due to Sid Wright, who gave up a great deal of his time to make numerous journeys to Portsdown, *Collingwood* and *Dryad* on behalf of authors unable to visit there themselves. His diligence in following up many queries, and his own extensive knowledge and experience of wartime radar have been invaluable.

Preface

The majority of the photographs and illustrations in this volume were provided by courtesy of the Defence Research Establishment, Portsdown; the Naval Historical Branch, Ministry of Defence; HMS *Collingwood*; HMS *Dryad* and the Defence Research Establishment, Malvern. Other examples were provided by the Imperial War Museum, Herr Fritz Trenkle, London News Agency Photos, Ltd (which company it has not proved possible to trace), E. B. Callick and Peter Peregrinus, Ltd. Permission to use this material, as identified in individual figure captions, is gratefully acknowledged.

Fred Kingsley has not only contributed two monographs to the second book but has been an exemplary editor for both volumes. Faced with a diverse collection of papers, some already published elsewhere, some in various stages of preparation, and a few not even started, he set about his thankless task with determination. It is mainly owing to his industry and application that the two volumes emerged in such good time after his appointment, and that its component parts were welded, with tact, persuasion and persistence, into a reasonably consistent whole. John Coales, Derek Howse, Basil Lythall, Harry Pout, Jack Shayler and Stewart Watson (in alphabetical order) acted as an informal advisory group, to which Alec Cochrane has actively contributed from overseas; Jack Shayler has been particularly conscientious in reading every monograph and providing constructive comments. Thanks are also due to Miss Carin Dean for patiently and expertly reproducing many of the original drafts to professional standards, and to Mrs Sheila Barker, who undertook several complex processing tasks with complete success.

Finally, the Naval Radar Trust is most grateful to the Ministry of Defence, Mr David Packard, and the Medlock Charitable Trust for major financial help, without which all the research and collection of archival material could not possibly have been carried out, nor the books published. Other valuable contributions were received from BICC plc, GEC-Marconi Ltd, the Royal Society and the Fellowship of Engineering, as well as many generous contributions from individuals, both Naval and civilian, who were involved in the developments during World War 2. Without their financial help and without the support and industry of so many wartime colleagues, who gave freely of their time and energy without reimbursement, this book could never have been completed.

Esher, Surrey   BASIL LYTHALL
1994   On behalf of the Naval Radar Trust

Reference

1. F. A. Kingsley (ed.), *The Applications of Radar and Other Electronic Systems in the Royal Navy in World War 2* (Macmillan, 1994).

Tribute
Cecil Horton: Father of British Naval Radar
Basil Lythall

The collections of articles in the present volume, and its companion,[1] on the early technical history of British Naval radar would not be complete without some recognition of the key part played by Cecil Horton. Almost every British seagoing radar that saw service in World War 2 – and several land-based offshoots – first came into being under his leadership; indeed, apart from a short gap, he was in charge of Naval radar developments from late 1937 until well after the war. It was very much owing to his drive and insistence that the Navy entered the war with some ships actually fitted with radar, and with many more sets on order. And throughout most of the great wartime expansion it was his hand at the helm that enabled such a catholic variety of talents to work so effectively together. In retrospect, his contributions to wartime radar, recognised with the award of CBE in the 1946 Birthday Honours, can be seen as a pinnacle in his long record of distinguished achievement.

Horton joined Signal School in 1921, and for many years led a small section working on radio direction-finding. During the 1920s the increasing use of short radio waves for communication with ships pointed to the need for high-frequency direction-finding (HF DF) at long range. Horton realised that this would only be possible in ships if the existing antennae could be replaced by rotating framecoils fitted high above the superstructure and rigging. By 1930 plans had been laid for developing equipment using remotely controlled coils at the top of the mast, and it is a great tribute to Horton and his tiny team that the Royal Navy led the world in ship direction-finding throughout the 1930s. No foreign ship at the Naval Review at Spithead in 1937 had a DF antenna sited where it could possibly have been used at frequencies above about 1.5 MHz, whereas HMS *Newcastle* could operate up to at least 22 MHz. And throughout the war that followed at least some German scientists still believed that HF DF on-board ship was not possible.[2]

Horton's programme included research on radio propagation, collaborating with and advising the Radio Division of NPL. It was largely because of this Admiralty interest that Watson-Watt and his colleagues were encouraged to work on radio propagation and ionospheric reflection phenomena in the early 1930s. This led to the birth of British radar when, following Watson-Watt's pioneering work in 1935, it was decided that the vital importance of radar to the RAF justified the setting up of an experimental establishment at Bawdsey dedicated solely to this field.

THE EARLY YEARS OF NAVAL RADAR

In contrast, the Navy decided that work should be pursued in an existing establishment already familiar with the seagoing environment, with its particular requirements and restrictions – entrusting it to HM Signal School, under the control of the Director of the Signal Department of the Admiralty. In August 1935 – almost exactly four years before the war – a small scientific effort was authorised for work on shipborne radar, but the Admiralty did not at first attach anything like the same importance to the new invention as did the Air Ministry. Radar became just another item, without any special priority, in a signals-oriented establishment. The scientific effort was much too small, spread over too many lines, and above all lacked the priority to obtain adequate supporting services. After two years a metric radar had reached only an early experimental stage, as yet with no indication that even the minimum operational requirements were likely to be met in a reasonable time scale. This, it should be recalled, was at about the same time that operational radars were actually being installed in German warships.[3]

Such was the sorry situation when, late in 1937, Horton was brought in to take charge, assisted by a handful of scientists also transferred from other work in Signal School. Perhaps more than anyone else at that time he appreciated the importance and potential of radar in the Navy, and how unfavourably the attitude towards it had compared with that of the RAF. To quote his own words ten years later, in an address to the Senior Officers' War Course: 'This establishment failed to see the significance of radar, and as late as 1937 it was a matter of great difficulty to get workshop and drawing office effort put on to it. The reason was always the same – other and more obvious demands took priority'.

Horton's immediate concern was how to concentrate his meagre resources towards getting practical systems to sea. This depended on the relative priorities of aircraft and ship detection, since there were grounds for believing that the former could be achieved sooner at a metric wavelength, whereas shorter wavelengths offered longer-term potential for detecting ships and low-flying aircraft. Agreement was reached to give first priority to an aircraft-detection radar, with any ship-detection capability it gave as a bonus. Horton was then able to devote a major part of his resources to the rapid improvement of the experimental metric radar (Type 79X) in which antenna, transmitter and receiver all underwent major changes. The effort remaining was concentrated upon experimental work for decimetric radars, and all other current lines of work were abandoned.

This concentration of technical effort was not in itself enough. Horton also needed the conviction and courage to fight an organisational battle to achieve a more equitable division of supporting services between communications and radar. To win required great determination, but win he did, and the outcome was a separate radar division under his leadership, able in its own right to demand more effort in design, production and ship-fitting. The results were dramatic. Within months not only had most of the significant shortcomings of the metric radar been removed, but two ships of the fleet had already been fitted with development models, and were reporting results of detection ranges against aircraft. Although there was no margin for any degradation from peak performance the results were enough to justify early ordering of production models. In parallel with all this the decimetric work had been put on a realistic basis to develop a series of gunnery radars, which entered the fleet from 1940 onwards.

During these pioneering years Horton's policy was to shield his staff as far as possible from organisational pressures and to allow them to concentrate on their work. He was constantly in touch with his section leaders, but never needed to call them formally together. Requests for equipment or resources would quietly be fulfilled, with no indication of the battles that had been fought to obtain them. Externally his dry, almost diffident manner belied a firmness of purpose and an ability for staunch resistance and pungent criticism when the occasion demanded. A typical example was his forthright rejection of Watson-Watt's advice at the Tizard Committee, despite the apprehensions of Admiral Somerville, that the Navy should limit itself to fixed antennae for shipborne radar.

THE WARTIME YEARS

The new radar organisation soon had to accommodate the large wartime build-up of staff, slow at first but then rapid. Its success may be judged by the fact that the next three years saw the development of the majority of the Naval radars that were to become operational in World War 2, including a second metric set (Type 281), the family of decimetric gunnery radars, the small-ship radars (Types 286/291) and the first centimetric sets.

By mid-1942 most of the radar work had been concentrated into a department at Witley. The programme had continued to expand, so that many more new families of radar systems and variants were being worked on at the same time. Horton appreciated the need for explicit planning and coordination; this had hardly needed formal definition in his tiny prewar group but had become progressively more important as the scale of activities increased. He introduced a novel organisation (see Monograph 1) in which the conventional 'component' divisions, each responsible for a specific technical area, were supplemented by three small but powerful 'equipment' divisions, each headed by a senior man and concerned respectively with Tactical Radars, Fire-Control Radars, and other systems, including IFF. Their responsibilities included overall planning of each individual system, defining the elements to be provided by each component division, ensuring technical compatibility between all components of the system, and monitoring progress. They were also the main points of contact with Naval application officers in the Establishment, and staff at similar level in Admiralty.

Horton's new organisation also had to be flexible enough to make good use of the catholic variety of scientific staff that continued to arrive in increasing numbers. Of course there were many young graduates, not only in physics and engineering, but in a number of other disciplines whose relevance to radar was less immediately obvious; there were also some more senior people with practical experience in industry, or with distinguished academic records in universities, and in addition there were contingents, temporarily in uniform, from allied nations whose countries had been overrun. The mainstream of development still lay in the four 'component' divisions concerned with the essential elements of a radar – antennas, transmitters, receivers and displays – but in the new organisation these were supplemented by a variety of other divisions concerned with particular techniques, measurements, or aspects of research. While the great majority of recruits were assigned to the mainstream divisions, others could be more profitably employed elsewhere. Some of the more senior men worked in small specialist groups to develop techniques for use in future systems. Others were able to take a broader look; for example a division consisting of a small group of mathematicians and theoretical physicists (several of whom subsequently became FRS) was effectively given a free hand to look at fundamental issues affecting radar performance, including features of the physical environment in which it had to work.

These arrangements defied orthodox precepts; for example the division of responsibility between equipment and component divisions was not sufficiently well-defined, and so tended to be influenced more than it should have been by personalities in one or other camp. Again Horton had up to 15 people reporting directly to him, which could have made it difficult to maintain adequate coordination and control. Nevertheless the arrangement worked remarkably well. It may well not have made the best use of every individual, but to attempt this would probably have been self-defeating because of problems of rank and personality. It certainly made the fullest combined use of the available talent, as indeed it was designed to do, and it continued essentially unchanged throughout the War.

The scale of activities now demanded regular and somewhat large meetings of all the senior staff, at which Horton acted very much as 'first among equals'. There was much vigorous discussion and airing of views, but Naval and civilian colleagues all played as a enthusiastic and happy team, and this informal but workmanlike attitude permeated the whole department. Effectiveness and output were remarkable, and for many of us the high morale and intimate, happy-family atmosphere of those days have never really been equalled. Of course we were all stimulated by the fight for national survival, but the spirit also owed much to Horton's ability to bring out the best from this great variety of human talents. The raw university recruit, the distinguished academic, the seasoned engineer from industry, though some of them might see him only rarely, all warmed to his friendly and unassuming manner, as indeed the pre-war scientific stalwarts had done before.

THE POST-WAR RESTRUCTURING

After a period in Admiralty headquarters Horton returned to ASE to become Chief Scientist of the whole Establishment, responsible for communications and allied subjects, as well as radar. After the end of the war the numbers diminished as many wartime recruits returned to their chosen careers in universities or industry, but ASE was still very much larger than its prewar counterpart, with many by now experienced people opting to remain with the scientific civil service.

A new era was beginning. It was time to adapt to a more measured progress, but capable of easy expansion should emergency arise in the next five years – as indeed it did in the shape of the Korean War. Within the Admiralty the vision was still that of a 'blue-water' navy, and project demands were becoming more elaborate and complex, particularly as the guided weapon began to overtake the gun as the Navy's future main armament. At the same time, financial and other administrative controls were returning to their former stringency after the relative flexibility of wartime. Horton recognised that demands on the Establishment would undoubtedly greatly exceed the supply, and stressed the need to focus on a limited number of important projects whilst preserving a degree of high-quality scientific work towards the future.

Early in 1947 he introduced a fundamental change from previous practice by introducing an organisation based on project groups. Within each group individual projects would be executed by largely self-contained teams, able to conduct their own applied research to establish the necessary techniques, and then to produce their own engineered experimental models for sea trials. From the start each team was to include design/development engineers – seconded from the development department, but reporting to the project leader – including the man who would subsequently be responsible for taking the project through its development stages. A project coordination party was charged with system coordination, forward planning and assessment studies.

Project leaders now had not only the authority but most of the resources necessary for their task, rather than having to be customers of the various techniques groups, each committed to supporting several projects. There was also a more effective combined use of experimental and design staff. Horton had to live for the time being with the traditional division between the development department and the experimental organisation, both of which now reported to him, but this was the first move towards their eventual integration.

Another novel feature was a centralised Post-Design Services Division, which served to free the project groups from continuous repercussions from the past. It was able to take an independent look at complaints against existing designs and was empowered to approve and implement *essential* changes; more ambitious modifications advocated by enthusiastic customers (and sometimes by the original designers!) had to fight for priority with the ongoing programme. Resources for basic research, design and development, inspection and test, ship-fitting and so on were also under separate centralised control, each reporting directly to the top.

It cannot be said, however, that this visionary new structure was permeated by the same vital spirit that existed at Witley in the early days of the war. Of course there was no longer a threat to national survival, and the Establishment was still inconveniently dispersed over a number of sites, but there were other difficulties. For example the project groups did not report to Horton himself but to the so-called Project Coordination Party, which then reported through the Deputy Chief Scientist. Between them the heads of project groups were responsible for by far the largest and most important part of the Establishment's programme, yet to them the Chief Scientist became a somewhat remote figure. Perhaps the organisation had become too large for Horton's particular style of management. He certainly preferred to work quietly and informally with a limited number of colleagues, rather than as a highly visible leader figure impressing his personality on the whole Establishment.

Whatever these shortcomings, Horton had undoubtedly again put the Establishment on the right track, along which it subsequently travelled with conspicuous success. The basic structure continued unchanged for 15 years, throughout his term of office and that of his successor, and proved robust enough to adapt to many changes of circumstance – even including the absorption of another establishment (AGE) in 1959. His decision to go firmly for a project-oriented organisation set an example that many others were subsequently to follow. Even more than twenty years later Horton's basic precepts were still being rediscovered in other organisations, and propounded as new and fundamental truths.

In 1951 he left ASE to become Director of Physical Research at the Admiralty, succeeding Sir William Cook, who had become Chief of the RNSS. He found the translation to headquarters uncongenial, and before long left to take up a new and successful career in industrial research, where he became a director of a well-known organisation manufacturing chemicals for agricultural and medical use.

Cecil Horton possessed a rare ability to maintain the respect, enthusiasm and affection of his colleagues while allowing them free rein to exploit their creative talents and giving them unquestioning support when necessary. He never ceased to encourage closer understanding, cooperation and mutual respect between his scientific staff and the Naval officers with whom they had to work. After the war he emphasised that morale and leadership were just as important in the new Royal Naval Scientific Service as in the Navy itself, though in a subtly different way, and urged the Navy to pay as close attention to the one as to the other. Throughout his long career with the Admiralty he always took his responsibilities very seriously, and especially during those pioneering days of radar must frequently have suffered from frustration. Fortunately he was able to find solace in his long and happy marriage, and in his abiding love of music, shared also by his wife who had a charming contralto voice. He himself was a most accomplished performer with the violin (see Frontispiece), of truly professional standard, having been invited by Eda Kersey to play in her quartet, which was very highly regarded at the time. The musical evenings the Hortons arranged from time-to-time were a source of spiritual refreshment to fortunate colleagues as well as to the hosts themselves.

Acknowledgements

I am particularly indebted to the late A. W. Ross for his discerning recollections of Horton in the early days of radar before the war. Thanks are also due to D. S. Watson, J. F. Coales, and other former colleagues, for helpful comments and advice.

References

1. F. A. Kingsley (ed.), *The Applications of Radar and Other Electronic Systems in the Royal Navy in World War 2* (Macmillan, 1994).
2. F. A. Kingsley, op cit., Monograph 6 by P. G. Redgment.
3. D. Pritchard, *The Radar War* (Patrick Stephens, 1989), p. 190 et seq.

Development and Installation of British Naval Radar – Some Significant Milestones

1928		HM Signal School applies for first patent on Radio-Location in name of L. S. Alder.
1935	Feb.	Watson-Watt demonstrates detection of aircraft by radio.
	Sep.	Admiralty instructs HM Signal School to start development of radar.
1937	May	Preliminary trials of metric radar completed. Research on 1200 MHz begins.
1937	Sep.	Development of warning radar (to become Type 79) settles on 43 MHz.
1938	Feb.	Decision taken to develop equipment on 600 MHz using pulsed triodes.
	Mar.	Type 79X, first experimental radar, installed in HMS *Saltburn*.
	Aug.	Type 79Y, first operational radar, with 20 kW output, installed in HMS *Sheffield*, and in HMS *Rodney* in October.
1939	Aug.	Type 79Z, with 70 kW output, installed in HMS *Curlew*. Full production started, leading to a total of about 100 sets
	Dec.	Development of Type 281 started on 90 MHz.
1940	Feb.	Trials of 600 MHz rangefinder at AA Range, Eastney.
	Apr.	HM Signal School instructed to design and produce 200 sets of Type 282 (600 MHz).
	June	Sea trials of 600 MHz radar in HMS *Nelson*. HM Signal School instructed to design and produce sets for fitting on all main armament and high-angle directors. 700 sets ordered.
		Type 286 (RAF 200 MHz ASV radar with fixed masthead antenna) started to be installed in large numbers in destroyers and smaller ships.
	Oct.	Type 281 installed in HMS *Dido*. Full production started, leading to a total of about 80 sets.
	Nov.	Signal School party visits Swanage to assemble copy of 'breadboard' TRE 3000 MHz (S-band) radar in a trailer, followed by preliminary trials against naval targets using TRE experimental equipment.

	Dec.	Decision to proceed immediately with the design of a 10-cm radar for convoy escorts. Trials of first production gunnery sets in HMS *King George V* (Type 284) and in HMS *Southdown* (Type 285).
1941	Mar.	Trials of first prototype 5 kW S-band naval radar (Type 271X) in HMS *Orchis*. Twelve prototypes completed, and a further 12 in hand. Initial production order placed for 100 sets. Versions for destroyers and large ships (Types 272 and 273) followed in July. First Type 79B, with single antenna, fitted in HMS *Hood*.
	Apr.	First multiple installation in a capital ship. Type 281 and eleven 600 MHz sets installed and commissioned in HMS *Prince of Wales*.
	May	First Type 290, interim replacement for Type 286 with 50 kW output, installed in HMS *Aurora*.
	July	Mobile trailer NT271X at Dover for coast defence. Resulted in Army conversion as CD No 1 Mark 4.
	Sep.	32 escort vessels at sea with Type 271; orders increased from 150 to 350. Experimental development in hand for higher power (70 kW) version (Type 271 Mark 4, to become 271Q and 273Q), also for yet higher-power (500 kW) version (Type 272/273 Mark 5, to become 276/277). Development begun for Type 274 S-band main-armament gunnery set of similar power.
	Nov.	Prototype Type 271/272/273P delivered; order for 1000 sets.
	Late	First installation of Type 281B with single antenna.
1942	Early	Development started of S-band gunnery set for high-angle directors (Type 275)
	Apr.	Work started on close-range auto-follow gunlaying radar on 10 000 MHz (X-band) (Type 262).
	May	Trials with prototype 271Q (70 kW S-band) in HMS *Marigold*, followed in July by 273Q in HMS *King George V*.
	Aug.	First installations of Types 284P and 285P, with beamswitching for blind fire and common antenna for T/R.
	Late	Initial work towards new fighter direction radar (Types 294, 295).
	End	First fitting of Type 291, final replacement for Type 286 with 100 Kw output.
	End	Types 271/272/273P: delivery of 1000 sets complete.
	Dec.	Development contract placed on EMI for Type 262.
1943	Mar.	500 KW S-band radar (Type 277T) installed in trailer cabins for coastal defence.

	Apr.	Trials of seagoing version (Type 277X) in HMS *Saltburn*. Trials of Type 276 in HMS *Tuscan* followed in November.
	Mid	PPI displays start to be installed in large numbers on most warning radars, reaching 5000 by war's end.
	Aug.	Trials of prototype Type 293 S-band target-indication radar in HMS *Janus*.
1944	. . ..	Progressive installation of Action Information Centres in most classes of ship.
	Mar.	Trials of first production Type 277 in HMS *Campania*, followed by extensive installation of Type 277 in the Fleet.
		Revised development plan for fighter direction radar (Types 980/981).
	Aug.	First installation of submarine radar Type 267W (Type 291 with additional X-band facilities) in HMS *Tuna*.
	Late	Type 274 S-band gunnery radar installed on main armament directors in large ships.
	Late	First installation of Type 262 X-band close-range blind-fire radar.
1945	Early	First installation of Type 275 S-band high-angle gunnery radar.
	Feb.	General installation of Type 268 X-band warning and navigational radar in Coastal Forces.
		Type 293M began to replace Types 276 and 293 for target indication.
	Apr.	Types 277P and 293P began to replace Types 277 and 293M.
	Mid	Introduction of Type 281BQ, with addition of continuous antenna rotation.
	Late	First installation of Type 960, replacement for both Type 79 and Type 281, in HMS *Vanguard*.

Notes on the Contributors

J. F. Coales, CBE, F Eng, FRS. Born in 1907, Professor Coales joined Admiralty service in 1929 after graduating from Cambridge University. After original work on radio direction-finding, he transferred to developments in ultra-short wavelength radar and communications in 1937. During World War 2 he was in charge of research and development of Naval gunnery radar. Subsequently he returned to Cambridge as Professor of Electrical Engineering. He was elected President of the Institution of Electrical Engineers for 1971–2 and Honorary Fellow in 1985. Similar posts were held in international organisations concerned with automatic control systems and instrumentation. Awarded an OBE in 1945 and a CBE in 1974, he was elected F Eng in 1967 and FRS in 1970. Currently he is Emeritus Professor of Engineering at Cambridge University.

C. A. Cochrane, MA F InstP, C Eng, MIEE. Born in 1919, C. A. Cochrane graduated from Glasgow University in mathematics and natural philosophy. He joined HM Signal School in August 1940 and was almost immediately involved in developing and introducing into Naval service the world's first operational centimetric wavelength radar. This was the Type 271 anti-submarine radar, operating on 3000 MHz, for installation on convoy-escort vessels. He continued to work on the development of centimetric warning radar for the duration of the war. In the early postwar years he did further work on microwave developments in an industrial research laboratory, where he was the inventor of the Cassegrain antenna for radar applications. Subsequent appointments were as Director of the Tube Investments Technological Centre, and later as Head of Division of Pollution Control in the Environment Directorate of the Organisation for Economic Cooperation and Development in Paris.

F. M. Foley, BA, MSc, MIEE. F. M. Foley graduated from Trinity College, Dublin, in 1937, followed by a year researching the physics of very thin metal films. He joined the Valve Division of HM Signal School in 1938, first working on the design of transmitter valves, later on receiving-valve applications and the achievement of very high valve reliability. In 1948 he transferred to the Seaslug radar project, working on the development of displays and ranging circuits. In 1954 he became Head of the Valve

Division; he later aided the development of thin-film microelectronics. From 1970–2 he was in charge of the Civil Marine Navigation Aids Division, before becoming Assistant Director, Post Design Services, in the Establishment.

H. D. Howse served at sea in the Royal Navy throughout World War 2, latterly as a specialist navigator. In 1958 he took early retirement as a Lt-Cdr, and became a curator at the National Maritime Museum at Greenwich from 1963 until 1982. He was Clark Library Professor of the University of California, Los Angeles, for the academic year 1983 to 1984. His published works include: *The Sea Chart* (with Michael Sanderson), 1973; *Greenwich Time and the Discovery of Longitude* (1980); and *Radar at Sea – The Royal Navy in World War 2* (1993). He was awarded a DSC and an MBE for his services at sea in World War 2 and the Korean War, respectively.

F. A. Kingsley, BSc, C Phys, F Inst P, C Eng, FIEE (editor) joined HM Signal School from Birmingham University in July 1941. Until 1945 he was engaged mainly in electronic-warfare projects, including technical planning of radar countermeasures activities as part of the Navy's contribution during the assault phase of the Normandy invasion in 1944. Postwar he was engaged in original radio-propagation research, electronic-warfare concepts and communications-systems developments. He became Head of the Communications Division of the Admiralty Surface Weapons Establishment in 1961, with the primary task of modernising the Royal Navy's ship and submarine communications. During this period he served on a number of inter-Service, NATO and CANUKUS communications systems Working Parties. He was a member of the original Space Research Committees of the Royal Society, and of a Cabinet Office Committee on Satellite Communications. In 1965 he was appointed as an Assistant Director in Central Staffs, Ministry of Defence.

R. A. Laws, MBE, BSc, FIEE, RNVR. Originally qualifying as a chartered accountant in 1939, R. A. Laws was mobilised as a member of the RNV(W)R in August of that year. He was commissioned as a Paymaster Sub-Lt in October 1940, and then appointed to the Special Branch, RNVR for radar duties in January 1941. He was mentioned in dispatches in 1944. On demobilisation in 1946 with the rank of Lt-Cdr, he joined the BBC research department. He gained his BSc from Sheffield University in 1950. He then held engineering appointments with Metropolitan-Vickers, GEC, CEGB and Marconi Underwater Systems. He was appointed FIEE in 1946 and awarded an MBE in 1986.

B. W. Lythall, CB, MA. B. W. Lythall joined HM Signal School in 1940 after graduating from Oxford University, working initially on the development of the first operational centimetric radars. He then worked on microwave systems throughout the war, subsequently specialising in antenna design. In 1953 he moved to the Admiralty Research Laboratory to develop new methods of underwater acoustic detection. In 1957 he was appointed Assistant Director of Physical Research at the Admiralty. From 1958 he was Deputy Chief Scientist at the Admiralty Signal and Radar Establishment. In 1964 he was appointed to the Admiralty Board as Chief Scientist, Royal Navy, serving until 1978. He was also Deputy Controller of the Navy for Research and Development until 1971, when he became Deputy Controller, Establishments and Research, in the Procurement Executive. In 1978 he became Director of the NATO Saclant ASW Research Centre at La Spezia. He was awarded a CB in 1966.

A. M. Patrick, CEng, MIEE. In 1939 A. M. Patrick was mobilised as a member of the RNV(W)R and served initially in the South Atlantic. Commissioned in 1940, he gained experience as a radar officer during ship-fitting duties at HM Signal School and at Sherbrooke House, Glasgow, before being appointed Port Radar Officer, Rosyth, in 1941. In 1944 he became Base Radio Officer, Sydney, Australia. Here the roles of Port Radar Officer and Port Wireless Officer were combined for the very first time. Shortly before the war ended he was Fleet Radar Officer, British Pacific Fleet, with the temporary rank of Commander. Subsequently he served in HMS *Collingwood* on planning and personnel duties connected with the formation of the new Electrical Branch of the Royal Navy. From 1946 to 1952 he headed the newly formed Naval Radar Section of the Marconi Company, before serving as Assistant Electrical Engineer-in-Chief (Trials) in the Royal Canadian Navy from 1952 to 1955. He later held successive senior marketing appointments with Decca Radar Ltd and the Plessey Co.

B. G. H. Rowley, MA, C Eng, FIEE. After graduating from Oxford in 1939, B. G. H. Rowley joined HM Signal School to work on radar development. He was seconded to the Staff of C-in-C, Home Fleet, from May 1940 till April 1942, at Scapa Flow. He then served as a Lieutenant (Special Branch) RNVR in the USS *Gleaves* on North Atlantic convoy-escort duty as Radar and Sonar Observer. In late 1942 he was posted as Staff Radar Officer, British Naval Staff, Washington, DC. Postwar appointments included that of US resident representative for Marconi's Wireless Telegraph Co Ltd, New York (1950–4), Manager of the Company's Maritime Division, England (1954–7); with the Canadian Marconi Company, Montreal (1957–60); with the English Electric Company, London (1960–3); with the North East Training Council

(1963–5); as Careers Adviser and Industrial Liaison Officer, Woolwich Polytechnic (1965–9); with the International Labour Organisation in Chile (1969–70); and with the University of Manchester Careers Advisory Service (1971–82).

J. S. Shayler, BSc, FIEE. J. S. Shayler graduated from Manchester University in 1938 and subsequently spent his whole career in defence research and development. He worked on Naval radar from 1938 to 1948 at HM Signal School (later Admiralty Signal Establishment). After this he was involved in sonar at the Underwater Detection Establishment; millimetre-wave CW radar at the Services Electronic Research Laboratory; aids to the operation of aircraft from carriers at the Naval Air Department, followed by automatic landing of aircraft at the Blind Landing Experimental Unit, both being located at the Royal Aircraft Establishment, Bedford; RAF and Army telecommunications at the Ministry of Defence; and finally as Head of Defence Research and Development, British Embassy, Washington, DC. He retired in 1978.

PART I
Radar Equipment Developments, 1935–45 (Monographs 1 to 6)

Editorial Note

Professor J. F. Coales is the senior survivor of the pioneering pre-war radar team at HM Signal School, and the Elder Statesman of wartime decimetric Naval gunnery radar systems. He is therefore in a unique position to recall the historical background that led to the early Naval radar experiments, the personalities involved, the successes and the failures. His Monograph 1 thus provides an overview of Signal School's/ASE radar research and development up till 1945. In addition it presents a detailed account of the development of decimetric Naval gunnery radar, for which Professor Coales was directly responsible.

The succeeding Monographs 2 to 6 provide the detailed substance of the associated research, the development of special valves, and accounts of the development of many individual radar systems, and associated IFF, written by scientists directly concerned.

Monograph 1
The Origins and Development of Radar in the Royal Navy, 1935–45 with Particular Reference to Decimetric Gunnery Equipments
J. F. Coales

SUMMARY

Pioneering Naval interest in radio communications at the turn of the century – ships equipped for 'Wireless Telegraphy' in World War 1. Experimental section initially established in HMS *Vernon* – transferred to HM Signal School in 1919 – work on all forms of signalling, including possibilities of Infra Red. In 1928 L. S. Alder took out a provisional patent for Naval radar, but this was never implemented. By 1935 Signal School well established in Training and Experimental Departments – Naval Officers and Civilians work closely together. Early radiolocation experiments against aircraft targets by Watson-Watt disclosed to senior Admiralty representatives – consequent establishment of small, secret group in Signal School, for investigations using metric wavelengths for Naval applications. Valve Section in Signal School provided high-power sources for *all* British metric radar requirements. Attempt by Watson-Watt to gain control of Admiralty radar research – rejected by Admiralty. Results achieved in first Naval research unimpressive, leading to major reorganisation of effort in 1937 under C. E. Horton, with a small increase in staff and resources. Successful trials of 7.5-m equipment at sea in 1938 led to equipping *Sheffield* and *Rodney* before the war. Further developments on improved sets in 1939 for long-range air warning, and low-level warning on aircraft and ships. Initial development of 50-cm equipments, and provision for gunnery fire-control. Problem of detection of U-boats, particularly at night – birth of Naval 10-cm radar using high-power cavity magnetron – successful and rapid development of integrated 10-cm set for rapid installation in convoy escorts. Further development of 50-cm sets, for blind-fire applications in particular – growth of the integrated-weapon-system concept. Developments in guided-missile systems against aircraft targets.

INTRODUCTION

Following the successful demonstration by Watson-Watt and Wilkins in February 1935 of the potential of radar for the detection, and ultimately location, of distant aircraft targets,[1] Admiralty decided to initiate an *independent* research programme forthwith for ship and shore applications. This decision was based on three premises:

- Naval requirements, and the associated severe environmental conditions of use, were markedly different from those of the RAF and the Army.
- The Navy already possessed a well-developed organisation and structure for the design, procurement, testing and installation of radio equipment capable of operation in the stringent shipborne environment.
- At the HM Signal School at Portsmouth there existed the *only* facilities in the UK for developing transmitting valves using silica envelopes, suitable for generating very high powers without the need for water cooling, a severe restriction for shipborne installations.

THE RESEARCH BACKGROUND, 1915–35

Captain H. B. Jackson (later Admiral of the Fleet Sir Henry Jackson, KCB, FRS) initiated a Wireless Section in HMS *Vernon*, the Navy's shore-based Torpedo and Mining Establishment, in 1904. H. A. Madge, BA, was the Civilian-in-Charge, and Commander F. C. A. Ogilvy the Experimental Commander. Jackson had first experimented with Wireless Telegraphy (W/T) in *Defiance* in 1895, and had later demonstrated its importance to the Navy in the Fleet manoeuvres of 1899. By 1913 all HM Ships were fitted to transmit and receive W/T signals.

By the start of World War 1 *Vernon* incorporated a small Experimental Department, in addition to providing training in the maintenance and operation of W/T equipment, afloat and ashore. In early 1915 it became evident that the W/T Branch needed to be expanded. Amongst new recruits were E. V. Appleton (later Sir Edward Appleton, FRS), Professor C. L. Fortescue, B. S. Gossling, BA, F. Brundrett, BA (later Sir Frederick Brundrett, Chief of the Royal Naval Scientific Service, RNSS), C. E. Horton, E. J. Grainger, and Lt. H. Morris-Airey, MSc (later Superintending Scientist.)

In April 1917 a new Department of Torpedoes and Mining (DTM) was formed from Staff in the Directorate of Naval Ordnance (DNO), whilst a Signal Division (DSD) was formed in the Admiralty, under the control of the Deputy Chief of the Naval Staff (DCNS). The respective responsibilities of the two Admiralty Departments and HM Signal School were defined in a Memorandum issued in April 1918. Briefly,

- DSD was responsible for advice on all questions involving Visual, W/T (including Telephony (R/T)) and Sonic Telegraphy (S/T) policy. He was to keep in close touch with DTM on all aspects of communications policy, including technical aspects.
- DTM was responsible for all matters connected with the design and development of W/T, S/T and Electric signalling equipment. He was to keep in close touch with DSD on all matters related to the suitability for development, and subsequent application, of all equipment for the above forms of communication.

Later, in 1919, the W/T Department was transferred from *Vernon* to HM Signal School in Portsmouth Barracks, responding to the requirements of DSD, Admiralty. Thus, in addition to its training commitment, Signal School (SS) now incorporated an Experimental Department under Commander J. F. Somerville (later Admiral Sir James Somerville). The original staff included four Naval Officers, approximately 35 Civilian technical staff under H. A. Madge, 27 draughtsmen and 30 clerks (Figure 1.1). In January 1921 Madge was replaced by H. Morris-Airey, with a complement of five senior engineers, and 62 engineers, in three grades. G. Shearing was later appointed as Head of the Experimental Department, a position he was to hold until 1941. A Valve Research Laboratory was established in Signal School in about 1918 under B. S. Gossling, primarily to develop high-power silica-envelope transmitting valves (see Monograph 3). Gossling was succeeded by T. E. Goldup early in 1920. The latter developed a silica valve that dissipated a power of 10 kW. This proved to be a critical component for the first British radar transmitter, some 16 years later. Goldup was succeeded by H. G. Hughes in 1923.

Professor Fortescue, who had been engaged originally on development of W/T reception techniques in *Vernon*, was succeeded by B. Hodgson in Signal School in 1920. He was himself followed by W. F. Rawlinson in 1923. Experimental work on W/T transmitters had started in about 1920 under W. Ure, and in 1922 trials were carried out in *Antrim* using transmitter Type 36.

Also in 1920, work was started on distant radio-control of *Agamemnon* acting as a target ship for full-calibre shoots. This development was operational in 1922, under the guidance of Lt Cdr Boles, C. R. Evershed and H. Noble. A later improved system was developed by Evershed and J. D. S. Rawlinson for the remote control of *Centurion*, the Mediterranean target ship for many years. This led to a system for the radio-control of motor torpedo boats in 1926, and later to the 'Queen Bee' radio-controlled aircraft, developed at RAE, Farnborough for Anti-Aircraft (AA) practice.

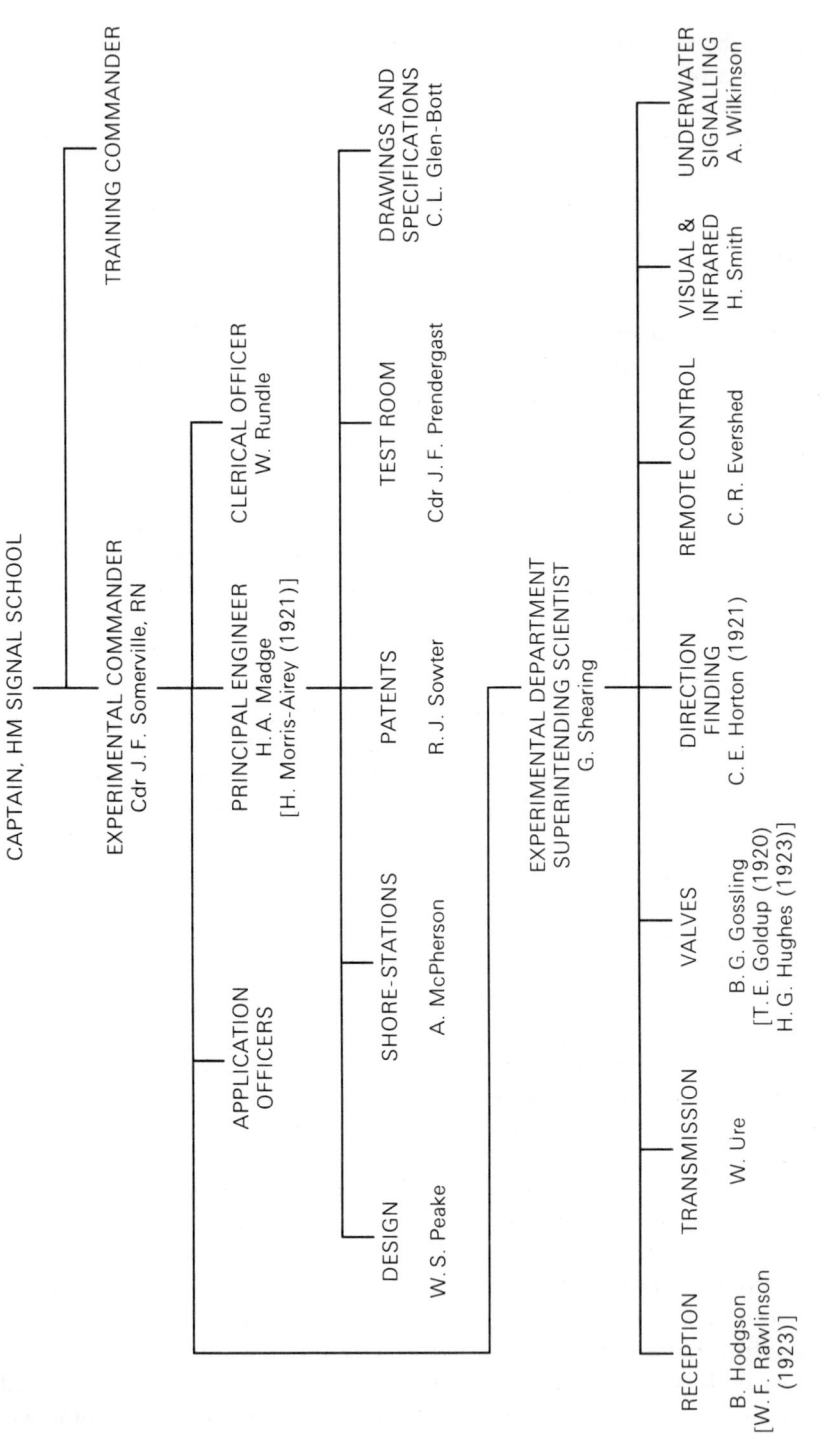

1.1 Organisation of the Experimental Department, HM Signal School, 1919

The latter was largely due to the initiative and persistence of Commander Henry Cecil RN, a signals specialist and Application Officer in HM Signal School in the 1920s and 1930s.

C. E. Horton, who was to play such a prominent role in the development of Naval radar from the mid-1930s onwards, joined Signal School in 1921 (see the Tribute in this volume). He was joined by C. Crampton in 1923, and by J. F. Coales (the present author) in 1930. At that time this small group led the world in the development of High Frequency Direction Finding (HF DF) in ships, and in particular in warships, for the next decade and a half. Watson-Watt, then Head of the Radio Division in the National Physical Laboratory (NPL), accompanied Horton on a long cruise in *Yarmouth* to investigate the nature of atmospheric noise in radio receivers at HF, and radio propagation via the ionospheric layers in the upper atmosphere. As a result of this Watson-Watt's group at NPL continued research into these topics during the 1930s. Horton became a member of the principal NPL Advisory Committee, and the present author a member of an Advisory Committee on HF generation, propagation and DF.

Apart from developments in radio and visual techniques of communication in HM Ships, in the early 1920s a section was set up in Signal School to investigate the possible use of Infra Red signalling. Thus, at this time, the Experimental Department of Signal School was organised in a total of seven groups, an arrangement that would continue almost unchanged until the advent of radar development in 1935. The Heads of the seven groups remained in position until the work on radar necessitated a major reorganisation in 1937.

Thus by 1935, when radiolocation (later RDF, and finally Radar) research was initiated in the United Kingdom,[2] HM Signal School was well established in the Portsmouth Naval Barracks HMS *Victory* for:

- training of all Signal Officers and W/T Branch Ratings in the operation and maintenance of all communications equipment, ashore and afloat, in support of the Naval world-wide network; and
- an Experimental Department under a Commander (S), RN, for the development, evaluation, procurement and testing of W/T equipment for Naval use.

The Experimental Department consisted of 43 professional engineers and scientists, plus about 20 technical assistants, supported by workshops, a drawing office and ancillary staff. Six Naval Signal Officers (known as Applications Officers) were attached to the Experimental Department to advise the civilian staff on Naval requirements, and for the organisation of experimental sea trials using the Signal School tenders: *Concord* (a light cruiser), *Sardonyx* (a destroyer), and submarine *L22*. Assistance was also provided by some serving Petty Officers and Ratings. Relations between the Civilian Staff and the attached Naval Application Officers were very cordial, and based on mutual respect.[3]

Proposals for new project developments in Signal School, based on discussions between the appropriate Naval Application and Civilian Staffs, were submitted to DSD, Admiralty, for consideration and, if agreed, to the Third Sea Lord. as Controller of the Navy, for approval. Development work would then be initiated in the programme of the Experimental Department at Signal School, in close association with the relevant Naval Application Officer, representing the operational interest. Any difficulties arising were resolved by Shearing and the Experimental Commander.

As a particular project progressed, design engineers and workshop staff produced experimental equipment for evaluation, including sea trials in one of the Establishment's allocated vessels, *Concord*, *Sardonyx*, or Submarine *L22*. Following successful sea trials (incorporating equipment modifications, if necessary) a ship-fitting programme would be determined in Admiralty, both for new-construction ships and older categories designated for modernisation in HM Dockyards. Manufacturing drawings were then produced in Signal School, and appropriate commercial firms approached who were capable of meeting the production programme required to fulfil the agreed ship-fitting schedule. Production Staff in the Establishment collaborated with Admiralty's Director of Contracts to place the necessary orders on the selected firms.

As production models of equipment became available from industry, these were tested in Signal School and, if accepted, despatched to the appropriate HM Dockyard or commercial shipbuilders for installation, in accordance with 'Fitting-Out' data prepared in the Establishment. Meanwhile the Instructional Department of Signal School prepared 'Operating and Maintenance' instructions for the equipment and initiated training for Naval personnel. Finally, acceptance trials would be carried out in the presence of the relevant Application Officer, who would write a report on the suitability of the equipment for acceptance to the Captain, Signal School and DSD, Admiralty. In some cases it was necessary to seek the approval of the Board of Admiralty to make modifications to the structural design of certain ships, to ensure satisfactory performance of a new equipment in operational use. For example, for optimum performance of HF DF, a top-mast installation was required. This involved top-weight, stability and windage considerations, plus questions relating to the safety of personnel during maintenance. These were resolved with the Directorate of Naval Construction. It will thus be apparent that HM Signal School possessed a powerful range of skills and resources in 1935, adequate to pursue the new challenge of the invention of radar independently of the effort being devoted to the RAF and Army developments at the Bawdsey Research Station.

In the period 1915 to 1935, reviewed in the present section, there was a remarkable portent to the future invention and development of 'radiolocation' (radar) for Service applications. This was due to the foresight of L. S. Alder, one of the Signal School scientists, who originated a provisional Patent Specification dated 1 March 1928, which quite precisely presaged radiolocation (Provisional Specification No. 6433/28). The following is an extract:

(a) Improvements in and relating to methods and means for determining positions, directions, or distances of objects by wireless waves, applicable to navigation and for the location of dangerous objects, or of enemy craft.

(b) Lodged by we, James Sacheverell Constable Salmond, Captain, Royal Navy, and Leonard Stanley Batar Alder, MSc, both of HM Signal School, Royal Naval Barracks, Portsmouth, Hampshire, both British subjects.

(c) The invention relates to methods and means for the employment of the reflection, scattering or re-radiation of wireless waves by objects as a means of detecting the presence of such objects, and of determining the position, directions or distances of such objects. According to this invention, apparatus which may be placed on land or in a ship or in aircraft may be employed to generate wireless waves and to observe or detect their reflection, scattering or re-radiation from a surrounding object or objects, or a nearby object such as shipping, icebergs, natural land features, or the surface of land or sea below an aircraft, or from suitable reflectors constructed for the purpose in known positions, as, for example, at points dangerous to shipping, the entrances to harbours, or at aerodromes and landing places for aircraft, etc.

With hindsight it is remarkable that this proposal was not followed up at the time (see Monograph 2). At that early date, perhaps it appeared impractical to obtain useful detection ranges on ships or aircraft, using available types of equipment and components, which would be superior to the usual ranges achievable by a ship's look-outs.

THE FORMATIVE YEARS, 1935-7

The first successful British demonstration of the possibilities of radiolocation (radar) in detecting a distant aircraft was made by Watson-Watt and Wilkins, of the Radio Division of the NPL, in February

1935.[4] This was in response to a request from the Tizard Committee, set up to consider the urgent problems of recommending an effective technique for the Air Defence of Great Britain, based on recent scientific advances.

On 18 March 1935 the Director of Scientific Research (DSR) at the Admiralty, C. S. (later Sir Charles) Wright attended the fourth meeting of the Tizard Committee, accompanied by A. B. Wood of the Admiralty Research Laboratory (ARL); G. Shearing, Superintending Scientist in HM Signal School; and Cdr N. G. Garnons-Williams, representing the Director of Naval Ordnance. At this meeting the Admiralty representatives first heard of the technical possibilities of RDF. Subsequently Wright kept in close contact with Watson-Watt's experimental work at Orfordness, and arranged for A. B. Wood to be sent there for two months to keep him informed of progress. At Signal School only the Experimental Commander and Shearing knew of the RDF development work at Orfordness. However three members of the Orfordness research team (A. F. Wilkins, L. H. Bainbridge-Bell and E. G. Bowen) visited Signal School in April 1935 to arrange for the production of high-power silica-envelope valves, supposedly for ionospheric research. At the same time Bainbridge-Bell discussed the design of directional antennae for short radio-wavelengths (less than 20-m) with Horton and the present author.[5]

On 7 July 1935 Shearing, accompanied by some Naval Officers, visited the experimental RDF site at Orfordness, where detection ranges of up to 40 miles were then being achieved on aircraft. Shortly afterwards Shearing sent R. A. Yeo, a member of W. Ure's Transmitter Section, X2, at Signal School, with A. B. Wood to Orfordness to study current developments.

The new Experimental Commander at Signal School, F. J. Wylie, together with Shearing, attended a meeting held by DSD, Admiralty, on 2 August. As a result of this meeting the Third Sea Lord (also Controller of the Navy) decided that Signal School should initiate work on RDF as soon as possible.[6] Treasury approval was then sought for an increase in Signal School's experimental complement by one Scientific Officer and one Experimental Assistant, Grade II. In December these posts were filled by E. M. Gollin and W. P. Anderson, respectively.

Yeo remained at Orfordness from 2 September to 9 October, studying the Air Ministry and Army developments. At a meeting called by DSD on 11 October, at which (inter alia) DSR, the Captain and also the Experimental Commander of Signal School, and Yeo were present, a programme of development was agreed, which required an increase in the Staff complement of three Scientific Officers, three Assistants and an additional liaison officer at Orfordness.[7]

Accommodation was now found at the Royal Marine Barracks, Eastney (Portsmouth) for Yeo, Gollin and Anderson, the party remaining an offshoot of the Transmitter Section, X2, under Ure. Dr C. F. Bareford joined the party in June, followed by H. M. Bristow and W. H. Pritchard in November. By the end of 1936 there were still only five Officers and two Assistants working on RDF in Signal School, plus one Officer (W. S. Eastwood) seconded to Orfordness. This work was then classified 'Most Secret', with the consequence that access to relevant radar information was restricted to the following personnel in Signal School, apart from Yeo's party: the Captain, Signal School; the Experimental Commander, F. J. Wylie; the Application Commander, the Hon. J. B. Bruce; the Superintending Scientist, G. Shearing; F. Brundrett, Assistant to Shearing; and W. Ure, Head of the Transmitting Section. Thus, whilst Yeo's section worked in isolation at Eastney, Signal School's Experimental Department as a whole continued with conventional developments to equip the Fleet with the most advanced communications and direction-finding equipments practicable at that time, unaware of the activities of Yeo's group.

At the meeting on 11 October 1935,[8] tentative requirements were stated for an aircraft-warning range of 60 nautical miles and a ship-warning range of 10 miles, with precise location at 10 and 5 miles, respectively. (A nautical mile is approximately 2000-yds, as defined in the Glossary. Subsequent reference to miles in this Monograph implies measurement in nautical units.) It was suggested that the operating wavelength should be less than 25-m in use at Orfordness, and the shorter the better. Signal School was therefore directed to:

- develop the equipment operating on 1.5-m used for the current aircraft Homing Beacon, and
- investigate the utility of other wavebands, both longer and shorter than 1.5-m.

These two experiments were to be carried out in parallel. However it appears that no serious attempt was made to adapt the 1.5-m Homing-Beacon equipment for RDF application, and an operating wavelength of 4-m was chosen as alternative, on the basis that an antenna for any longer wavelength would be too large to install in a warship. It will be recalled that knowledge of the reflecting properties of both aircraft and ship targets to incident radiation was negligible at that time. In the case of the Orfordness equipment, the operating wavelength of 25-m was chosen on the basis that the wings of a typical bomber-aircraft target would provide the major component of reflected energy to the RDF set. Hence initial choices of operating wavelengths tended to be relatively long, supposedly to produce the best echo returns from the target.

In fact the choice of 4-m for the operating wavelength for the first experimental Naval RDF was unfortunate. This was because the design of available silica-envelope valves incorporated very long lead-filled seals, with the result that the valves only operated efficiently at wavelengths longer than about 7.5m (as in the Orfordness equipment). Yeo and Anderson designed a receiver, which they took to Orfordness for evaluation. They then modified a high-power shipborne communications transmitter, Type 7DX, which used silica-envelope valves similar to those supplied to Orfordness for the 25-m pulse equipment there. The experimental equipment was installed in a hut at Eastney in July for tests. Its immediate derivative, designated as Type 79X, was manufactured in the Signal School Workshops and installed in the Trials Tender *Saltburn* in the autumn of 1938. During trials in mid-December, aircraft were reliably detected at about 17-miles range – a promising start. Yeo's team also conducted experiments using various types of antenna, and operation on different wavelengths (including 23-cm equipment), and with different techniques of modulation.

A. B. Wood was assigned to Signal School from ARL in December 1936 to investigate the problems of generating cm-waves for RDF applications. Meanwhile the Controller of the Navy expressed the view that the detection of aircraft was of 'primary and immediate importance, whilst detection of surface targets was "vital"'. In response Signal School informed DSD, Admiralty, on 16 February 1937[9] that it would:

- pursue the RDF work on 4-m wavelength with maximum (available) effort;
- discontinue the work on 1.5-m wavelength, since there was no indication that this would give better results than 4-m;
- exert pressure on the new work exploiting the 23-cm wavelength;
- place a contract with GEC Research Laboratories for a communication set, Type 73X, operating on 60-cm wavelength; and
- propose placing a contract on industry for equipment operating on 20-cm wavelength for both communications and RDF applications.

In consequence Yeo's group continued to work during most of 1937 on the Type 79X RDF equipment operating on 4-m wavelength, whilst Gollin and Bareford commenced work on a 23-cm wavelength equipment on a site at Southsea Castle. Wood continued work on Infra Red (IR) target detection, since the relative performance capabilities of RDF and IR in operation against ship targets were not yet known. Wood was joined by O. L. Ratsey in 1937, and was involved in the investigation of the generation of wavelengths shorter than 23-cm. There was, however, little contact with Yeo's group at Eastney.

Meanwhile the Air Ministry Research Establishment (AMRE), Bawdsey, which now incorporated a large group from the Army Signals Research and Development Establishment (SDRE), were being very successful. For example, using the 25-m wavelength equipment, detection ranges of 100 miles plus were being achieved on aircraft targets. As a result the first RAF Chain Home (CH) Station was established in March 1937. E. G. Bowen, at AMRE was engaged in the successful development of both an airborne intercept (AI) system for use against aircraft targets using a 1.5-m wavelength, and also a variant for airborne detection of ships (ASV).[10] The latter system had been used to detect ships of the Home Fleet during exercises in murky weather on 5 September. Similarly the Army Group at Bawdsey (Pollard, Butement, Chivers and Friend) were developing gun-laying equipment operating on 1.5-m wavelength.

In contrast Signal School's achievement of a reliable detection range of only about 15 miles on an aircraft, and a single observation of 3 miles on a ship, were very disappointing in relation to the Staff Requirement. Range could not be measured at that time (although a technique for doing so was being investigated). However the bearing accuracy using the experimental 23-cm equipment was about $\pm 1°$. This fact led to the view that work on 23-cm equipment 'was required to provide the most satisfactory means of detecting low-flying aircraft and surface craft'. The overall position of Signal School's activities was subsequently reviewed at an important meeting on 8 September 1937. This was attended by DSD and DSR from Admiralty and several representatives from SS,[11] including Cdr B. R. Willett (Experimental Commander Designate), Captain A. J. L. Murray (newly appointed Captain, Signal School), Cdr H. M. A. Cecil (Application Officer for RDF), Shearing, Ure, Yeo, Gollin, Bareford and C. R. Evershed, who had taken over the administration of the Experimental Department from Brundrett.

Up till this time no special priority had been accorded to the experimental Naval RDF work in relation to other activities in progress in Signal School's Experimental Department. DSD said that this was because he had originally thought it necessary for the radar work to produce good results before asking the Board for unlimited facilities to further developments. He now believed that this decision was a mistake, and that other experimental work in SS should give way, if necessary, to the radar work.

The meeting took a number of important decisions. Development of Type 79X was to be of the 'highest priority' towards achieving the required warning range of 60 miles on an aircraft, and that subsequent sea trials were to be carried out in a cruiser. Additional facilities and extra staff (comprising four Scientific Officers, five Assistants and six Laboratory Assistants, plus one Senior Scientific Officer in the Transmitter Group) were agreed. A 'comprehensive work programme' was defined, with a number of staff to be allocated to each of the existing five radar work items (4-m; 1.5-m; 60-cm; 23-cm; spark). However, of the extra staff allocated, only two Laboratory Assistants were to be specifically allocated to Type 79X. (Two of the four Scientific Officers were for 'general analytical work'.) Further, the combined programme still had no designated leader, save the Head of the Transmitting Group, Ure, who was still occupied with other experimental programme responsibilities.

DSD's and DSR's subsequent submission to the Board on 28 October 1937[12] differed significantly from what had been agreed at the earlier Signal School meeting. The bid for extra staff was cut drastically (one Scientific Officer, one Technical Officer and so on), and it was 'proposed internally to release one Principal Scientific Officer from some less important duties to give close attention to the supervision of Type 79X'. Whilst it is not clear precisely what happened in the intervening weeks, it appears probable that pressure was put on SS not only to 'ginger-up' the Type 79X work under a senior officer, but also to bolster the effort applied on radar research from its own resources, at the expense of other work.

The Controller agreed to the scaled-down staff proposals very quickly, with the result that a major reorganisation of the radar work in SS was put in hand in November 1937, under the new Experimental Commander, B. R. Willett (see below).

It was during this period that continued pressure was being exerted on the Admiralty to centralise all research work on RDF at AMRE, Bawdsey, under Watson-Watt. This pressure was firmly resisted by the Controller of the Navy and DSR, Sir Charles Wright. The AMRE group was not in a strong position to counter the Admiralty's case (see the Introduction to this monograph) since AMRE was totally dependant on Signal School for the supply of the high-power silica-envelope valves needed for its research. Thus research and development activity on radar for Naval applications remained with Signal School throughout the war (and subsequently), in spite of a later renewed attempt to gain control by Watson-Watt when he became Director of Communications Development (DCD) at the Air Ministry.

THE 1937 REORGANISATION OF NAVAL RADAR RESEARCH

The Captain, Signal School, called a meeting in November 1937 to initiate the changes called for in the organisation of Naval radar research. Apart from Commander Willett, this was attended by Shearing, Ure, Horton and others. The event was unusual in that the Captain did not usually concern himself with the affairs of the Experimental Department, under Willett. In the event, the most important recommendation of the meeting was that Horton would take over responsibility for radar development in a new grouping, to be known as R Department. Figure 1.2 shows the organisation of the Experimental Department in Signal School in September 1937, just prior to the reorganisation, and Figure 1.3 the new arrangements for the conduct of radar development under Horton, in November. Yeo was replaced by A. A. Symonds as Head of R1 Division, responsible for the Eastney group. This now included O. L. Ratsey, following the appointment of A. B. Wood as Chief Scientist of the Admiralty Mining Establishment. The present author, who had been a member of Horton's DF Section since late 1930, became Head of R2 Division, responsible for *all* SS work on wavelengths less than 1-m, whether for RDF, radio direction finding or communications. A. W. Ross, who had originally joined the DF Section under Horton in September 1936, became Head of R3 Division to lead a small team to develop antenna systems for Type 79X. Yeo returned to the Transmitter Section in Signal School, whilst Symonds and Ratsey moved to Eastney. The author divided his time between the accommodation in RN Barracks and the Southsea Castle site, where Gollin and Bareford were working on the 23-cm set, whilst Ross was established at a field station at Nutbourne to conduct his antenna experiments in an interference-free environment. Gollin, Bareford and Burtt worked under the author in the DF laboratory in the RN Barracks, when not at Southsea Castle.

Horton's modus operandi was to discuss current technical problems with the personnel directly concerned, determine a course of action, and then leave it to the Staff involved to implement it. Thus Symonds and Ross began work on a new version of Type 79, operating on 7.5-m wavelength, whilst the author supervised all the work (whether RDF or communications) going on in Signal School involving wavelengths below 1-m, and in addition was responsible for trials of a television system, produced under contract with EMI, in HMS *Iron Duke*. At the same time he gave serious consideration to the future use of cm-waves for radar applications, bearing in mind that the 23-cm results were not very promising. Through his responsibility for the development of 60-cm wavelength communications equipment for the Navy at the GEC Research Laboratories, contact was established with E. C. S. Megaw, the leader of the group involved in these developments. The latter became a life-long friend and, through the television trials, he also became closely involved with A. D. Blumlein, C. O. Brown, Cork, Connell and others of EMI, who constituted the most professional electronics team in Britain at that time. The author profited enormously from these associations with Megaw and his colleagues at GEC and also with engineers at EMI.

Megaw was at that time the leading British worker on magnetron oscillators, and was investigating a complete communications system on 60-cm wavelength. Although unable at this stage to discuss RDF applications with Megaw for security reasons, it was possible for the author to form an appreciation of the possibilities of exploiting

EXPERIMENTAL DEPARTMENT
SEPTEMBER 1937

EXPERIMENTAL COMMANDER: Comdr F. J. Wylie RN

TECHNICAL SECRETARY
TO C. R. Evershed
T1 H.A. Brooks

EXPERIMENTAL DIVISION DIRECTOR
XO G. Shearing

W/T RECEPTION AND WC/TELEPHONES
- X1 W. F. Rawlinson
- X11 L. S. B. Alder
- X12 S. J. Moss
- X13 H. E. Hogben
- X14 P. T. W. Baker
- X15 J. W. Clarke
- R. L. A. Borrow

DIRECTION FINDING
- X4 C. E. Horton
- X41 C. Crampton
- X42 J. F. Coates
- X43 M. J. Gates
- X44 A. W. Ross
- X45 F. Briggs
- X46 D. H. Toller-Bond

SC/S and V/S
- XG H. Smith
- XG1 E. G. Hill
- XG2 (Miss) M. G. Marsh

DESIGN OF APPARATUS
- X7 R. L. Randall
- X71 S. E. Trigle
- X72 S. H. Trippe
- X73 R. R. Kent
- X74 G. W. Harris
- X75 G. Bishop
- X76 J. G. Farrow
- X77 F. G. H. Lewis
- X78 W. S. Offord
- X79 J. J. A. Mason
- X710 T. C. Roberts
- X711 J. G. Story

W/T TRANSMISSION
- X2 W. Ure
- X21 E. J. Grainger
- X22 J. E. Sheldrick
- X23 A. E. R. Ballard
- X24 C. Matthews
- X241 K. W. Blake
- X25 W. J. R. Merren
- X26 A. M. Hardy
- X27 J. D. S. Rawlinson
- X28 C. H. Webb
- X29 J. C. W. Drabble
- X210 R. A. Yeo
- X211 E. M. Gollin
- X212 W. P. Anderson
- X213 A. A. Symonds
- X214 C. F. Bareford

PRODUCT DIVISION DIRECTOR
PO W. S. Peake

SPECIFICATIONS, PURCHASE AND PROGRESS
- P1) G. D. Dewar
- P4)
- P10 E. S. Russell
- P40 D. W. S. Challans
- P41 G. W. A. Birkett

INSPECTING, TESTING AND CALIBRATING
- P2 T. H. Baines
- P20 A. G. Akehurst
- P21 E. W. Penny
- P22 G. C. W. Lane
- P23 H. E. Christie (Test Room)
- P24 F. H. P. Hoolahan
- P25 A. Dooley
- P26 A. J. P. Warlow
- P27 G. A. Terry
- P28 E. A. Barnsdale
- P29 R. McMillan

EQUIPMENT & FINANCE DIVISION DIRECTOR
DO C. L. Glenn-Bott

EQUIPMENT PROVISION PROGRESS & STORE LISTS
- D1 R. Savory
- D11 P. L. Ta Bois
- D12 A. S. C. Phillips
- D13 E. J. Ridge

EQUIPMENT LAY-OUTS & FITTING INSTRUCTIONS FOR SHIPS
- D2 Comdr R. F. Pitcairn
- D21 R. H. Garner
- D22 W. Austin Smith
- D24 A. L. Foort

JOB PROGRESS & DRAWING OFFICE
- D4 N. Pemberton
- D41 W. J. B. Holleyoak
- D42 E. Trevaskis
- D43 J. G. Barrett
- D44 G. V. Jenkins
- D45 C. O. Baverstock
- D46 W. Crofton

APPLICATION DIVISION DIRECTOR & DEPUTY FOR EX. CDR
AO Comdr G. R. Waymouth

DIRECTION FINDING
- D/F1 Comdr Hon J. B. Bruce
- D/F2 Lt F. W. B. Edwards

CO-ORDINATION OF W/T DEVELOPMENT
- A01 Lt Comdr J. D. M. Robinson

REVISION OF RESERVES OF W/T STORES
- A02 Tel. Lieut. R. W. Craig

RECEPTION Wa/T SUBMARINES R.N.V.R. R.N.W.A.R.
- A1 Lt. J. M. Villiers

TRANSMISSION. PORTABLE & MARCONISETS M.T.Bs. TRAWLERS. RESERVE FLEET. R.F.As, A.M.Cs, etc.
- A2 Lt J. W. McClelland
- A21 Tel. Lt W. E. Moss

CLERICAL DIVISION HEAD OF DIVISION
CO W. A. Rundle

MAIN OFFICE
- C1 H. Oliver

COSTING & ACCOUNTANCY
- C2 C. G. Ellis

SPECIFICATION AND PERSONNEL OFFICE
- C3 A. H. Childs

SHORTHAND TYPING POOL
- C4 (Miss) M. Duckworth

SPECIAL DUTIES
- A8 Comdr Hon H. M. A. Cecil

SHIP FITTING PW/TO. PORTSMOUTH
- A9 Comdr A. W. Loveband
- A91 Warr Tel. H. Bolton
- A92 Warr Tel. G. Raffery

19

X215 W. S. Eastwood
X215 J. Davis
X217 W. H. Pritchard
X210 H. N. Bristow

VALVE CONSTRUCTION
X3   H. G. Hughes
X32 C. O. Pringle
X32 A. L. Kirkham
X33 D. T. O'Dell
X34 J. F. Spilling
     W. S. Butt
     C. H. Morey
     N. H. Foot
     Miss Jollin

STANDARDS & L/F SIGNALLING
X8   A. Wilkinson
X81 R. E. Blakey
X82 A. W. Short

RESEARCH
XR1 A. B. Wood
XR2 O. L. Ratsey

INSPECTING AT CONTRACTORS' WORKS
P3   E. E. Tipper
P31 P. G. M. Hebert
P32

EXPERIMENTAL PRODUCTION & LIAISON WITH DOCKYARD PRODUCTION
P5   H. Noble
P50

IN CHARGE OF WORKSHOPS
P51 L. Fisher

D47 T. E. Bond
D48 H. G. Jolliffe
D49 H. R. Verry (Photographer)

EQUIPMENT LAY-OUTS & FITTING INSTRUCTIONS FOR SHORE STATIONS
D5   A. McPherson
D51 E. N. Lee

REMOTE CONTROL, POWER A.C. SUPPLIES. LAY-OUTS & AERIAL RIG
A3   Lt G. M. Bennett

VH/F. D/C. TENDERS INTER-SERVICE LIAISON
A4   Lt R. T. Paul
A41 Comd Tel. W. A. Sayers

SHORE STATIONS & SHORE ESTABLISHMENTS
A5   Lt Comdr C. J. W. Ryanson
A51 Tel. Lt C. C. Nash

STORES LIAISON & RESERVE OF STORES
A6   Comdr H. H. Rawlings
A61 Warr Tel. G. E. Bassett

SURVEY OF STORES
A7   Comd Tel. A. Wheeler

A93 Warr Tel. H. V. Drury
A94 Sig Bosn H. Coward
A95 Sig Lt G. Barton

CENTRAL STORES OFFICER
R1   G. S. Attrill

The following are shown in either the S.R.E. Dept Establishment List or the Navy List as serving in H.M. Signal School in 1937 but are not included in this organisation chart.

ASSISTANTS II
R. H. J. Brown
G. F. Shepherd

TEMPORARY ASSISTANTS III
H. Minns

LABORATORY ASSISTANTS
P. Maloney   P. L. H. Burgess
J. W. Webber   J. A. Mitchell
E. G. Banham   P. S. Ford
C. R. Gilmore   D. F. Small
W. H. Fielding   B. A. Howlett
J. C. McLaren   F. Eeles
H. Beckett

1.2   Organisation of the Experimental Department of HM Signal School in September 1937

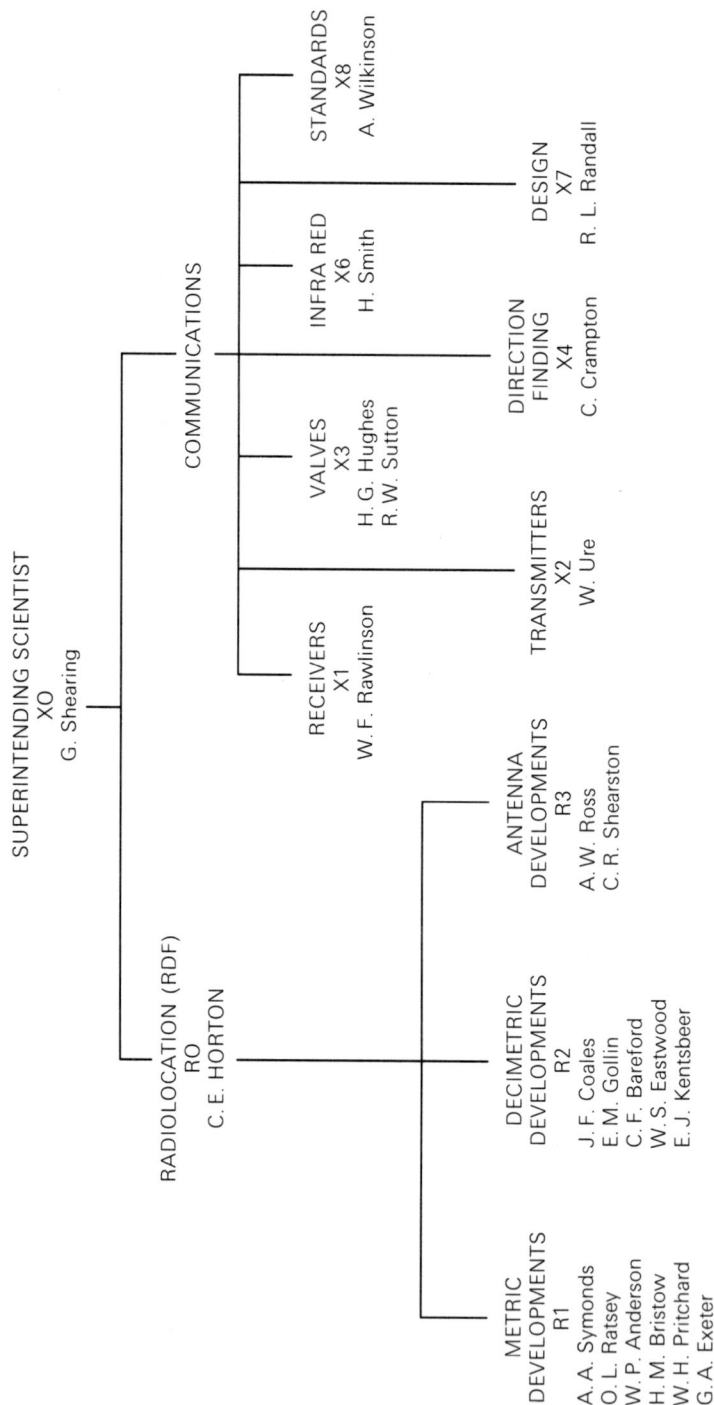

1.3 The November 1937 reorganisation of research (R Department)

wavelengths less than 1-m from Megaw's very wide knowledge of the field. Eventually, having observed the current practical difficulties in the use of magnetrons, the author decided to use high peak-anode voltages in a conventional triode pulse transmitter, operating on 50-cm wavelength. Horton agreed to the abandonment of the work on 23-cm wavelength, and to the application of future effort to the development of RDF on 50-cm wavelength for ship detection and gunnery applications. This was quickly approved by the Experimental Commander, Signal School, and endorsed by the Admiralty. It is interesting that a similar decision was reached by the developers of German Naval RDF at about the same time,[13] reflecting the similar standards of valve development in that country.

PREPARING FOR WAR, 1938–9

At the beginning of 1938 the highest priority was allocated to the development of RDF systems operating on 7.5-m wavelength for long-range aircraft warning, and 50-cm wavelength for surface ship detection and gunnery applications. Increases in staff were approved, and the necessary support facilities in Signal School were further developed. The clouds of war were obviously gathering, and a sense of urgency prevailed in Signal School, which quickly generated a spirit of enthusiasm. The foundations for equipping the Fleet with effective RDF equipment had been laid, and appropriate plans were put in hand.

Symonds' group at Eastney, including Anderson, Ratsey, Bristow and Pritchard, continued development of the experimental 7.5-m equipment. This was the shortest wavelength at which the NT57 silica valves employed in the transmitter would operate efficiently. The transmitter generated a pulse power of 20 kW at a pulse length in the range 8–30 microseconds, at a prf of 50. The receiver was a detector and 'straight' (that is, non-superheterodyne) amplifier, associated with a simple A-scan display of echo-amplitude versus target range. Ross's group at Nutbourne developed a common design of rotatable DF antenna array, to be used individually by both transmitter and receiver, which was suitable for mast head mounting in a ship. This, of course, involved two separate masts for the system. A full account of the complete development work is given in Monograph 4.

This experimental equipment was developed and manufactured using only Signal School resources, before being installed in *Saltburn* for sea trials in 1938. In these trials the transmitting and receiving antennae were installed on the main and foremasts, respectively, in what was to become the standard arrangement in the Navy until common antenna working was developed.

Following successful initial trials in *Saltburn* (see Monograph 4), decisions were taken at the Admiralty on 25 March 1938 as follows:

- Two equipments of the experimental type fitted in *Saltburn* (to be known as Type 79BX) should be installed in two ships of the Home Fleet as soon as possible.
- Thorough sea trials of an improved Type 79X should be carried out in the cruiser *Dunedin* in mid-1939.
- All the work involved should be carried out in Signal School.

The estimated requirements in resources to effect this programme were approved in due course, and the ships designated for the first sets, Type 79BX, were the cruiser *Sheffield* and the battleship *Rodney*.

An important consideration in the development of the original experimental Type 79 RDF was to determine whether horizontal or vertical polarisation would be the most effective for detection of aircraft targets over the sea surface. There were two facets to the problem. On the one hand, in common with the Bawdsey Group, it was at that time thought that horizontal polarisation would give larger target returns from aircraft at metric wavelengths, the principal reflected components being from the wings. On the other hand was the question whether, in practice, vertical polarisation would provide a more effective transmission path at low angles of elevation over the electrically conducting sea surface than horizontal polarisation, as indicated by theory. Ross therefore conducted experiments with Type 79X in *Saltburn* to investigate these effects. He found that, for some unknown reason, vertical polarisation was not as effective as expected, and in addition resulted in increased unwanted clutter returns from the sea surface. This latter effect had not been expected. Arising from the results of these trials, horizontal polarisation was therefore adopted universally for Naval radar (unlike the German Naval equivalents at the time).[14]

The inevitable lobes in the radar vertical-plane coverage diagrams over the conducting sea surface slope upwards more steeply in the case of a low, rather than a high-sited antenna. Also, these lobes slope upwards more steeply for a long, rather than a short wavelength of operation. There are obvious restrictions on the permissible height of a radar antenna in a ship. The *Saltburn* trials provided the essential clue that adequate radar coverage in the vertical plane against low flying aircraft targets could only be achieved by the use of much shorter wavelengths in the radar system, such as the dm or cm wavebands.

The completion dates for the refits of *Sheffield* and *Rodney* were August and October 1938, respectively. Thus from April onwards activity in Signal School was intensive to manufacture the equipments (now designated as Type 79Y), install them and set them to work in the ships concerned. This was achieved successfully. In sea trials *Sheffield* reported detection of aircraft at 30 miles for 3000-ft altitude, 40 miles for 7000-ft altitude and 53 miles for 10000-ft altitude. Following this success, it was decided to proceed immediately with the manufacture of 40 more equipments for an extensive ship-fitting programme, commencing in July 1939. In order to expedite this programme J. D. S. Rawlinson, who had wide experience in commissioning high-powered Naval W/T Stations in the Far East, was detailed as Officer-in-Charge, Eastney, in January 1939.

Some teething troubles had been experienced during the Type 79Y trials, and these had to be overcome before going into production. The principal ones involved the very rapid development of an entirely new modulator by D. S. Watson and the introduction of a stable superheterodyne receiver, based on that developed by A. W. Ross as part of the equipment at Nutbourne for the measurement of antenna polar diagrams. Also, the introduction of thoriated-tungsten filaments into the silica-envelope transmitting valves at just this time resulted in the output power being raised from 20 kW to 70 kW. The improved radar system was then designated as Type 79Z.

The first of the improved Type 79Z equipments was installed at Eastney for trials in July 1939. Excellent detection ranges were recorded on aircraft targets, as follows:

- 40 miles at 3000-ft altitude
- 70 miles at 10000-ft altitude
- 90 miles at 20000-ft altitude
- 120 miles at 30000-ft altitude

Thus the requirements originally stated in 1935 were now met.

Another model of Type 79Z which was also made in the Signal School workshops, was installed in the Anti-Aircraft (AA) cruiser *Curlew*, in August 1939. Soon after the outbreak of war, this ship took station at Scapa Flow to augment the surveillance for approaching aircraft, since coverage by the RAF in this area at the time was provided only by a small mobile set. *Curlew* consistently detected aircraft flying at 1000-ft altitude at ranges of over 60 miles.

EARLY DEVELOPMENTS IN 50-CM RADAR EQUIPMENTS

It was evident that metric wavelengths were more suitable for air warning applications than decimetric waves (1-m to 10-cm, incorrectly referred to as centimetric waves at the time). The immediately obvious task for the author's R2 group was to develop RDF for gunnery applications, using wavelengths in the decimetre band to exploit the possibilities of improved range and bearing determination on distant targets. The problem lay in the generation of a sufficiently high transmitter output power on these wavelengths.

The experimental FM CW radar transmitter made by Gollin used a CW10 split-anode magnetron developed by Megaw at the GEC Research Laboratories, but the available power output was only of the order of 200 mW. A search of the literature established that, of all the possible ways of generating decimetric waves, the split-anode magnetron, producing at most a few watts output, was the best available. As stated above, the problem was discussed in detail with Megaw, since he had achieved a power output of a few watts CW in his Naval communications project. Taking into account the earlier experience on the 23-cm RDF project, and that of Lorenz in setting up a cross-Channel communications link, it appeared to the author that the longer wavelengths (60-cm) offered the most promising possibilities for RDF development. Since it was very problematical whether useful output powers could be achieved by pulsing available magnetrons with higher anode voltage, the author decided to initiate RDF development using the shortest operating wavelength that could be generated using available triode valves. A subsidiary reason for this decision lay in the very poor performance of Bareford's receiver operating in the 23-cm band, coupled with the fact that it had not been easy to make a satisfactory receiver for the 60-cm communications project.

The Western Electric company in the USA had recently developed a valve known as the E316 ('Doorknob') valve capable of producing a few watts on 3-m wavelength. It seemed plausible that, with pulsed operation using a higher anode voltage, it might prove possible to achieve reasonably high peak-power outputs at operating wavelengths of less than 1-m. After calculations of the transit times of the electrons in the E316 using different anode voltages, it was found that the valve would almost certainly oscillate at 600 MHz (50-cm wavelength) with a 1 kV anode voltage applied. In February 1938 Horton agreed that future development work should take place on a wavelength of 50-cm, the 25-cm project being abandoned.

During a visit to Signal School from E. G. Bowen of AMRE, Bawdsey, he was informed of this decision. Bowen never visited Signal School again, and so was not acquainted with the Naval 50-cm equipment whose development, by mid-1939, would almost certainly have been more suitable for his airborne radar developments than the 1.4-m wavelength then in use. In this connection, Sir Bernard Lovell records that some experimental work was done on 50-cm wavelength at AMRE, in the interest of reducing ground returns.[15]

In April 1938 H. Calpine joined Signal School from Standard Telephones and Cables, Ltd, where he had been engaged on fundamental research on thermionic valves. He replaced Gollin, who had been transferred to the author's section in 1937. Calpine immediately commenced design on a pulsed transmitter incorporating two of the Western Electric E316 valves separated by a quarter-wavelength concentric line, the valve anodes being connected to the outer conductor and the grids to the inner. Anode voltages of abut 3 kV were applied from a thyratron modulator, using a concentric-cable delay-line to produce microsecond pulses. This arrangement produced a peak-power output of more than 100 watts. Meanwhile Bareford developed a superheterodyne receiver that incorporated a commercial 45 MHz TV receiver as its IF amplifier. The output was fed to a simple A-scan CRT display on a tube of 5-in diameter.

It was quite evident that a radar operating on 50-cm wavelength would have a relatively short range-capability, most suitable for gunnery applications. In this case several equipments would be needed in the same ship. The associated antenna system would require a narrow beamwidth for target discrimination purposes, but also a small swept radius. These are normally incompatible requirements. For example, the simplest form of highly directive antenna system would have utilised a half-wave dipole at the focus of a parabolic reflector. However this would have needed to be some 20-ft in aperture to produce the antenna gain and directivity required in the RDF application. The author therefore devoted considerable effort to the investigation of the possibilities of developing a three dimensional Franklin-beam type of antenna array, which would be highly directional, and maximise the use of the volume swept by the array in rotation. A preliminary experiment with an array of 24 half-wave dipoles provided useful gain. However there were formidable problems in adjusting the phases of the individual dipole elements – owing to the difficulties of determining the mutual impedances between dipole elements in different parts of the array – with the test equipment available at that time. Attention was therefore turned to the Yagi ('fishbone') type of antenna, so familiar today for domestic TV reception.

The radiation pattern of Yagi's original antenna possessed relatively large side-lobes in addition to the main lobe, and these would have to be suppressed for the Naval RDF application. This antenna used a single reflector element spaced behind a half-wave-driven dipole, plus up to 20 director elements spaced in front of the dipole, all of the same length. The author, however, considered that by using different lengths for the directors and hence varying the mutual impedances between elements, it might be possible to reduce the unwanted side-lobes and thus increase the gain of the array. He started to investigate this experimentally at the Nutbourne field station. When W. F. Drury joined his Section, he continued the work until R. V. Alred joined a little later. He succeeded in reducing the magnitude of the offending side-lobes and in marginally increasing the Yagi antenna gain, some 20 years before the same technique was applied to domestic TV antennas. However the magnitudes involved were not sufficiently great to warrant further work on directors of differing lengths at that stage. R. V. Alred showed that the unwanted back-lobe of the Yagi could be greatly reduced by replacing the single reflector element behind the driven dipole by a cylindrical reflecting screen. Structural support for the antenna array proper was provided by installing tie rods from the corners of this reflector to the far extremities of the 'fishbone' array. In due course this type of array became the most widely fitted type for 50-cm Naval RDF.

At the time of the 50-cm development, one of the greatest problems lay in designing a satisfactory type of receiving system. Suitable triode valves to provide a good RF detector, much less an amplifier, were not available. The most promising solution lay in developing a superheterodyne receiver with a diode mixer, with the autodyne as a possible alternative. After first trying a superheterodyne, Bareford developed an autodyne receiver, using a harmonic of a 200 MHz oscillator. This was designated the P14 receiver. It was used in all RDF system trials up till the middle of 1940, and also for demonstrations at the AA Range at Eastney. However the performance of the P14 was not really satisfactory and it was later replaced by a superheterodyne receiver P16, designed by J. Croney (see below).

Meanwhile the author was conducting negotiations with the Pirelli Company for the production of a low-loss feeder cable using polythene insulation. Such a cable possessed a much higher protection against insulation breakdown when the peak power was increased than available air-insulated types, which were susceptible to moisture ingress.

In the autumn of 1938 the author was shown by R. Le Rossignol at the GEC Research Laboratories a new design of a small triode valve capable of providing an output of several watts CW at a frequency of 100 MHz. This valve had a finned cylindrical anode with a copper/glass seal, and a grid brought out through another copper/glass seal at the side. It immediately occurred to him that if the grid and anode connections could be made concentric, the valve would be ideal for Calpine's transmitter, Since it was designed for an anode voltage of 5 kV. since the GEC Laboratory staff were not then cleared for access to RDF techniques, it was not possible to discuss a redesign of the valve with them, and on return to Portsmouth the author discussed the problem with Horton. He explained that a request for redesign could only be because it was required for application at a much higher frequency, using a tuned concentric-line as the oscillating circuit. It was not possible, therefore, to disclose the particular function for which the redesigned valved was required without infringing the security restriction concerning RDF. Since a successful redesign could well increase the peak power output at least 10 times, Horton agreed to ask DSR Admiralty (Sir Charles Wright) to give permission for the GEC Research Laboratory representatives (who were already cleared to Secret level), to be included on the 'Most Secret' list of people having knowledge of radar. Ultimately Sir Charles gave his permission for disclosure of the requirement to GEC, who developed the modified valve, later designated as the E773. He felt, however, that other valve manufacturers must also be given the opportunity to compete with GEC, and arising from this, the 'Coordination of Valve Development' Committee (CVD) was organised to coordinate the efforts of other valve manufacturers. (There still remains controversy as to whether CVD stood for Coordination of Valve Development or Communication Valve Development, in an attempt to preserve secrecy.) Because of its origins CVD remained for more than 20 years under the aegis of the Admiralty, in spite of its activities on behalf of all the Service Departments. Ultimately CVD was concerned not only with triodes, but also with magnetrons, klystrons, travelling-wave tubes and other devices that became essential components of the later microwave radar system[16](see Monograph 3).

The impetus given to valve development by the CVD organisation eventually also benefited the 50-cm radar receiver programme. This occurred in 1941, when STC developed a grounded-grid triode whose use in an RF preamplifier greatly reduced the inherent noise level generated in the receivers (see Monograph 3).

The first 50-cm transmitter using the redesigned triode valves in the quarter-wave coaxial-line circuit, the anodes and grids being connected to the outer and inner conductors, respectively, developed peak powers of more than 1kW. There was, however, great difficulty in extracting the generated power and feeding it to the antenna effectively. This problem was brilliantly solved by Calpine, who conceived the idea of mounting the oscillator in a tuned cavity, or 'Hohlraum', which incorporated a dipole output feed situated at an appropriate position in the cavity. Calpine designed such a cavity resonator, comprising a copper cylinder about 12-in diameter and 18-in high. Although first tried using DET 12 valves Calpine set up the resonator with the new GEC E773 valves on 19 November 1938. By April 1939 the transmitter provided a peak-power output of over 600 watts using an anode voltage of 4600 V. This was raised to 1.2 kW by June – almost certainly a 'World first' in exceeding 1 kW output at 50-cm wavelength.

The increase in transmitter power output itself raised difficulties in the design of a suitable modulator to provide pulses in excess of a kilowatt at the 5000V level. Discussions were initiated with the BTH research Laboratories (L. J. Davies and H. de B. Knight), which resulted in a very successful programme of research on the development of high-voltage thyratrons. Up to this time, thyratrons had been low-voltage, very high-current devices for use in arc welding, the reverse of that required in a radar modulator.) This culminated in a design capable of producing peak powers of more than 5 kW.

Early in 1939 all the components of the 50-cm radar were assembled at the Southsea Castle site for trials using *Sardonyx* as target. Good echoes were obtained at a few hundred yards range which, although not impressive, provided a more successful performance than that achieved on 23-cm wavelength. Plans were therefore made to carry out trials with the equipment installed in *Sardonyx*, in June 1939. The antenna system, comprising an assembly of four driven dipoles, each with its own rod reflector, was of only moderate gain. However good echoes were received using the Nab tower as Target, at ranges up to 4 miles. Later, Yagi antennas of superior performance to the dipole array were developed, and a more powerful transmitter was produced for further system trials in October.

At this time, there were still only just over 20 qualified engineers and technicians working on radar in the Signal School laboratories. In spite of war clouds gathering on the horizon in July 1939, Signal School still had to remind the Admiralty[17] that work on RDF took second place to that on communication equipment for new-construction ships! (see Monograph 2). The Captain's Memorandum to DSD Admiralty delineated new lines of development required on RDF work, and outlined the increases in staff and the facilities required to implement such a programme. It was anticipated that the laboratory staff engaged on RDF, plus the ancillary grades of designers, draughtsmen and mechanics, would need to be doubled. Also, there would need to be a significant increase in the administrative and production staff associated with the radar work, to make new equipments available to the Fleet.

These matters were, therefore, considered at the Second Naval Radar Panel Meeting on 27 July[18] under the Chairmanship of ACNS. (the first meeting of the Panel had been held on 16 December 1938). The Panel was informed that *Sheffield* and *Rodney* were already equipped with Type 79Y, and that *Suffolk* and *Curlew* would be soon; also that *Valiant*, plus the AA Cruisers *Curacoa*, *Capetown* and *Colombo* would be equipped by the end of the year. The names of ships were decided for the next 30 equipments coming from production, at the rate of three or four a month, from mid-1940. Priority within that group was accorded to AA Cruisers, but no mention was made of aircraft carriers. All new-construction and large-reconstruction ships (which included the *Illustrious*-Class Carriers) were to be prepared for fitting. One Army GL1 fire-control set was to be fitted in *Carlisle* by the end of 1939, and it was forecast that another 200 GL equipments of some form would be required. There was also a need for the supply of some 600 Air-to-Surface (ASV) equipments for the Fleet Air

Arm, although it had been agreed earlier that these would be obtained from the Air Ministry. The latter would also be asked to supply 2000 Identification, Friend or Foe (IFF) equipments, whilst Signal School developed an IFF set for use in ships and merchantmen (see Monograph 6). Finally, the priorities laid down at the First Meeting of the Panel on 16 December 1938 were reviewed, with the following result:

Priority 1   Long-range warning of aircraft (already largely met by Type 79Y, but still capable of improvement, particularly in relation to the liability of DF by an enemy).
Priority 2   (a)   Ship identification, from air and ship.
(b)   Range and bearing capabilities for surface gunnery.
(c)   Range and bearing facilities for long-range AA gunnery.
Priority 3   Warning of approach of aircraft at 20 – 25 miles range, and of ships at 6–10 miles range, on the frequency least liable to radio-direction finding by an enemy.
Priority 4   Continuous location of aircraft by bearing and elevation, for searchlight control (SLC).

The glowing reports of the performance of Type 79Y in *Sheffield* and *Rodney* had not been lost on Vice-Admiral Sir Bruce Fraser, the Third Sea Lord and Controller of the Navy. Immediately after the second meeting of the Naval RDF Panel he sent a Memorandum[19] to ACNS, DSD and DSR stating that, if RDF developed as expected, it would revolutionise the Navy's methods (of waging war). He asked for radar development to be given the first priority, and requested that DSD and DSR present their proposals and requirements by 18 August so that the necessary facilities could be provided as soon as possible. On 10 August a Memorandum from their Lordships was sent to C-in-C Portsmouth stating that 'It had been decided that the development and supply of RDF for Naval purposes is to be extended, accelerated, and given the highest priority'.[20] This was less than one month from the date at which the United Kingdom declared war on Germany.

WARTIME DEVELOPMENTS

Shortly after the outbreak of war on 3 September 1939, Signal School's requirements for large increases in staff and resources to be provided for radar development were quickly implemented. The Valve Section, which had already been strengthened in 1938 by the appointments of T. J. Jones, F. M. Foley and R. W. Sutton from industry to meet the demands for silica-envelope valves for both the Navy and the RAF, was given a greatly increased number of technicians (see Monograph 3). The Radar Design

Staff recruited W. D. Mallinson, C. E. Fenwick (a New Zealander) and M. J. Jones, and the Production Department took in W. A. P. Wykeham, all from industry. N. Shuttleworth, from ARL, took over as Head of the Design Department, and B. Hodgson (who had headed the original Receiver Section in 1919) came back from retirement to assist. In order to meet the requirement for new experimental and developmental models for fitting in ships before full production could be implemented, the Signal School workshops were expanded and additional Naval Artificers were provided. New buildings were erected at Eastney, and a number of buildings in Southsea (including an old school) were commandeered. The Nutbourne Antenna Group was expanded by the inclusion of G. E. F. Fertel, S. H. Falloon and others early in 1940. H. E. Hogben, who had joined the Receiver Section of Signal School in 1927, was transferred to Eastney, and was soon followed by J. Croney and L. A. Moxon, new recruits from industry.

Brundrett (the 'power behind the throne' in the Experimental Department of Signal School since the early 1920s) had moved to DSR headquarters in 1937 to recruit scientists and engineers, and allocate them to the most appropriate Naval establishments. By this means he recruited E. B. Moullin (Reader in Engineering Science at Oxford University) plus two of his students, J. R. du Parcq and A. G. Bogle (from New Zealand); L. B. Turner from the Engineering Department at Cambridge; and S. E. A. Landale from Edinburgh (where he had been Technical Director of his family's Brewery – Youngers).

Meanwhile Rear Admiral Bruce Fraser, Controller of the Navy, recalled Vice-Admiral Sir James Somerville for special duties in Admiralty, to provide coordination and development of RDF applications for the Navy. (Somerville had been invalided from the Navy only a few months earlier with suspected pulmonary tuberculosis, but had made a complete recovery.) A former Signal Officer and the first Experimental Commander of Signal School from 1917 to 1920, Somerville used his knowledge, drive and seniority to press the Navy's case for RDF, and to galvanise the Navy, the Dockyards and the relevant sections of industry involved. In particular he had the fullest confidence and support of the First Lord of the Admiralty himself – Winston Churchill. Somerville was largely responsible for the rapid equipment of ships with radar during 1940 and 1941. He was described as 'the foster-father of Naval radar' by Watson-Watt.

At this time Watson-Watt (now Director of Communications Development at the Air Ministry, a cover for his radar activities) made a renewed bid to centralise all RDF research under his own control. This would have involved moving the Naval experimental work from Signal School to Dundee, where the combined RAF and Army research activity had been temporarily dispersed from Bawdsey for security reasons.

Neither Somerville nor Sir Charles Wright, DSR, would agree to this proposal, and the Board of Admiralty accepted their view. Nevertheless Watson-Watt still continued to try to assume responsibility for valve research and development. However, because the main preponderance of valve development at that time was either in Signal School or in industrial firms controlled by CVD (under Admiralty sponsorship), their Lordships would only agree to permitting Air Ministry establishments to undertake such valve research as was required for their own radar development. A strengthened Valve Section was dispersed to the H. H. Wills Laboratory at Bristol University under R. W. Sutton, in the interests of relative safety from air attack. This was in response to a letter from Churchill, following his visit to Signal School as First Lord of the Admiralty on 21 September 1939, to the Controller, ACNS, Somerville, DSR and DSD calling attention to the risk of having all radar-research facilities concentrated in such a vulnerable locality as Portsmouth.

Another ploy by Watson-Watt at this time was to persuade the Chief of the Air Staff to ask the Admiralty to release Horton to become his deputy. Ultimately Churchill vetoed the proposal, and Horton was promoted to Superintending Scientist at about the same time that Willett, the Experimental Commander in Signal School, was promoted to Captain. The Air Council then recommended that DSR Admiralty should be responsible for the allocation of all new staff to the different Service research establishments. Brundrett, the Assistant DSR, Admiralty, was appointed to give this effect. In particular recruitment of new professional staff for Signal School was entirely in his hands.

THE NEED FOR A SURFACE DETECTION CAPABILITY

The ship-to-air warning (SA) Type 79 radar operating on 7.5-m wavelength was now well established with the order to fit 40 equipments in large ships in 1940, both those in Dockyard for refits and in new-construction ships such as the *King George V*-class, the *Illustrious*-class carriers, *Fiji*-class heavy cruisers, *Dido*-class light cruisers, and *Hecla*-class depot ships. However, the need remained for an equipment capable of detecting surface ships and low-flying aircraft, defined at the meeting of the Naval RDF panel on 16 December 1938. Obviously, shorter wavelengths than 7.5-m would be necessary, and at the outbreak of war a decision on what this should be was still pending.

The preferred operating wavelength was 50-cm for such a Ship-to-Ship (SS) equipment, on which the author's group had been working since 1938. However sea trials in *Sardonyx* on 14 October 1939 were disappointing in the SS role, whilst on low-flying aircraft at only 200-ft altitude the ship-to-aircraft (SA) performance gave a maximum range of

5000 yards. In fact the antenna system used was somewhat makeshift, and considerable losses were experienced in the air-insulated feeders due to damp ingress.

Somerville, Willett and Horton then decided that, as an interim solution to the SS problem, a ranging panel adapted from the Army's GL1 set should be added to the existing 7.5-m equipments (that is, SA, Type 79), together with a shorter (3-microsecond) pulse, and a new design of receiver. A range accuracy of 50-yds out to 14,000-yds was expected, but bearing accuracy would be relatively poor at $\pm 10°$ if reliance were placed on simple beam-swinging techniques, under hand control, to assess the point of maximum echo. However, with the use of a simple operator aid (see Monograph 4), this could be reduced to about $\pm 2°$

It was also decided that a new SS set, to be designated as Type 281, should be developed to fulfil the requirements as:

- a long-range aircraft-warning set (WA),
- an AA range-finder, and
- a main-armament range-finder,

although it was accepted that in operation these functions were mutually exclusive.

J. D. S. Rawlinson's group at Eastney set to work with vigour to implement the various proposals. Trials with the modified Type 79 were conducted in the AA Cruiser *Curlew* in March 1940, and in consequence all the existing Type 79s (except those in *Rodney* and *Sheffield*) were to be modified, being reclassified as Type 279.

The group then devoted much effort to the development of the new 'general-purpose' set, Type 281. In order to increase both detection range (particularly on surface targets) and bearing accuracy, it was essential to use a shorter operating wavelength. Fortunately in mid-1939 the Valve Section in Signal School had introduced new, shorter seals (developed by Mullard Co, see Monograph 3), which enabled efficient power generation down to a wavelength of 3.5-m. A larger-diameter silica envelope than that used in the NT 57 was adopted, and this enabled a larger thoriated-tungsten filament to be used, resulting in a peak emission of at least 100 amps (see Monograph 3). Using a pair of these valves (designated NT 86), J. S. Shayler developed a transmitter producing a pulse power of 350 kW at 15 microseconds pulse length, and 1MW at 1.7 microseconds pulse length. S. E. A. Landale developed the high-power modulator required, and L. A. Moxon the associated superheterodyne receiver. An antenna with a higher gain than that for Type 79, with a narrow horizontal beamwidth, was designed by A. W. Ross. The overall equipment operated in the 86–94 MHz band, and was now designated Type 281. A detailed description is given in Monograph 4.

The first experimental model of Type 281 had been produced in the Eastney laboratories by May 1940, and the first development model was operating there in time to plot the first air raid on Portsmouth, on 11 July 1940, on its approach from the French coast. Shipborne trials were carried out in the cruiser *Dido* in October 1940, as a result of which the Admiralty placed a production order for 36 sets.

Early in the war it appeared that use could be made of the Army's Gun Laying (GL1) anti-aircraft radar, developed at Bawdsey, in attempting to solve the Navy's SS problem. The Army set[21] operated on 3.5-m wavelength. J. D. S. Rawlinson's Eastney group undertook the adaptation, to be designated as Type 280, for fitting in *Carlisle* by the end of December. This involved design of a new antenna system and rotatable mounting by Ross' group at Nutbourne, for installation on the main and foremasts of *Carlisle*, plus design of a new display system at Eastney. The antenna pedestals were installed at Devonport dockyard in December, and a number of different types of antenna system were taken in *Carlisle* to Malta in February 1940, by O. L. Ratsey and G. E. F. Fertel, for trials evaluation.

In these Malta trials, aircraft were readily detected with accurate ranges (50-yds), and bearings to $\pm 2°$ for the gunnery system when there was no interference from unwanted land echoes, or when the aircraft target was not too low. Radiation patterns of the antenna systems in both the vertical and horizontal planes had been calculated, and were confirmed during the trials. Thus estimation of target height was good, and gave the high-angle (HA) director an elevation bracket to search. *Carlisle* subsequently served with distinction in the Norwegian campaign in 1940, both for guiding the guns of ships in company (not at that time radar-equipped), and also for directing fighters on to attacking aircraft. As a result it was decided to fit six more ships with this equipment (Type 280), starting in August 1940.

THE SMALL-SHIP RADAR REQUIREMENT, 1940

The evacuation of the Army from Dunkirk in May and June 1940 brought home to the Admiralty the urgent need for radar in small ships, such as destroyers and mine-sweepers. Such a set, which would need to be compact, would be required in large numbers. Fortunately Bowen at AMRE had developed a small airborne ASV radar operating on 1.4-m wavelength,[22] which seemed capable of modification for shipborne use. The task was given to Bogle, who had originally been engaged in attempting to adapt the RAF's IFF sets for shipborne use. The RAF radar was just in quantity production. It provided a peak-power output of 7 kW, at a pulse length of 1.5 microseconds and a repetition rate of 200.

The equipment, designated Type 286, was very quickly adapted for shipborne use, incorporating a light weight antenna system designed by Falloon at Nutbourne (see Monograph 4 for dull details). The whole equipment was then installed in *Ambuscade*, at Portsmouth, for brief trials, which showed a worthwhile performance. Falloon was then delegated to design a rotatable antenna system for masthead mounting, suitable for future ship-fitting. The improved system, known as Type 286P, was quickly in production, and was fitted in ships from February until July 1941. It was then replaced by a new set, Type 291.

THE REQUIREMENT FOR GUNNERY AND FIRE-CONTROL RADAR (50-CM)

Meanwhile the Director of Naval Ordnance had asked Signal School about the possible provision of a range-finder system for use against dive-bomber targets at ranges up to 5000-yds. The idea was to combat enemy dive-bombers by firing the main armament of battleships and cruisers at a predetermined, preset range, the shell fuses having been set to the flight time in accordance with the range rate established by the radar.

It was therefore decided to carry out trials of the 50-cm radar at the AA Range at Eastney using two of Alred's design of improved Yagi antenna mounted on a close-range director, normally associated with a multiple pom-pom gun mounting (see figures 1.4 and 1.5).

The transmitting and receiving antennae were connected by coaxial feeder to the associated equipments situated in a nearby office. Successful trials were carried out in which low-flying aircraft were detected at about 5000-yds range, and tracked on a CRT A-scan down to a few hundred yards. Following a demonstration to representatives of DNO – Captain K. Eddon, RN, DDNO(F), and the Director of Training and Staff Duties, Cdr. SW. Roskill, RN, (DTSD) – in February 1940, it was agreed informally to proceed with the project. A formal directive from Admiralty to proceed with the production of 200 sets, now designated Type 282, in accordance with a Staff Requirement raised by DNO, was received in Signal School in April 1940.

Quite clearly, in order to meet all these requirements Signal School would need a great deal of help from industry. Fortunately E.V. Appleton, now Secretary of the Department of Scientific and Industrial Research (DSIR), had already enlisted the help of some of the leading electronics firms for the development of RDF. Early in January 1940 he attended a meeting in Signal School with representatives of GEC, the Marconi Company and STC Ltd. GEC was already collaborating closely with Signal School on the 50-cm work, and the secondment of staff from

1.4 Close-range director with experimental 600 MHz Yagi antennae at the AA Range, Eastney (© Defence Research Agency)

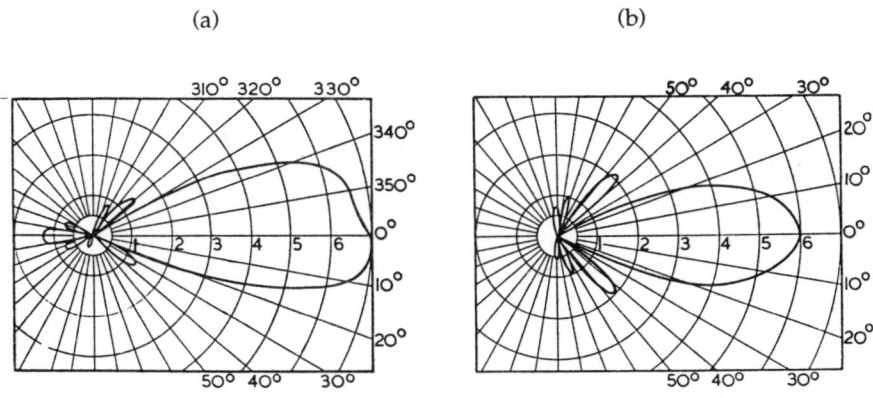

1.5 (a) Horizontal polar diagram of the Yagi antenna installation; (b) vertical polar diagram of the Yagi antenna installation (© Defence Research Agency)

Marconi (N. E. Davis, S. H. H. Falloon, G. N. Coop) and STC also added stimulus to the meeting. G. M. Wright, who had earlier been Chief Engineer of the Marconi Company, joined Signal School as head of the Radar Countermeasures Group,[23] and O. Böhm (formerly holding a key position in the Telefunken Company before fleeing from persecution in Nazi Germany) joined Ross' antenna and propagation group at Nutbourne. Wright's younger son, Paul, also joined Sutton's valve group at Bristol.

Cooperation with industry was further extended by the BTH Company (which was already involved in the development of suitable thyratrons for modulating the 50-cm transmitters) in seconding Fossey and some other staff to assist Mallinson with the design of equipment for large-scale production, and I. Yorke to assist with power systems. On the production side, Aeronautical and General Instruments (whose proprietor, Mr Joseph, was already well known in Signal School) was awarded the contract to produce the 40 models of Type 79Z required for fitting in the Fleet in 1940. Willett and Brundrett also arranged for Messrs Allen West of Brighton to become an outpost of Signal School, to help with design and manufacture of prototypes and development models based on experimental equipment devised in the laboratories. Mr Paddick of Allen West spent most of his time over the next few years in Signal School to provide close liaison with the works at Brighton.

DEVELOPMENT OF NAVAL GUNNERY AND FIRE-CONTROL RADAR SYSTEMS

At this time none of the designs for the individual component units for 50-cm radar had been finalised, yet within six months plans had been laid for the production of 200 complete equipments. Nevertheless it was essential to expand the 50-cm group to 10 officers and six assistants. Ratsey was transferred from Eastney, and new recruits H. W. Pout, J. du Parcq and E. F. Daly were drafted into it, together with W. D. Mallinson and G. W. Jones from the design staff. The enlarged R2 Division was necessarily transferred to new accommodation – a disused school in Onslow Road, Southsea. It was, however, evident that much design assistance and the preparation of manufacturing drawings would have to be entrusted to the appropriate manufacturing firms.

It was clear that the author's 50-cm group (still numbering only about 20) could not possibly complete the development of transmitter, modulator, receiver, display unit, antennae, feeders and test equipments in a form suitable for large-scale production, and operation and maintenance at sea. Arrangements were therefore made for GEC to manufacture the transmitters, a special unit being set up at Wembley to manufacture the E773 valves; for BTH to design and manufacture the modulators and associated 500 c/s generators; and for Marconi to design and manufacture the receivers.

At this time the design of suitable CRT display units presented a difficulty. The only types readily available presented A-scan on a 5-in diameter CRT. However, to meet the Staff Requirements, a range accuracy of ± 5-yds on a linear scale was thought necessary if the regenerative tracking were to generate a sufficiently accurate range-rate. In addition the operator would have to be capable of following the moving target-echo smoothly and accurately in the heat of battle. The author reached the conclusion that this performance could only be achieved using a 12-in diameter CRT screen. At that time A. C. Cossor Ltd was one of the few manufacturers of 12-in CRTs in the country and there was no knowledge whatsoever of how linear an A-scan would be on such tubes. L. Bedford and his team at Cossor made some experimental tubes that were found to have a sufficiently linear A-scan – 30-microseconds duration (equivalent to 5000-yds range) at 500 times per second. Cossor developed a display console with a sloping screen, fitted with a mechanical cursor and controlled by hand. The space above the rack was for the associated receiver, manufactured by Marconi.

Design engineers under S. E. Trigle and G. W. Harris now undertook the engineering development of Alred's adaptation of the Yagi directional antenna, which reduced unwanted side-lobes. It will be recalled that this exploited a cylindrical aluminium reflector instead of the single rear-reflector element, supported in a cast aluminium frame so designed that two or more could be mounted on a 2.5-in diameter rod above the fire-control (FC) director. It was capable of rotation in elevation by the director layer, so that the antennae always pointed directly at the target. The spar supporting the antennae elements was ultimately made in tapered form from 'permali' (teak vacuum-impregnated with bakelite), which was strong and rigid enough for the supporting tie rods (from the corners of the cylindrical reflector to the ends of the element spar) to be dispensed with. This antenna-system design was retained throughout the war (it was still in service in the Israeli Navy in 1967!).

During the initial trials difficulties had been experienced due to the ingress of moisture into the antenna feeders. Messrs Pirelli-General then developed a polythene-insulated coaxial cable, whose electrical characteristics at 50-cm wavelength were determined by Professor Willis Jackson at the Manchester Institute of Science and Technology. The results were very promising, and this development was adopted. As a result it became possible to mould the feeder-cable termination and the central ends of the associated dipole element into a small block of polystyrene, which also served for mounting the dipole on the supporting spar.

It remained to provide an adequate impedance match between the antenna array (about 50 ohms) and that of the coaxial feeder (about 70 ohms), to avoid loss of transmitted power and of received signal, respectively. This was achieved using a quarter-wave coaxial stub, consisting of a length of 21 SWG copper wire as inner conductor in an outer conductor of 3/8-in copper tube, also moulded integrally with the dipole mounting block. The whole dipole assembly, including matching stub, tail leads and watertight connectors, was manufactured by STC Ltd.

In May 1940 the design of a radar receiver of acceptable sensitivity and stability for the 50-cm equipment was discussed at a meeting at GEC, Wembley. The super-regenerative type was clearly unsuitable for the Naval application, and Bareford's autodyne P14 receiver, had a very high noise factor. At this point the author (in some despair) suggested that a superheterodyne incorporating a crystal-diode mixer should be investigated. Opinion at the meeting considered that the conversion factor would be too low. (This was, in fact, incorrect. Barely one year later the work of H. W. B. Skinner of the Telecommunication Research Establishment – TRE, formerly AMRE – on crystal detectors resulted in this becoming the preferred radar receiver detector, operating in both the 9-cm and the 3-cm bands – (see Monograph 5). Nevertheless J. Croney, who joined Signal School in 1940, was asked to develop a superheterodyne receiver incorporating a diode mixer, followed by a 6 MHz IF amplifier available from commercial TV equipments. This receiver, designated P16, was tested at the AA Range at Eastney in the autumn of 1940 and was found to possess a signal/noise performance ratio some 6 db better than Bareford's P14.

A meeting was held on 20 May 1940 at BTH Rugby (A. L. Whiteley and K. J. R. Wilkinson) to finalise the design of the modulator for the 50-cm transmitter. This incorporated the BT9 thyratron and a length of Pirelli-General polythene-insulated cable type PT20 acting as a delay-line (developed by Ratsey at Signal School). The modulator was required to operate at up to 21 kV – much higher than ever before – using a thyratron whose geometry required further research, by H. de B. Knight. Also, because of the unusual wave form of the current required from the 500 c/s generator, BTH undertook development of a special voltage regulator, because, in a ship's generator, speeds cannot be controlled very accurately. This problem was resolved by A. L. Whiteley, of BTH. Later it was found that a radio-interference suppressor was also necessary in the mains supply.

Meanwhile, from April 1940 onwards, the author's group was working late each day (including weekends) to complete the development of the 50-cm set in order that design details could be finalised and manufacturing drawings prepared by the six major firms involved in production . Effort was also required for the preparation of equipment-installation instructions for the Dockyards, and technical handbooks for the operators.

In May, impressed by the results of the close-range trials at the AA Range, Eastney, with the prototype Type 282 equipped with Yagi 'fishbone' antennas on a pom-pom director, DNO enquired whether 50-cm range finders could be provided for high-angle (HA) directors, and for the main armament directors in capital ships and cruisers. Visits were therefore paid to the Experimental Department of the Gunnery School at *Excellent*, the technical staff of DNO and the Vickers Armstrong, Rose Brothers and Laurence Scott firms. As a result it was concluded that five or six 'fishbones' (Yagis) could be fitted side by side above these HA Directors. However a subsequent visit to Elswick, on the Tyne, to examine main armament directors resulted in the decision to develop cylindrical parabolic reflector systems for the antennae. It seemed that these could be up to 2.5-ft high and up to 24-ft long, incorporating an array of 24 driven dipoles (see Figure 1.6).

Such an array (known colloquially as 'pig-trough') would be expected to produce a horizontal-plane beamwidth of 5°, and a vertical-plane beamwidth of 38°, for tracking of surface targets (see Figure 1.7). (The corresponding figures for Type 282's twin-fishbone antenna were 38° and 35°, respectively). Separate 'pig-trough' antennae would have to be mounted one above the other for transmission and reception, respec-

1.6 Experimental 24-Dipole tapered array for Type 284 undergoing field trials
(© Defence Research Agency)

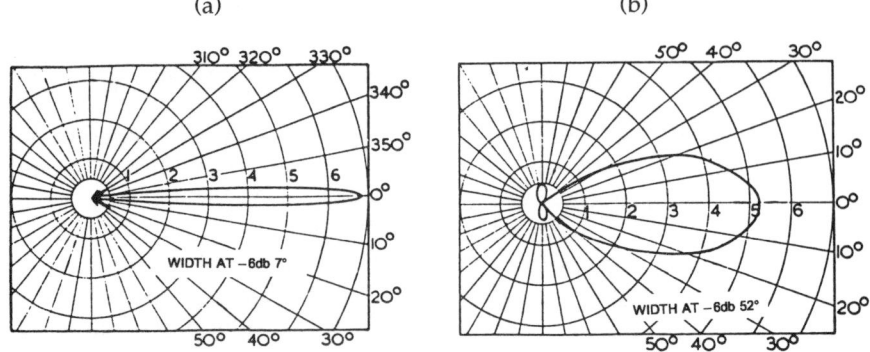

1.7 (a) Horizontal polar diagram of 24-dipole 'pig-trough' array; (b) vertical polar diagram of 24-dipole 'pig-trough' array (© Defence Research Agency)

1.8 Office equipment of Types 282, 284 and 285 (600 MHz); modulator and receiver on the right, transmitter in resonant cylinder centre, display with range finder on the left (© Defence Research Agency)

tively, the increase in power gain each way being expected to increase the tracking range on a given target by a factor of 2.6.

Trials of this type of antenna arrangement were conducted in June in *Nelson*, using an antenna array only 10-ft wide, in conjunction with Type 282 office equipment (see Figure 1.8). Ranges of 18 miles were obtained on ships in convoy, and 11.5 miles on the destroyer *Ambuscade*. It was later realised that these ranges were somewhat optimistic due to abnormal radio-propagation conditions, influenced by the prevailing weather. Nevertheless Admiralty took an immediate decision to fit 50-cm sets, designated as Type 284, in all battleships and cruisers and Type 285 sets, with five or six 'fishbone' arrays on all high-angle and range-finder directors. In consequence production orders for the basic set were increased from 200 to 900. The first prototype equipment was planned for fitting in the battleship *King George V*. For HA anti-aircraft gunnery using Type 285, antennas would have to be fitted to a variety of directors of different shapes and sizes, involving further development work.

Alred continued experimental work on 'pig-trough' antennae, the reflector aperture being 2.5-ft high by 21-ft long, fitted with 24 dipoles in line. This arrangement was made up of three units bolted together, each 7-ft long and incorporating eight dipoles. It was found experimentally that the side lobes in the vertical plane were smallest, and the antenna gain greatest, when the dipoles were 25-cm from the apex of the reflector rather than at the focus at 13.5-cm. To reduce the unwanted side lobes in the horizontal plane, which could give rise to spurious echoes from elements of the ships' superstructure, Alred was able to show theoretically that there should be no such side lobes if the power distribution across the dipoles from the centre outwards followed the binomial expansion coefficients. Clearly it would be difficult to feed all the individual dipoles with different levels of power, and further there appeared to be advantage in reducing the number of groups of dipoles. Having established the optimum focal length for a reflector with a 75-cm aperture, and the best position for the line of dipoles, Alred experimented with the distance between them and found that this could be increased to 40-cm without loss in antenna gain. Thus the number of dipoles was fixed at 24, fed in groups of four, with a power distribution between the six groups in the ratios 1:4:10:10:4:1. With this arrangement the side-lobe radiation would be only 1 per cent of the main beam. A successful solution was then found to the problem of designing an antenna feeder system that would divide the total available power in the quoted ratios without introducing mismatch reflections at the various feeder junctions (see Figure 1.9). All dipoles and junction boxes had to be absolutely watertight and damp-proof.

In addition to this work on development of the main armament director antenna system (Type 284), work proceeded to determine how a

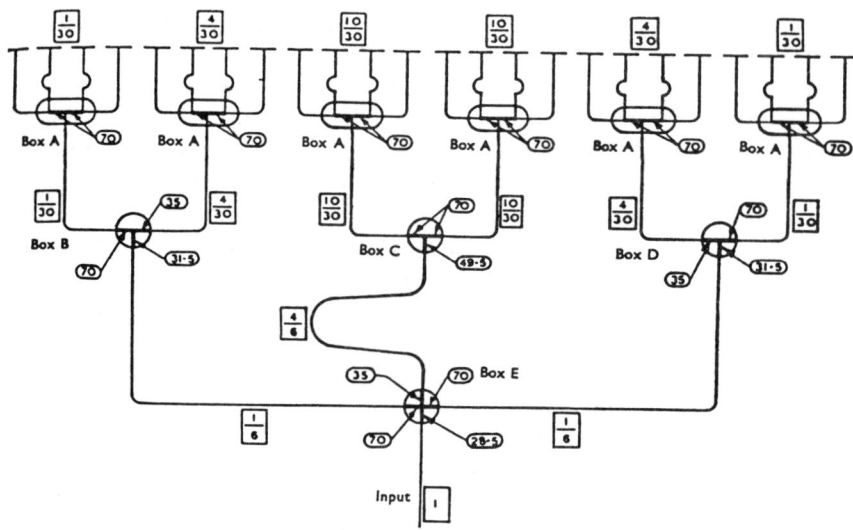

1.9 Diagram of 24-dipole array, showing how power is partitioned (© Defence Research Agency)

1.10 Type 284 Yagi array on the LA Director in *Suffolk*, with Type 285 Yagi array on the HA Director, further aft (© Defence Research Agency)

number of 'fishbone' antennae could be mounted on the HA director (Type 285). For the older cruisers it was decided to use two arrays of three 'fishbones' each, mounted side by side on the directors (Figure 1.10). Since some of the larger ships might require up to 11 sets operating simultaneously in close proximity, there was the strong possibility of mutual interference between them, but fortunately Alred quickly found that when the spacing was optimum for two or three 'fishbones', with the dipoles in line, the main lobes of the antenna radiation patterns were still symmetrical, and that the unwanted side lobes could be further reduced. Mounting the transmitting and receiving antennae side by side had little adverse effect, and interference with other antennae 20-ft distant was negligible. The two-'fishbone' arrangement gave a gain of 3 db over a single one, and a three-'fishbone' 5 db.

TRIALS OF MAIN ARMAMENT AND HIGH-ANGLE DIRECTOR RADARS

The first Type 284, with two 21-ft 'pig-trough' parabolic reflector antenna systems was installed in *King George V* for trials early in December 1940. This set worked immediately it was switched on, and when at sea the next day tracked the cruiser *Norfolk* to a range of 19000-yds. The first Type 285, equipped with five 'fishbones' mounted on a range-finder director, was also installed in December in *Southdown*. After initial troubles with faulty thyratrons, the set worked well. Croney's P16 receiver was found to be about 6 db more sensitive than the P14, although its tuning was liable to drift. This installation detected the small cruiser *Phoebe* in the bows-on aspect at about 20000-yds, and had no difficulty in tracking the ship during manoeuvres at ranges between 10000 and 20000-yds. Surprisingly, shells in flight could be followed to some distance. For example, 5.25-in shells fired at a battle practice target at 15000-yds range could be followed out to 5000-yds. Shell-splashes could also be detected. Radar operators later experienced some alarm when echoes from *approaching* enemy shells could be seen on their A-scan displays! However trials against low-flying aircraft were disappointing, reliable detection ranges of the order of 10000-yds being achieved rather than the 15000-yds expected.

Thus, by the end of 1940, 50-cm gunnery radars were being widely fitted in HM ships – Type 284 in large ships on the main surface directors (LA); Type 285 in several classes of ship on the long-range HA directors (and also for surface targets in destroyers); and Type 282 for close-range AA directors in all classes of ship. All this was achieved using one basic equipment with many variants of antenna system. Finally, Type 283 was introduced in 1942 to permit the main armament of the larger ships to lay down a barrage against air attack. (The Signal Officer of a destroyer acting as part of the screen of a Mediterranean convoy described verbally to a colleague how the accompanying battleships fired such a barrage, using the main armament, against a flight of low-level Italian torpedo planes approaching the convoy out of the dying sun. After the explosions, nothing was left in the air.)

THE BIRTH OF NAVAL 10-CM RADAR

Sir Charles Wright, DSR at the Admiralty, had been keenly interested in the potential capabilities of cm-waves for radar from the earliest days, in view of the narrow antenna beamwidths that would be achieved. This was later to prove of great importance in both Naval and airborne applications (see Monograph 5). However, in the early stages of the war no suitable high-power cm-wave sources existed for these potential applications.

Fortunately the CVD Committee arranged for an Admiralty-sponsored contract to be placed on the Nuffield laboratories of the Physics Department at Birmingham University, under Professor M. L. E. (later Sir Mark) Oliphant, for the development of high-power transmitting valves for 10-cm radar applications. Initially this work embraced research on both klystrons and magnetrons as there was no immediate evidence to indicate which type had the greatest potential for further development. However, by February 1940 J. T. Randall and H. A. H. Boot had evolved the first resonant-cavity magnetron design capable of producing high power at about 10-cm wavelength.[24] Similarly R. W. Sutton, of Signal School's Valve Section at Bristol University, developed a reflex-klystron valve suitable for radar receiver application. Also H. W. B. Skinner, of the Air Ministry's Telecommunications Research establishment (TRE) group (ex AMRE, Bawdsey and Dundee, now located at Swanage), evolved a practical crystal-mixer device – fulfilling the author's idea expressed about one year earlier.

By inter-Service agreement the initial research on 10-cm wavelength radar was the responsibility of the airborne radar group at TRE, Swanage, primarily to develop an AI equipment for night-fighter application to replace the current 1.5-m set. A. C. B. (now sir Bernard) Lovell and W. E. Burcham observed the first 10-cm echoes from an aircraft on 12 August 1940.[25] Shortly afterwards they installed a 10-cm set overlooking Swanage Bay, with which good echoes were obtained from neighbouring land and various surface craft. Following the success of the experimental equipment at Swanage, E. C. S. Megaw, at the GEC Research Laboratories, undertook full development of the Birmingham magnetron for production.[26] At this time the author was still responsible for all centimetric (below 1-m wavelength) developments in Signal School, and so made it his business to keep abreast of all these developments. He was in almost daily touch with Megaw, on account of the work going on at Wembley for the 50-cm gunnery sets, and was kept fully informed by him of the progress of the 10-cm developments. As a result of this, very careful consideration was given in Signal School to the possibilities of developing gunnery radar sets for the Navy operating on 10-cm rather than the current 50-cm wavelength. It was, however, apparent that even with a crash programme not enough 10-cm equipments could be produced in time to fulfil the vast ship-installation programme for 50-cm up to the end of 1941, involving nearly 200 outfits. It was therefore decided to increase the overall production of 50-cm sets to 900, and to embark on the development of improvements to them, whilst investigating the 10-cm position. However, in the autumn the author visited Skinner at Swanage to review TRE's 10-cm work, and to determine the possibilities of exploiting 10-cm wavelength radar as a replacement for the Navy's 50-cm sets. He reported to Horton accordingly, on the success Skinner's group was having with its 10-cm equipment.

Sometime in October 1940 a visit was paid to Swanage by Commander H.W. Fawcett and Lt-Cdr E.C. Bayldon, both of the Admiralty's Anti-Submarine Warfare Division, who expressed themselves keenly on the application of 10-cm for submarine detection.

Horton and Captain Willett then attended a demonstration of the TRE 10-cm land-based set at Swanage on 8 November 1940. Lovell[27] records that their response was initially pessimistic, from the Naval viewpoint, but that Skinner's group strenuously advocated fitting a 10-cm equipment in a destroyer for anti-submarine (A/S) applications. Signal School reacted with speed, bearing in mind the urgent requirement to equip convoy escorts in the Atlantic with an equipment capable of detecting surfaced U-boats by night in the attacking role, when they were free from Asdic detection by the escorts.

On 11 November a further trial of TRE's experimental set against the submarine *Usk* was carried out from a site some 250-ft above sea level in the presence of Horton and Captain Willett (and very probably representatives of the Admiralty Staff). In the stern-on position *Usk* was seen out to about 7.5 miles, with strong returns up to about 6 miles. With only the conning tower exposed above the surface, ranges of about 4.5 miles were achieved. Lovell's diary records that this demonstration 'seemed to shake Willett and Horton pretty much'[28]. The equipment was later moved to a position on the beach, where it was demonstrated that target echoes could readily be obtained from a small motor boat (see Monograph 5). These results led to the decision that a Signal School team should assemble a copy of the TRE equipment mounted in a mobile trailer vehicle for easy transfer to the laboratories at Eastney (see Monograph 5).

S. E. A. Landale headed a small group taken from J. D. S. Rawlinson's staff at Eastney, which was ordered to TRE Swanage on 21 November to assist in the centimetric radar development work there, learn the new techniques and then start work on the development of centimetric radar at Eastney (see Monograph 5) for installation in convoy escorts for U-boat detection. The other members of the group were C. A. Cochrane, J. Croney and C. Owen.

The TRE equipment used a magnetron providing some 5 kW of power output, a Sutton klystron local-oscillator, and Skinner's hand-made sillicon-crystal mixer. The antenna system consisted of Lovell's design of 3-ft diameter parabolic reflectors with dipole elements at their focal points, mounted side by side.

Having regard to the shipborne environment, involving continuous pitch and roll in typical Atlantic conditions, it was clear that the narrow vertical beamwidth of the parabolic reflector antennae used in the experimental model would be unsuitable for convoy escorts without stabilisation. Lovell therefore suggested[29] the use of a parabolic cylinder format, closed in the vertical by parallel horizontal plates, with dipole feed at the focus in the aperture plane. This was the 'cheese' format, which was to become one of the basic standards in future Naval applications. On 18 December the trailer antennae were changed to a 'cheese' design, having an aperture of 6-ft x 9-in. A test run from the trailer on the beach at Swanage against a small motor-boat gave a range of three miles, which was very satisfactory in view of the low antenna height involved (see Monograph 5).

On the following day the trailer disappeared 'mysteriously' from Swanage and reappeared at Eastney, whilst TRE's Senior Storekeeper was left with an accounting problem! For some time it continued in use as a test-bed at Eastney for upgraded components. It was later used to test the first 'strapped' magnetron delivered to the Navy, the greatly increased power-output melting the antenna feed!

THE APPLICATION OF 10-CM RADAR IN THE U-BOAT WAR

1941 opened in Signal School with the Staff at fever pitch with excitement and enthusiasm – despite the fact that everyone had worked all possible hours, with no leave being granted in 1940. Clearly this could not go on, so Admiralty decreed that Staff should have one day per week off, and take a whole week's holiday in 1941.

The excitement was generated by a number of successes. The return of Landale's group to Eastney with a 10-cm radar mock-up in the trailer appeared to promise ultimate fulfilment of the Navy's requirement for an effective set in convoy-escort vessels. Trials of Type 281 in the cruiser *Dido* had been highly successful, and approval had been given for fitting 20 ships as soon as models could be produced. The Type 79Xs based at Eastney and Fort Wallington had proved their worth in tracking the first air raids on Portsmouth. Finally, the author's group had returned to its laboratories in Onslow Road, Southsea, after successful trials of Type 284 in *King George V* and Type 285 in *Southdown*.[30] These gunnery sets were coming off the production line at the rate of several per week and being installed in ships at more than 10 per month. It appeared that radar could provide the solution to the Navy's most demanding and urgent problems – to combat the dive-bomber and the defeat the U-boat in the convoy war.

The immediate requirement was to develop the 10-cm equipment into a form capable of withstanding sea going conditions in small convoy-escort vessels, as fully described in Monograph 5. Briefly, less than four months from the date of commencing to equip the experimental trailer at Swanage the first prototype Type 271X of a batch of 12 was completed in the laboratory workshops. This was sent for fitting in the corvette *Orchis* for the sea trials on 25 – 26 March 1941. In these trials a submarine on the surface was detected at ranges up to 5000-yds. When only the conning tower was exposed detection was at 2800-yds and at periscope depth at 1500-yds, using relatively unskilled operators.

With the success of the *Orchis* trials it was decided to manufacture a second batch of 12 prototypes of Type 271X in Signal School Workshops. This was intended to maintain momentum until a suitable manufacturer could commence production in larger quantities. The firm chosen was Allen West of Brighton, who were given a contract to produce 100 sets in six months. Assistance from design engineers of the firm was offered in order to expedite the preparation of the necessary manufacturing drawings. One of their senior engineers, Mr Paddick, now spent most of his time at Eastney, and constituted the liaison between Landale's party and the works at Brighton.

Installation and commissioning of the new corvette radar was supervised by a few civilian officers from Signal School, assisted by a team of RNVR Officers and Ratings, specially trained for the purpose. A complication was that, although the transmitter and RF section of the receiver could be mounted on the back of the respective antennae, the receiver local-oscillator had to be tuneable by the operator in the radar office. Because rotatable waveguide joints had not been developed at that time, concentric cable had to be used between the antenna and the associated receiver. Since this cable had very high losses at 3000 MHz (10-cm wavelength), it was decided that the transmitting and receiving antennae, together with their rotatable pedestal, should be fitted on the roof of the radar office and controlled by the operator within it. It was therefore decided that the radar office, complete with a perspex lantern on its roof to protect both the antenna system and the transmitter, should be installed as a complete unit in a suitable position in the ship whilst in dock between sorties, instead of requiring structural modifications that would have to await a Dockyard refit. Later, arrangements were made for W. H. Smith (Engineers) Ltd to assemble all the equipment into prefabricated offices, complete with perspex lanterns (Type 271P). With the radar installed and tested, the assembly could then be transported to the appropriate ports for immediate installation. By the end of July 1941 some 25 escort vessels had either been fitted or were in the process of being fitted with Type 271X. One of the first prototypes was sent to the USA as part of the US – UK technical-exchange agreement. For a full account of these developments, see Monograph 5.

Meanwhile, with the advent of waveguide technique the antenna dipole source and associated concentric cable were replaced by a waveguide-fed horn.

When higher-power magnetrons became available, the use of the experimental trailer at Eastney was abandoned. However a successful experimental Coast-Defence (CD) radar with an output power of 500 kW was developed and installed in a rotatable trailer in the summer of 1942. This was used for an extended series of trials against aircraft targets between December 1942 and February 1943 (see Monograph 5).

PARALLEL DEVELOPMENTS

Whilst this exciting new 10-cm development was in progress, no relaxation could be permitted on the programmes started in 1940. The manufacturing prototype of Type 281 had been completed and was being commissioned in *Prince of Wales*. Production equipments were expected in February, to be installed and commissioned at the rate of one per month. A new rotatable antenna had been designed at Nutbourne to replace the original fixed-array system for Type 286. Also, a new transmitter intended to provide 100 Kw peak power (but actually achieving 50 kW, see Monograph 4) was developed in the hope that destroyers would have a capability comparable to that of Type 281 in big ships, allowing for inevitable limitation in size and weight in the smaller ships. The experimental model of this set, Type 290, was complete by March to the point where a contract for 300 sets could be placed. By this time fitting-out bases had been established in all the principal home ports, plus a number abroad, and some 200 Type 286 sets had been installed. All this was possible only because since the outbreak of war the technical staff at Eastney had been increased to 50, and that at Nutbourne to 15, exclusive of design staff (see Monograph 4).

Meanwhile, for the 50-cm group the most urgent task was to provision the 11 sets required for installation in *Prince of Wales* (one Type 284 for surface gunnery, four Type 285 for AA gunnery and six Type 282 for close-range gunnery). This was a consequence of the decision in June 1940 that all FC directors, whether main armament, high angle or close range, should be fitted with 50-cm sets for range finding. Similarly, cruisers would be provided with five sets (one Type 285 and four Type 282). The question arose, inevitably, of avoiding mutual interference between the various equipments when operating simultaneously in a given ship. It was obviously too difficult for the individual sets to operate on different frequencies – for example in provisioning individually tuned antennae – quite apart from the associated maintenance problems. It was therefore decided to operate the individual sets in a given ship in time-sequence. This was just possible because the total go-and-return transit time for the maximum ranges required by the 11 sets in *Prince of Wales* was less than the repetition period of 2000 microseconds involved with the 500 c/s prf. Thus, it was decided to synchronise the sets to their 500 c/s supplies, using two six-phase generators in tandem with a 16° phase-difference between them, to provide a total of 12 separate phases. The twelfth phase was utilised to operate an equipment testroom. Shuttleworth of Signal School, in collaboration with Mr Jack of BTH, perfected the design, the firm then producing the machines. It was soon found that a new type of voltage regulator was required, owing to the unusual waveform resulting from the very large pulses of current demand by the modulators, but A. L. Whiteley of BTH soon resolved this problem.

In January 1941 *Prince of Wales* was fitting out at Rosyth, in accordance with a programme allowing her to join the Fleet at Scapa early in April, for working-up exercises. The author took a party of six from his group to Rosyth, but by 1 March only the main armament director Type 284 was complete. Whilst the Type 285 offices were almost complete, the feeder cables to the antennae required installation. On the other hand the feeder cables for four of the Type 282s had been installed, but the offices were not nearly ready. During the next three weeks the party checked each element of all the sets and corrected many teething troubles. By 25 March seven of the 11 sets were working correctly. There were still problems with the 12-phase generator and associated voltage control equipment, which Whiteley of BTH resolved. Final trials were carried out during the working-up exercises (Calpine supervising Type 284, Hill the forward Type 285s, Hansford the aft Type 285s and Pout the Type 282s). The trials were quite successful, ship targets being picked up by Type 284 at 19000-yds and aircraft targets at 15000-yds by the 285s.

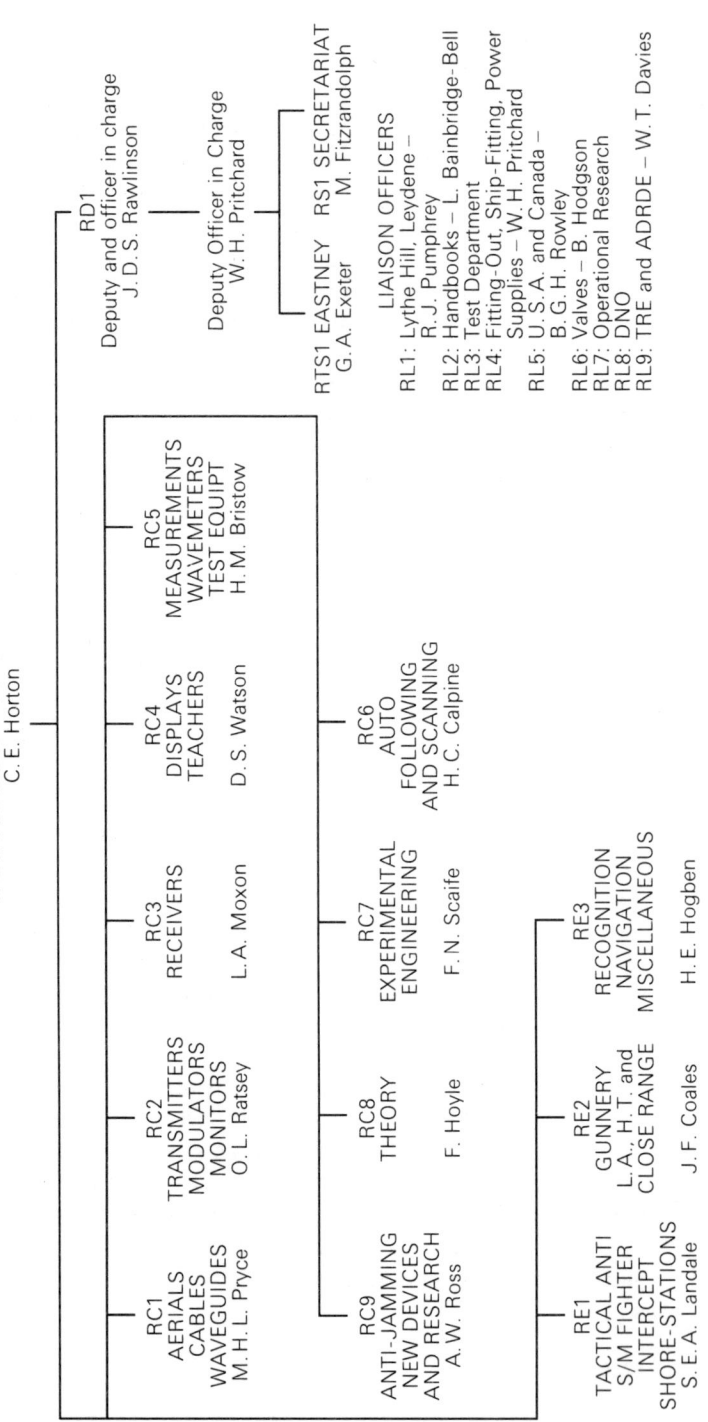

1.11 Organisation of Radar Department under C. E. Horton in October 1942 (© Defence Research Agency)

GUNNERY RADAR DEVELOPMENT AND IMPROVEMENTS, 1942 ONWARDS

An important reorganisation of the Radar Department under C. E. Horton took place in 1942 (see Figure 1.11). This is described in more detail in Monograph 2. Following the work on the *Prince of Wales*, the next requirement of the author's group was to determine target direction with an accuracy adequate for gunnery purposes, that is, a few minutes-of arc. The technique of beam-switching had been applied successfully to Type 281, giving a bearing accuracy of $\pm 0.5°$, but this was not good enough. On the big main armament director towers the 24-dipole array could be phased relatively easily to displace the beam by about 0.5° from the line of sight on one side, and then be switched similarly to 0.5° displacement on the other side. However, in cruisers and destroyers the largest antenna system that could be accommodated consisted of three double 'fishbone' arrays side by side. The only way to achieve reasonable bearing accuracy was to use the same array for transmitting and receiving, and to switch the phases of the feeds to the outer pairs (Figures 1.12 and 1.13). Suitable transmit/receive (T/R) switches to isolate the receiver from the transmitted pulses were available, but in this case the full transmitted peak power of 100 kW would have to be switched to the feeders. This was achieved by switching a quarter-wave line 25 times per second, first into one feeder and then into the other, by means of an oil-filled rotary capacity switch operated by a Geneva-cross mechanism.

This was developed by W. T. Davies and designed by G. W. Jones. By this means a bearing accuracy of better than five minutes-of-arc was achieved with Type 284 (21-ft antenna arrays), and better than 15 minutes-of-arc with Type 285P (six fishbone arrays). These accuracies were adequate for blind fire on surface targets and low-flying aircraft. However, for high-flying aircraft accurate elevation was required in addition. This had to await the installation of equipment operating in the cm-wave band, although trials of a Type 285 with an antenna array of 20 vertical dipoles were carried out at Fraser Battery, Eastney. This system proved to be too complicated to be reliable, and it was therefore decided to await the arrival of the 10-cm Type 275 set.

Improvements were also made in range-finding by developing a strobe that produced a notch in the A-scan trace. This was set on an echo by hand using a rate-aided hand wheel that kept the strobe on the target with the little further intervention by the operator if the target rate of approach was constant. This enabled accurate ranges and range rates to be fed automatically to the FC predictors, whilst the barrage computer ensured that the guns fired at the correct predetermined range to place shells near the target. *Duke of York* was the first ship to be equipped with this Auto-Barrage-Unit (ABU) facility when she joined the Fleet in

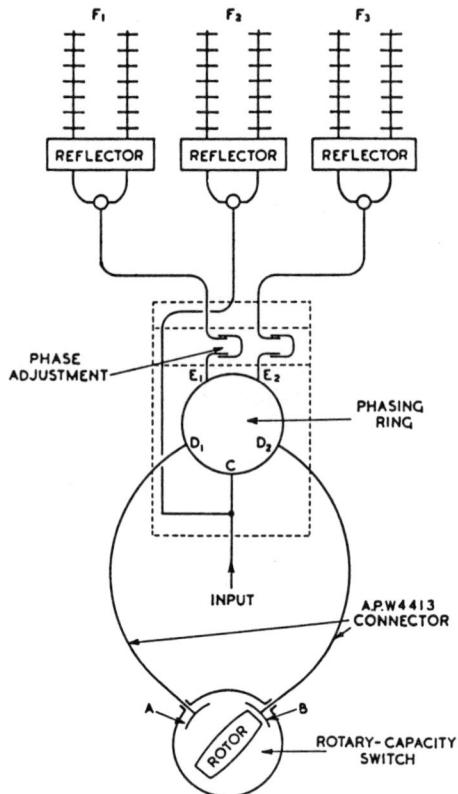

1.12 Diagram of 6-Yagi antenna array with rotary switch (© Defence Research Agency)

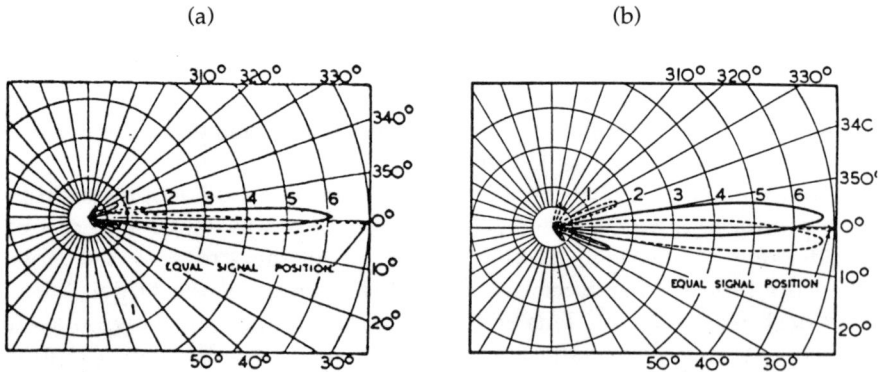

1.13 (a) Horizontal polar diagram of switched 24-dipole array; (b) horizontal polar diagram of switched 6-Yagi array (© Defence Research Agency)

January 1942. Later it used its Type 284P to score direct hits on the *Scharnhorst*, firing completely blind, at an early stage in the Battle of North Cape. This slowed her sufficiently for destroyers to catch and ultimately sink her.

By 1942, with the evolution of cm-wave radar well established, consideration turned to similar applications on these wavelengths: firstly on 10-cm wavelength to provide really accurate blind-fire facilities on both low-angle and high-angle directors for both surface and air targets; and high-angle directors for both surface and air targets: and secondly on 3-cm wavelength to provide the blind-fire facilities for the close-range Bofors gun-mountings. In the autumn of 1941 design of the principal radar components for Type 274 was initiated in a group at Eastney under Dr A. A. Symonds, in association with Dr Landale's group. Because gun systems were required to be at instant readiness against surprise attack, the long warm-up time of the thyratrons then in use as modulator valves was unacceptable. The solution was to develop an air-spaced, triggered spark gap capable of handling the high powers involved and possessing an adequate life. A satisfactory design allowed development to proceed, but the air gap was soon replaced by a sealed and pressurised inert gas-filled device. This was also used in the subsequent design for the high-angle FC tracking radar Type 275. It soon became apparent that the design of the radars for anti-aircraft application would have to go hand in hand with that of the gun-mounting, or FC director. This involved close liaison between the radar group in Signal School – now renamed Admiralty Signal Establishment and relocated at Witley, Surrey – and the designers of the directors and mountings in the Department of Naval Ordnance at Bath. For this reason the design of the cm-wave equipments took longer than the earlier warning radar systems, and in consequence, with the exception of Type 274 low-angle main-armament FC radar installed on the low-angle director of *Norfolk*, was operational in time to destroy a German convoy by blind fire at 11-miles range.

The Navy's most important roles in 1942 were to protect the shipping bringing troops and essential supplies from America; supplying our armies in Africa and the Far East; and getting much needed war material to Russia. Thanks to Hitler's obsession concerning the safety of his battlefleet, it was mainly with U-boats and shore-based aircraft that the Navy had to contend. However, the battleships and cruisers based in Norway were always a threat to the Russian and Atlantic convoys, which therefore had to be shielded from attack by cruiser escorts and the Home Fleet.

Action was joined between cruiser and destroyer forces on occasion to the North of Norway and in the Mediterranean, when radar played its part in providing air and surface warning and accurate range-finding for the guns. However it was the escort ships that were continuously in action, and for whom centimetric radar had come only just in time. Throughout the first four years it was a ding-dong battle with the U-boats, with the outcome always in the balance.[31] Even in 1942, after Coastal Command was equipped with long-range Catalina and Liberator aircraft fitted with centimetric radar, they and the convoy escorts shared the honours equally in the number of U-boats destroyed.

From 1941 until the end of the war a long series of modifications and developments were made to improve the capability of the original Type 271, and to provide variants suitable for fitting in a wide range of ship types, both large and small. A detailed account of these developments is presented in Monograph 5.

In the Far East a quite different type of naval war developed. The main naval actions were between fleets including strong aircraft-carrier forces, or in small operations within reach of shore-based hostile planes. It was due to lack of fighter cover that the *Prince of Wales* and *Repulse* had been lost off Malaya on 10 December 1941. From this time on defence against dive-bomber and airborne torpedo attacks became the became the highest priority. Types 279 and 281 provided excellent long-range warning of high-level air attack, the only problem being the loss in coverage in multilobed vertical polar diagrams due to reflection from the sea interfering with the direct wave.

In order to overcome this, from 1944 onwards large ships were fitted with both sets, the Type 279 antenna on one mast and the Type 281 on the other, so that the lobes of one polar diagram filled the gaps of the other (see Monograph 4). Further, a virtue was made out of this necessity by using the ranges at which an aircraft entered and left respective lobes to determine its height (see Figure 1.14). The other defect was that metric wavelengths were really too long a wavelength for detecting low-flying aircraft at long range, even with peak powers of megawatts. (Early in the days of radiolocation it had been established that on surface ships and low-flying aircraft the detection range increased only as the one-eight power of transmitted power.) By mid-1942 it was also clear that the higher the frequency, the more effective became the radar for low-angle detection, even with relatively small antennae. The 10-cm sets using such antennae, and transmitters whose peak powers were only a fraction of what could be obtained with the greatly improved magnetrons now in large-scale production, had proved their value for low-angle target detection. Following on Type 271, a set with 3-ft diameter paraboloidal reflectors designated Type 273 had therefore been developed for warning of surface ships, and low-flying aircraft in capital ships and cruisers. This gave some 30-miles detection range on low-flying aircraft, and filled the low-angle gap well enough for the guns to be brought to bear. However it was quite insufficient warning for aircraft carriers, which needed to fly-off fighters for their own defence.

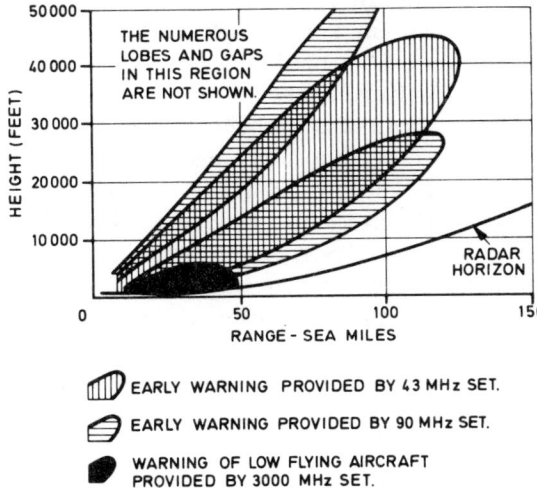

1.14   Vertical polar diagrams of antennae of Type 279 (43 MHz) and Type 281 (90 MHz), with that of Type 273 (3000 MHz) added   (© Defence Research Agency)

For reliable detection of low-flying aircraft, development was necessary on bigger antennae, higher transmitter powers and improved displays (see Monograph 5). By mid-1942 a transmitter and modulator delivering 500 kW peak power, in conjunction with common T/R operation had been developed, and a 4-ft 6-in diameter parabolic antenna (which would require stabilisation for roll and pitch of the ship) was planned. In the search role the antenna would rotate continuously, and target echoes would be displayed on a Plan Position Indicator (PPI), but when required for height finding the antenna could be remotely controlled in both bearing and elevation from the radar office. Trials of an early experimental set (Type 277X), which was not fitted with a stabilised antenna, were carried out successfully in HMS *Saltburn* in April and May 1943. Sea trials of the final design took place in HMS *Campania* in March 1944, following which, with the introduction of antenna stabilisation, it became the standard replacement for all 10-cm warning radars. These sets gave reliable detection of aircraft flying above 1000-ft at ranges in excess of 40 miles, and height finding out to 20 miles with an accuracy of $\pm 0.5°$. A shore-station version, Type 277T, installed in a trailer cabin, was produced and went into service in 1943 for coastal defence.

The PPI had been developed originally at TRE for Fighter-Direction (FD) applications in the RAF. In 1942 a model suitable for shipborne use was developed in ASE, and in 1943 this became the standard form of display for warning radars. At about this time a dark-trace projection tube known as the Skiatron became available (see Monograph 3), and this was used with a large, horizontal projection screen in the role of a plotting table for Fighter Direction.[32] In most ships the radar data was required in the bridge and plotting offices, both remote from the radar office.[33] 'Slave' PPIs were therefore soon developed, and fitted in them. Information had always been passed from the various ships in a fleet to the flagship and vice versa, first by visual signalling and later by radio, so that a comprehensive plot of the disposition of the fleet, as well as of any enemy or neutral shipping and aircraft in the vicinity, could be maintained in the flagship. The flagship would also receive information concerning enemy movements, broadcast by the Admiralty at specified times, and information would then be signalled to the other ships so that they could update their individual plots. Radar having increased the watchkeeper's field of 'vision' many times, the amount of information on these plots had increased enormously[34] and it was becoming more and more difficult to keep them updated. Further, they were in danger of becoming congested with too much raw data, much of which for some ships of the fleet would be unnecessary, if not redundant. The idea was therefore conceived of having a television-type plot that could be kept up to date in the flagship and transmitted by radio to the other ships. C. A. Laws and R. Benjamin started to develop a suitable information transmission and display system for this purpose in 1944 using analogue methods that interlaced the information in coded form on the fly-back of the CRT time-base. In the postwar years it was expected that the speeds and likely numbers of attacking aircraft in any possible future conflict would increase considerably, and to cope with this situation automation of the Action Information Organisation (AIO) was considered essential. After the war R. Benjamin evolved the concept of a digital system, which was fully developed by ASE, in association with industry. The system was able to store and display target tracks and other target data in symbolic form. Together with other aids to the Direction Officers it greatly increased the capacity and effectiveness of the AIO in both Weapon Allocation (WA) and Fighter-Direction (FD) applications.

From 1943 onwards the emphasis changed from being concerned with individual equipments designated to meet specific requirements, to developing information and weapons systems to enable individual ships and fleets to engage concerted attacks, either on the surface or from the air. The intercommunications within a ship and between the ships of the fleet had been developing with this in view ever since the days of Admiral Sir Henry Jackson, FRS. In the early stages of an action the tactics were coordinated by the flagship. The weapons in any particular ship were under the orders of the captain, through the remote-control office alongside the tactical plot on the bridge.[35] Before the advent of radar the assembly of information was too haphazard (and usually too late) during the course of enemy action for this coordination to be maintained for long. Sooner, rather than later, weapons crews would be told to operate independently. Under these conditions, when dive-bombing or torpedo-bombing attacks were pressed home by eight or more determined aircraft (as was repeatedly the case in the Far East) some would inevitably get through, with disastrous results.

It has already been described how Types 279 and 281 were used together to determine aircraft height, and the 10-cm sets were used for detection of low-flying aircraft and surface targets. Type 277, which provided height finding on aircraft targets, had now replaced Type 273. With the aid of PPIs and height finding equipment, the information gathered was fed into the tactical plot. The system was further improved immediately after the war had ended by the introduction of the Type 960 operating in the 3-m band to replace Types 279 and 281. The new system provided greater reliability and stability, together with long-range air warning and height-estimation facilities in aircraft carriers.

It will be recalled that, from 1941 onwards, battleships had carried 11 gunnery radars and cruisers up to six. It was anticipated that in a well-coordinated air attack, targets would have to be switched from long-range to short-range weapons. New targets would have to be allocated to weapons as earlier targets were either destroyed or went out of range, or sector. Thus the concept of a target indication set (TI) was born, made possible by the development of the 10-cm search radar and the PPI display. For this purpose a 'cheese'-type antenna was developed for use with the Type 277 transmitter and receiver. With a PPI display in the gunnery office this equipment could be used by the gunnery officer for allocating targets to the most appropriate weapons. This set was designated Type 293. There was no provision for height finding since the 50-cm sets were broad-beam in the vertical plane and could not follow accurately in elevation. However the 10-cm high-angle gunnery set Type 275 and the 3-cm close-range gunnery set (Type 262) under development would both give accurate elevation, and if they were guided on to the right azimuth they could then search in the vertical plane rapidly enough to locate the target, since both radars were stabilised against pitch and roll. As mentioned above, two 10-cm gunnery sets were developed, the Type 274 for surface gunnery and the Type 275 for anti-aircraft (see Monograph 5).

This led to the concept of a much more sophisticated Target Indication (TI) system.[37] In this, each weapon would have its own dedicated TI operator in the gunnery office, seated in front of a PPI. On this he could track the target allocated by the gunnery officer to his particular weapon, with a moving cursor. This required quite sophisticated information selection and transmission arrangements in order to avoid ambiguities and confusion. A system was agreed and specified in 1943 or 1944 (Type 292), and a contract for its development was placed with EMI. The first model was undergoing test at Hayes before the end of the war. After being updated to provide for the new surface-to-air missile systems, in addition to the conventional armament, it went into production in the 1950s and survived many changes in weapon systems into the 1980s.

By 1942 plans had been formulated by DNO to replace the multiple pom-pom gun mountings by a fully automatic twin-Bofors mounting, for which ASE had proposed a 3-cm automatic-following radar to provide range, bearing and elevation out to 10000-yds. This was to be integral with the gun mounting. Although a limited search scan was provided for conditions of poor visibility, both approximate direction and elevation had to be provided for these conditions, and so, with several mountings in the same ship, target indication was essential. This set, Type 262, was developed by EMI in close collaboration with the Gunnery Equipment Division of Admiralty Signal Establishment. A. K. Solomon and A. V. Hemingway – who with D. R. Logie were seconded to the author's RE2 Division from the Radar Research and Development Establishment (RRDE), Malvern, in 1943 – supervised the project. It was first fitted in the aircraft carriers *Ocean* and *Bulwark*, but neither of these ships were completed in time to see action in the Pacific.

The problem of target allocation in big ships continued to exercise our minds right up to the end of the war. Since the expected detection range of Type 980 was 50 miles, similar systems should have given adequate time for allocation of the targets, and for the guns to be brought to bear to develop adequate range, azimuth and elevation rates; but fading of the target echo reduced the reliable detection range to 33 miles, which was unacceptable operationally. Moreover, calculations showed that a technical specification to meet the Staff Requirement was barely attainable in 1945. Several experimental systems were investigated, but none gave the operational performance required. Success was only achieved several years after the war had ended, when magnetrons of much higher output became available and parametric amplifiers (which greatly reduced receiver noise-factors) had been developed (see Monograph 5). Table 1.1 summarises the principal characteristics of Naval gunnery radar sets.

In the summer of 1945 L. H. Bedford and Major Sedgefield at AA Command had designed a simple guided missile, and were building an experimental model. The Director of Naval Ordnance at Bath asked the author for an assessment of the possibilities. He deemed the possibilities to be good at the existing state of the art and proposed a programme of

Table 1.1 Characteristics of typical naval gunnery radar sets   (© Defence Research Agency)

Gunnery role	Original DM-Wave Sets			Improved DM-Wave Sets			CM-Wave Sets		
	CR	LA	HA	CR	LA	HA	LA	HA	CR
Modulator peak power, kW	200	200	200	400	400	400	1000	1000	160
Transmitter peak output, kW	25	25	25	150	150	150	500	400	30
Pulse length, microsecond	2	2	2	1	1	1	0.5	0.5	0.5
Pulse shape	Triangular, rapid rise							Square	
Pulse recurrence rate/sec	500	500	500	500	500	500	500	500	1500
Nominal frequency, MHz	600	600	600	600	600	600	3000	3000	10000
Antenna gain, relative to half-wave dipole	15	120	50	30	120	90	835	725	1380
Feeder loss (db)	3	3	3	3	3	3	3	1.2	0.4
Beam-switching loss (db)	–	–	–	3	3	3	4	3	2.2
Receiver noise-factor (db)	25	25	25	15	15	15	13	13	18
Intermediate frequency (MHz)	6	6	6	6	6	6	60	60	30
Receiver bandwith (MHz)	4	4	4	4	4	4	4	4	4
Average range error, yds	50	120	150	50	25	25	25	25	25

Notes: CR = close range; LA = low angle; HA = high angle

development possible in about two years. His assessment was forwarded to the Third Sea Lord, who sent it on to the Prime Minister. It arrived at the same time as a proposal from AA Command for the guided missile being developed by Bedford and Sedgefield. An Inter-service Committee was therefore set up to study the whole field of guided-missile design, target detection and following, launching, propulsion, guidance and control, the only aspect not within its terms of reference being warheads. Eventually, after much deliberation under the chairmanship of Sir Alwyn Crowe, it was decided early in 1945 to build a number of test vehicles that would be rocket propelled and beam riding. ASE was given responsibility for the target detection and following and the provision of the beam-riding signals, but not of the control system within the vehicle. At this time the ASE was already committed to develop an improved long-range anti-aircraft system that, because of the known difficulties of target indication, would itself be able to carry out a search operation over a limited solid angle before locking on to a particular target. For this purpose a large 6-ft diameter paraboloid with hornfeed reflected into it from a small disc was developed (on the principle of a reflecting telescope in reverse). The small disc could be slightly offset and rotated to give a nutating beam for accurate following. Alternatively it could be swept through appropriate angles in azimuth and elevation for searching. Of more importance when tracking, it was the direction of this small disc that was controlled to respond rapidly to target movement. Hence the big reflector, with larger inertia, could be allowed to lag a degree or two behind the large reflector and catch up later. This large antenna was, in the first instance, fitted with all its remote power controls on a two-axis mounting. Trials on aircraft of the development model, designated Type 901, were successfully carried out at Witley in 1945 and 1946. Special filtering and strobing systems had to be developed for accurate following, and because it was to be used for missile guidance a specially coded information system was devised which could not easily be interfered with by countermeasures.

This was almost certainly the first use of coded information in beam-riding systems. Later the large parabolic reflector was replaced by a lens antenna. The system, designated Type GMY4, went into service as the control system for the Navy's first guided missile, Seaslug, and remained in service for many years.

Meanwhile communications, DF, IFF and Electronic Countermeasures were pursued vigorously in ASE, although most of the research and development of IFF was carried out by TRE or in America (see Monograph 6). Modifications required for IFF applications in the Navy had to be carried out by engineers from ASE, who also had to be responsible for their counterparts in TRE, and were frequently seconded to work at Malvern or in the USA for considerable periods.

In this short account of the development of Naval Radar in the 10 years from 1935–45 it has only been possible to give a general picture of how the techniques evolved in an attempt to meet the ever-changing needs of the Fleet, and how they were then converted into operational equipment. Space has permitted only a brief description of some of the more important technical details. Fortunately the most important of them can be found described in papers presented to the Radiolocation Convention held at the Institution of Electrical Engineers in 1946,[38] and in several of the Monographs in this book.

It has not been possible to give an account of the rapid development of small centimetric sets and passive radar for use in small boats, nor of countermeasures developed for the invasion of Normandy, and fitted in landing craft and other boats.[39] One of these many and rapid developments was a 3-cm set for submarines. The design was such that a rotating antenna, mounted on its own retractable telescopic mast, made possible a PPI display. Later an antenna was added to the periscope to provide accurate ranging during daylight engagements.

Another such was the K-Band 1.25-cm radar with a novel scanning antenna, developed by the National Research Council Laboratories in Canada, copied by ASE and fitted as an auxiliary to Type 274 for observing the fall of shot in surface actions.

Space has not permitted much description of the organisation of ASE after the moves from Portsmouth to Haslemere and Witley, but the relevant organisation charts are given in Figs 1.11, 1.15 and 1.16. Towards the end of 1943 C. E. Horton was transferred to DSR Headquarters in the Admiralty, and his position as Head of the Radar Department was taken over by Dr S. E. A. Landale, with A. W. Ross as Deputy (see Figure 1.15). When Dr Shearing retired in 1941, G. M. Wright formerly Head of the Electronic Warfare Group[36] became Chief Scientist of ASE at the Headquarters in Haslemere. Dr H. Smith then headed the Electronic Warfare Group (see Figure 1.16).

Horton was to succeed Wright after the end of the war, when Wright returned to Marconi, Chelmsford, as Chief Engineer. By the time of the move to Witley the technical staff (officers and assistants) of the Radar Department had grown to 155 and the total complement at Witley was about 400, including Naval personnel. By the end of the war the number of officers and assistants at Witley had grown to 350. (The overall 'all-hands' total of personnel in ASE during the war, including clerical and administrative staff, is believed to have numbered about 2400.)

What has never been either recorded or published is an account of the quite remarkable efforts of all the experienced designers and ancillary staff who converted experimental models, of quite novel design, into well-engineered equipments. These equipments that had to be maintained and operated by seamen under action conditions in appalling seas

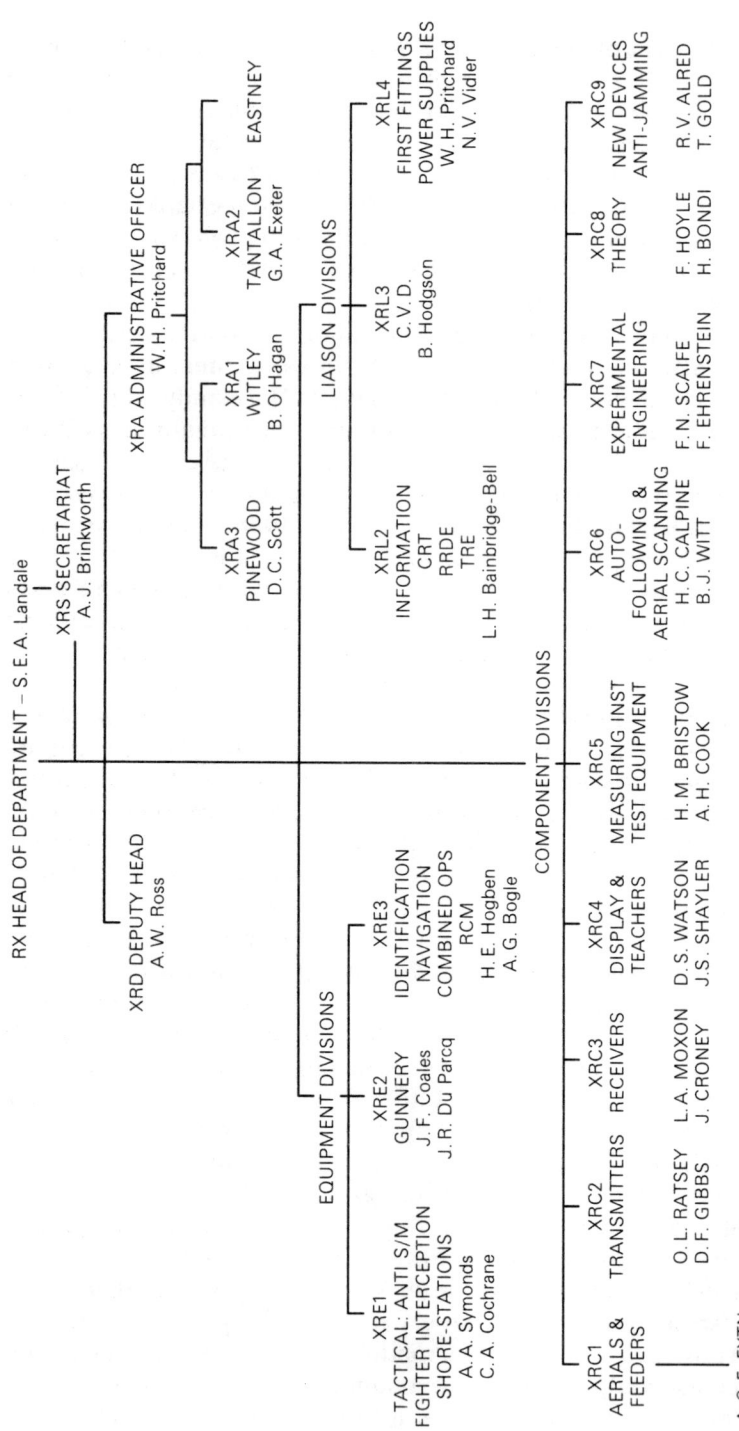

1.15 Organisation of Radar Department under Dr S. E. A. Landale in October 1943 (© Defence Research Agency)

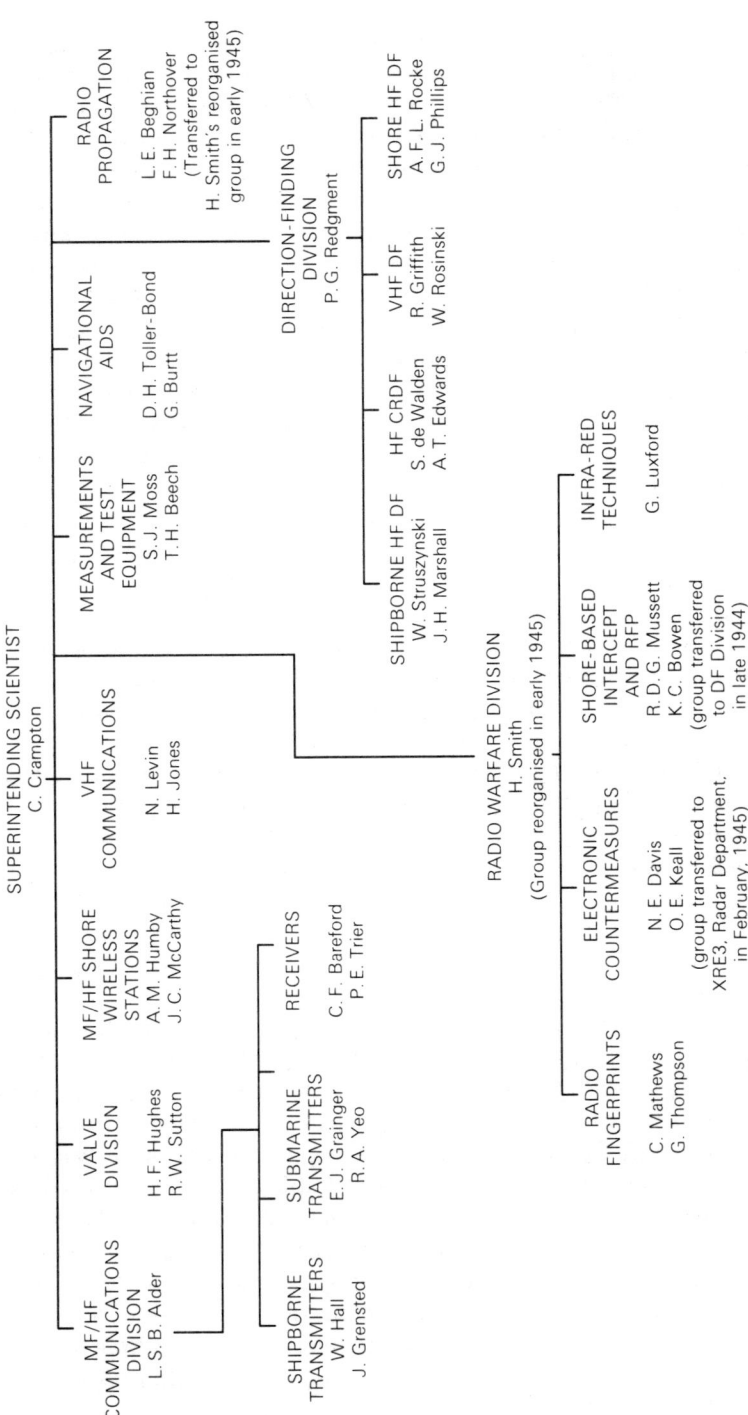

1.16 Organisation of the Communications Department ASE, in October 1944

for weeks, if not months, and thousands of miles from port. It is quite impossible to design equipment that will never break down under such conditions, so the Navy had to train officers and men who could not only operate, but also maintain and repair the new equipment at sea. HM Signal School had trained the signal officers and wireless operators to do this for Naval radio for 20 years but the advent of radar, of course, required an enormous and rapid expansion (see Monographs 7,8 and 9). It is remarkable how successfully this was achieved by the new HM Signal School, (HMS *Mercury* I) at Leydene, and the Radar Maintenance School, *Collingwood*, at Fareham. (ASE was HMS *Mercury* II.)

Even so the task of the operators and radio mechanics would have been impossible if the design and manufacture of the equipment had not been to rigourous standards. Experience had shown it to be necessary to withstand the buffeting met in ships at sea in bad weather, and due to gunblast in exercises and near-misses in action. Fortunately, at the outbreak of war HM Signal School had not only a team of very experienced mechanical and electro-mechanical designers, but also an excellent drawing and design office under N. Pemberton, a well-equipped workshop with experienced craftsmen under H. Noble, a procurement department under Mr Peake and a test department under Mr Christie, in which the equipment manufactured under contract was tested, and if necessary calibrated, before issue to the Dockyards for fitting in the Fleet. Up to the outbreak of war all radio equipment for the Navy was designed in HM Signal School, where full manufacturing drawings were made and proved by making development models in the experimental workshops. Only then were contracts for the manufacture of equipment placed with the established manufacturing firms by the Director of Contracts, Admiralty. Further, very close liaison was maintained between the staff of Signal School and the designers and engineers in the manufacturing firms, by means of which very good relations were built up.

It was thanks to these good relations that when the urgent need arose HM Signal School could proceed from a 'breadboard' model to a finished design to be produced in hundreds in a few months. The Establishment had the assistance of experienced engineers in so many radio and electrical firms, both large and small: GEC, BTH, Metropolitan Vickers, Marconi, EMI, Standard Telephones and Cables, A.C. Cossor, Allen West, Advanced Components, Aluminium Plant and Vessel, Pirelli-General and Telcon, to name but a few (see Appendix 4 to this volume). Of course designs and manufacturing drawings had now to be made in the firms instead of in the Establishment, but the firms seconded engineers and draughtsmen, and Admiralty engineers spent long periods in the firms. In this way the work was done in weeks where previously it would have taken months. This was engineering at its best.

Even so there were times when the urgency was such that there was not time to make manufacturing drawings and resort had to be made to making a development model as quickly as possible, and then copying it. It was then a case of 'all hands to the pumps', and young scientists and craftsmen, Engine-Room Artificers and radio mechanics in the experimental workshops to turning out the equipment, testing it and taking it to the ship at some port, sometimes completing the installation and finallly testing it while at sea.

It is impossible to praise too highly the skill and devotion to duty of the technicians, draughtsmen and artificiers who worked all hours throughout the air raids of the winter of 1940–41 to get those sets to sea, or to the wireless operators who kept new and strange equipment working in the appalling weather conditions of the North Atlantic. It is probable that never before, or since, has such a major technical advance been so widely and successfully deployed in such a short time, even in war.

References

1. R. A.Watson-Watt, *Three Steps to Victory* (Odhams Press, 1957).
2. Ibid.
3. J. F. Coales, CAC NRT ND63, His Majesty's Signal School, Portsmouth, 1919–1935. Soon after Coales joined Signal School he was a colleague of Lord Louis Mountbatten, at that time one of the Instructional Officers in the rank of Lieutenant Commander. A little later Lord Louis was driving his Rolls Royce to the Dockyard, and stopped to pick up Coales who was on his way to work. The (passenger) chauffeur descended to open the rear door for Coales, and Lord Louis moved to the rear also, whilst the Chauffeur took over the driving. This incident was subsequently repeated on a number of occasions, and typified the courteous and friendly relationships between Naval Officer and Civilian Scientist that was to persist into the future.
4. R. A. Watson-Watt, op. cit.
5. J. F. Coales, CAC NRT ND64, 'Naval Radar – the Formative Years'.
6. ADM 220/70, DSD to CSS, 13 August 1935.
7. Ibid.; Signal Directorate meeting (11 October 1935).
8. Ibid.; loc. cit. (11 October 1935).
9. Ibid.; CSS to DSD, Admiralty, 16 February 1937.
10. E. G. Bowen, *Radar Days* (Adam Hilger, 1987).
11. ADM220/70, CSS to DSD, Admiralty 16 September 1937.
12. Ibid.; DSD to DSR, Admiralty, No. SRE 1476/37 of 28 October 1937.
13. D. Pritchard, *The Radar War* (Patrick Stephens, 1989), p.32 et seq.
14. Ibid.;p.190 et seq.
15. A. C. B. Lovell, *Echoes of War. The Story of H2S Radar* (Adam Hilger, 1991), p. 25.
16. E. B. Callick, *Metres to Microwaves* (Peter Peregrinus, 1990).
17. ADM 220/70, CSS paper 'Memorandum on Expanded Effort in RDF'.
18. Ibid.; Naval RDF Panel Meeting, 27 July 1939.
19. ADM 220/83, Controller to ACNS, DSD, DSR, 28 July 1939.

20. ADM2 220/72, Their Lordships to C-in-C, Portsmouth, 10 August 1939.
21. L. H. Bedford 'The Development of Gun-Laying Radar Receivers Type GL Mark I, GL Mark I*. and GL/EF', *Journal IEE*, Vol. 93, Part IIIA (1946), p. 1115 et seq.
22. E. G. Bowen, op. cit., p. 42 et seq.
23. F. A. Kingsley (ed), *The Applications of Radar and Other Electronic Systems in the Royal Navy in World War 2* (Macmillan, 1994), Monograph 5 by F. A. Kingsley.
24. J. T. Randall and H. A. H. Boot, 'Early Work on the Cavity magnetron', *Journal IEE*, Vol. 93, Part IIIA, p. 182 et seq.
25. A. C. B. Lovell, op. cit., p. 41
26. W. E. Willshaw, 'The High-Power Pulsed Magnetron: Development and Design for Radar Application', *Journal IEE*, op.cit., p. 196 et seq.
27. A. C. B. Lovell, op. cit., p. 51 et seq.
28. A. C. B. Lovell, op. cit., p. 52.
29. A. C. B. Lovell, op. cit., p. 53.
30. J. F. Coales, CAC NRT ND85, 'Preparing for War', Appendix D.
31. G. Hessler, *The U-Boat War in the Atlantic* (HMSO, London, 1989).
32. F. A. Kingsley, op. cit., Monograph 4 by Cdr R. S. Woolrych, RN.
33. F. A. Kingsley, op. cit., Monograph 3 by Cdr A. E. Fanning, RN
34. F. A. Kingsley, op cit., Monographs 1 to 4 by H. W. Pout, Cdr A. E. Fanning RN and Cdr R. S. Woolrych RN, respectively.
35. F. A. Kingsley, op. cit., Monograph 1 by H. W. Pout.
36. F. A. Kingsley, op. cit., Monograph 5 by F. A. Kingsley.
37. F. A. Kingsley, op. cit., Monograph 2 by H. W. Pout.
38. Various authors; 'Radio-location Convention', *Journal IEE*, op. cit., contains a whole range of papers devoted to radar component and system developments by the Service Research Establishments and Industrial Associates in the period of interest, 1935 to 1945.
39. F. A. Kingsley, op. cit., Monograph 5 by F. A. Kingsley.

Monograph 2
Basic Science and Research for Naval Radar, 1935–45
B. W. Lythall

SUMMARY

Early visionary thinking by two Admiralty scientists – the concept of radar and of the resonant-cavity magnetron – regrettably not pursued. Following Watson-Watt's demonstration of RDF, limited Naval research initiated, but at inadequate priority. Reorganisation of RDF effort under C. E. Horton in 1937, leading to series of rapid development and production programmes – but still preserving some effort for supporting research. Reorganisation of Radar Department in 1942 – setting-up of research groups – recruitment of distinguished scientists, devoting attention to fundamental topics. Need for improved prediction of radar performance – theoretical studies of proposed systems – growing understanding of system parameters and environmental influences – analysis of evidence from sea trials – early methods of aircraft height-determination. Research on influence of environment – radio propagation – target reflection characteristics – noise – sea-clutter. Interference with the environment – anti-jamming research – 'Window' reflection characteristics – countermeasures. Examples of interaction between theory and experiment – antennae – transmission lines and waveguides.

EARLY HISTORY – AND TWO MISSED OPPORTUNITIES

One of the more striking features to emerge from Monograph 1 is the impressive speed at which families of Naval radar equipments were conceived, developed and brought into service. From 1938 onwards almost all the limited radar effort at HM Signal School had to be devoted to meeting the overriding pressures for operational hardware. Even so, a useful amount of more fundamental theoretical and experimental research was carried out, both by permanent staff and by wartime recruits, some of whom subsequently achieved high academic distinction elsewhere in other scientific fields.

As an example of the quality, if not quantity, of scientific support available to the Admiralty before the war, it is salutary to recall two examples of brilliantly creative thinking that were well ahead of their time. Regrettably, however, these proposals were not pursued. They relate to two of the most important milestones in the history of radar: Watson-Watt's original concept, in 1935, and the invention of the multiple-cavity anode magnetron in 1940. Both these ideas were conceived independently some years beforehand by Admiralty scientists at Signal School.

The Invention of Radar

It has been mentioned in Monograph 1 that L. S. Alder, of Signal School, proposed to the Admiralty in 1928 a scheme for the detection and location of objects by radio that quite specifically anticipated the invention of radar. It was never followed up, and even the request to proceed with a patent application was refused. Coales and Rawlinson[1] have suggested possible reasons why the idea may well not have seemed sufficiently important to the Signal School's parent department in Admiralty. Nevertheless it was also referred to the Directorate of Scientific Research, where it was rejected by the Patents Section without even consulting the Director, or his scientific staff. One can only speculate that had the latter been done, the idea would not have gone unheeded.

Seven years later Watson-Watt demonstrated radio detection of aircraft, and a dedicated Air Ministry Research Establishment (AMRE) was set up at Bawdsey. Within months the Admiralty had authorised independent work on Naval applications at Signal School, but it cannot be said that it accorded as high a degree of importance and urgency to this new invention as did the Air Ministry, or even the War Office. This might have been understandable in the light of what were then seen to be the main Naval priorities. Whatever the reasons, there is clear evidence that the deployment of scientific effort in the next two years was small, diffusely employed, and received inadequate supporting services because of the lack of priority in comparison with work on communications and DF.

The Multiple-Cavity Anode Magnetron

During this period A. B. Wood, a distinguished Admiralty physicist, was detached from ARL to work on radar at Signal School for a period of about 10 months. Ever since his earlier visits to Bawdsey in 1935–36 Wood had been advocating the advantages of higher frequencies. He realised that Gollin's work at Signal School on centimetric-wave systems was grossly limited by the low power generated by the split-anode magnetrons then available, and indeed by the highest power that might conceivably be generated in such a device (see Monograph 1). Wood turned his mind to ways of overcoming this problem. Amongst other possibilities he conceived the idea of a magnetron with a multiple anode, and sketched out two designs. In one of these he planned 'a cylindrical annular block having six or eight small cylindrical holes, each with a slit leading into a central cylindrical cavity containing a common central cathode'.[2] This appears to have been very similar to the concepts developed about three years later at Birmingham University by Randall, Boot and others, except that Wood perhaps originally thought of the anode as being included within a glass envelope, rather than itself becoming the main vacuum enclosure. Unfortunately, as Wood recalled in his memoirs,[3] his idea was received very dubiously by the Chief Scientist of Signal School, and he was told that the Valve Laboratory was too busy making large silica valves for the Fleet, and also for the Air Ministry radar chain stations. Again a most important idea was not pursued.

It was not until late in 1937, just before Wood left, that the Naval radar programme was at last put on a sounder footing. Several experienced scientists were transferred from other work in Signal School and a new Radar Department was formed under C. E. Horton, who was charged with getting radar equipments into operational service as soon as possible. This marked the real start of serious Naval radar development, and Horton was to remain in charge throughout most of the war. He was also to return soon after it to lead both radar and communications R & D into the post war era. One can only speculate that if the change had occurred earlier, or if Wood had stayed longer before leaving to become Chief Scientist at HM Mining School, his ideas would have fallen on more receptive ground, and strong efforts would have been made to have them pursued.

THE ENVIRONMENT FOR RESEARCH

1938–42

The creation of a new Radar Department under Horton set in train a period of remarkably rapid development and production programmes, which continued for at least five years. Despite the still limited resources, metric radars appeared in the Fleet before the outbreak of war, and decimetric sets soon after. By then a rapid build-up of manpower and resources was under way, but demands on them grew at least as fast. Hence the conventional sequence of research, experiment, development and manufacture had to be greatly telescoped. For the next few years whole families of systems were developed in rapid succession, each set in turn pushing technical possibilities to the limit. In parallel with each development there was a continual process of 'predevelopment' so that, as each set entered service, the technology for its successor was already being demonstrated. At the same time experimental work on new ideas to extend the state of the art was providing the foundation for further sets to follow.

This heavy concentration of resources on the overlapping development of successive systems did not prevent Horton from ensuring that, at all times, some effort was explicitly assigned to research. Whilst two Divisions of his new Radar Department (R1 under A. A. Symonds, and later J. D. S. Rawlinson, and R2 under J. F. Coales) focused on the development of metric and decimetric systems, the third Division (R3, led by A. W. Ross) was given responsibilities for general research, as well as the development of antennae for warning radars. Within a few years this research effort was to be greatly augmented. Among the many university graduates joining during the early days of the war, there were some scientists with distinguished records in mathematics and theoretical physics, who were better fitted to devote attention to more fundamental topics than to join the mainstream of experimental development. Among these were M. H. L. Pryce, F. Hoyle, H. Bondi, T. Gold, C. Domb, E. T. Goodwin, A. R. G. Owen and others; mention should also be made of E. B. Moullin and L. B.. Turner, who came from electrical-engineering faculties, and Otto Böhm, formerly of Telefunken, who arrived via Marconi. Many contributions were also made by the Allied groups from Poland, Norway, Holland and France.

1942–45

By mid-1942 the much-enlarged Radar Department had moved to Witley, and Horton had introduced a new organisation intended to achieve better planning and coordination of individual systems (see Monograph 1). Another feature was the creation of a number of specialist divisions to supplement the four basic technical divisions concerned with antennae, transmitters, receivers and displays. Among these were XRC8 Division, headed by Hoyle, with Bondi as Deputy. This was a small group of mathematicians and theoretical physicists (at least four of whom subsequently became FRS), which was given a more-or-less free hand to look at fundamental factors underlying the performance of radar, including features of the environment in which it had to work. Ross moved to Witley and later took charge of another new Division (XRC9) working on anti-jamming (A/J) and miscellaneous research. Pryce took over the antenna Division at Nutbourne, while maintaining close contact with Hoyle and his section.

THE PATTERNS OF RESEARCH

Research Themes

Because of the small effort available, particularly in the prewar years, it was necessary to concentrate on specific issues. Before deciding on the programme for each topic, surveys were made of the state of the art by studying available papers and visiting other activities in the UK and the USA. If any general pattern of work is to be distinguished it would have two threads. The first was a continuing need for better performance prediction, in order to account satisfactorily for the observed behaviour of existing radars, and to predict that of new concepts. Although this was helped by a steadily improving ability to measure internal parameters, it also required a better understanding of limitations imposed by the external environment, both natural and man-made. Hence, among the topics studied were the propagation of radio waves in the atmosphere, particularly over the sea; the reflecting properties of various targets; the characteristics of sea clutter and noise; and the effects of potential enemy interference. Each topic required both theoretical analysis and experimental research to acquire relevant data. As well as these more fundamental issues it was also necessary to make sure that practical problems of particular concern to the Navy, for example the need to site antennae at a great distance from the radar office, were properly understood.

The second thread is perhaps best described as a sort of 'ad hoc' consultancy service. This grew up between the theoreticians on the one hand, and the scientists and engineers developing new systems on the other. For example, design problems might require extensive theoretical analysis and computation, or a new device exploiting a fresh scientific principle would benefit from help in the interpretation of mathematical theory. There were many and varied instances, too numerous to mention; some typical examples are described in the final section on antennae and waveguides

The Need to Predict Performance

During 1938–39 some kind of yardstick with which to compare trials results from the first metric radars was badly needed. Ross carried out a theoretical analysis of the expected performance of Type 79, which he later repeated for Type 281 (see Monograph 4). He was able to take account of radio propagation from his work described below, but he had to make simplifying assumptions. These concerned the reflecting properties of the target, which he assumed to behave like a dipole reflector, and the detection capabilities of the human operator, where he assumed that, with an A-scan display, there was no integration by eye and brain. The errors in these assumptions were quite likely to have been in opposite senses; at any rate the analysis seemed to produce fairly reasonable predictions. Predictions were similarly made in R2 Division for the 50-cm gunnery radars; in practice, propagation considerations were of little significance for the close-range AA sets, but the performance of the longer-range surface gunnery sets was enhanced by the meteorological conditions prevailing in 1940 in a manner that was not at the time fully appreciated (see pp. 76–9).

It was to be several years before these relatively crude methods could be improved significantly. Much of the story of how this became possible will emerge later. The task became more and more difficult with the move to higher and higher frequencies; for example radio propagation was found to be more dependent on local meteorological conditions, and echo strength much more sensitive to small changes in target aspect.

Gradually the measurement of system parameters became more accurate, system behaviour more explicitly analysed and the environmental factors better understood. The performance of a hypothetical new system could be predicted with more confidence, and during the middle years of the war this became a central task of XRC8 Division. Requirements received from the Admiralty were each subjected to a performance assessment by establishing the figure-of-merit that would be needed to realise them. If a figure was unattainable within the state of the art, this had to be stated at the outset; if it was thought to be possible, a series of trade-offs would be investigated to determine the most economical solution. An example was the Staff Requirement in 1943 for new target-indication (TI) system; Hoyle and Pryce were very concerned about its practicability and pressed for relaxations.

In April 1944 Hoyle made an analysis of the observed performance of all existing Naval warning radars against aircraft. Wherever possible, known or measured values were inserted in the 'radar equation' and empirical values deduced for the other factors for each family of radars of similar type (that is, the metric warning radars Types 79 and 281, and the scanning centimetric radars Types 276, 277 and 293). This could then be used as the basis for predicting the performance of new systems in each family. Hoyle proposed that some of the outstanding uncertainties should be resolved by measurement of performance against a spherical target of known echoing area, for example a balloon with a metallised coating.

However the tools for prediction were still far from perfect. For example they did not indicate explicitly how the performance ought ideally to depend on the number of pulses falling on the target, an important factor in scanning radar design. (At the time work in the

Display Division (XRC4) was establishing that echo visibility was proportional to the square root of the number of pulses falling on the target, and for Naval metric and cm-wave radars it did not depend on where the echo was on the PPI, or on the beamwidth of the radar. In fact the CRT screen, human eye and brain together integrated signals remarkably well compared with an ideal observer.)

More seriously perhaps, the prediction process was only applied to establish the maximum possible range of detection. Whilst this might well be achieved in practice by a radar pointing continuously at the target, it would have been highly optimistic to expect it to be consistently obtained with a rotating-beam radar, sampling a limited sequence of rapidly fluctuating echoes on each revolution. Indeed, as we shall see later, XRC8 had already been measuring fluctuations of successive individual aircraft echoes, but because of their potential importance in countermeasures to 'Window', these results were not made available to other workers in the Radar Department to help in estimating the degradation of performance to be expected in scanning system.

Much more direct evidence had been accumulated by C. A. Cochrane in 1943 and 1944 from a series of trials at sea with Types 276, 277 and 293, the first shipborne centimetric warning radars capable of continuous antenna rotation, and fitted with PPI display. The maximum ranges achieved had been used for XRC8's analysis, but as Cochrane observes in a recent review: 'Unlike the metric waveband radar, first detection range on successive runs . . . was quite variable. Also there was no apparent regular variation associated with . . . interference lobes in the envelope of vertical coverage' (see Monograph 5). It was found that the results could be characterised by a *maximum possible range* of first detection, which occurred on very few of the trial runs, and a *minimum expected range* of first detection, by which an incoming aircraft was virtually certain to have been detected, and could be reliably tracked. By mid-1944 it was established that the ratio of these two ranges was remarkably consistent, at about 1.5 to 1. Cochrane and Domb subsequently developed an analysis based on empirical probability distributions for echo strength and sea reflections (see Monograph 5).

The evidence from these trials, together with those on fluctuations of aircraft echoes, was too late to influence the initial assessments of performance for proposed new aircraft warning and direction radars (Types 294/5, 990/981, and so on). Estimates were probably based on maximum possible detection ranges and were therefore optimistic, apart from the height-finder Type 981 when pointing continuously at a target. It is not clear how soon the implications of scanning loss were generally realised in the Establishment. It was not until after the war that a more realistic view of the performance of Type 980 finally prevailed, not surprisingly to the dismay of the Naval Staff.

Early Methods of Height Estimation

Typical detection ranges with Type 79 were large enough to allow interception of incoming aircraft by fighters, provided that the enemy's height was known approximately. Fighter direction really became feasible with the arrival in August 1940 of the new carrier *Illustrious*, fitted with the most modern Fulmar fighters and Type 79. However, early attempts had been made by several other ships and enquiries were soon received from the Fleet asking if there was a linear relation between detection range and target height that could be used to determine the latter. Ross prepared a reply based on exploiting the vertical polar diagram formed by interference between the antenna and its mirror image reflected in the sea surface. He explained that the range of detection was related, though not linearly, with aircraft height above the optical horizon, from which the height above the sea surface could then be determined. He pointed out that this could only be an approximate estimate, because of factors such as the operating efficiency of the radar and the type of aircraft. Use of the minima of the vertical lobe pattern should provide an estimate less dependent on such factors, but would only be possible at shorter ranges.

The method was developed by Hoyle, who had joined Ross at the end of 1940. He was also asked to look into the performance of Type 281, particularly as regards high-flying aircraft, which might escape detection in the lowest lobe if operational efficiency was poor. Hoyle devised an ingenious graphical presentation that proved of great operational value. The vertical lobe pattern was plotted on a rectangular grid of detection range versus height above the optical horizon, and superimposed on this were curves of constant aircraft heights above sea level. (In this presentation the earth's surface curves downward from the origin.) The aircraft height could then quickly be read off by eye, avoiding the need for the ship's radar officer to calculate the effects of the earth's curvature. It was soon followed by an even simpler presentation in which true heights were drawn as horizontal lines. The radar horizon then turns upwards, as do the lobes of the polar diagram.

Reports on the effectiveness of this method soon became increasingly unfavourable. It was realised that not only were the nominal performances of transmitter, receiver, antennae, feeder and so on inadequately known, but the performance of individual sets was often well below full efficiency and could vary considerably from time to time. The solution was to provide each individual set with a direct calibration of performance by flying in friendly aircraft at known heights. Hoyle followed this up by attending post-refit trials to check that the calibration method did in fact work successfully. He refined his presentation to include detection contours corresponding to various levels of performance degradation below the calculated standard. The ship could then choose an appropriate set of contours based on the most recent calibration and any subsequent changes to the set. The most irksome feature was the need for repeated calibration, which was often limited by availability of aircraft. A significant step forward occurred in July 1943 when the first performance meter for Type 281 was developed, so that performance changes since the previous calibration could be monitored, but this did not become widely fitted until the end of the war.

It is not known where the idea of utilising antenna interference patterns to obtain information on aircraft height first occurred, but Type 79 must have been one of the earliest applications – perhaps the first. In the absence of specialised British heightfinding radars it remained in widespread use throughout the war. Following initial help from the scientists, and building on Hoyle's tables and graphical aids, practical techniques were developed largely through the initiative, enthusiasm and ingenuity of individual radar officers and ratings, especially in aircraft carriers. To achieve and maintain a reasonable standard required constant training and practice and depended heavily on the experience and skill of the operators. Nevertheless by the end of the war this inherently rather crude process reached a surprising degree of refinement. By 1945 the British Pacific Fleet was able to make a unique contribution to joint operations with the US Navy by providing long-range height estimation, as a complement to the superior American capability in low cover and in numbers of true heightfinding radars.[4]

The advent of high-power centimetric waves held out promise of a more direct and accurate means of obtaining height by using a narrow vertical beam. In 1943 considerable work was done on an air warning set to provide continuous azimuthal cover with simultaneous heightfinding. Some ingenious ideas conceived for the warning antenna (Type 295 and later 980) are described later. In 1944 it was decided to develop a separate set for heightfinding, but for various reasons this did not emerge, as Type 981, until well after the war. The American 10-cm heightfinding radar SM-1 was fitted in some British Carriers towards the end of the war. It was invaluable up to ranges of about 50 miles, beyond which the old methods of heightfinding with the metric sets still had a role to play.

RESEARCH ON THE RADAR ENVIRONMENT

Radio Propagation

A knowledge of radio propagation is important, both to predict the performance of a radar and to estimate the distance at which it can be intercepted. During 1938–39 Ross conducted a study of available papers on propagation, and subsequently carried out measurements of field strengths at and beyond the optical horizon, for both horizontally and vertically polarised waves. These trials were carried out in an MTB in the Channel, and had to end when the ship came uncomfortably near to patrol-boat fighting near Calais. (The superheterodyne receiver Ross used in these experiments was subsequently adapted for use in 79Y and 279, in preference to the 'straight' model originally developed for the radar – see Monograph 4.) Ross presented a unified review of contemporary theory in May 1940, pointing out that the situation was still unsatisfactory as regards vertical polarisation. The review also included information on ground and sea reflections, referred to in a later section.

A better understanding was needed of the diffraction of radio waves around the earth, and how the field strength would be expected to fall off beyond the optical horizon. In 1941 Pryce started a brilliant attack on this problem. The classical theory of the diffraction of electromagnetic waves by a sphere had been evolved by G. N. Watson, and developed for radio waves by T. L. Eckersley and others. Pryce noted that the approach was unnecessarily complex for a sphere as large as the earth, where only the curvature near the transmitter was relevant to the solution. This yielded a substantial simplification of analysis, in which the key function was the Airy Integral of complex argument.

Previous work had involved a new and cumbersome calculation for each fresh case considered, so that there were few worked examples from which to build up a more general understanding. Pryce's elegant solution was particularly suitable for horizontal polarisation, for which it could be expressed in a standard form, any specific case being treated simply and quickly in terms of two given parameters. Calculation of the relevant values of the Airy Integral was undertaken by Domb, who had recently joined Hoyle at Nutbourne. Dr J. C. P. Miller of Liverpool University provided initial guidance. Formulae and graphs for practical use were presented as a Monograph, which was widely used on both sides of the Atlantic. This paper was subsequently published in the literature.[5] For vertical polarisation the dielectric properties of the ground are relevant, and more extensive calculations were needed; these were undertaken by HM Nautical Almanac Office. In these days of ubiquitous computing power it is difficult to appreciate the skill, mathematical ingenuity and sheer hard work involved in this project using only the mechanical desk calculators then available.

Whilst this work was in its early stages, reports were coming in from the Fleet of unexpectedly long ranges of detection with the recently introduced centimetric radars. Diffraction theory would predict that these wavelengths should be virtually cut off at the horizon, although metric waves would penetrate somewhat farther. Yet the reports were suggesting exactly the opposite, that centimetric waves had penetrated far beyond the horizon, indeed much farther than would have been expected with the longer wavelengths.

Pryce and Hoyle concluded that, if these reports were to be believed, the effects must be due to abnormal atmospheric refraction. Under normal conditions the refractive index of the atmosphere decreases gradually with height because of the decreasing amount of water vapour present. However it was difficult to see how this alone could account for the effects observed, until they both realised that there could be significant ducting effects if the water-vapour content varied sharply within one or more rather thin vertical layers. Pryce is reported as recognising this as being analogous to optical mirages when he was lying observing the horizon one lunch time on Eastney beach. Hoyle recalls that his thoughts were jolted along similar lines by the shock of seeing Owen fall off the roof of a wooden hut at Nutbourne, contriving to land with each foot squarely in separate buckets of sand provided for fire fighting!

Whatever the background for these flashes of inspiration, a search by Hoyle of the meteorological literature produced a series of examples of unusual water-vapour distributions in the lower atmosphere that, on first examination, seemed to offer hope of a satisfactory explanation. The ideas were therefore put to the test by calculating the propagation for a selected series of sample conditions. The results indicated that this theory could give a reasonable explanation of the effects observed.

Hoyle and Pryce reported their findings in a paper in 1941, concluding that anomalous propagation conditions ('anaprop') were more likely to happen at wavelengths of 25-cm and below. They recommended an experimental investigation, and a programme was planned in June 1941. This started in September, jointly by Signal School and GEC Research Laboratories. Hoyle was the Admiralty sponsor, and E. C. S. Megaw was the leader for GEC. The experimental results of Megaw's team at GEC were published as various ASE reports, and Domb and Pryce also applied their diffraction analysis to some of these.

The Army had also observed strikingly large ranges, in the hot summer of 1940, at 50-cm radar stations on the South Coast, and at 10-cm stations during the following summer. Others also suffered from the remarkable meteorological conditions in 1940. In the summer anaprop conditions were encountered during trials with the first 50-cm Naval gunnery radars, tending to give somewhat optimistic forecasts of performance. In addition, in the prevailing ionospheric conditions of the preceding winter associated with sunspot maximum activity, second-trace echoes, presumably from far-distant land, were observed on Type 79 in *Curlew* based at Scapa Flow. D. S. Watson recalls that, with headphones connected to the radar receiver, signals from American police radios could clearly be heard by the ship's staff!

The surprisingly long ranges at centimetre wavelengths led to the planning of a concentrated Inter-Service attack on long-range propagation and its relation to meteorological conditions. The resulting programme, under the auspices of the Ultra-Short-Wave Propagation Panel of the Ministry of Supply's RDF Applications Committee, chaired by Sir Edward Appleton, used equipment provided by the Admiralty and sites operated by the Army. Organisations involved included SRDE, ASE, NPL, the Meteorological Office, the Naval Meteorological Service and GEC. The sites were originally chosen on the South Coast, but had to be abandoned for security reasons early in 1942. The main investigations were then conducted at sites on the West Coast. It was difficult to find suitable sites for the longer oversea paths in which ASE was principally interested. Eventually it was decided to use the top of Mount Snowdon, which gave wide arcs of view across Cardigan Bay and the Irish Sea. The Snowdon Mountain Railway was commandeered, together with the Café at the top, where Bondi lived for several months. Goodwin was installed at Aberporth, and measurements were made of propagation at 10-cm and 3-cm across Cardigan Bay. A Type 277 radar was subsequently installed for sea-clutter experiments described later.

This programme of theoretical and experimental work continued throughout the war and was a good example of wartime scientific cooperation between the Services. Many of the findings were reported at the IEE Radiolocation Convention in 1946, including a comprehensive summary of the experimental results by Megaw.[6] The scale of effort may not have been justified solely on grounds of investigating anaprop, but it provided a much improved understanding of how propagation was affected by atmospheric, terrestrial, marine and other phenomena. These included the behaviour of fading, which was important to communication systems as well as radar. Although not reported at the time, it was the results emerging from the Cardigan Bay trials in 1943 that led Megaw to look further into long-range propagation via the mechanism of forward scattering in the troposphere, which was to become so important for military communications and intelligence gathering. After the war Megaw, who by then had moved to ASE, pursued this in much greater detail.[7] He developed a comprehensive theory of scattering by atmospheric turbulence[8] complementary to work by Booker and Gordon in the USA.[9] His posthumously published theory[10] won the Annual Premium prize awarded by the IEE.

Another form of beyond-horizon propagation was observed at metric wavelengths, caused by partial reflection from temperature-inversion layers that commonly occur between 2000 and 10000-ft in anticyclonic weather conditions. A theoretical analysis of this was made by L. E. Beghian and F. H. Northover of the Communications Department, but unfortunately the associated successful experimental results were not fully analysed before the run-down of staff at the end of the war. (This is the phenomenon that today periodically affects TV reception by abnormally long-distance transmission from other transmitters operating on a similar frequency.)

Target Reflection Characteristics

By the 1942 the greatest uncertainty in predicting performance had become the reflecting properties of various targets, particularly aircraft. Hoyle decided that it was time to find out what really happened when an aeroplane reflected a single pulse. This required a range-gating device, supplied by the receiver group, and a system developed by Gold that could draw photographic film across the display tube at a sufficiently high rate (say 10-ft per second) to separate the images of successive pulses. (Gold first experimented with a commercial device, which apparently proved quite unworkable until he removed many 'useless' parts. With the experience thus gained he then had a much-improved camera made up in the ASE workshop. Gold later developed other high-speed photographic techniques. He used continuous film processing to implement an idea for discriminating moving targets from stationary echoes by subtracting the signals of one scan from those of the next. Dr Tricker of ASE Extension, Bristol, had already demonstrated the concept using conventional photographs to give a display in slow time ('Tricker's Flicker'), but Gold was interested in a system operating in real time. The apparatus he devised, with a continuous spray process for developing and fixing, was considerable technical achievement and the forerunner of large-scale commercial systems after the war.)

A few initial trials were carried out at Nutbourne, followed by some runs with a Hampden bomber at Eastney. It was soon found that the variations of echo amplitude were much more pronounced over sea than over land. The equipment was then assembled in a trailer and taken to a cliff-top site near Bedruthan Steps in Cornwall. Here, targets could readily be obtained using aircraft flying from the nearby RAF Station at St. Eval. After several alarming incidents with this highly experimental equipment, successful results were obtained. These confirmed that there were indeed remarkable variations in the intensity of the received signals. As had been expected, they would often fluctuate a few times per second. However they were also found to change violently within only a few pulses, so that the 'instaneous' echoing area, at least at 10-cm wavelength, could change decisively in less than a hundredth of a second.

This work had a number of significant consequences. Firstly it suggested the possibility of discriminating aircraft echoes from those of Window, which would not be expected to suffer such violent changes.

Gold's subsequent pursuit of this is described later. Secondly, because the potential implications were assumed to be very sensitive, the results were kept quiet and not disseminated to other workers in ASE. Otherwise the work might well have contributed earlier to a more thorough understanding of air warning radar and a more realistic prediction of the performance of projected sets. Even as late as 1946 Cochrane was still unaware that any pulse-to-pulse records were available in ASE, although he had found some from ADRE.

Thirdly, the results of these experiments questioned the assumption that antenna rotation rates should be kept low enough to allow some gain from integrating the effect of a reasonable number of successive pulses falling on a target. In 1944 Gold conducted trials at Tantallon with Types 277 and 293. He showed that much higher rotation speeds could be employed with little loss (for example only 1 db when the rotation rate was trebled), because detection depended more on peaks in the instantaneous echo strength of the target than on the integration of pulses received in a single scan.

Noise

In all early radars, the background against which echoes had to be detected was dominated by noise generated within the receiver. Throughout the war the main effort towards improving the performance of receivers was devoted to reducing their 'noise-factor' – the degree to which performance was degraded compared with an ideal system which introduced no noise. A comprehensive survey of the principles used in wartime developments was published in 1946 by L. A. Moxon.[11]

The receiving antenna had not originally been thought of as a significant source of noise, but by about 1943 this was thrown into question when new low-noise RF stages were developed for the receivers for Types 79 and 281 (see Monograph 4). These had both demonstrated a 5 db improvement in noise factor in the laboratory, but in trials at sea there was no improvement with Type 79 working at 40 MHz, and only 3 db with Type 281 on 90 MHz. Moxon and his colleagues established that the cause was 'cosmic noise', extra terrestrial radiation arriving from all celestial directions, although with a marked increase in intensity near the plane of the Galaxy. The noise received is more or less constant (unless the antenna has a narrow beam and is pointing at the galactic equator), and at these frequencies this is the main factor controlling the useful sensitivity of a receiving system. Another important ASE contribution was the very first determination of the variation of cosmic noise level with frequency. It was based on simultaneous measurements at 40, 90 and 200 MHz made over a 24-hour period at Nutbourne and Witley. This indicated, inter alia, that cosmic noise decreased rapidly with frequency, remaining barely significant at the upper end of the frequency range. A resume of the work was published by Moxon after the war, and a brief summary of the properties of extra terrestrial noise was given in the paper.[12]

Measurements were also made of the increase of noise observed with narrow-beam antennae pointing in the galactic plane, for various longitudes. Although the practical relevance to radar was only in 'worst-case' situations, this phenomenon was perhaps the main feature of general scientific interest at the time. ASE's contribution, apart from confirming results of other workers, lay in observing the variation with frequency of the extra noise when looking at the hottest spot on the galactic equator.

This early work in the embryonic science of radio astronomy was complementary to a major effort at TRE, for example by J. S. Hey and his colleagues. This led after the war to the well-known schools of radio astronomy set up by Bernard Lovell at Jodrell Bank, and Martin Ryle at Cambridge. It is interesting that the wavelengths chosen for those first Naval radars lay exactly in the required 'window' to reveal these effects – they were just short enough to penetrate the ionosphere, but long enough for cosmic noise to be sufficiently strong to be revealed by the best receivers.

Theoretical work on noise and similar random processes was also being developed from first principles by several workers in the USA and the UK. Perhaps the best-known general statistical approach remains that of S. O. Rice at the Bell Telephone Laboratories.[13] In 1943 Bondi made a formal analysis of how the probability distribution of the various sources of (Gaussian) noise generated in the first stages of a receiver would be influenced by the subsequent circuits, particularly the characteristics of the second detector. It had been long known[14] that there was little difference between linear and square-law detectors in respect of discrimination between strong and weak signals. Bondi's analysis showed that this similarity also applied to the correlation of the output noise. In principle, Hoyle was now able to allow for the characteristics of actual receivers in his general theory of performance prediction.

Soon afterwards Domb was engaged on a theoretical analysis of echoes from rain clouds. It had previously been assumed that echoes from individual water droplets would be in random phase. However Hoyle had pointed out that this might not always be justified, and that in certain conditions the contributions from many particles might be cophased. Domb's formal analysis showed that, in most practical cases, the assumption of incoherent scattering would be justified because at longer wavelengths, which are more favourable to coherent scattering, the individual echoes are so small that the total echo is likely to be undetectable. He also showed that his analysis could be applied to the distribution of noise from the 'shot' effect, and from this derived some of the results[15] previously obtained by Bondi and Rice by a different method.

During these years the underlying principles involved in distinguishing wanted signals from a statistically varying background were becoming better understood, although in a rather piecemeal manner. It was not until after the war that the picture could be seen in a fuller perspective, with the publication in 1948 of C. E. Shannon's classic paper on the Theory of Communication – better known now as 'Information Theory'.

Sea Clutter

Ross's prewar experiments had included trials in *Saltburn* using a Type 79 fitted with a vertically polarised array. He argued that this would give better results than horizontal polarisation against targets on the surface. In fact the idea was thwarted by the much higher sea clutter with vertical polarisation, probably the first time this effect was appreciated. It was realised that a much more directional antenna would be needed, and this would really only be practicable at higher frequencies. As a result all Naval radars for many years to come were to use horizontal polarisation (see Monograph 1).

For the next few years there was little systematic research on sea clutter at ASE, although work was going on elsewhere in the UK (especially for airborne systems), as well as at The MIT Radiation Laboratory in the USA. Pressure of systems work at ASE had restricted attention to observations that could be made in the course of other trials, and to 'ad hoc' methods of alleviating the effects of sea clutter in existing radars. However in 1944, using a Type 277 installed at the top of Snowdon, Bondi conducted a series of trials across the Irish Sea to establish the detectability of low-flying aircraft in clutter, and to examine the properties of sea clutter in general.[16] These appear to be the first quantative studies of clutter from a rough sea, experimental studies in the US having been largely confined to the calmer waters of Boston Bay. Bondi was assisted by H. Dahl, the senior member of the Norwegian group at ASE, who later worked on microwave-measurement techniques and developed the 'preplumbed' crystal mixer mentioned below.

Measurements were made of the mean power of the clutter return and its variation with range, for a variety of sea conditions. The dependence on wave height was also established. Gold's high-speed photographic apparatus was used to investigate the scan-to-scan characteristics, including the correlation time. Estimates were made of optimum antenna heights for detection of low fliers, and the likely performance of AEW radar.

In April 1945 Gold (who had by now joined XRC9) and Renwick commenced another series of trials at Seaford, using three different heights representative of shipborne installations. Both 10-cm and 3-cm radars were used. Again the variation of clutter with range was measured, and successive single scans were photographed. In their report Gold and Renwick stressed the importance of a very precise AGC system to match the envelope characteristics of the clutter. They concluded that devices such as flicker displays, which could exploit the scan-to-scan correlation of clutter, would not provide enough improvement to justify the complexity. More elaborate possibilities for the future were to use two antennae at different heights, and eventually to use a Doppler system for tracking fast targets.[17]

Ross then made a review and analysis of all the available wartime results. He found that the Seaford trials, which covered a range of conditions not explored experimentally in the USA, were in reasonable agreement with theory developed at MIT. He concluded that although reasonable predictions of sea clutter could be made for horizontal polarisation in S- and X- Bands, and for moderate seas, there was still too little known about clutter with vertical polarisation, especially outside these bands. In addition to the precise AGC system already mentioned, Ross proposed a number of other features to be considered in future sets. These included the use of longer wavelengths, higher prf, and range-gating with high-pass video filters. Soon after the war a further programme of measurements was put in hand, including extensive trials from high coastal sites in the Isle of Wight.

INTERFERENCE WITH THE ENVIRONMENT: ANTI-JAMMING RESEARCH

Electronic Jamming

At the beginning of the war the pressure was naturally to exploit the new technique of radar as rapidly as possible, without much regard to the possibility of disruption by the enemy. However, as early as 1941 an internal paper (author unidentified) speculated on what jamming methods the enemy might employ, and suggested possible measures to combat them. Ross was certainly giving thought to this in 1942. Early in 1943 he made proposals for a broad anti-jamming (A/J) policy. These included general measures to be taken with all current radars; specific suggestions for individual metric, decimetric and centrimetric sets; and recommendations for a number of features to be incorporated in future systems. It was realised, however, that little could be done about the true noise jammer, if and when it appeared, without a new philosophy of radar design.

There followed a period of steady application, in which R. V. Alred played a prominent part. Modifications or additions were made to individual sets, and various additional anti-jamming measures tried out. By careful attention to detail it was found possible to achieve large improvements with relatively small modifications to the design of existing receivers. Particular attention was paid to the frequency response at 40 db below peak, in order to see through modulated jamming, as well as to the avoidance of unnecessarily low saturation levels. Suitable specifications were then drawn up to be met by all new receivers.

Two special measures were also explored. The first was 'back-biasing', whereby an increase in the mean amplitude of signal, caused for example by jamming, was made to cause a reduction in overloading of IF stages in the receiver. A technique for reducing overloading of displays was subsequently proposed and developed by Alred into the 'logarithmic' receiver. Eventually this became almost universally adopted after the war, particularly as an anti-clutter device, and for applications in tracking radars.[18]

The second device, which proved more successful in the short term, was the video-frequency filter unit, proposed and developed by Ross. The receiver was operated at low gain, in order to avoid saturation by jamming signals, and the gain was restored in the video amplifier after jamming modulation frequencies had been removed by a specially designed filter. Ross ordered four units for use during the Normandy landings in June 1944. Alred tried out the first in *Belfast*, with successful results against the interference encountered. (This interference seems to have been more from our own radio communications than from German jamming!)

Up to this time there seems to have been very little communication, let alone interaction, between the anti-jamming division at Witley and the radar-countermeasures division at Haslemere. This may have been because of the very high security attached to the work of the latter. There was closer collaboration later in the war. This included trials of British jammers against British centimetric radars, and a joint study of Japanese centimetric radar, and the countermeasures to be adopted against it, as the British Naval presence in the Far-Eastern Theatre developed.

Non-Electronic Jamming: Window

By 1944 Window was considered one of the most serious forms of jamming. Wartime radars were not able to exploit the Doppler effect to discriminate between fast-moving aircraft and drifting Window. About all that could be done quickly was to reduce the pulse duration, and to introduce a differentiation circuit with a time-constant equal to the pulse length. Gold was interested in other possible ways of discriminating between aircraft and Window echoes, and conducted trials to measure the polarisation effects of Window. It was found that Window reflected a considerably greater proportion of cross-polarised energy than did aircraft. Here, in principle, was a method for discrimination which was roughly explored in a second series of trials. Goodwin analysed the earlier results to deduce the distribution of orientations in a cloud of Window particles, and hence the mean echoing area.

Goodwin's theoretical analysis also showed that the characteristics of echoes from a fully developed Window cloud should be very similar to noise, whereas Hoyle and Gold had already found that aircraft echoes fluctuated much more rapidly. Gold decided to investigate the possibility of discriminating against Window by electronically storing the outputs of individual range scans, or short groups of scans, and subtracting them from the output of the subsequent group. He was, of course, also interested in such methods for discrimination of moving targets, as a much more compact and convenient alternative to the method using rapid photographic processing mentioned earlier.

The signals needed to be stored for, say, a hundredth of a second, which in 1943 – 44 was not yet possible by digital means. An analogue solution had to be adopted, whereby the information was transformed into acoustic waves and transmitted through a medium having an appropriate sound velocity. Mercury was chosen as the most suitable substance. Gold's work was the precursor of a quite different application immediately after the war, when the mercury delay line, as it came to be called, was used in the memory system for the first British digital computer, EDSAC 1, at Cambridge.

The components of a mercury delay line, especially the crystals used for input and output, had to be kept very clean. It was judged that the best solvent was pure alcohol. Hoyle recalls that research was duly carried out on other properties of this fluid, especially the ease with which palatable liqueurs could be produced by the simple addition of sugar, water and fruit juice. Perhaps this is why it has not been possible, so far, to find any report of the outcome of Gold's experiments!

Radar Camouflage and Echo Enhancement

An associated line of work in XRC9 Division was the study of means of reducing the echoing area of sensitive targets. It was believed (correctly) that the Germans were somewhat more advanced in this field, probably because of pressure to reduce the detectability of the newly-developed submarine Schnorkel exhaust. It was decided to experiment with radar-absorbent materials, firstly to assess their potential benefit to the enemy, and secondly to investigate uses such as coating parts of ship superstructures to reduce side lobes, or the absorption of leakage radiation from transmitting antennae.

It was also pointed out at this time that very considerable reductions in the echoing area of ships could, in principle, be achieved by proper attention to design, particularly in avoiding large plane surfaces and the occurrence of corner reflectors. Not surprisingly such suggestions were received unsympathetically by the Director of Naval Construction, and indeed continued to remain so for very many years to come. It was not until several decades later that such ideas began to be applied in practice, for example in Russian warships for some years now; in the design of the current Type 23 frigate for the Royal Navy; and in the future concept of 'stealth' ships (and aircraft) for the US Navy.

Of more immediate application was the deliberate artificial enhancement of radar echoing area. A particular case occurred in Operation NEPTUNE. A major seaborne attack on Le Havre was simulated by anchoring 15 small barrage balloons in groups of three, each containing internal metallised corner-reflectors to produce echoes comparable to the large bombarding ships present. Also, four minesweepers each flew two balloons at different heights (so that their echoes would fade independently, an idea of Ross), patrolling to the northward of the bombarding ships. The CO of *Roberts* reported that, when the bombarding ships were obscured by smoke, the German shore-batteries engaged the radar balloons. The minesweepers reported eight near misses, whilst the bombarding ships were not fired at. Whilst the evidence is not conclusive, it would appear that the anti-radar balloons were successful in drawing the limited fire of the shore-batteries on to themselves.[19]

INTERACTION BETWEEN THEORY AND EXPERIMENT

This final section describes some examples of the interaction between theory and experimental technology in two particular fields, antenna development and the related subject of transmission lines and waveguides. The examples of waveguide work have been selected to illustrate responses to specific Naval problems, in particular the Navy's pioneering use of very-high-power centimetric sets, and the peculiarly Naval need to fabricate and install lengthy runs of waveguide under the conditions obtaining in a Naval dockyard. They do not include a great many other waveguide achievements at ASE, to say nothing of the very great deal of pioneering work on waveguides done elsewhere, for example at TRE.

Antenna Developments

Ross's first design for the Type 79 shipborne antenna required a compromise between high gain and high radiation resistance. The latter was required to achieve a bandwidth adequate to transmit the narrow pulse and to ease matching problems with the feeder lines. Both factors were influenced by the spacings between the various radiating and reflecting elements of the dipole array, but changes that increased one factor generally reduced the other. To arrive at an informed choice it was necessary to calculate not only the gain of the array, which was relatively straightforward, but the radiation resistance as well. Fortunately Ross was able to draw on the extensive work on the latter by G. H. Brown of RCA,[20] leaving as the main task the development of suitable impedance meters to check the results experimentally.

As antennae became more sophisticated, so the calculations of radiation pattern and impedance became more complex. Moullin and L. Reynolds carried out a series of calculations relating to the antennae for 50-cm sets being developed by Alred. These included, for example, the effects of parabolic and V-shaped reflectors on a variety of radiators, from single dipoles to side-by-side arrays of Yagi antennae. In the course of this work Alred himself originated the idea of using a binomial power distribution across the elements of an array to produce a radiation pattern with no side-lobes.

Common antenna working for warning radars required the development of a transmit/receive (T/R) switch to enable the sensitive receiver to be isolated during the transmitted pulse. A conventional diode was used in Type 281, but for Type 291 a gas switch was developed by G. E. F. Fertel, who had come to Nutbourne from the Admiralty Valve Section at Bristol University. He added a small quantity of water vapour to the gas in order to reduce the deionisation time, and hence the minimum range of the radar. Later he experimented with the gas mixture in T/R switches developed at the Clarendon Laboratory and TRE for the 10-cm sets, in which a gas discharge was set up across the throat of the resonating chamber taken from a reflex klystron. Rumour has it that Fertel maintained that the only source of neon of the requisite purity was in the old neon lamps to be found in London Underground trains, which were indeed mysteriously vandalised at about this time!

Otto Böhm was formerly Director of Research at Telefunken, but had left Germany for the UK in the 1930's. He was interned at the beginning of the war, but fortunately was released to join ASE in the summer of 1941, where he played a key role in the work on centimetric radar.[21] He was also concerned with the difficult problem of antennae for IFF transponders, which needed an omnidirectional radiation pattern but could not be sited at the top of the mast to obtain an all-round view (see

Monograph 6). Domb recalls that Böhm's own calculations were relatively simple but that, with remarkable perception, he used empirical approximations to obtain reasonable solutions to very complex problems, refining them by experiment as necessary. For the more involved mathematical calculations he was assisted by L. Lewin, but always seemed to know intuitively if the calculations were correct, or if an error of significance had occurred.

At least of equal importance to Böhm's personal work was his position of unofficial doyen of young mathematicians and antenna engineers at Nutbourne throughout the war. To quote Hoyle 'there was no young turk at Nutbourne who would take a serious step without having a word with "Doctor Böhm" about it. It was significant that everybody instinctively used the title. Nor was the respect for Böhm confined to the more mathematical end of the Nutbourne staff. It extended to the mechanics in the workshop, for whenever Böhm designed a piece of equipment the job never had to be done twice'.

In 1942–43 Böhm proposed an approach to the increasingly important problem of heightfinding on low-flying aircraft, in which a single receiving antenna with two different feeds would provide two different vertical receiving patterns, whose responses could be compared. Gold made some early tests of these ideas using two separate receiving antennae, but the elevation accuracy achieved was less than expected. A theoretical analysis by Owen suggested a number of possible reasons and recommended further trials with the two receiving feeds fitted to the same antenna, as originally proposed. The ideas were not apparently taken further.

By now another ingenious scheme was being studied for the warning antenna for Type 980, proposed as a replacement for Type 295 when the expected increase in magnetron power did not materialise. The aim was to obtain high gain by using a large vertical aperture, and at the same time achieve the required elevation coverage, normally requiring a small vertical aperture, with low gain. The antenna was a vertical stack of three 'cheeses' whose upper and lower elements were each fed through a waveguide phase-shifting device. The central sections of the two phaseshifters were continuously rotated in opposite senses, so that the narrower vertical radiation pattern of the combined antenna was rapidly and repeatedly swept in elevation within an envelope shaped like the broader patter from a single 'cheese'[22] F. N. Scaife and Hoyle subsequently pointed out that rotary phasechangers could also be used for vertical stabilisation of the antenna; perhaps one of the earliest ideas for 'electronic' stabilisation.

As early as the end of 1943 experimental phasechangers had been developed, and the principle successfully demonstrated working with an antenna. The antenna was then developed for Type 980, and under Type number 982 was eventually fitted in *Eagle*, and other carriers, after the war. However it was superseded, in later fittings of Type 982, by a mechanically much simpler antenna with no vertical scanning losses. This used a horizontal linear waveguide array feeding a specially-shaped cylindrical reflector, to provide the required vertical coverage with only a small reduction in gain. Hence those ingenious but rather cumbersome and inefficient antenna techniques of 1943 saw only limited service.

Transmission Lines and Waveguides

Type 79 used a long, open-wire transmission line of about 400-ohms impedance. This was chosen to reduce losses from wet spacers; to reduce the danger of flash-over; and to facilitate matching at both transmitter output and receiver input. Long distances between antennae and office equipment had long been a feature of Naval radio. In 1936 – 37 Coales and Ross had studied long transmission lines with particular regard to the effective Q at the near-end resulting from various impedances at the far end. Applied to Type 79 this determined the matching requirements at the antenna, as already described, and the limits on its radiation resistance in order to preserve the shape of the transmitted and received pulses. It was also well known that, in the worst case, the coupled circuits of the transmitter and the long mismatched transmission line could cause 'frequency pulling', an effect to be encountered again later with long waveguides.

Coaxial cables offered attractive advantages over open lines but, as frequencies and powers increased, problems had arisen in finding dielectrics with acceptable loss and resistance to electrical breakdown, which could also withstand a salt atmosphere and corrosive funnel gases. The first cables had braided metal outer conductors and a rubber dielectric, later superseded by polythene. In 1940 a cable with solid copper conductors and a magnesium-oxide core became available from Pyrotenax Ltd (see Monograph 1). It had been developed as a fireproof cable for electrical power transmission, but proved to have acceptable loss at metric wavelengths, provided that measures were taken to prevent ingress of water to the highly absorptive filling. This problem was solved successfully and the cable was used in the 214 MHz radars for smaller ships (Types 286 and 291), as well as in later versions of Types 279 and 281 (see Monograph 4).

Almost immediately afterwards the first centimetric radars were entering service, using braided cables with a flexible polythene core to reduce the losses at the higher frequency. There then occurred a phenomenal rate of increase in the power available at 10-cm, resulting first from a more than tenfold improvement in efficiency from magnetron 'strapping', and soon afterwards by the use of robust secondary-emission cathodes. Within months the requirement was to handle peak RF powers of 70–80 kW and this was soon increased to 500 kW – nearly a hundred times greater than that available in the first Type 271. The mean power increased in proportion, and as Cochrane relates in Monograph 5, when the first strapped magnetron was used in the Type 271 experimental outfit on Eastney promenade it promptly melted the coaxial cable to the antenna! The problem had already been expected for the 500 kW set, and because the waveguide was as yet unproven, experiments had been made with large-diameter Pyrotenax cable. It was fortunate that at just about this time the waveguide did become a practical proposition. Otherwise the greatly increased RF power could not have been handled without undue loss. A waveguide feed was quickly adopted for Type 271Q, and for all subsequent sets.

Unlike the previous developments in coaxial cables, the waveguide was not a direct extension of current technology. It arose from a renewed interest in 1936 in the USA, in an almost forgotten branch of elctromagnetic-wave theory originally developed by Lord Rayleigh. The American work, at Bell Telephone Laboratories, had shown that at sufficiently high frequencies waveguides should be more efficient than coaxial lines. By 1941 the theoretical basis was well established and available in the literature[23] but engineers were not yet used to thinking in terms of solutions to Maxwell's field equations. Practical experience was still very limited, especially of systems that would transmit high powers without breakdown, or that could be mass-produced in a sufficiently accurate but rugged form.

Examples of waveguides to handle high power in the 10-cm band were first made by the Polish group working at Eastney. Their efforts were just in time for suitable devices to be developed to couple the new high-power magnetrons to a waveguide. Much theoretical analysis was done by S. Kuhn, sometimes in collaboration with Bondi, particularly on transitions between waveguides and coaxial lines, and on waveguides with various dielectric loading. Soon many different centimetric sets were being developed at the same time, each requiring new and better waveguide components. Again, theoretical help was available when needed. For example Dahl recalls the advantages of being able to consult with the theoretical group when he developed the first 'preplumbed' crystal mixer for Naval radars. This was deliberately designed to leave the *minimum* number of adjustments available to the operator, and so reduce the risk of gross mistuning.[24] M. Surdin, of the Free French group, and A. H. Cook each contributed to the theory of waveguide directional couplers first invented by Bethe in the USA.[25]

Meanwhile, under the leadership of Pryce and S. H. Falloon, the Antenna Division at Nutbourne was developing techniques to solve another peculiarly Naval problem. The low power loss offered by the waveguide had enabled the antennae of 10-cm warning sets to be sited as high as possible, while the transmitter remained on deck. However the wide variety of ships to be fitted meant that waveguide runs for each ship could not be factory-produced under strict quality control, but would have to be assembled and installed by dockyards (see Monograph 5). Nutbourne designed a kit of standardised mass-produced components, such as die-cast bends and connecting flanges. These could be assembled by simple techniques into waveguide runs tailor-made for each ship. Whilst the performances of individual components were not up to the best 'hand-made' standards, they were adequate to allow Type 277 to be fitted with runs of 100-ft or more, containing perhaps 15 – 20 corners. After some training the dockyards were able to assemble runs with an acceptably low failure rate, as Types 276, 277 and 293 came to be fitted in a very wide variety of ships.

It soon became clear that the free use of waveguides was not without its own difficulties. As early as 1942 problems had arisen because the transmitting magnetron could change its frequency as the rotating antenna beam encountered obstacles in its path that reflected energy back into the transmitter. Considerable effort was put into investigating the design of the output circuit and tuning controls, as well as adjusting the internal coupling loop of the magnetron during manufacture. However the problem was regarded as serious enough for work also to be put in hand to investigate various ways of automatically adjusting the frequency of the local-oscillator to follow the changes in magnetron frequency.

The difficulties were exacerbated when the magnetron was more than about 5–10-ft from the antenna. Even with the limited lengths of precision-made waveguide to be used in the new gunnery sets (Types 274 and 275) it was found that the transmitted frequency could jump at random from pulse to pulse, regardless of what was 'seen' by the antenna. Ironically enough the mulitple-anode magnetron, having had its own complex mode structure cured by 'strapping', had now become part of another multimode system through being coupled to the antenna by a waveguide many wavelengths long. The causes were analogous to the frequency pulling already encountered with long transmission lines in metric systems. It was found necessary to 'tune the frequency away' from those values at which frequency jumping occurred. 'This was done by moving the waveguide piston of the output circuit, which allowed the frequency to be adjusted by a few tens of MHz at substantially constant power. There were stable frequency 'levels' over this range; the number of levels depended on the electrical length of the waveguide, but not on the match of the antenna or the 'pulling figure' of the magnetron. Frequency jumping only occurred near a transition between two levels.

A more complex effect, with which even local oscillator AFC would have been unable to cope, was found to occur when the waveguide length exceeded about 40–50-ft, as it did in many Type 277 installations. It was first interpreted by W. T. Davies and R. H. Kay, who realised that the length of waveguide was now so great that the magnetron could settle down to oscillate at a particular frequency before reflections from the antenna arrived back to cause a continuing change of frequency throughout the pulse, and a transmitter spectrum with significant energy outside the pass band of the receiver. Domb[26] made a theoretical analysis to account for this frequency modulation and computed typical examples. B. W. Lythall and colleagues made experiments to understand this phenomenon, and its relationship to frequency jumping and other problems. They eventually showed how all could be avoided by a single tuning procedure.[27] A long waveguide system needs to be adjusted so that the reflected energy does not cause a change in frequency when it arrives at the transmitter. The solution is the same as that for maximum stability against frequency jumping, but requires much more precise adjustment. (In fact it really needs the aid of a spectrometer, but for various reasons this did not become available in service until after the war, in Type 980 and its successors. It is remarkable that so many sets managed to be operated successfully without its aid.)

Acknowledgements

The great majority of the numbered references in the text are to items published in the literature, or which are available in the Public Records Office. At least as many more relevant papers have been found in collections not released to the general public at the time of writing, and I am grateful to various Defence authorities for allow access to them. Thanks are due in particular to the Director of the former Admiralty Research Establishment (now part of the Defence Research Agency) for the opportunity to sift through a multifarious assortment of archival papers, and to John Briggs, Librarian, for his assistance during many visits to Portsdown. I am also most grateful to several former colleagues for their personal memoirs. A compendium listing additional reference information material from these various sources is to be deposited in the Naval Radar Trust Archives at Churchill College.[28] Copies of some of the papers listed are also held in the Trusts Archives; for the rest the compendium indicates where the source material was found.

Finally I am most grateful to many colleagues for their helpful comments on earlier drafts of the text.

References

1. J. F. Coales, and J. D. S. Rawlinson, 'The Development of Naval Radar, 1935–45', Parts I and II, CAC NRT, ND1 and ND2.
2. A. B. Wood, 'From Board of Invention and Research to the Royal Naval Scientific Service', CAC, NRT.

3. A. B. Wood, Ibid.
4. H. D. House, *Radar at Sea: The Royal Navy in World War II* (Macmillan, 1993). Appendix C, by the author of the present Monograph, describes the development of Height-Determination techniques more fully, and contains several more references.
5. C. Domb and M. H. L. Pryce, 'The Calculation of Field Strengths Over a Spherical Earth', *Journal IEE*, Vol. 94, Part III (1947), Pg. 326 et seq.
6. E. C. S. Megaw, 'Experimental Studies of the Propagation of Very Short Radio Waves', *Journal IEE*, Vol. 93, Part IIIA (1946), Pg. 79 et seq.
7. E. C. S. Megaw, 'Scattering of Electromagnetic Waves by Atmospheric Turbulence', *Nature*, Vol. 166, (1950), Pg. 1100 et seq; 'Waves and Fluctuations', *Proc. IEE*, Vol. 100, Part III, (1953), Pg. 1 et seq.
8. E. C. S. Megaw 'Fundamental Radio Scatter Propagation Theory', IEE Monograph No. 236R, Vol. 104C, (1957), Pg. 441.
9. H. G. Booker and W. E. Gordon, 'A Theory of Radio Scattering in the Troposphere', *Proc. IRE*, Vol. 38, Pg. 401 et seq.
10. E. C. S. Megaw, Fundamental Radio Scatter Propagation Theory'. op. cit.
11. L. A. Moxon, 'Noise Characteristics of Radar Receivers', *Journal IEE*, Vol. 93, Part IIA (1946), Pg. 1130 et seq.
12. L. A. Moxon 'Variation of Cosmic Radiation with Frequency', *Nature*, Vol. 158 (1946), Pg. 758 et seq.
13. S. O. Rice, *Bell System Technical Journal*, Vol. 23 (1944), Pg. 282 et seq.
14. F. E. Terman, *Radio Engineers Handbook* (McGraw Hill, 1943) (which also has other references).
15. C. Domb, *Proc. Cambridge Phil. Soc.*, Vol. 42 (1946), Pg. 245 and Vol. 43 (1947), Pg. 587 et seq.
16. PRO ADM 220/227, H. Bondi, MK38/45, 'Summary of Results of Snowdon Trials'.
17. PRO ADM 220/229, Gold and Renwick, XRC9/46/8, 'Wave Clutter Experiments at Seaford, April – October, 1945.
18. R. V. Alred, 'Logarithmic Receivers', *Journal IEE*, Vol. 95, Part III (1948).
19. PRO ADM 239/367 (CB 004 385 A & B), 'Report by the Allied Naval Commander-in-Chief, Expeditionary Force, on Operation Neptune', Part VI, Section II, paras 81 – 84.
20. G. H. Brown *Proc. IRE*, Vol. 25, Part 1.
21. O. Böhm, 'Cheese Aerials', *Journal IEE*, Vol. 93, Part IIIA, (1946), Pg. 45 et seq.
22. L. G. H. Huxley, *The Principles and Practice of Waveguides* (Cambridge University Press, 1947), Pg. 102 et seq.
23. H. R. L. Lamont, *Waveguides*, Methuen Monographs on Physical Subjects (1942).
24. H. Dahl, 'A Pre-Tuned Frequency Changer for the 10-cm Region', *Journal IEE*, loc. cit., Pg. 280 et seq.
25. M. Surdin, 'Directive Couplers in Waveguides', *Journal IEE*, loc. cit., Pg. 66 et seq.
26. C. Domb *Proc. Roy. Soc.*, Vol. LIX, (1947), Pg. 958 et seq.
27. B. W. Lythall, 'Frequency Instability of Pulsed Transmitters with Long Waveguides', *Journal IEE*, loc. cit., Pg. 1081 et seq.
28. CAC NRT ND 151 (Appendix) Compendium of Additional Reference Material for 'Basic Science and Research for Naval Radar, 1939 – 45'.

Monograph 3
Valve Developments for Naval Radar Applications, 1935–45
F. M. Foley

SUMMARY

Formation of Signal School's Valve Laboratory in 1917 – early history – particular contribution to technology of transmitting valves. Application of high-power valves in Watson-Watt's first radar experiments – subsequent adoption of these valves for the RAF CH Early-Warning Chain – also for the first Naval metric wavelength radars. Later wartime valve developments that proved essential in radar equipment – silica output-power valves not suitable for mass-production, and limited in operation to an upper frequency of about 200 MHz. Introduction of metal-to-glass valve technology – decimetric-wave triodes for Naval 600 MHz gunnery radars. Invention of resonant-cavity magnetron at Birmingham University – subsequent development by GEC Research Laboratories as a very-high-power oscillator. Pulse modulators – hard valves, thyratrons and spark-gaps – limitations. Low-power RF valve developments. Duplexer valves for metre and microwavelengths. Crystal-mixer developments. General-purpose valve developments. Cathode-ray tubes, including PPI and Skiatron developments.

INTRODUCTION

The first part of this Monograph reviews the early history of the Valve Section of HM Signal School, starting in 1917, and the particular contribution it made to the technology of transmitting valves. This was to play a vital role in the formative years of British radar development prior to World War 2, initially in Watson-Watt's pioneering radar experiments and the subsequent development of the CH Early-Warning (EW) Radar Chain, and then in the independent evolution of metric wavelength radar for the Royal Navy.

In the second part of the Monograph a brief account is presented of later wartime developments, including the resonant-cavity magnetron and the reflex klystron, and the many categories of valve and thermionic devices that came to be necessary as radar technology advanced.

EARLY HISTORY OF THE VALVE SECTION, HM SIGNAL SCHOOL

Development of Naval transmitting sets began about 1917, as referred to in Monograph 1. The first output valve, the T1, had a glass envelope, and was rated at 150 watts anode dissipation. In the Signal School annual report for 1917 it was noted that 'these valves have the electrodes mounted within glass bulbs which at present limit the output power for which a valve can be designed'. It went on to note that 'fused-quartz had been considered but that, with the present experience of working silica, it seemed unlikely that a satisfactory seal could be made'.

Nevertheless work on silica valves must have been started soon after this. At the end of World War 1, Signal School had a strong valve group under B.S. Gossling, with S.R. Mullard the leader on silica valves. He had been seconded from the Royal Flying Corps, but it seems certain that he had previous experience with fused silica from his time at the Edison Swan Co, which developed quartz arc lamps. This view is supported by the fact that in May 1919 he took out a patent for the silica valve in his sole name, and not jointly with the Admiralty. Later, in October 1919, he resigned to set up the Mullard Co with the express purpose of manufacturing silica valves for the Navy.[1]

Undoubtedly the Thermal Syndicate Co had an important role in these developments. They had been manufacturing quartz tubing and bulbs for arc lamps, and as early as 1913 a subsidiary had taken out a patent on a quartz-to-metal vacuum seal. This was the lead seal later used in all silica valves up to 1939. (It is surprising that the Signal School scientists appear to have been unaware of this seal until so late.) In 1919 Thermal Syndicate made parts for the experimental valves. They obviously had other skills because, in early 1920, they were given the task of assembling the first valve, Type T22 using electrodes made by the new Mullard Co. The second sample was assembled by Signal School later in the year, presumably using technicians trained by Thermal Syndicate. The wide capability of Thermal Syndicate is borne out by a request from Signal School in 1924 that they should take on complete valve manufacture in competition with the Mullard Co, who had a monopoly at the time.[2].

Over the next few years a series of patents on constructional details was taken out by other members of the Signal School group, all jointly with the Admiralty. A list of all the relevant patents is included in Reference 3.

Between 1919 and 1922 there were major staff changes in the Signal School group. B.S. Gossling went to GEC, T.E. Goldup and B. Hodgson joined the Mullard Co, and others moved within the HMSS Establishment. By 1924 the group was reduced to H.G. Hughes and a few technicians, and was not increased again till the late 1930s. A staff list is included in Appendix 1 to this monograph.

In these early years Thermal Syndicate were developing silica valve-part fabrication. The first valves had been in 6-in diameter corrugated envelopes, but in 1923 one of their scientist, F. Reynolds, developed a new 8-in envelope, 30-in long. Signal School noted that this should allow construction of a design rated at 100 kW. In fact the 6-in envelope was the largest ever used for operational types. Valves using this envelope were used only in shorebased sets. Most types were in 90-mm or 100-mm diameter straight-sided envelopes.

The first sample of the original silica type, the T22, assembled by Thermal Syndicate using Mullard-made electrodes, was sent to Signal School to be pumped. Unfortunately the envelope collapsed on the pumps. Thermal Syndicate redesigned the envelope and a second sample was assembled in Signal School. This too failed on the pumps. A third attempt in September 1920 was successful. This type had an anode dissipation of 10 kW.

The speed of evolution of electronic valve technology over the past 70 years has been so great that the silica valve, invented in 1920, has long been obsolete and the technology is a lost art. Nevertheless at two points in time the silica valve was extremely important. In its earliest years (1920–24) it played a major part in increasing radio-transmitter output power. Later, in 1935, it was the power source in all Watson-Watt's aircraft-detection experiments, the only one which met all the requirements, both technical and in security. Silica valves were then used operationally in some of the RAF's CH Early-Warning (EW) Radars, and in all the Navy's metric radars throughout the war.

HM Signal School, Portsmouth, was the driving force behind development of the silica valve. The Establishment was among the leaders in radio research after World War 1, and it is not surprising that the laboratory included a valve-development facility. It had been realised as early as 1917 that the transmitter power limit was set by the glass-envelope material. The technology of silica-valve manufacture was very difficult, but was gradually developed. The new valves produced a huge increase in transmitter power.

The present monograph puts on record a broad description of the technology, and of the valves and their use in radio and radar. No other record appears to remain. The technology is particularly part of Naval electronics history, and the author is one of the few surviving silica valve designers.

SILICA AS A VALVE-ENVELOPE MATERIAL

The common feature of all transmitting valves is, of course, a high power-dissipation in the electrodes. This power must be lost mainly through the envelope. Until the development of metal-to-glass technology, where the heat could be transferred by direct conduction to a cooling fluid, all heat loss was by radiation through the glass or silica bulb. Neither material has high transmission in the infrared, and the bulb absorbs this radiation. Comparing the two materials, fused silica transmits heat up to a wavelength of about 7 microns, while even boro-silicate glass cuts off at about 3 microns. For a particular electrode temperature a silica bulb will therefore absorb less, and will run at a lower temperature. At the same time the softening point of silica is about 1000°C, compared with about 600°C for glass, so a silica bulb can be allowed to operate at a much higher temperature than a glass one.

A further advantage of silica is its low coefficient of thermal expansion. This is only about 10 per cent that of glass, so it is much less likely to crack due to temperature variations or thermal shock. It is also much stronger mechanically than glass, and so better able to withstand the shock and vibration found in Service environments. Further, it is a very good material electrically. It has high resistivity, four or five orders of magnitude higher than glass. It has high dielectric strength, and a dielectric constant about half that of glass.

This is an impressive list of advantages, but there are disadvantages also. It is chemically inert, making it very difficult to devise vacuum-tight seals with metals. It is a far more expensive material than glass. The actual 'working' is more difficult, requiring higher working temperature, and it takes longer than glass. Valves were therefore relatively expensive, although the high initial cost was largely offset by the fact that they could be repaired simply and cheaply in the event of filament burn out or electrode damage.

SILICA-VALVE TECHNOLOGY

Silica-valve technology was developed in Signal School at the end of World War 1 with the help of the Thermal Syndicate Co, both in the supply of parts and instruction in the art of silica working. It was based on simple manual construction and assembly. Later the same methods were used by Mullard and other companies for factory production. These manual methods were adequate for the small numbers needed for Naval wireless between the wars. However the technology was not amenable to mass-production methods. The huge increase in demand for all types in 1940 for radar applications could only be met with difficulty, and the use of unacceptably large numbers of highly skilled technicians. This certainly contributed to the demise of silica valves in favour of easily mass-produced metal-to-glass types.

The technology for all types was essentially the same: hand-made electrodes were assembled and mounted in a set of fused-silica components made by Thermal Syndicate. The lead-out seal rods were first bound to electrode ribs. The whole envelope was then sealed together, first the main body flanges, then the lead-out seals and filament-tensioning spring tube. An oxyhydrogen torch is necessary to fuse silica. The completed unit was then sealed to the pumping unit. The detailed stages are described below.

The Electrode System

The anode, usually cylindrical, was formed by winding molybdenum tape 2-mm × 0.1-mm on a wooden mandrel, and weaving longitudinal strips of the same tape into the winding to produce a basket structure. Three or four molybdenum ribs were also woven in to make the structure more rigid, and also to leave tails to locate in the envelope and make a connection to a lead-out seal. This type of construction allowed new or modified designs to be made up very quickly. It was also better thermally than a solid cylinder, in that a significant proportion of the filament radiation passed through the basket's interstices and did not contribute to the temperature rise of the anode.

The grid also was wound on a wooden mandrel, on which a lightly-engraved spiral was cut. The grid winding of molybdenum wire, typically 0.2-mm or 0.3-mm diameter, was wound in the spiral over three or four molybdenum ribs, and each turn was laced to the ribs with fine molybdenum wire. The ends of the ribs were located in narrow tubes fused to the end-caps, one rib being bound to a lead-out seal. In some of the much later designs the grid spiral was spot-welded to the ribs.

Until 1939 the filament was always of pure tungsten, usually in hairpin form, but in a few types as a number of parallel wires. The filament ends were bent for insertion in holes drilled in the lead-out seal rods, to which they were then bound. The filament was tensioned by a spring at the loop end of the hairpin. Some experiments had been done in 1922, apparently successfully, on a thoriated-tungsten filament, but surprisingly it was never applied to an operational type.

In 1939 the need for very high peak-emission currents in radar valves led to new research on thoriated tungsten. T.J. Jones, in Signal School, developed a system that would withstand the high processing and operating temperatures in silica types. Before evacuation the filament was heated in naphthalene vapour, tungsten carbide being formed to a depth of about one third of the filament diameter. Thorium is relatively tightly bound to tungsten carbide, and so gives an electrically robust emitting surface.

The Silica Parts

Each valve type used an individually specified set of silica parts, comprising a cylindrical body, two approximately hemispherical end-caps with sleeved openings to take the lead-out seals and, in some cases, brackets or spacers to locate the electrodes. On one end-cap there was also a central tube to take the filament tensioning spring, and a tube for connection to the vacuum pumping system. Both body and end-caps were flanged to simplify sealing them together.

There were three standard body diameters, 90-mm, 100-mm and a corrugated cylinder 150-mm diameter. Body length was chosen to fit the specified electrode lengths (Figure 3.1). In practice, a new valve was designed to use as far as possible existing body and cap parts, possibly with modification to the seal opening positions. Figure 3.2 shows a typical bulb with end-cap fitted.

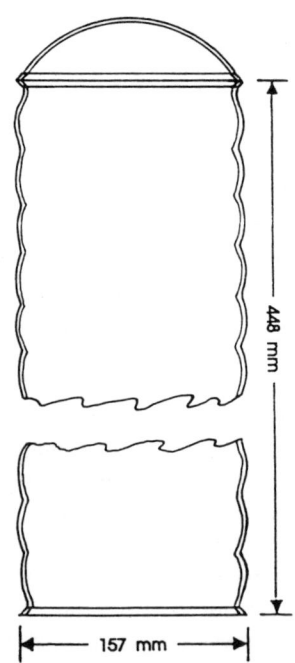

3.1 Corrugated envelope for NT22B (© Defence Research Agency)

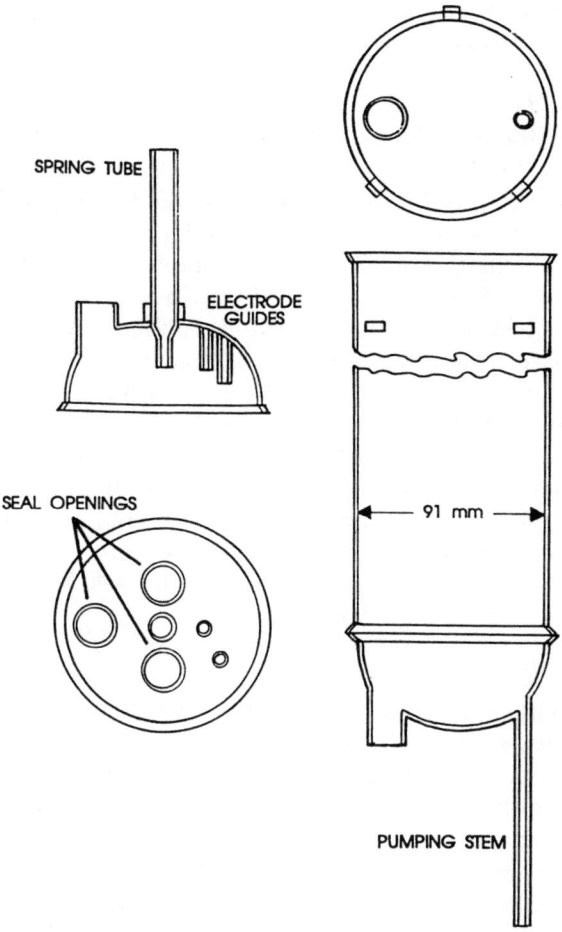

3.2 Typical bulb with end-cap fitted (© Defence Research Agency)

It has been noted above that the Thermal Syndicate Co was the sole parts manufacturer. Details of their manufacturing processes were never disclosed, but it is understood that a moulding process using graphite moulds was employed.

The Lead-out Seals

The formation of a vacuum-tight seal with a metal was the major problem in silica technology. The coefficient of thermal expansion of silica is several orders less than for metals, and also it is chemically very inert, with the result that there is no chemical bonding at the interface. Three schemes to overcome the difficulty have been used, as follows:

The Metallic Lead Seal

Historically this was the first solution, having been patented by Sand and Reynolds of the Silica Syndicate (later a part of Thermal Syndicate) in 1913 for use in quartz arc lamps. However the Signal School scientists appear to have been unaware of it till after the war.

In this scheme (Figure 3.3) a silica tube 6-in long and 0.5-in bore was shrunk on to a 3-mm tungsten rod over a length of about 2-in near its mid-point. Held in the vertical position, molten lead was poured into the annular space around the tungsten rod, and a flexible copper lead was set in the open end of lead to make an external connection. At the inner end, a molybdenum rod was butt-welded to the tungsten for connection to an electrode. The system was vacuum-tight because the metallic lead was sufficiently plastic to take up thermal expansion changes in the tungsten.

There were two disadvantages – the seals required strong air cooling in use to prevent the lead melting, and they were necessarily long, which limited a valve's use to frequencies below about 40 MHz. Their maximum current-carrying capacity was 25 amps.

The Strip Seal

In this scheme a thick-walled silica tube was fused and squeezed on to a feather-edged strip of molybdenum 5-mm wide, 0.2-mm thick and about 2-in long, with connecting wires welded to the ends. The very thin molybdenum had such a small overall thickness change on heating that the seal remained vacuum-tight. Its maximum current-carrying capacity was only a few amperes so it did not have many applications in transmitting valves. It seems to have been used only in one small split-anode magnetron.

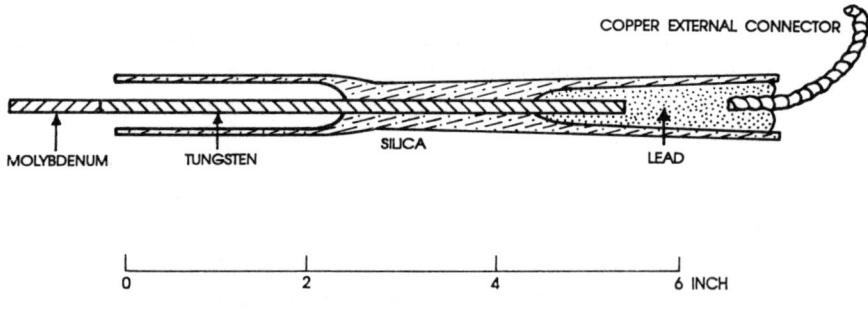

3.3 Lead seal (© Defence Research Agency)

The Graded-Glass Seal

In this design a thin sleeve of a hard glass was fused on to a 3-mm tungsten rod over a length of about 1.5-in (Figure 3.4). To this sleeve was sealed a short length of glass tube, of 20-mm bore, the glass having a slightly lower expansion-coefficient than the sleeve. Two further annular rings were sealed to this tube, each with a successively lower expansion coefficient. Finally a length of silica tube was sealed to the last ring, which could be sealed directly into the bulb openings. As with the lead seal a molybdenum rod was butt-welded at one end, and a copper connector brazed to the other end. The current-carrying capacity was 30 amps. It was only about 2-in long, less than one-third the length of the lead seal, making possible much higher frequency operation.

This design was invented by Philips in Eindhoven in 1939, and independently by GEC a little later.

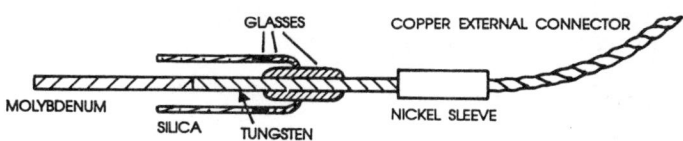

Figure 3.4   Graded-glass seal (© Defence Research Agency)

Valve Assembly

First the electrodes (formed as described above) were cleaned, by dipping into hot sodium nitrite, and then washed thoroughly. Each electrode was then bound or riveted to a seal rod, and the set of electrodes fitted in the valve body, and the end caps sealed to the body. Finally the seals were fused to the sleeves in the caps. Unlike glass, silica does not readily flow together, and the sealing operations involved pushing the heated surfaces together with a small, spade-shaped molybdenum tool. When the sealing was complete the surface around the fused area was found to be covered with a hard white deposit, a lower oxide of silicon, and this had to be 'burned off' by playing an oxidising flame over the surface.

Processing

There were three main processing stages:

- The valve was baked for four hours at 1050°C with hydrogen streaming through the valve to remove oxides from electrodes and seals.

- The valve was sealed to the pumping system – a single-stage mercury diffusion pump backed by a rotary pump, and with a liquid nitrogen trap – and was vacuum-baked at 1000°C till the pressure was less than 10^{-5}-mm mercury.
- The anode and grid were heated by electron bombardment to a temperature well above that at which they would operate and held till the pressure reached a 'sticky' vacuum on the Macleod gauge. The valve was then sealed off.

In valves with a thoriated-tungsten filament the process was modified. After the hydrogen-bake the filament carbonisation was carried out, and the electrode bombardment was replaced by eddy current heating of anode and grid.

In general, radar valves operated at very high voltages, which often resulted in flash-arcs between grid and anode. To avoid this, immediately after seal-off a controlled arcing, or 'spot-knocking', was done by applying a high voltage with a high resistance in series to prevent the arc current reaching a damaging level.

Valve Repair

A frequent cause of valve failure, particularly with pure tungsten-filament types, was filament burn-out. The electrode assembly was designed in such a way that a valve could be cut open with a narrow carborundum wheel through one of the body/end-cap flanges, thus allowing removal of the electrodes. It was a simple matter to replace the filament or a damaged electrode, reseal and pump the valve. This represented a big cost saving over that of a new valve.

SILICA VALVES FOR RADIO TRANSMITTERS

The first silica valve for the Navy, the T22 output power valve, was designed by the Signal School early in 1920. Clearly there were as yet no valve technicians in Signal School, for the first sample was assembled by Thermal Syndicate using electrodes made by the new Mullard Co. The valve was about 6-in in diameter and 3-ft long, with an anode dissipation of 10 kW. It was sent to Signal School for pumping, but unfortunately collapsed on the pumps.

Presumably, Thermal Syndicate had been training Signal School staff in silica-working techniques because the next sample was made in Portsmouth using a strengthened envelope – probably the corrugated design used later for all the largest types. On this sample the flanges cracked during sealing. A third sample was completed successfully in September 1920.

There is no record of the T22 actually operating in a transmitter, but a slightly modified design, the T23, was used in the long-wave (4000-m) transmitter at Pembroke W/T Station. It produced about 8 kW in the antenna from a pair of valves, which was three times the power achieved with glass valves. The set also used a pair of U21 silica rectifiers.

Figure 3.5 shows one of these very-high-power valves being sealed off the pumps in Signal School in 1922.

3.5 H. G. Hughes and T. E. Goldup sealing-off a silica valve, 1922 (© Defence Research Agency)

Valves for short-wave use were made by Mullard in early 1922 using the standard large envelope, but with very short electrodes; oscillation was obtained down to 6-m. The Naval short-wave set 7H, using silica valves, operated at 10-m and was fitted in *Hood* and *Repulse*. A similar set, the 7K, was fitted in some aircraft.

Silica valves were used in all Naval transmitters during the next 15 years. A number of types, most of them in smaller envelopes (about 4-in diameter) than the early 6-in corrugated design, were introduced for ship and shore transmitters. Ship sets in the early days (1922) seem to have been largely experimental, for the study of propagation at different wavelengths between 1000-m and 5000-m. Much shorter electrode structures were designed for experiments at 15-m (1938) but the lead seals (the only design available) were inherently long, and set a lower limit to the achievable wavelength. In all, 21 silica types were standardised for communication sets; of these 19 were triodes, and two were pentodes. Ten rectifiers were also standardised.

It is worth noting here that, in 1932, after discussions with E. C. S. Megaw at GEC Research Laboratories, a split-anode magnetron, with four anodes in a small silica envelope (NT52) was developed for a homing-beacon transmitter for aircraft carriers. Opposite anodes were internally connected and brought out to a Lecher-wire system, and oscillation was obtained down to about 1-m.

SILICA VALVES FOR RADAR

The Start

It was perhaps, in a way, fortuitous that silica valves played such a vital role in the earliest days of radar. When the first experiments on aircraft detection were set up at Orfordness in 1935, silica-valve technology was indeed one of the most advanced available, although this was not the decisive reason for its choice.

Suggestions that aircraft might be detected by radio waves had come from various sources, starting with Alder's patent application in 1928 (see Monograph 1). No experiments were done in the UK, however, till those by Watson-Watt's team from the Radio Research Station in 1935. The same team had been studying reflections from the ionosphere using a short-pulse transmitter on 50-m. The output valves in that set were silica types, designed and made by Signal School. This transmitter was adopted for the aircraft-detection experiments, which were successful and expanded, and Signal School was charged with making a number of additional valves. By this time the chance of war with Germany seemed likely, and aircraft-detection research was initiated, and made highly secret. The fact that the transmitting valves came from a secure Service Laboratory was a vital factor in continuing to use silica valves. As late as May 1938 A.P. Rowe, of the Air Ministry Research Station, Bawdsey (AMRE) wrote to DSR, Admiralty, noting that 'we are completely dependent on HM Signal School for sealed-off valve research.' Aircraft- and later ship- detection experiments were started in Signal School in 1936, using the same valve types as AMRE, Bawdsey.

The first type used at Bawdsey was the triode NT41. It had been designed for a CW radio transmitter and was certainly not optimised for a short-pulse mode. Initially, processing changes were made to allow higher-voltage working, and in use the filament was over-run to increase the peak emission and hence the output power.

Special Developments

Full radar development was started at AMRE, Bawdsey, following a visit by the Tizard Committee in June 1935. In July Signal School were asked to undertake the development of special valves. The record reads:

'Signal School to produce six experimental silica valves, on the lines of the NT41 modified as required for the voltage conditions (5kV but worked up to 12 kV). Of these 2 to be ready by September 1, and 4 by October 1. Also 2 NT41A experimental valves to be supplied immediately. These were for 4-metre to 8-metre trials; for 25-metre to 50-metre trials, 2 NT41 valves to be supplied. Also consideration to be given in Signal School to possible modification to NT34 as regards peak emission. Three experimental valves to be made up for test by November 1935'.

Completely new designs for radar were in development by early 1937. Obviously there was still a shortage of valve-development engineers because, in January 1937, Signal School advised Bawdsey that they did not have staff to work on the higher-powered valves they had requested (This was surely due to bureaucracy, no doubt coupled with a lack of appreciation of the importance of radar). The first type designed specifically for pulse radar, NT57, was used by both Bawdsey and Signal School in experimental work, and in early operational sets.

Early in 1939 NT57 was modified by the introduction of Jones's thoriated-tungsten filament, and also of the new, short, graded-glass seals. There was an immediate increase in power output, by a factor of four, to 70 kW from a pair, and the possibility of operation at twice the frequency. Bawdsey used this type NT57T (Figure 3.6) to drive another silica valve, the tetrode NT77A, in a mobile radar, and it was the output valve in the Naval sets Types 79Z and 279 (see Monograph 4).

In the summer of 1939 a series modulator triode and an output triode were designed specially for the new Naval 3-m set, Type 281. The

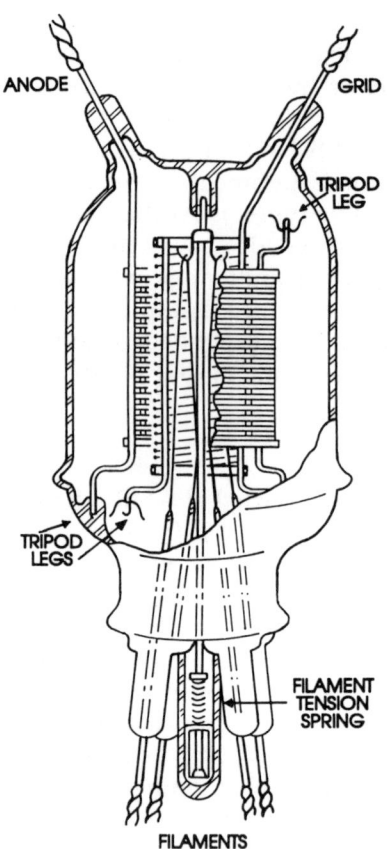

3.6 The NT57T silica valve (© E. B. Callick and Peter Peregrinus Ltd)

modulator, NT78, had serious development problems due to overheating of the grid by back-radiation from the anode, the resulting high grid emission causing flash-over during hold-off. Replacement of the basket-anode by one of carbonised nickel, a high-efficiency radiator, was tried, but was difficult to process and tended to poison the filament. Finally the whole electrode structure was increased in diameter, holding the spacing the same, and using a basket anode. The larger anode-radiating area and the loss of more filament radiation from the structure ends lowered the anode temperature sufficiently to ensure a low grid temperature, and negligible grid emission.

The output valve (NT86) was a short-structure triode. A pair of NT86s driven by four NT78As in parallel gave a peak output in the short-pulse mode of 1 MW. A number of these sets were in operational use, very reliably, till well after the war.

In 1939 the RAF, using a huge number of mobile radars, needed large-scale production of NT77As. This requirement was difficult to satisfy using only the manual production technology of silica valves. GEC, by this time privy to the radar secrets, were asked to develop a metal-to-glass equivalent, and the silica valve was phased out. The Navy continued to use them.

In 1941 an attempt was made to develop a much smaller silica triode to work at 240 MHz in a small-ship radar. This was the NT88. The electrode structure and envelope were made as short as the very-high-temperature silica sealing allowed, and samples were produced. On test in the radar the output was very variable from valve to valve. It was found that valves were operating on the steep part of the efficiency/frequency curve, and that minor parameter variations were causing wide output variation. Such variability was clearly unacceptable, and the design was abandoned. This attempt showed the design limitation of silica technology for high-frequency operation, and undoubtedly spelled its demise.

The End of the Era

One further type, a very-high-power series modulator, the CV313, was developed successfully. It was a modulator tetrode with a very low amplification factor (4.5), and could therefore pass very high current without driving the grid positive. Further, by careful alignment of the grids and by taking precautions to obtain a smooth grid surface, it was able to hold off very high voltages without arcing. Three of these in parallel were used to switch the 2 MW magnetron in the Naval Type 992 TI set, which went into service about 1950. They continued to operate reliably till the much smaller hydrogen-thyratron replaced them during an equipment modification in the 1960s.

This was the last silica valve used operationally for either radio or radar by the Navy. For some time after that the Mullard Co continued to market several types for industrial RF heating applications, but the silica valve is now a lost technology. A list of the valve types is given in Appendix 2 to this monograph.

OUTPUT POWER VALVES

Metric-Wave Valves

In the early Naval radars the power source in every case was a silica valve, as discussed above. They had serious limitations. Firstly, their design technology was not adaptable to mass production. Secondly, and of increasing importance as operating frequencies were increased, was the low upper-frequency limit of operation, due to the necessarily long seal and lead lengths – about 200 MHz.

By 1939 metal-to-glass seal techniques had been developed and refined, and metal-to-glass equivalents of the silica transmitting valves were in development. These could be mass-produced, and soon replaced their silica counterparts in the RAF's CH sets. The Navy made no changes, and continued to use the silica versions for the life of the shipborne sets throughout the war.

Decimetre-Wave Valves

The 1938 decision that the first series of Naval gunnery radars should operate at 600 MHz meant that a completely new range of output valves was required (see Monograph 1). Clearly silica valve technology could not meet the need, and a design based on metal-to-glass techniques would have to be developed by Industry. In fact, at that time, work on aircraft detection by radio waves had not been disclosed outside the Services. Industrial know-how was obviously needed, and the decision to acquaint at least one company with the principles of radar was therefore unavoidable. GEC was the first company chosen (see Monograph 1). The GEC Research Laboratory was already working on small RF copper-to-glass valves, and were therefore asked by Signal School to develop a valve to operate at 600 MHz, with high pulse-power output.

A design was already available that was capable of generating a few watts of CW output at frequencies above 100 MHz: this triode was used as the basis for the new pulse valve. The result was a range of triodes, developed by J. Bell *et al.*,[4] which came to be known as 'Micropups'. The electrode assembly was cylindrical, with the anode cylinder forming part of the envelope, and the grid lead was brought out concentrically at one end. The cathode lead was concentric at the other end. The cathodes in early valves were composed of parallel thoriated-tungsten wires forming a cylinder. At that time GEC demonstrated that oxide cathodes could give very high emission current in short pulses, without deterioration, and all the later 'micropups' were fitted with cylindrical oxide cathodes. The first of these valves used by the Navy was the NT99 (Figure 3.7). A pair of NT99s gave 200 kW peak output at 600 MHz. These valves were difficult to replace in equipments, and the concentric-line circuit was difficult to tune (due in part to the exposed position of the equipment on the gun director – see Monograph 1). This led to the development of a pretuned valve-circuit unit that could be simply and quickly plugged into the equipment.

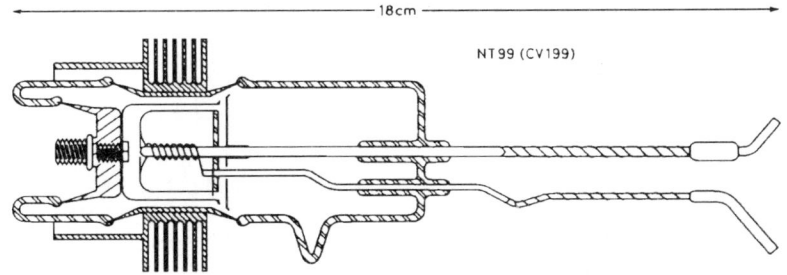

3.7 A 'micropup' triode, NT99 (© E. B. Callick and Peter Peregrinus Ltd)

Microwave Valves

By 1938 it was already clear that much higher frequencies for radar applications would have big advantages, particularly in allowing narrower beamwidths using antennae of practicable size. E. G. Bowen at AMRE suggested a first target of 3000 MHz.[5] Whilst triodes had been made to oscillate at this frequency, the power output was negligible, and clearly could not be increased significantly.

The problem was solved by the invention of the resonant-cavity magnetron by J. T. Randall, and then investigated experimentally with H. A. H. Boot at Birmingham University early in 1940[6] (This research was initiated at the request of DSR, Admiralty). Their valve was a crude laboratory device with six cavities. In operation, the circular electron motion around the cathode excited resonance in the cavities. Power was extracted by a loop in one cavity. The experimental magnetron was then passed to Megaw at GEC Research Laboratories to be developed and engineered for production. He accomplished this successfully, and also added a 'fat' oxide cathode, an essential factor in achieving the necessary high peak-output powers.[7] In 1936 Megaw had concluded that a larger-diameter cathode than that currently being used in split-anode magnetrons would increase the output power significantly. In a visit to SFR, Paris, in June 1939, Megaw made this suggestion to H. Gutton. Gutton made up some split-anode magnetrons with big cathodes, achieving a tenfold power increase. Megaw therefore felt well justified in fitting his resonant-cavity valves with such cathodes. At about the same time it had been found that oxide cathodes would withstand high voltage and high cathode loading. Megaw therefore fitted a fat oxide cathode in his development.

An early design, the E1189 (later adopted as NT98), is shown in Figure 3.8. It had eight cavities, machined in a copper block, about 40-mm thick, around a central space in which the oxide cathode was installed.

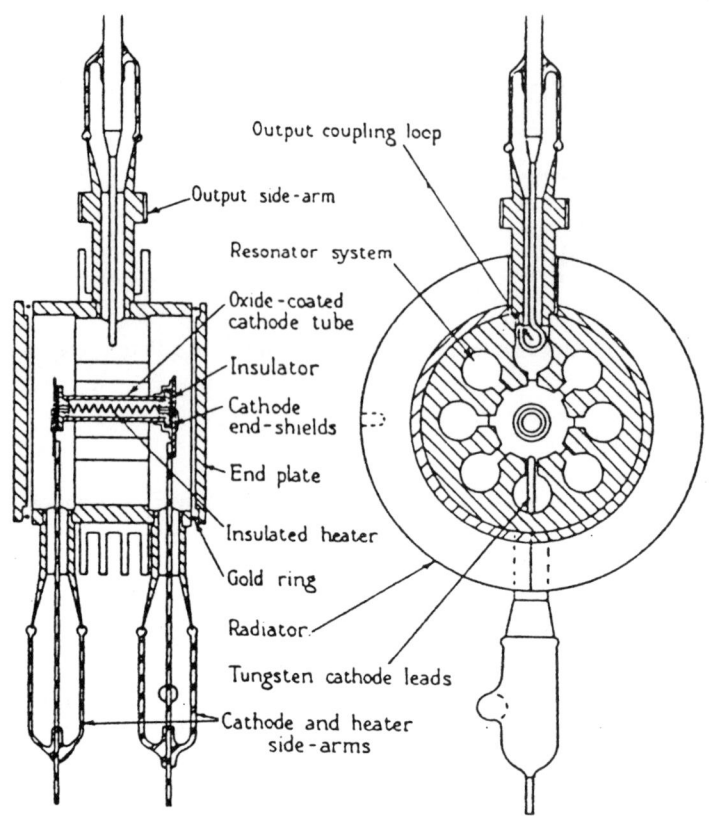

3.8   An early magnetron, NT98 (© Defence Research Agency)

The valve was mounted between the poles of a high-power magnet. Some later designs were supplied 'packaged', that is, complete with their own magnets permanently mounted. With the basic valve there was some instability of the oscillation due to a lack of phase-locking between the oscillation in individual cavities. H. J. Sayers, working with Randall at Birmingham University, solved the problem by cavity 'strapping'.[8] In this system wire links were fitted across the gaps of alternate cavities (Figure 3.9)

A number of frequency variants of each basic type were introduced, made by small changes in cavity dimensions. This enabled ships in company to operate on different frequencies without mutual interference. Early types gave peak-power outputs of about 10 kW, but with improvements (including strapping) this was later increased to about 500 kW. Much later designs, using higher-power cathodes, gave peak

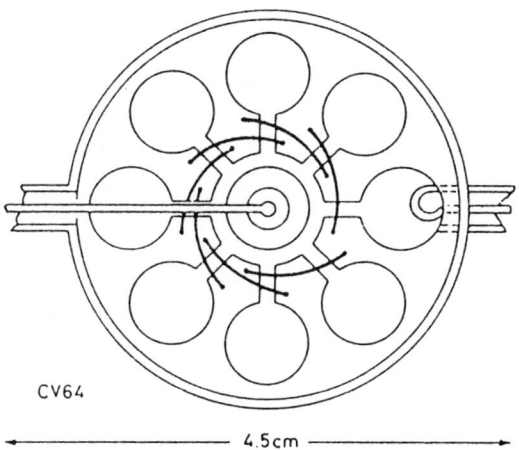

3.9 Magnetron cavity strapping (© E. B. Callick and Peter Peregrinus Ltd)

outputs up to 2 MW, but these were not in service until after 1945. The vital need for more and more microwave radars, both in ships and aircraft, and the need to have an alternative source in case of factory loss due to enemy action, made it essential to give details of the Birmingham/ GEC work to another contractor. BTH was therefore introduced to the magnetron-development work.

The 3000 MHz band (S-band) was almost the only one used by shipborne radars during the war, and large numbers of S-band magnetrons were made by both GEC and BTH. The microwave shipborne radars were Types 271, 273, 276, 277 and 293 and some of the early models of the gunnery sets Types 274 and 275 (see Monograph 5). Development of magnetrons for the 10 000 MHz band began in 1942. The fire-control set Type 262, and the submarine periscope radar Type 267 using this frequency (magnetron CV214) were fitted just before the end of the war. During this period experimental valves were also developed to operate at up to 30000 MHz, but did not go into service.

Initially magnetron blocks were made by conventional drilling and machining, but this was relatively slow and costly. In America, Raytheon tried a process of stacking stamped-out copper segments, but soon a cold-hobbing process was adopted universally. This produced accurate results and was a cheap method for mass-production.

Up to the end of the war the magnetron was the only power source used in microwave radars. In fact work on power klystrons had preceded the cavity-magnetron development at Birmingham. Professor M. L. E. (later Sir Mark) Oliphant made experimental klystrons, based on the work of the Varian brothers at Stanford. Further research was carried out both at the Signal School Bristol laboratory and at Oxford University. By 1941 Bristol had achieved a peak output of 1 kW at 3000 MHz. However these devices were far more costly to make, and had a much lower power capability than the cavity magnetron. Also, the relatively unsophisticated radars of that period did not require wideband frequency agility, which is probably the klystron's most valuable property. Furthermore, being amplifiers rather than self-oscillators, the overall circuitry would have been more complex. The result was that, up to the end of the war, the power klystron work was not taken beyond the experimental stage.

PULSE MODULATORS

Hard Valves

All the Naval metric radars used triodes to switch on the output oscillators.[9] In Type 279 the large glass triode NT87 was in series with the push–pull oscillators, and the positive pulse on its grid caused it to conduct and 'pull down' the oscillator cathodes, starting oscillation. In Type 281 there were four silica triodes Type NT78A, in parallel in a similar circuit, driving a pair of silica-triode oscillators NT86. (The NT78A has been discussed in more detail above.) Particular design characteristics of these valves were high pulse-current availability and high anode hold-off voltage rating. Several other more advanced designs were developed later, but only the NT100, made by GEC (a copper-glass tetrode with air-cooled anode block), was used by the Navy. This was in the early microwave set Type 271.

Thyratrons

Small thyratrons were used in the low-power stages of some early modulators but the 600 MHz gunnery sets were the first to use a high-power type as the output stage (see Monograph 1). This was the mercury-filled valve CV22, designed by Knight at BTH.[10] Mercury thyratrons, with their very low voltage drop when conducting, were ideal modulators, but the achievement of high hold-off voltage required the development of complex baffle-grid structures (Figure 3.10).

The CV22 was rated at 50 amps and hold-off voltage of 20 kV. Later, H. T. Ramsay at GEC developed the CV12, rated at 200 amps and 15 kV, which was used to switch the magnetron CV76 in the Type 277. This valve also had a complex baffle grid. Whilst these two valves (and other mercury-filled types) were very efficient switches, they tended to have rather short and variable lives. This was due to diffusion of mercury

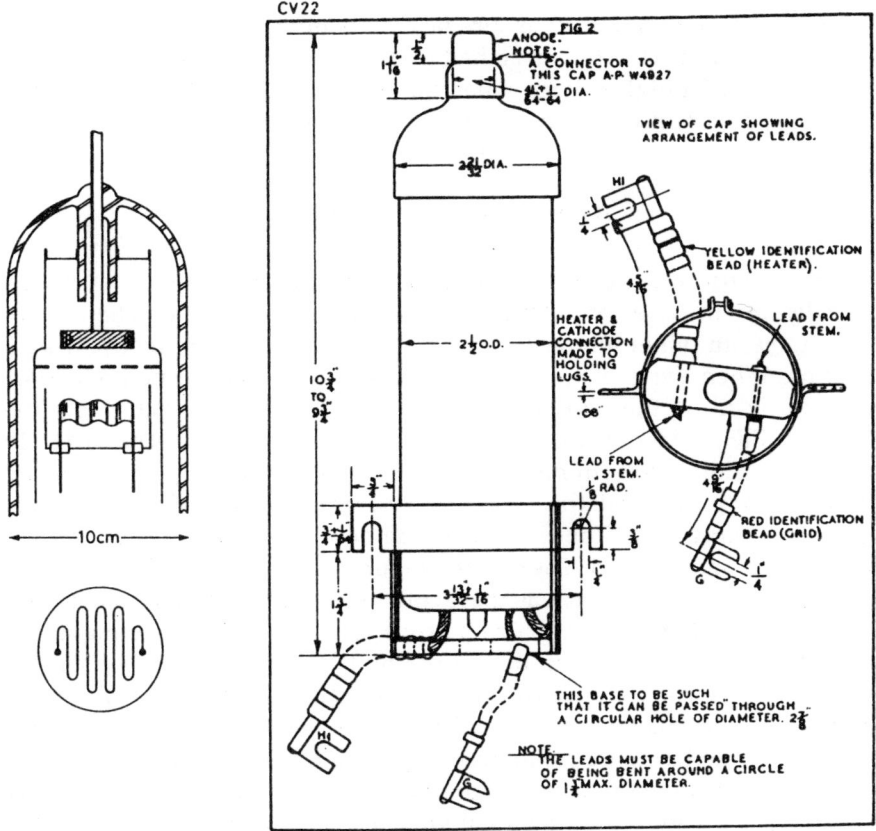

3.10 High-power thyratron modulator, CV22 (© Defence Research Agency)

vapour through the oxide cathode coating, where it amalgamated with the nickel base and caused loss of adhesion of the coating. This was particularly the case under conditions of ship vibration and shock. No solution to this problem was found, and thyratrons had serious limitations till the development of the hydrogen thyratron after the war.

Spark Gaps

Some very early 600 MHz gunnery sets used an open spark gap between adjustable tungsten electrodes (see Monograph 1). In use, the spark caused considerable electrode erosion, and part of the daily maintenance routine was to clean the electrodes and adjust the gap.

Sealed-off spark gaps were also developed. These had three tungsten electrodes, one acting as a trigger. The filling was an inert gas mixture, with some water vapour to ensure fast quenching of the discharge at the end of the pulse. The discharge had a low impedance, and the main gap could hold off a high voltage when not triggered. It is doubtful if sealed spark gaps were used in any shipborne radar modulator.

LOW-POWER RF VALVES

For the metric wavelengths a number of commercial types, developed initially for the prewar television market, were available. These gave acceptable amplifier performance, at least at the lower metric wavelengths. When Type 79 was in development the Ediswan pentode SP41 (standardised as the VR65) was in production, and for later sets the Mullard EF50 (standardised as CV1091) became available. The SP41 HF receiving triode (forerunner of the CV91) was probably the ultimate design on a 'pinch' seal: the electrode system was advanced, but the leads were inevitably long. The EF50 used the new pressed-glass base, which gave much shorter lead lengths (see Figure 3.17). One other design was important for a time. This was the 'acorn', so called because in shape and size it resembled a large acorn (figure 3.11), with leads brought out radially through the mid-section of the bulb. It was made in both triode and pentode forms. An 'acorn' triode was used in both the first RF stage, and the local oscillator stage in the Type 279 receiver. Although 'acorns' would operate to higher frequencies than the EF50, they were unreliable and expensive to manufacture. Consequently they were not used in later sets. Type 281 had two RF stages, both using EF50s, but later Marks of the set added a preamplifier using a disc-seal, grounded triode, CV53 (discussed in detail below). The use of this preamplifier doubled the signal-to-noise factor and gave a range increase of about 20 per cent.

It is interesting to note that, for all the early radar research, the valve development was concentrated on improving the transmitting valves (particularly their output power), and the significance of receive noise on overall performance appeared to be overlooked initially (see Monograph 2).

The best conventional receiving valves were satisfactory at VHF frequencies, but certainly would not operate at 600 MHz. In parallel with GEC's work on power valves, C. N. Smyth at STC worked on low-power disc-seal triodes, to be used in concentric line circuits. The electrode system was planar (Figure 3.12).

The grid was 10-mm square, with wires 0.05-mm diameter on a pitch of 0.165-mm. The interelectrode spacings were less than 1-mm. The grid was mounted in an opening in a copper disc across the middle of the tube, and the anode and cathode leads were brought out concentrically at either end. The achievement and accurate maintenance of such small

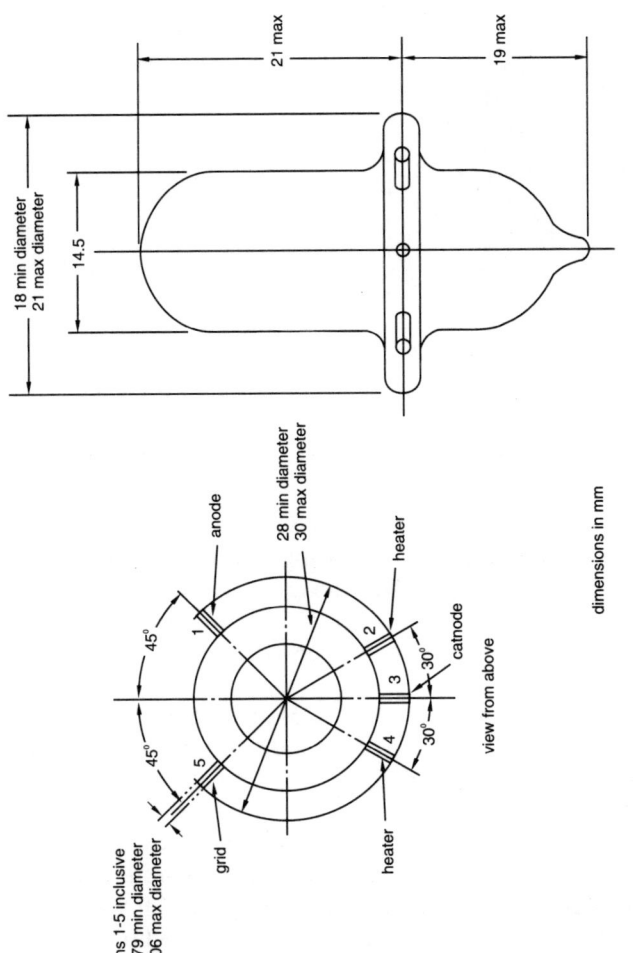

3.11 Outline of an 'acorn' valve

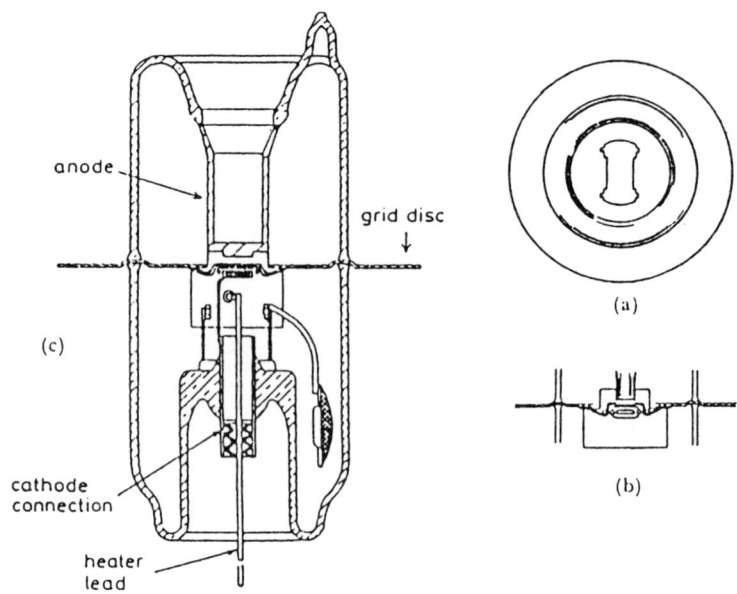

3.12 600 MHz amplifier triode, CV53 (© Defence Research Agency)

dimensions, when the valve was operating, required exceptional design and manufacturing techniques. The design was successful and gave a power gain of 16 db, with a 4 MHz bandwidth and a noise factor of 3 db at 600 MHz in a grounded-grid, concentric-line circuit. This was the CV53.

An oscillator version of the CV53 was also made. In this a small amount of capacitive feedback from anode to cathode was mounted inside the valve. As a local oscillator it was tuneable over the range 500 MHz to 750 MHz.

The performance of CV53 was tested up to microwave frequencies, but at 1700 MHz it had fallen off, so that there was no useful gain. The CV53 design represented the limit of triode technology, and it was clear that an amplifier to work at 3000 MHz would have to be based on some different principle of operation.

No low-power signal amplifiers were available for the microwave bands, and in the receivers the input signal went direct to the mixer. The development of the reflex klystron (Figure 3.13) by R. W. Sutton, of Signal School, as a microwave local-oscillator was of equal importance with that of the cavity magnetron in the practical achievement of microwave radar (he received the Physical Society's Boys' Medal for this development). In this tube an electron beam is velocity-modulated in passing through a

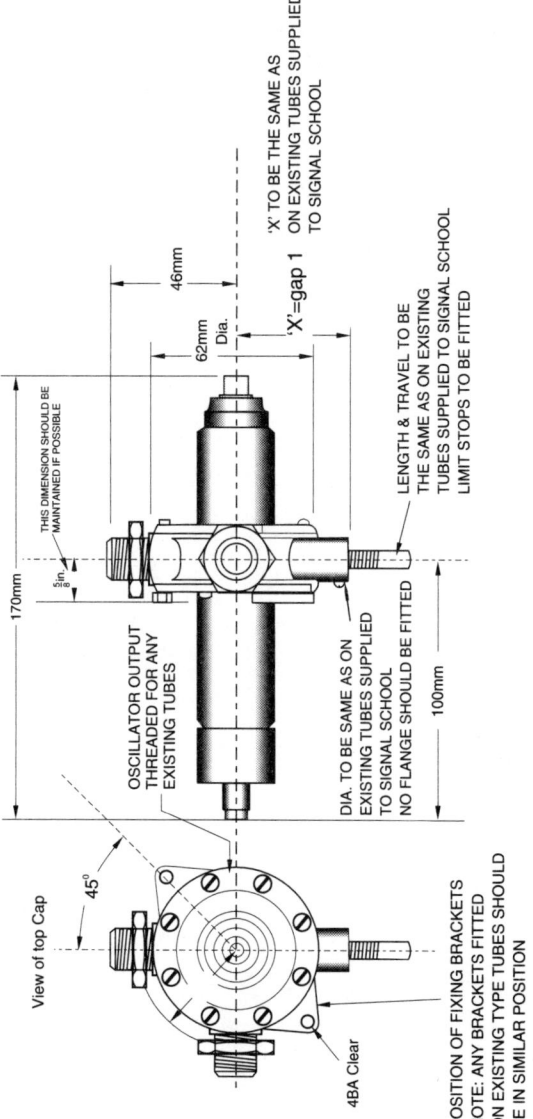

3.13 Local-oscillator klystron, NR89 (CV35) © Defence Research Agency

cavity resonator, and is then reflected by a negative electrode to make a second in-phase transit through the resonator. Power is taken by a loop in the resonator. The resonator was completed by external annual cast sections and was tuneable by stubs screwed into the castings, so that a single valve could be tuned to any particular magnetron frequency in the band.[11] A later tube is shown in cross-section in Figure 3.14.

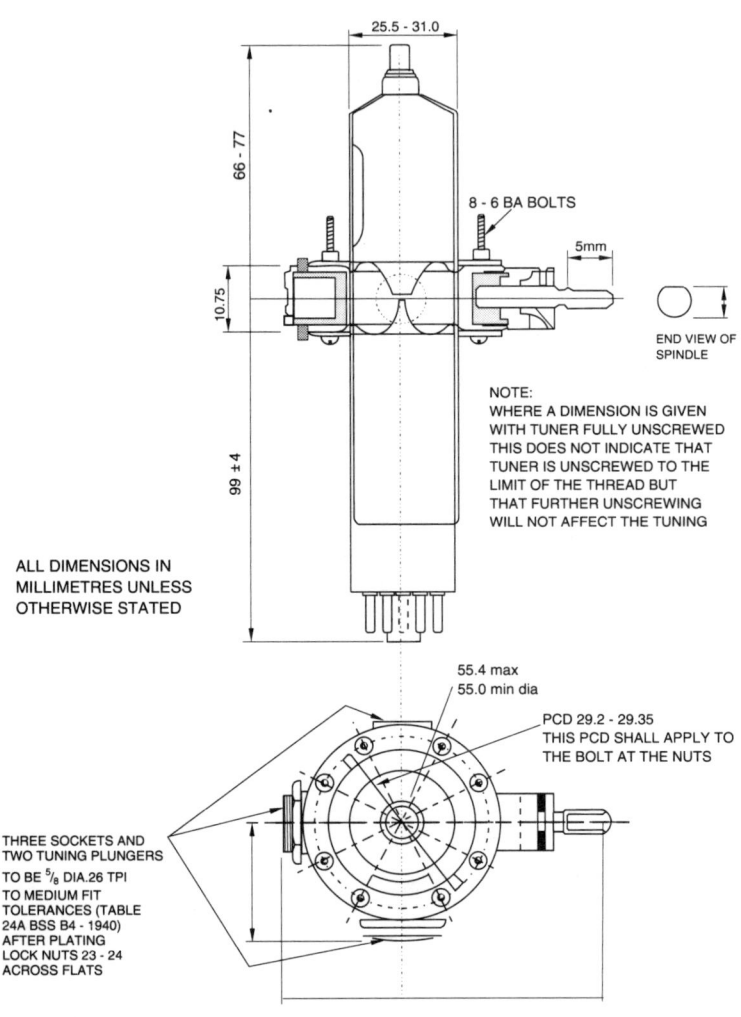

3.14 Cross-section of local-oscillator klystron, CV67 (© Defence Research Agency)

The original S-band valve was the NR89, later engineered for production as the CV35: later still it was scaled down for other frequency bands.

DUPLEXER VALVES

Metric Valves

The introduction of single T/R working for transmission and reception (first with Type 279) made it necessary to isolate the transmitter and receiver circuits. The receiver had to be protected from the high-power transmitted pulse, and then the received pulse had to reach the receiver without attenuation. In the first metric set, Type 79, this was achieved by connecting six small diodes in parallel (a suitable single diode was not available) across the receiver input, a system invented by H. J. W. Reeves of Signal School. These presented a low-impedance path for the transmitted pulse and when suitably back-biassed, allowed the received pulse to reach the receiver unattenuated.

When the higher-power set Type 281 was developed, providing an output of 1MW, this arrangement could not handle the power, and a new system had to be found. In early models an open spark gap was connected across the receiver input, backed by the parallelled diodes. The spark gap passed most of the transmitted power and the diodes took the rest. Later a special high-current, low-impedance diode, the CV94, was developed by GEC, and this handled the whole power output without the need for a spark gap.

Microwave Valves

In microwave systems the duplexer function was the same, but its performance was particularly critical because the receiver input was the crystal mixer, which was extremely sensitive to any overload. Two cells were used, a TR cell (Transmit–Receive) and a TB cell (Transmitter–Block) arranged as in Figure 3.15. Both cells were basically gas-filled resonators whose conductance was raised instantaneously when the transmitter pulse appeared and fired the discharge. After the pulse the discharge was quickly quenched by the presence of water vapour in the cell. The valve used in Types 271 and 277 was the CV43, a gas-filled Sutton tube with minor modifications.

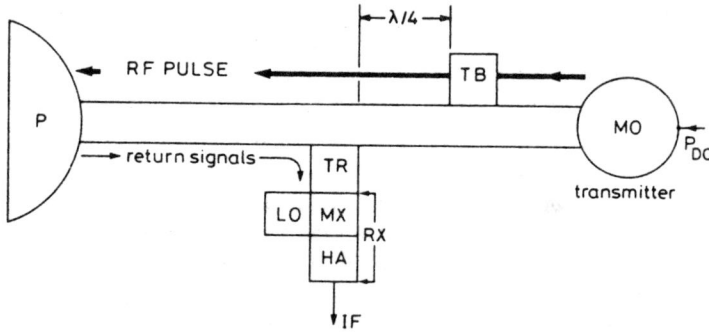

3.15 Duplexer arrangement (© Defence Research Agency)

MIXERS

The metric wavelength sets used the 'acorn' pentode NR54 as the mixer, whilst the 600 MHz sets used the small diode, VR78. Later a small planar diode was developed by GEC specially for mixer applications, but it does not appear to have been used in any Naval sets.

For microwave frequencies no valves were available as mixers, and the point-contact silicon diode, invented by H. W. B. Skinner of TRE in 1940, was always used[12] (see Monograph 5). The design of Skinner's 'crystal', a tungsten 'cat's whisker' in contact with a small, polished silicon chip mounted in a wax-filled glass bulb, was necessarily a laboratory device. It required a 'mysterious' forming process by tapping to achieve a good back-to-front impedance ratio, an empirical measure of a good mixer! A small number of these hand-made crystals were made by ASE (formerly HM Signal School) for the first Type 271 sets (see Monograph 5). A mass-production design was obviously essential, and this was achieved by T. H. Kinman at BTH using a ceramic capsule with brass ends, one of which could be screw-adjusted to 'form' the mixer[13] (Figure 3.16). The design was simple and robust and was universally adopted, not only in

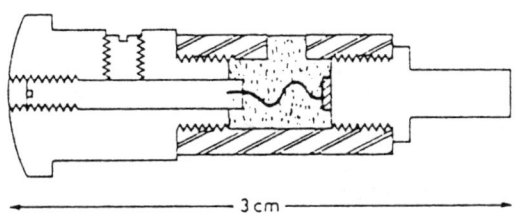

3.16 The microwave crystal mixer, CV101 (© Defence Research Agency)

Britain but also in America. Further improvement was later achieved by using aluminium-doped silicon.

GENERAL-PURPOSE VALVES

This category includes the many low-power valves used in receivers, displays and all signal-processing circuitry. As radars became more complex these applications became more numerous and involved. They required the specification of previously unused and uncontrolled screen-grid and suppressor-grid characteristics in pentodes.

From about 1939 onwards receiving-valve technology was virtually revolutionised, not only leading to superior characteristics and more robust mechanical designs, but also to more efficient manufacturing methods. Before then the electrode structure was mounted on a 'pinch' seal with long leads to a bakelite base, which had to be cemented on to the bulb after it had been pumped. Electrode structures were crude, with wide spacings, giving low gains and long transit times, the overall result being low performance. The pressed-glass base, pioneered by Philips, shortened the leads and reduced interelectrode capacities. This allowed more automated production. At the same time much closer interelectrode spacings were achieved, so that design figures-of-merit were increased, and relatively conventional designs such as the EF50 could operate at over 100 MHz. As the degree of complexity of radars increased, so the volume of equipment increased, and the problem of available space (especially in aircraft) became critical.

There was therefore an urgent need to minaturise the components. The problem for ship equipment was less pressing, but standardisation to achieve maximum overall production efficiency in the industry was vital (at one stage the UK's production capability was 45 million valves per year, but the Service's demand was 88 million: the shortfall had to be made good by imports from the USA). Thus the Navy also adopted miniature types. Only a few types had been designed on the old, large, pressed-glass base, but on the miniature seven-pin base (and a similar nine-pin version) a large range of triodes, pentodes, double triodes and rectifiers was produced. These types were not only advanced electrically, but were also very robust mechanically. Figure 3.17 shows the formats of the three types.

CATHODE-RAY TUBES

The first radars were essentially simple range finders, needing only very simple displays known as the A-scan, range being measured by the

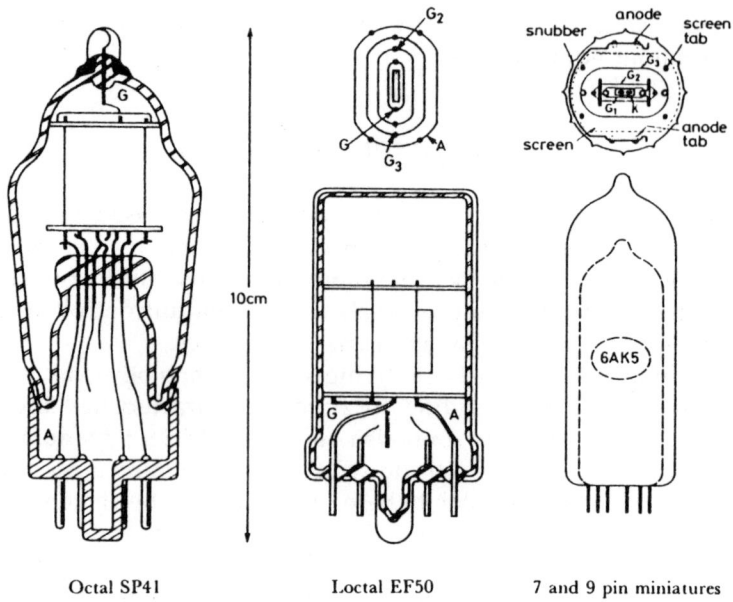

| Octal SP41 | Loctal EF50 | 7 and 9 pin miniatures |

3.17 Receiving valve outlines (© Defence Research Agency)

distance of the echo from the origin along a linear trace. This display generally used a 6-in electrostatic focus and deflection tube with a single-phosphor screen, giving a bright green trace. A typical tube was the VCR97 (Figure 3.18), used almost universally by all the Services. In the early metric sets the A-scan was the only display fitted. In most later sets, where the main display was more complex, an A-scan was provided as a subsidiary.

A continuously rotating antenna was introduced with the microwave sets. This now provided bearing information also, which had to be displayed. The single biggest display advance was the invention of the 'Plan Position Indicator' (PPI) display, showing both bearing and range. In this the range trace, with its origin at screen centre, rotates in synchronism with the antenna and remains visible for the duration of a rotation, the phosphor having a long afterglow. Electromagnetic deflection, and usually focus, by external coils was used. The tubes were usually of 12-in or 9-in diameter. The main developments were in the phosphors, whose afterglow had to be matched to the antenna rotation time. Early screens used two layer phosphors, one layer giving a short, bright, blue flash, which excited the second, long-afterglow, yellow–green phosphor. Later single-layer fluoride phosphors with orange glow and almost ideal afterglow characteristics were developed and used universally.[14] A typical tube was the VCR516 (Figure 3.19).

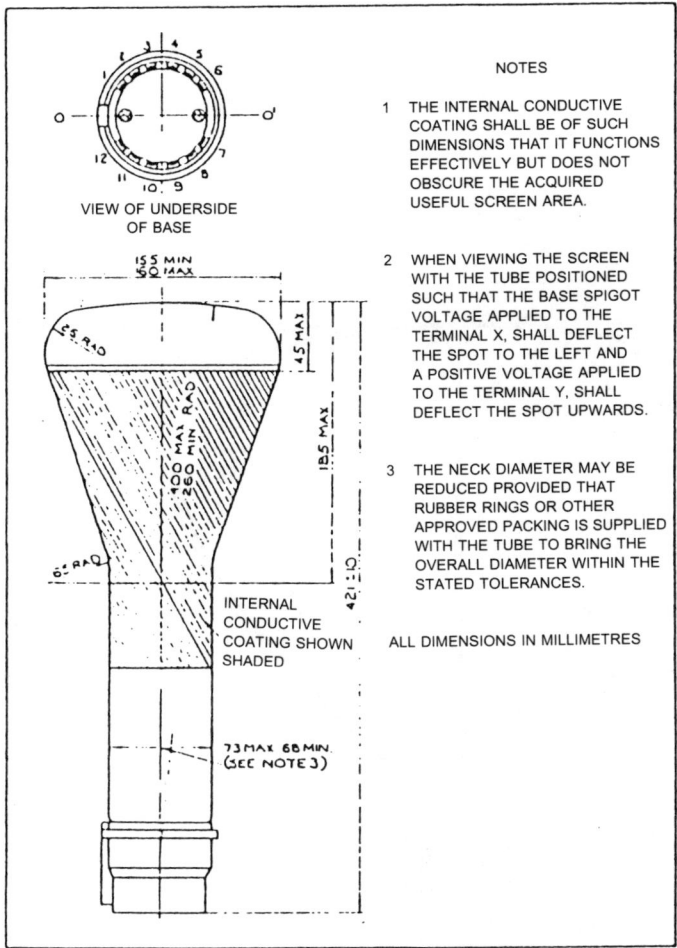

3.18 Electrostatic focus and deflection CRT (© Defence Research Agency)

One fundamentally different tube was developed by Tricker and King at Signal School, Bristol. This was the Skiatron dark-trace tube.[15] The electron beam produced a dark trace on the white phosphor that persisted indefinitely, until it was irradiated with infrared radiation. This screen was used on only one tube type, the NC17 (Figure 3.20), which was fitted in the Naval plotting table display. The tube itself had a flat 3.5-in diameter screen, and an optical projection system was used to give the 24-in table display. This type of screen was liable to permanent burning by the electron beam, and was difficult to operate. It was not used in any other operational equipment.

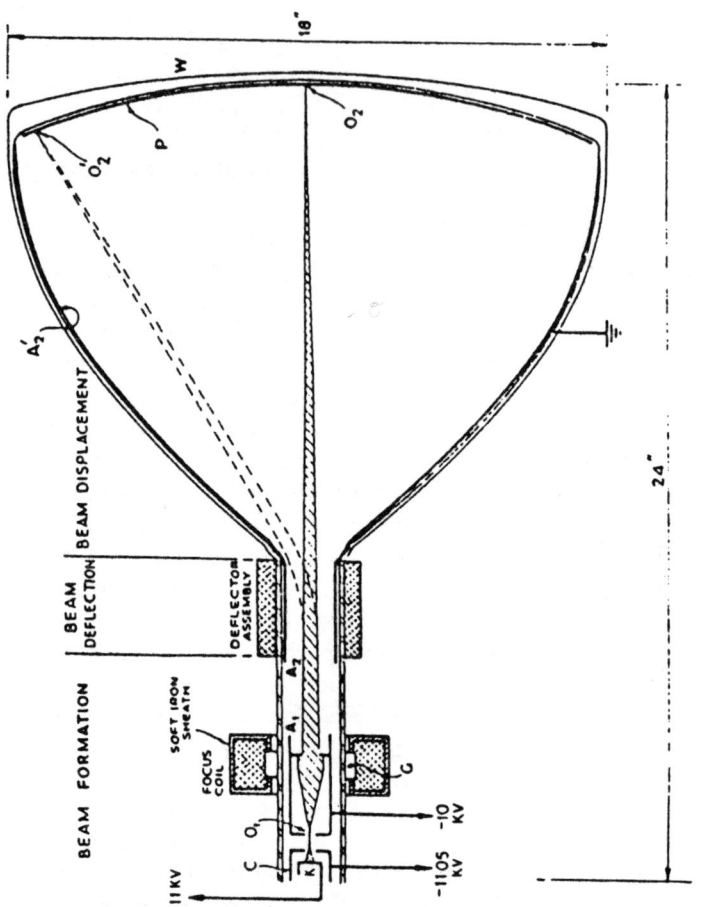

K—cathode. *C*—control electrode, also used for beam current modulation. $A_1 A_1'$—first anode. $A_2 A_2'$—final anode. *G*—gap in focus coil magnetic shield. O_1—first focus of the beam. O_2—second focus of the beam.

3.19 12-in electromagnetic focus and deflection CRT (© Defence Research Agency)

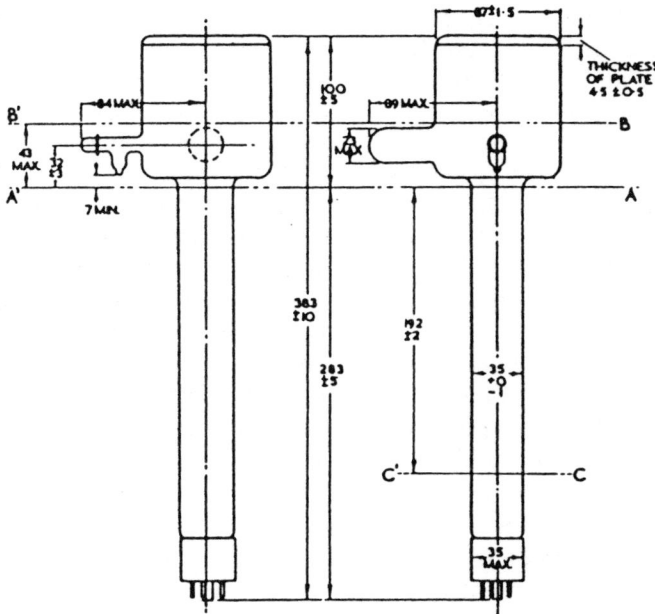

3.20 The Skiatron tube for optical projection, NC17 (© Defence Research Agency)

APPENDIX 1: H. M. SIGNAL SCHOOL STAFF WORKING ON VALVE DEVELOPMENT UP TO 1945

B. S. Gossling	Scientist	Head of Group, 1918. Joined GEC in 1919.
H. Morris-Airey	Scientist	Later, Principal Engineer (1921).
S. R. Mullard	Scientist	Founded Mullard Co in 1919.
T. E. Goldup	Scientist	Joined Mullard Co in 1922.
B. Hodgson	Scientist	Joined Mullard Co in 1922.
A. K. Macrorie	Capt. RN	Superintendent HM Sig. Sch., 1918
L. G. Preston	Capt. RN	Superintendent HM Sig. Sch., 1920
G. Shearing	Scientist	Later Chief Scientist, HM Sig. Sch.
J. Scott Taggart	Scientist	Ultimate Career not known.
H. G. Hughes	Scientist	Head of Group, 1924 – 54.
A. L. Kirkham	Technician	Joined in 1921.
J. F. Spilling	Technician	Joined in 1932.
C. O. Pringle	Scientist	Joined in 1936.
D. T. O'Dell	Technician	Joined in 1937.
J. S. Shayler	Scientist	Joined in 1938 (for four months. Transferred to Radar Division).
F. M. Foley	Scientist	Joined in 1938. Head of group 1954.
T. J. Jones	Scientist	Joined in 1938.
B. S. Gold	Scientist	Joined in 1940.

The following scientists joined later, mainly at the outbreak of war, and were involved in development of klystrons and other special types. From June 1940 this team worked at the Wills Laboratory, Bristol University.

R. W. Sutton	Scientist	Joined in 1939. Director SERL, 1946.
T. J. Buchanan	Scientist	Joined in 1939.
A. M. Reith	Scientist	Joined on outbreak of war.
J. Thomson	Scientist	Joined on outbreak of war.
R. A. R. Tricker	Scientist	Joined on outbreak of war.
G. P. Wright	Scientist	Joined later.
P. G. R. King	Scientist	Joined later.

In addition to these scientists, both groups had a number of assistants, and silica and glass workers. There were also groups working on valve application problems and on valve specifications, led by A. Sczaniecki (engineer, Polish Army) and J. W. Webber (engineer), respectively.

APPENDIX 2: SILICA VALVE TYPES

Type	Category	Filament V (v)	Filament A (a)	i (a)	Anode V (kV)	Anode W (kW)	Ampl. fact.	Mut. cond. (ma/v)	Seal type	Bulb diam. (mm)	Application
NT22B	Triode	27	48	5	14	15	32	10	Lead	150	Shore-station long wave comm. trans. output
NT22C	Triode	27	48	5	14	15	55	4.2	Lead	150	Shore-station long wave comm. trans. output
NT23B	Triode	12.5	18.5	0.9	12	2.5	26	2.4	Lead	100	Shipborne long wave comm. trans. output
NT24	Triode	16.5	28	2	12	4	180	2.2	Lead	100	Shipborne long wave comm. trans. output
NT30	Triode	27	48	5	14	15	75	5	Lead	150	Shore-station long wave comm. trans. output
NT31	Triode	16.5	28	2	12	4	80	2	Lead	100	Shipborne long wave comm. trans. output
NT32B	Triode	23	47	4.25	12	4.5	55	4.5	Lead	100	Shipborne med. wave comm. trans. output
NT33	Triode	16.5	28	2	12	4	80	2	Lead	100	Shore-station med. wave radio comm. trans. output
NT35	Triode	29	52	5	14	15	20	12	Lead	150	Shore-station med. wave radio comm. trans. output
NT41A	Triode	9	20	0.9	10	1	18	1	Lead	100	Shipborne med. wave output first radar output (Bawdsey)
NT43	Triode	21	20	1.5	10	3.5	18	3	Lead	100	Shipborne med. wave radio trans. output
NT45A	Triode	10	20	0.9	10	1.25	37	2	Lead	100	Shipborne med. wave radio trans. output

(continued overleaf)

Appendix 2 continued

Type	Category	Filament V (v)	Filament A (a)	i (a)	Anode V (kV)	Anode W (kW)	Ampl. fact.	Mut. cond. (ma/v)	Seal type	Bulb diam. (mm)	Application
NT46R	Triode	15	40	3	10	3.5	18	3	Lead	100	Metric radar output
NT52	Magnetron	3	6		2	0.2					Radio homing beacon (1.5-m)
NT54	Triode	28	20	2.5	10	4.5	36	5.5	Lead	90	Comms. trans. power absorber
NT57	Triode	15	48	3.6	10	1	16	2.4	Lead	90	Radar output
NT57D	Triode	15	48	5	10	1.75	16	3.2	Graded Glass	90	Radar output
NT57T	Triode	9	35	18	10	1.75	16	3.2	Graded Glass	90	Radar output (Thor. tungst. fil.)
NT59A	Pentode	8	20	0.9	10	1.75		3	Graded Glass	90	Shore-station radio
NT69	Pentode	13	40	2	10	2.5		3	Graded Glass	90	Comm. trans
NT77	Tetrode	10	68	70	4			7.5	Graded Glass	90	Radar output (RAF)
NT78A	Triode	10	65	50	50	4	18	20	Graded glass	90	Radar modulator
NT84	Triode	12	40	2	100	1.5	4.5	5.5	Graded Glass	90	Radio shore-station trans. absorber
NT86	Triode	10	100	90	50	2	11	10	Graded Glass	100	Radar output
NT90	Triode	33	20	1.5	6	6	25	5	Graded Glass	90	Comm. trans. absorber
CV313	Tetrode		11.5	68	55	50	4.5	32	Graded Glass	100	Radar modulator
NU22	Rectifier	19	19	1.4	40	4			Lead	100	Shore-station comm.
NU23	Rectifier	27	46	5	40	15			Lead	150	Shore-station comm.
NU24	Rectifier	22	52	4.5	25	4			Lead	100	Shipborne comm.
NU25	Rectifier	19	19	1.8	40	4			Lead	100	Shipborne comm.
NU26	Rectifier	22	18	1.8	28	0.5			Lead	90	Shipborne comm.
NU28	Rectifier	28	20	2.5	28	2.5			Lead	100	Shore-station comm.
NU29	Rectifier	29	52	5	40	3.5			Lead	100	Shore-station comm.
NU30	Rectifier	14	10	0.8	60	1.8			Graded Glass	90	Radar comm.

Acknowledgements

I must acknowledge great help, particularly on older archival material, from The Library at the Defence Research Establishment, Portsdown; the museum at HMS *Collingwood*; Mr J. A. Winterburn, formerly of The Thermal Syndicate Co; Mr R. Davis and Mr Sowan, formerly of the Mullard Co; and Mr E. B. Callick and Peter Peregrinus Ltd for permission to use diagrams from the book *'Metres to Microwaves'*, published in 1990.

References

1. B. S. Gossling, 'Thermionic Valves for Naval Uses', *Journal IEE* (1920), p. 670–97; B. S. Gossling and M. Thompson 'Development of Valves for Wireless', *World Power* (April 1925); Sir B. A. Surtees Paget, 'History, Development and Commercial Uses of Fused Silica', *Jour Roy Soc Arts* (January 1924); PRO File ADM 220/171, Includes very early photographs of silica valves.
2. Ibid PRO Files AVIA 7/352, 353, 430, 431, 433, 434 and 435. Included much information and correspondence on the development of early radar valves for AMRE, Bawdsey; PRO File AVIA 7/715 Gives data on the unsuccessful attempt to make a silica valve to operate at 240 MHz.
3. S. R. Mullard, 'Thermionic Valve with Cylindrical Anode mounted in a Silica Bulb', Patent 149076, May 1919; 'conductors in a Lead-Seal System', Patent 170096, July 1920; 'Electrode Mounting System', Patent 170097, July 1920. A. K. Macrorie 'Electrode Support systems in Silica Envelope', Patent 170955, August 1920; 'The Molybdenum Basket-Anode', Patent 170953, August 1920; 'Electrode Mounting on a Frame in a Silica Envelop', Patent 170954, August 1920. H. Morris-Airey *et al.*, 'Connections from the Anode to a Seal', Patent 170955, August 1920. L. G. Preston *et al.*, 'Assembly Method allowing a Silica Valve to cut open for Repair', patent 178898, December 1920; 'A filament Tension Spring', Patent 177270, December 1920; H. J. S. Sand and F. Reynolds, 'Improvements in Production of Gas-Tight Seals between metal and Vitreous Materials', Patent 23854, October 1913.
4. J. Bell *et al.* 'Triodes for Very Short Waves', *Jour. IEE*, vol. 93, Part 3A (1946), p. 833.
5. E. G. Bowen, *Radar Days* (Adam Hilger Ltd, 1987), p. 29.
6. J. T. Randall and H. A. H. Boot, 'The Cavity Magnetron', *Jour. IEE*, loc. cit., p. 928 et seq.
7. E. C. S. Megaw, 'High-Power Pulsed Magnetron: A Review of Early Developments', *Jour. IEE*, loc. cit., p. 977 et seq.
8. W. E. Willshaw et al., 'The High-Power Pulsed Magnetron: Development and Design for Radar Applications', *Jour. IEE*, loc. cit., p. 985 et seq.
9. O. L. Ratsey, 'Radar Transmitters. A Survey of Developments', *Jour. IEE*, loc. cit., p. 245 et seq.
10. H. de B. Knight, 'The Development of Mercury Vapour Thyratrons for Radar Modulator Service', *Jour. IEE*, loc. cit., p. 189 et seq.
11. E. B. Callick, *Metres to Microwaves* (Peter Peregrinus Ltd, 1991), p. 91.
12. Sir B. Lovell, *Echoes of War. The Story of H2S Radar* (Adam Hilger Ltd, 1991), p. 38.
13. B. Bleaney, 'The Crystal Valves', *Jour. IEE*, loc. cit., p. 184 et seq.

14. G. F. J. Garlick, 'Cathode-Ray Tube Screens for Radar; their Development and Measurement for Intensity Modulated Displays, especially H2S', *Jour. IEE*, loc. cit., p. 167 et seq.' 'Cathode-Ray Tube Screens for Radar', *Jour. IEE*, loc. cit., p. 815 et seq.
15. P. G. R. King, 'The Skiatron or Dark-Trace Tube', *Jour. IEE*, loc. cit., p. 171 et seq.

Monograph 4
Royal Navy Metric Warning Radar, 1935–45
J. S. Shayler

SUMMARY

Admiralty decision to investigate RDF applications independently – provisional Staff Requirement for warning set against aircraft and ships, October 1935. Early experimental work on 4-m wavelength disappointing – reorganisation of radar research under Horton in 1937. New development on 7.5-m for shipborne installation – choice of horizontal polarisation – successful development of Type 79. Fitting in *Sheffield* and *Rodney* in autumn 1938 – further improvements leading to Type 79Z, first installed in AA cruiser *Curlew* in August 1939. Improvements in silica-envelope transmitter valves giving higher peak power – introduction of ranging panel, Type 279, in September 1940 – the anti-DF facility – development of common Transmit/Receive operation in Type 279B. Anti-jamming filter – discovery of effect of galactic noise on long metre-wave radars. Technique for height determination – development of performance meter – development of improved warning set Type 281 for aircraft and surface targets in the 90 MHz band. Development of new high-power 90 MHz transmitter valve at Signal School – development of split-lobe DF to give improved bearing accuracy. Introduction of common Transmit/Receive operation into Type 281B. Problem of long-range detection of high-flying aircraft – introduction of PPI display in Type 281 – introduction of continuous antenna rotation in Type 281BQ – height estimation using dual installation of Type 79Z and Type 281. Development of Type 286 in the 200 MHz band for small ships – introduction of common Transmit Receive (T/R) Antenna system – replacement of fixed-sector antennae by rotatable unit – crash programme for Type 290, leading to problems – overcome in Type 291, gradually replacing Type 286, for small craft. The ultimate wartime metric radar development for large ships, Type 960, in 1945.

INTRODUCTION

As described in Monograph 1, following Watson-Watt's successful demonstrations of the potential of radiolocation, later known as RDF, in the Air-Defence role, the three Services considered the mechanisms for further development. The Air Ministry and the Army initially joined forces under Watson-Watt at Orfordness, and later, Bawdsey. The Navy, however, decided to remain independent, and to use resources at its HM Signal School for RDF investigations. On 13 August 1935 DSD informed Signal School of the Controller's decision to start work on the Naval applications of 'detection and location of aircraft by wireless methods'.[1]

At a meeting held in Admiralty early in October 1935, a provisional Staff Requirement was formulated for the following performance:

- Aircraft: warning of approach 60 miles; precise location 10 miles.
- Ships: warning of approach 10 miles; precise location 5 miles.

Note: these ranges are expressed in nautical miles, approximately 2000-yds per mile (see Glossary). All subsequent references to miles in this monograph imply nautical miles.

At this stage there was no hard evidence to indicate whether such ranges were practicable. They were based largely on current concepts of the application of gunnery as a means of ship defence. Warning ranges were judged to be the minimum to allow gun crews to 'close up', thus removing the necessity for keeping them continuously at action stations. Precise location ranges were judged to be the minimum at which the guns commenced tracking targets, in readiness for opening fire. Only later was it realised that warning ranges should be sufficient to allow aircraft to be launched from carrier decks in time to perform a successful interception. Even later it came to be appreciated that, following such a warning, a continuous flow of information was required in a form suitable for the conduct of controlled fighter interception from the ship.

The above mentioned directive of 13 August 1935 from DSD to the Captain, Signal School, to commence RDF investigations resulted in the appointment of R. A. Yeo to lead a small team. In 1936 an extension of Signal School was set up at the Royal Marine (RM) Barracks, Eastney. The site was ideal since it directly overlooked the sea and was in an enclosed area (an old fort), where it was straightforward to impose tight security. A single-storey two-roomed brick building was erected in 1936. Together with a wooden hut as the Superintendent's office, this sufficed until 1939. Two tubular masts (made of steel tubes and clamps, as used for scaffolding) were built to support antennae at simulated mast height. In 1939–40 other two-storey buildings were added to provide laboratory space of 10000-sq ft and roof area of 7000-sq ft. Pylon towers, each 100-ft high with large platforms at the top to simplify work on antennae, were built to replace the tubular masts.

Various experiments were performed during the first year at Eastney, of which only little information has been found. By the autumn of 1937 work on warning RDF had settled on a frequency near 40 MHz, and three element Yagi antennae were being used. At this time no special priority was attached to RDF within Signal School, the main emphasis remaining on wireless telegraphy (W/T) and direction finding (DF) (see Monograph 1).

Towards the end of 1937 responsibility for RDF was transferred to C. E. Horton, who was already in charge of DF (see Monograph 1). Although no extra priority for RDF was then allocated, Horton did all he could to obtain the necessary financial support and resources to produce equipment of operational use to the Navy. It is very largely due to him that such a strong base for Naval RDF had been achieved by the start of the war when, at last, full priority was allocated from the Admiralty.

TYPES 79X AND 79Y

An experimental shipborne RDF set, Type 79X, was fitted in Saltburn *for sea trials in early 1938. The results were so promising that two similar sets, Type 79Y, incorporating some improvements, were made in Signal School workshops. One was fitted in the cruiser* Sheffield *in August 1938, the other in the battleship* Rodney *in October 1938. They worked well, and were not replaced until September 1941 for* Rodney *and July 1942 for* Sheffield.

In late 1937 the decision was taken to produce an experimental RDF set suitable for withstanding the shipborne environment and undertaking seagoing trials, the objective being to meet the Admiralty Staff Requirement referred to earlier.

The preceding laboratory work had investigated a frequency near 40 MHz; in fact 43 MHz was chosen for Type 79X. The highest available frequency was required for compact antennae with high gain and narrow horizontal beamwidth (to aid DF). These had to be light enough (together with the mounting required to turn the array) and of sufficiently low windage for them to be mounted at the mast head. On the other hand the frequency had to be such that components and technology were immediately available to enable both a transmitter and receiver to be designed and engineered. Laboratory tests had shown that the highest frequency at which the silica valves then available would generate power efficiently was about 40 MHz. Also, the BBC Television Service (which had been transmitting since 1936) used 41.5 and 43 MHz for sound and vision, and this provided confidence that an RDF equipment could be engineered successfully at about these frequencies.

The antenna team at Nutbourne quickly designed an array to meet the shipborne requirements. To improve the gain over the three element Yagi ('fishbone') antenna, which had so far been used for laboratory tests, and yet keep weight and windage within acceptable limits, a four-dipole array was used, with two driven and two passive elements.[2] Horizontal polarisation was chosen. The two driven elements were mounted a half-wavelength apart, one above the other, each with a passive reflector spaced 0.2 wavelengths behind. This provided a gain of about 7 db, compared with about 4 db for the Yagi, or a combined transmission – reception gain of 14 db compared with 8–9 db.

Horizontal polarisation was originally chosen for RDF because it was thought that, at these long wavelengths (11-m for RAF CH stations and 7.5-m for Type 79), better reflection would be obtained from the wing-span of target aircraft. However the Navy was very interested in the detection of surface targets as well as aircraft. It was realised that, theoretically, vertical polarisation should give a higher field-strength close to the earth's surface. Consequently, when the complete Type 79X was fitted in the Signal School's trials sloop *Saltburn* early in 1938, A.W. Ross conducted some tests using simple horizontal and vertical antennae. He found that vertical polarisation produced a loss in performance because of the increase in returns from the sea (sea clutter, a previously unsuspected phenomenon) compared with horizontal polarisation. As a result horizontal polarisation continued to be used for Type 79, and in fact for all other Naval metric sets developed throughout the war.

The array was fed by a twin open-wire feeder of two wires spaced 2.5-in apart by moulded insulators, fitted every 2-ft along the run. The impedance was about 400 ohms. Power was fed to the two driven dipoles by a similar open-wire feeder, the two wires being splayed apart to be attached to the dipoles at the appropriate impedance points. At the junction of the feeder line up the mast and the two short feeder lines to each dipole, a section of shorted line was also attached of length appropriate to provide proper impedance matching.

The four dipoles were supported on a mast, which was mounted on a pedestal so that it could be rotated by $\pm 190°$. Continuous rotation was impossible because of the feeder line, which had a section made of flexible wires to allow the plus/minus half revolution. A drawing of an array is shown in Figure 4.1. Similar arrays and pedestals were used separately for both the transmitter and receiver.

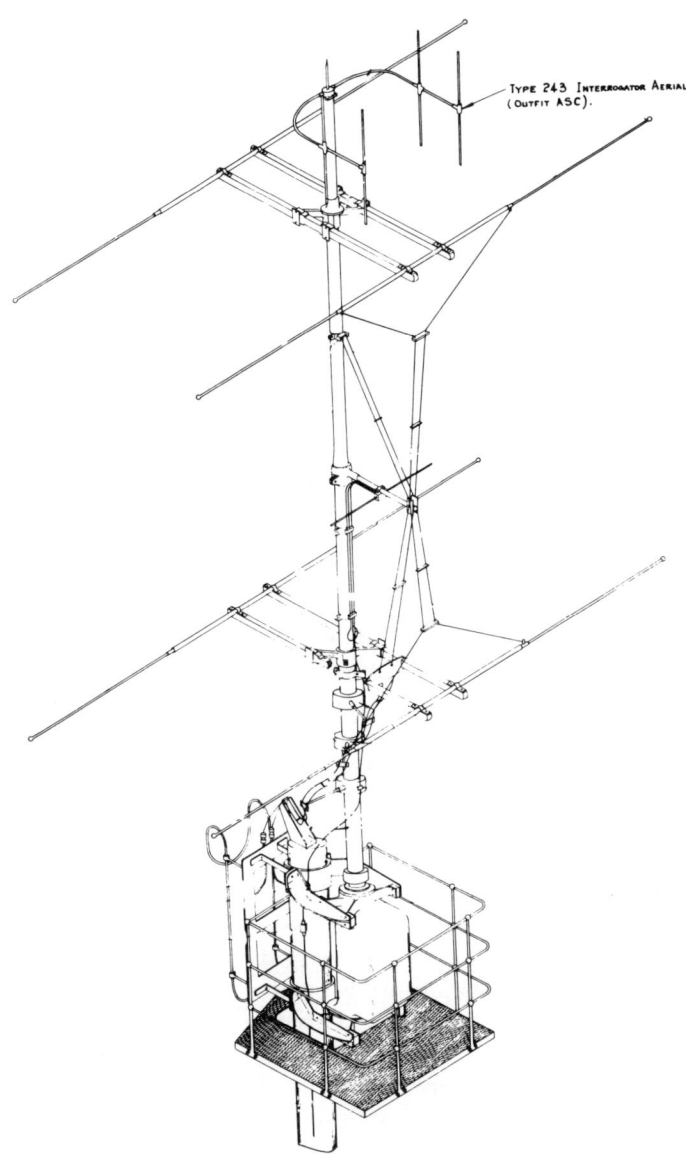

4.1 Type 79/279 Antenna Array, with Type 243 IFF (scale: Type 79 dipole approx. 12-ft long) (© Defence Research Agency)

In the pedestal was a Selsyn torque motor coupled to the antenna mast through a 90-to-1 step-down geartrain. A control unit was housed in the receiving office beneath the receiver and display. It consisted of a handwheel coupled through a step-up geartrain to two Selsyn torque transmitters, each independently connected electrically to the torque motors in the two pedestals. Thus by turning the handwheel the two antennae were turned in step.

Maximum torque was transmitted when the receiving Selsyn lagged the transmitting Selsyn by 60°, but because of the 90-to-1 gearing in the pedestal there was a maximum lag of only twothirds of a degree between the bearing indicator in the office and the antenna masts. When the antennae were stationary the error was much less, but since the combined horizontal beamwidth was about 60° the lag was negligible.

In passing, it is worth noting that this array was so successful as a compromise between the various conflicting needs of the Naval environment that it was used throughout the life of Types 79/279. It was also the basis of design of antennae for later RDF sets on higher metric frequencies, namely Type 281 on 90 MHz and Type 291 on 214 MHz. Considerable credit is due to the team under A. W. Ross for this near-perfect design, so early in the history of Naval radar.

Prior to and during the war, ships' main power supplies were on 220 V DC. To provide a more flexible AC supply, Type 79X had its own motor generator giving 230 V 50 c/s. Why 50 c/s was chosen is unclear, other than that presumably no reason was seen at the time to move away from experience gained from the domestic supply, and components designed for it. For simplicity, the whole function of Type 79X was locked to the mains, and consequently the pulse repetition rate was 50 c/s. This supply frequency was retained throughout the life not only of Types 79/279 but also Type 281.

The transmitter used two NT57 triode valves in push–pull (see Monograph 3). These were silica-envelope valves with a molybdenum basket anode, a molybdenum wire grid and a bright pure tungsten filament. The origin of the design is unclear. It may have been a result of the maximum amount of filament that could be accommodated in a standard silica envelope. The new requirement for RDF was, of course, high peak emission combined with modest anode dissipation, and so special valves had to be produced. They were designed and made in the Valve Section at Signal School.[3] The silica envelope had the advantage that it could be opened in the laboratory, so damaged valves could be repaired. Also, water cooling was not required, a complication that was not acceptable on-board ship.

No record has been found of the circuits of either the transmitter or modulator, but the transmitter is believed to have been similar to the later model Type 79Y. In this case the main resonant circuit was a Lecher-line in the anodes of the push–pull arrangement, with power to the antenna being tapped off this circuit via isolating condensers. The grid circuit of each valve was a three-turn coil of 1-in diameter, and the cathodes were earthed. A negative bias was supplied to the grids so that the valves were normally non-conducting.

The modulator generated a positive pulse to overcome the negative bias on the transmitter grids. The pulse was derived by squaring the 50-cycle mains waveform in a distorting amplifier, differentiating this to provide a sharp rise and exponential-decay triangular pulse, and squaring this up in another distorting amplifier. The pulse length was a nominal 15 microseconds, probably variable from 8 to 30 microseconds. The transmitter produced a power of about 20 kW during this pulse. The voltage for the transmitter of 20 kV was derived from a half-wave rectifier from the mains.

The receiver was 'straight', that is, had no superheterodyne stages. It consisted of a series of low-gain RF stages using 'acorn' valves coupled together, and feeding a final output detector. To improve stability, the 'acorn' anodes were supplied with 50-cycle AC, the transmitter firing at the peak of this waveform.

The display was an A-Scan (range versus echo amplitude) on a CRT. The scan was derived very simply by applying a 50-cycle mains current to horizontal deflecting coils, thereby vastly over-scanning the CRT face and displaying only a portion of the linear part of the mains sine wave. The transmitter firing was phased so that the groundwave appeared at the left-hand side fo the CRT. The CRT itself had about a 5-in diameter face, with a green non-afterglow phosphor. Focusing was achieved by having a small amount of inert gas in the CRT, and adjusting the beam current by varying the cathode filament current until optimum focus was achieved.

The whole of the equipment described above was rapidly engineered in Signal School Experimental Workshops into a form suitable for ship-fitting.

The complete Type 79X was installed in the Signal School trials sloop *Saltburn* early in 1938. Trials were completed rapidly, and good results obtained on an aircraft (type unknown). The performance was promising enough for a further two sets to be manufactured within Signal School as rapidly as possible. These were for fitting in two operational ships in order to obtain full in-service experience. The sets were designated Type 79Y, and were made in the workshop as copies of Type 79X, except possibly for some improved engineering. The first was fitted in the cruiser *Sheffield* in August 1938, and the second in the battleship *Rodney* in October 1938. Both gave satisfactory performance, and in fact were used until replaced by Type 281, in September 1941 in the case of *Rodney*, and July 1942 for *Sheffield*. Typical detection ranges on aircraft were 30 miles at 3000-ft, 40 miles at 7000-ft and 53 miles at 10000-ft altitude.

Early experience with these two sets was so good, and Naval Staff were so impressed, that it was decided to arrange for the production of 40 sets for extensive ship fitting. The first of these sets was to be ready for fitting in a ship by July 1939, having completed shore trials. However inevitable problems came to light. Firstly the 'straight' receiver showed signs of instability. Secondly, the grid modulator of the transmitter sometimes failed to cut off the transmitter at the end of the pulse, thereby significantly increasing the mean power. This created havoc with several of the meters, some of the components and the anode dissipation of the transmitter valves. The reason was believed to be the grid emission in the transmitter valves presenting too low an impedance to the modulator.

Additionally, any increase in transmitter power would be readily acceptable, and by coincidence a means of achieving this became available just at this time. Thus some rapid laboratory work was undertaken at the beginning of 1939 to improve these items for introduction into the production equipment.

TYPE 79Z

An improved equipment, Type 79Z, was developed for further ship-fitting. This had higher transmitter power, a more reliable modulator, and a more stable and easier-to-manufacture receiver. Again, the first two models were made in Signal School workshops before bulk production started in industry. The first set was fitted in the anti-aircraft cruiser Curlew *in August 1939. Production sets started to be fitted about May 1940, eventually totalling about 100 sets.*

Late in 1938 the Valve Section of Signal School introduced a technological breakthrough in their silica RDF valve design, namely thoriated-tungsten filaments[3] (see Monograph 3). With no basic redesign of the valve, other than a change of filament material, the peak emission was immediately increased some five times. The NT57 valves used in Type 79Y were quickly converted and redesignated NT57T, providing a peak emission of 20 amps compared with 4 amps previously. At the beginning of 1939 laboratory tests started using a pair of these valves in the Type 79 transmitter.

At the same time redesign of the modulator was required to eliminate the failure to cut off, as in Type 79Y. It was decided to introduce cathode modulation, that is, to earth the transmitter valve grids and insert a valve between their cathodes and HT negative which, when pulsed on, would pass the transmitter current. A large glass-enveloped STC triode, Type

4279A, was found that would handle the transmitter current, and this was used as the series modulator valve in the transmitter cathode circuit. The pulse to drive the grid of this modulator was derived as follows.

A pulse generator was housed in the receiver office, close to the receiver and display. This produced a push–pull square pulse of ± 5 volts by distorting and differentiating the 50-cycle mains waveform. It could be varied in length by between 8 and 30 microseconds, and varied in phase with reference to the mains supply, by which means the groundwave could be adjusted to the left-hand end of the A-scan, itself derived from the mains. The pulse was transmitted via a two-core cable to the modulator in the transmitter office.

In the modulator the push–pull pulse was fed into a balance-to-unbalance transformer with a step-up ratio of five to one. The positive pulse so derived was amplified by two pentodes in series to produce a positive pulse of 100 volts or more. Since the second pentode had to carry current all the time in order to be cut off by the pulse and produce a positive pulse at its anode, its screen was supplied with mains AC in order to reduce valve dissipation. This positive pulse was fed via a cathode-follower to the grid of the large series modulator valve. This rapidly assembled modulator functioned extremely well and was used throughout the life of Types 79Z and 279.

The transmitter RF circuit remained similar to that already described for Types 79X and 79Y, namely a Lecher-line in the anodes and coils in the grids of the push–pull circuit, the cathodes of the NT57T oscillator triodes being earthed at RF. Minor changes were made to the DC circuits of the grids and cathodes to accommodate the new cathode modulator. A pulse power of 70 kW (in contrast with 20 kW for Types 79X and 79Y) was obtained from this combination of high-emission transmitter valves and new modulator.

An absorption wavemeter to check the transmitter frequency, and a 1-in CRT monitorscope to check transmitter and modulator waveforms were built into the top of the modulator panels.

In addition to the tendency to instability in the original receiver, it was difficult to manufacture and to maintain. Thus it was decided that it must be replaced for wider Service use. Fortunately a laboratory model of a superheterodyne using 'acorn' valves in the RF stages had been designed at Nutbourne for carrying out antenna measurements. This was copied in the workshops and became Receiver P11, to be used with Types 79Z and 279 throughout the war. The only laboratory modification before engineering in the workshops was to increase the bandwidth to 140 kHz in order to handle the Type 79Z pulses.

The P11 receiver had an impedance-matching input stage, an 'acorn' pentode RF stage, a triode local oscillator, all separately manually tuned, and a pentode mixer stage. This was followed by a three-stage pentode IF amplifier centred on a frequency of 5 MHz, a full-wave double-diode detector, a video amplifier and output cathode-follower. The input stage must have been greatly overloaded by the transmitter pulse, but never seemed to suffer.

Another change was introduced into production versions of Type 79Z. In order to make jamming slightly more difficult and to reduce mutual interference between ships sailing in company, an operational frequency band was allocated of 39–42 MHz. The tuning of both the transmitter and receiver was able to cover this band. However the antenna design was narrow band, and so, four sets of dipoles were produced of dimensions matched to each frequency. Thus, once an antenna was installed, that ship's frequency was determined and unalterable.

The antenna control unit was unchanged from Type 79Y except for the introduction of a small mechanical feature, borrowed from DF practice, to improve the accuracy of measuring the bearing of a target. Because the beamwidth of the combined antennae was about 60°, the peak could only be judged to about ±10° when looking at the echo amplitude on the A-scan. Now in the centre of the antenna training wheel was a bearing indicator, which showed the antenna position by a pointer relative both to true North and to ship's head. A foot pedal was provided that, when depressed, rotated this pointer at only half the antenna speed. To take an accurate bearing, the antenna was turned off-target until the A-scan echo was reduced to about half amplitude. The pedal was then depressed and the antenna rotated through the echo peak to half amplitude on the other side. The operator then read the bearing off the pointer, which had bisected the antenna rotation, and then released the pedal so the pointer jumped to its position in line with the antennae. By this means a bearing could be obtained to an accuracy of some 5° on an echo of reasonable strength.

As with Type 79Y the above improvements, together with some engineering lessons learnt from Type 79Y, were introduced into a complete Type 79Z produced by the experimental workshop for tests at Eastney (see Postscript to this monograph). These were completed by July 1939. Very acceptable detection ranges on aircraft of about 40 miles at 3000-ft height, 70 miles at 10000-ft, 90 miles at 20000-ft and 120 miles at 30000-ft were achieved. Also, during the winter of 1939-40 the first example of anomalous propagation was experienced; owing to ionospheric condition, echoes from the previous cycle, at a range of some 1600 miles, overlapped the wanted echoes.

The first Type 79Z set was installed in the anti-aircraft cruiser *Curlew* in August 1939. Shortly afterwards *Curlew* was stationed at Scapa Flow to provide aircraft-detection cover for the Home Fleet, since at that time the RAF's Home Chain RDF stations did not cover that area, only a small portable system having been provided there.

At the same time as this first set was being manufactured in Signal School workshops, the Drawing Office produced a full set of engineering drawings. Contracts were placed with industry to manufacture 40 equipments rigourously to these drawings. No flexibility was allowed for industry to modify the design to their own production methods, or to introduce greater cost-effectiveness (a term not then invented and a concept not alive in Signal School). Nevertheless this *modus operandi* provided extremely quickly equipment that withstood shipborne operating conditions, and produced very acceptable operational results for the duration of the war. Figure 4.2 shows the equipment fitted in the transmitter office and Figure 4.3 that fitted in the receiver office.

The production sets started to be fitted from about May 1940 onwards, and altogether some 100 ships were fitted with Types 79Z or 279 during the war.

4.2 Rear of Type 79 equipment in the transmitter office (left-hand unit, EHT rectifier; centre unit, transmitter; right-hand unit, modulator) (© Defence Research Agency)

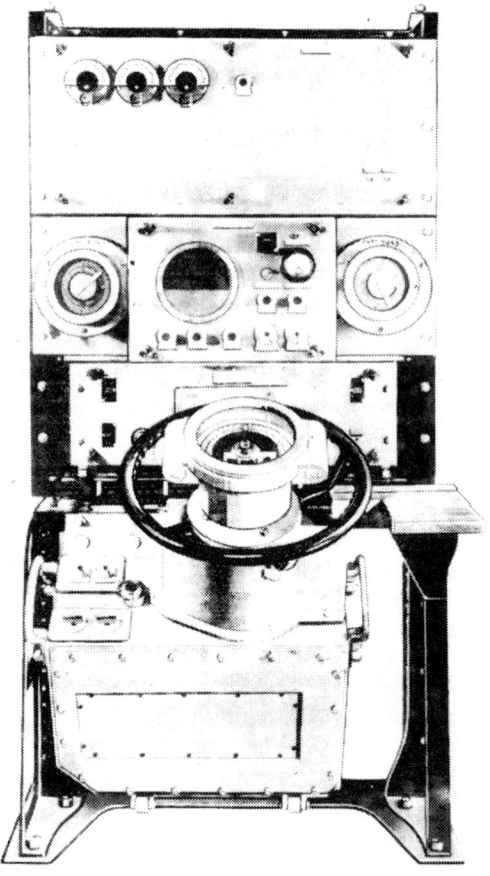

4.3   Type 79 Equipment in the Receiver Office (© Defence Research Agency)

TYPE 279

To improve the accuracy of range measurement available for gunnery (optical ranging was of limited accuracy), an Army accurate ranging panel was added to Type 79Z, then being designated as Type 279.

Early experience from Type 79Z caused the Naval Staff to realise that, in addition to its warning potential, it had the ability to provide an accurate target range and range-rate, parameters that are available with only limited accuracy from optical range finders. In consequence they asked for this feature to be added rapidly to Type 79.

The team at Eastney found that an acceptable ranging unit had been developed by the Army RDF research team for their GL1 equipment (Gun Laying Mark 1). This was bought as a unit and added on to the equipment in the receiving office. The complete RDF set including this feature was rechristened Type 279. Its introduction into service is thought to be by the conversion of the Type 79Z in *Curacao* in September 1940.

The ranging unit consisted of two CRTs. one for ranging and the other for bearing. Both were supplied with exponential A-scan timebases that vastly overscanned the tube, so only part of the scan could be seen. The total scan was shifted horizontally by the potential derived from a precision wire-wound, oil-immersed potentiometer with an exponential law matched to that of the scan, thus displaying a different small part of the total scan on the CRT display. There was a precision calibration procedure, and ranging was performed by adjusting the potentiometer so that the leading edge of the echo lined up with hairline on the ranging tube. If a moving target was kept on the hairline, an ingenious mechanical system produced a range-rate. Another winding on the potentiometer moved the bearing-tube A-scan in step so the same echo appeared in the middle of the screen. However, in Type 279 the antenna horizontal beam was too wide for accurate bearings, so reliance on this had to remain with the gunnery optical system. The unit gave accurate ranges (within 50-yds if the setting up was accurate) on echoes between 2000-yds and 14000-yds.

In order to provide a better echo on which to range, and to provide better range discrimination during ranging, the pulse generator was modified to provide a pulse length of 3 microseconds (in addition to the normal pulse lengths for warning). In consequence, in order to handle this pulse and provide a steep leading edge for accurate ranging, a second receiver (the P12) was mounted on top of the ranging unit. This receiver had a bandwidth of 1 MHz instead of the 140 kHz of the P11. Its input circuit, RF stage, local oscillator and mixer were 'acorn' valves with circuits similar to the P11, but the IF consisted of four stages of EF50 valves to provide the increased bandwidth. A single diode second-detector was used. The single video amplifier and output valve was also an EF50.

The EF50 was just becoming widely available in the UK. It was a general-purpose RF pentode, entirely metal-screened in a glass envelope, with a metal – glass seal in the base for its connector pins. It had been developed by Philips in Holland, and just before the fall of that country in 1940 Philips handed over full production information to the UK with full permission to use it. It soon became ubiquitous in RDF circuits developed by all the Services, and was allocated the Service number VR91.

In practice it is doubtful whether the added ranging facility was much used operationally, since the general warning information could rarely be dispensed with in order to provide precision ranging on a single target. In addition, RDF sets specifically designed for gun-laying were beginning to make their appearance (see Monograph 1). Later in the war, when Type 279 was used mainly for warning and for fighter direction, it is believed that this ranging equipment was no longer fitted.

A small addition to Type 279 was the provision of an anti-DF facility. It was common operational practice to impose radio silence so that the enemy could not determine one's ship's position by DF. However there was an even stronger need for warning information, and so a Morse key was introduced for the use of the receiver operator. When pressed, this activated the pulse generator for only half a second, thus transmitting some 25 pulses. A switch was provided to cut out this feature. In order to allow the operator to read the received signals, a CRT with an afterglow screen was fitted in the warning A-scan position. This had a two-layer screen with a blue flash and green afterglow of about 10 seconds persistence. Because of the wide azimuth beamwidth, the operator could pulse the transmitter once every 30° in azimuth and still obtain reasonable warning cover. Again it is not certain how much this feature was used operationally.

One final modification became available about this time, that is, during 1941. This was a more versatile timebase for the warning display. It is not certain that this was tied specifically to Type 279 since the new unit was housed in the same chassis as the old one and could therefore be fitted retrospectively. It is believed that, in addition to being supplied with all new Type 279s, it was also replaced in all Types 79 and 279 already installed.

The new unit used the same gas-focused CRT as the old one (either with an afterglow or non-afterglow screen), but introduced a triggered timebase instead of the original mains sinusoid scan. This had the advantage of easily providing different scan speeds. It proved useful when the accurate ranging unit was either eliminated or little used, by providing a selection of scan speeds on the warning tube appropriate to air warning, fighter direction and surface-vessel warning and tracking.

The timebase consisted of two valves, an EF50 and another pentode with a top-cap anode connector capable of withstanding a high voltage. The EF50, and the screen and grid of the other pentode were connected as a 'one-shot' multivibrator, and this provided a brightening pulse fo the CRT. When the multivibrator fired, the second pentode was cut off, and a condenser from its anode to earth charged exponentially through its anode load. Since the supply voltage to this anode was 1100 volts, a sufficient amplitude was obtained to overscan the CRT screen via the X-plates, the comparatively linear part of the exponential being displayed.

Scan speed could be changed by selecting different condensers in the anode CR circuit.

Two further EF50 valves provided calibration pips, the first being an LC oscillator triggered and gated by the multi-vibrator, the second distorting the sine wave produced by the first into sharp pips at 5-mile spacing.

To recapitulate briefly, Type 279 had separate receiving and transmitting antenna arrays appropriate to the frequency allocated to the ship. The transmitting office contained an EHT rectifier, a modulator and a transmitter identical with Type 79Z (Figure 4.2). The separate receiver office had an antenna control unit, a pulse generator unit, a warning A-scan display unit and a receiver unit with a narrow bandwidth (Figure 4.3). Some installations also had an accurate ranging unit and a wideband receiver in the receiver office.

A word may be appropriate about the RDF operators, one in the transmitter office and one in the receiver office. Apart from routine, and occasionally panic maintenance, the transmitter operator had a quiet life, only having to monitor some meter readings. His main problem was living with a heat-generating equipment in an office often inadequately ventilated! In contrast the receiver operator had a busy time. He had continuously to rotate the antennae, and since the Selsyn system provided no power amplification but only torque transmission, the operator had to contribute the appropriate muscle power. This was not difficult in calm conditions but could be very tiring in windy conditions, owing to the windage presented by the antennae. Additionally he had mentally to remember the echoes he was tracking, so that he could recognise a new one, and to track them against the True North compass-corrected bearing ring remembering that, when the ship was manoeuvring, the bearing of the echoes would change relative to Ship's Head. When there was action, he had a challenging time both physically and mentally.

TYPES 70B/279B

All the early RDF sets had separate transmitter and receiver antennae, thus requiring masthead positions on two masts. These positions were also required for other radio services, and so a Transmit/Receive switch was provided in the RDF set to allow the transmitter and receiver to use a single antenna. The first fitting was in Hood *in March 1941.*

As more sets were introduced into the Fleet from 1940 onwards it soon became clear that the demand for two masthead positions for transmitter and receiver arrays was causing a major difficulty for other W/T, DF and later RDF equipments, all of which needed masthead antenna positions. Work was therefore started at Nutbourne in late 1940 to develop a diode switch and matching transformers to provide single antenna working for Type 79/279. It compromised a network of cables interconnected between the single antenna, the transmitter, the receiver and a diode. In operation, when the transmitter was switched on so that the diode conducted, the transmitter was matched to the antenna and the receiver was isolated. When the transmitter was off and the diode not conducting, the receiver was matched to the antenna and the transmitter was isolated. Since there was no time to develop a single diode to pass the required transmitter current, six receiver-type diodes were connected in parallel.

It was decided to base the design on concentric Pyrotenax cable instead of the open twin-transmission line originally used to feed the arrays in Type 79/279. This was done to provide soundly engineered and stable matching transformers, and to ease the cable run up the mast. A system, shown schematically in Figure 4.4, was devised and a first experimental model was installed in *Hood* during a refit in March 1941. Sea trials took place in April 1941, and these confirmed the soundness of the design. Thereafter the system was put into production and used in subsequent installations. This single antenna equipment was designated Type 79B or 279B.

Pyrotenax had been designed originally as a mains power cable capable of withstanding very high temperature (even fire) conditions. In addition to being robust, it was found to have very low RF loss and to be capable of handling high RF power. The concentric cable had a solid copper inner and a copper outer sheath, the insulator being magnesium oxide. The only problem was that this was very hygroscopic. If water was absorbed, the insulation resistance dropped dramatically and RF losses increased to such an extent that performance was seriously affected. In order to protect the cable, all junction boxes (of which there were about a dozen in the system) were filled with bitumen. The seal was apt to deteriorate under seagoing conditions, and before resealing the last foot or two of cable had to be dried out by heating it with a blowlamp. This was very difficult at the top of a 150-ft mast in both wind and rain! But apart from this tedious maintenance problem the system was very robust, worked well and freed one mast for other applications.

Other items were added as they became available during the remainder of the war, mostly during 1943. One addition was a set of anti-jamming filters, introduced between the receiver output and the feed to all the displays. Selection could be made between two different high-pass filters, intended to eliminate the beat frequency between the jammer and the RDF set, and differentiating circuits to reduce interference if long-pulse jamming were employed.

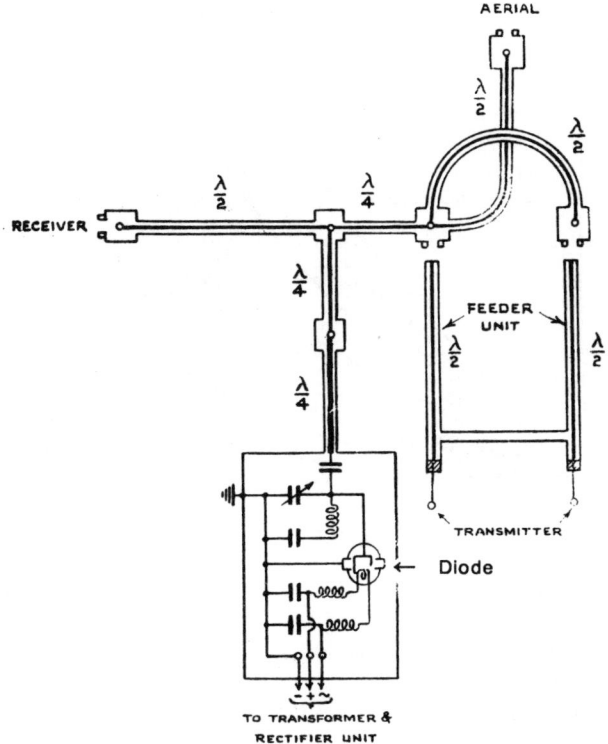

4.4 Schematic of Type 279B T/R switch (© Defence Research Agency)

A frustrated introduction was that of a low-noise, grounded-grid RF stage, to be inserted at the input to the receiver. In the laboratory this improved the receiver noise figure by about 5 db. However, during tests on the first installation in a ship no improvement at all could be obtained. This led to the discovery of galactic noise, and the study in a more academic climate of galactic and cosmic noise (see Monograph 2). As a result the introduction of this device to Type 79B/279B was cancelled.

The final item to be added was a test set, first introduced in late 1943. The background to this was height estimation (HE) of the aircraft target. It was realised, and observed from the very first installations of Type 79, that a series of vertical lobes would be formed by interference from sea reflection (Figure 4.5). It was soon appreciated (see Monograph 2) that an estimate of aircraft height could be obtained from the range of first detection of an aircraft, and that this estimate could be refined by observing the echo as the aircraft closed in range, noting the range of first maximum, first minimum and so on. Valiant attempts were made to use

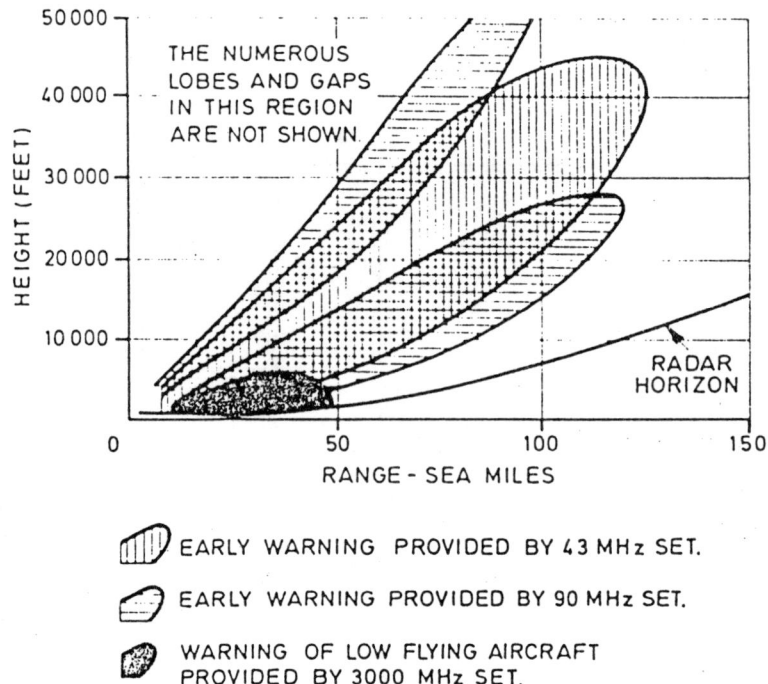

4.5 Vertical lobe structure, Type 79/279 and Type 281/281B (© Defence Research Agency)

this system operationally, particularly in a fighter-interception environment, where any indication of enemy height was of great assistance to the intercepting fighter. However rather mixed success was obtained, categories of high, medium or low being achieved, but more accurate estimation being somewhat variable. It was eventually realised that the success of the scheme depended critically on the performance of the RDF set. If its performance varied (as it easily could in practice) from that when the height-calibration flights were made, then errors would be introduced into the height estimation.[4]

Consequently the Naval Staff raised a Requirement for a Performance Meter. This was worked on during the second half of 1942 and first installed operationally in mid-1943. It consisted of a small test dipole mounted in the centre of 79B/279B array feeding a tuneable RF circuit containing a vacuo junction via a concentric feeder down the mast. This gave a measure of the transmitter power. Alternatively, with the transmitter switched off a signal could be fed to the dipole from a very-low-power test oscillator, and this could be used to check receiver sensitivity. The unit was mounted in the receiving office and had a circular slide-rule on its front panel, which could be used to combine the readings from the transmitter and receiver to obtain an overall-efficiency figure. If this figure varied, then the height-versus-range graphs could be appropriately adjusted, or better still remedial action could be taken to restore the RDF set to peak performance. In practice, the war in Europe was over by the time the Performance Meter was widely fitted in the Fleet. It is not certain how successfully it was used operationally by our ships in the Pacific.

TYPE 281

Despite the immediate reasonable capability of Type 79Z for air warning, it provided only limited performance on surface targets. The Naval Staff soon asked for even greater ranges against aircraft, and for a significantly improved performance against surface targets. Hence this dictated the use of a higher frequency (about twice that of Type 79Z was chosen). This itself provided a higher gain antenna with a narrower horizontal beamwidth, for about the same topweight. A significantly higher transmitter power was also required. This led to the development of Type 281, two preproduction models made in Signal School workshops being installed in Dido *in October 1940 and in* Prince of Wales *in January 1941. Production followed in February 1941, some 80 equipments being fitted during the war.*

Thoughts about an improved warning RDF set must have started to germinate during the late stage of engineering the Type 79Z, about mid-1939. The success with Type 79X and Y had provided a firm basis for the usefulness of RDF and proved its worth for aircraft warning and tracking, and indicated a potential for detection and tracking of surface targets.

As a result of discussions between the Naval Staff and C. E. Horton, the new set was to give maximum possible detection range of both air and surface targets, combined with as accurate as possible bearing and range information for the tracking of both types of target to aid gunlaying.[5] Gunnery was still seen as the main method of ship defence.

Translated into technical terms, this demanded a higher frequency in order to achieve a higher antenna gain and narrower horizontal beam to improve bearing accuracy; and also to increase the field strength close to the sea to improve performance against surface targets. It also demanded maximum transmitter power. Pulselengths of about 15 and 2 microseconds were chosen, the first to give good detection range, and the second to allow reasonable range discrimination. This would enable an estimate to be made of the number of aircraft in a group, and hopefully allow individual information to be obtained on ships operating in a group.

High power at a frequency of about twice that of Type 79Z was available by this time because of the introduction of much shorter sealed leads on silica valves (see Monograph 3). So 90 MHz was chosen, not only so that a higher-gain antenna could be designed with about the same topweight and windage as that of Type 79Z, but also a design that had beam-switching facilities. This required symmetry about the vertical centre-line of the array. Beam-switching was introduced to provide much improved bearing accuracy (see later). At the same time it was considered that this frequency required components and techniques that were within existing technology, an essential prerequisite in view of the very short development time required.

It seemed to be assumed from the start that the same mains frequency and prf, namely 50 c/s should be used, and the maximum possible pulse transmitter power should be aimed at. This was presumably because the basic radar equation had by this time been derived, and it was realised that detection range varied as the fourth root of the pulse power (eighth root for targets detected well off the peak of the vertical lobe, for example low-flying aircraft and surface targets). But the importance of post-detector integration and the role of mean radiated power rather than peak pulse power was still not understood. Thus concentration focused entirely on maximum pulse power.

A specification for a new high-power 90 MHz transmitting valve was received by the Signal School Valve Section from the RDF Transmitter Section at Eastney during August 1939. Its main features were:

- Peak emission of 100 amps minimum;
- conductance to be such that with $V_a = V_g = 1500$ v, the anode current to be at least 25 amps;
- amplification to be about 10;
- anode-grid capacity not to exceed 10pF;
- anode-to-grid to withstand 60kV; and
- all connections to electrodes to be as non-inductive and non-resistive as possible.

Samples were quickly produced by the Valve Section (see Monograph 3), and transmitter experiments started in December 1939. Experiments started at the same time on the modulator, antenna and turning gear, receiver and displays.

The transmitter valves, designated NT86, used a larger silica envelope than the NT57T, with larger-diameter grid and anode to provide room for the three-hairpin thoriated-tungsten filament. A pair was installed in a push - pull circuit, using Lecher-line circuits in the anode, grid and cathodes. The anodes used a quarter-wavelength, open-circuit line, provided with HT via chokes, of such a length that the voltage node was inside the valve. The grids, earthed at DC, used a short-circuited half-wavelength line, the voltage node also being inside the valve. The cathodes used a short-circuited line to optimise the phase between cathode current and anode voltage. Power was tapped off the grid lines to feed the antenna. Filament power was provided by making the cathode Lecher-line of 1-in copper tubes threaded with a heavy-duty conductor pairs to feed the two filaments of the transmitter valves.

As a result of the success of the Type 79Z cathode modulator, it was decided to use a similar system for Type 281. However, because of the much higher peak current demanded by the NT86 transmitter valves, and the fact that the modulator valves had to pass this current at an anode voltage dictated by the NT86 mutual characteristics of only about 3kV, special high conductance valves were required. These were developed by the Valve Section in the form of silica-envelope triodes designated NT78 (Monograph 3). To obtain the high conductance, each valve was about twice the length of the transmitting valve. Four were used in parallel.

These valves in parallel were connected directly between the transmitter cathode circuit and the EHT negative. Their grids were normally biassed beyond cut-off, and they were driven as follows. A pulse generator consisting of two pentodes generated a steep leading-edge pulse by squaring and differentiating the 50-cycle mains waveform. The positive pulse was applied to the grid of a normally cut-off Thyratron mercury vapour-filled valve. The Thyratron anode was supplied from the 50 c/s mains, so that the gas current was deionised and quenched during the negative half cycle. The anode also contained two selectable pulse-forming networks, which, when the Thyratron was fired at the peak of its anode voltage, produced a suitable pulse to modulate the grids of the series modulator valve. One pulse-forming network contained a single L and C and produced a 1.7 microsecond pulse, the other was a three element network producing a 15 microsecond pulse. These were the first pulse-forming networks used in an operational radar.

The combination of this modulator and the transmitter produced a pulsepower of 350 kW at an EHT of about 15 kV during the 15 microsecond pulse, and in excess of 1 MW at an EHT of about 28 kV during the 1.7 microsecond pulse. This was the first time a 1 MW pulse power had ever been achieved. The long-pulse power was limited by the anode dissipation of the NT86s, while the short-pulse power was limited by the peak emission available from the NT86s. Another limitation was found empirically. The same twin open-wire transmission line was used to feed transmitter power to the antenna as was used in Type 79Z. This could just carry the RF voltage generated by the 350 kW without flashing over, but sparked over at only a slightly higher power with a long pulse.

However it successfully withstood the 1 MW at 1.7 microsecond duration because it provided insufficient time for the arc to build up. This was not predicted, and was a bit of luck for the engineering of the equipment.

As for Type 79Z, an absorption wavemeter and a small CRT for waveform monitoring were included in the transmitter panels. Figure 4.6 is a drawing of the Type 281 equipment in the transmitter office.

Separate transmitter and receiver arrays were used, as in Type 79Z. The basic array consisted of two scaled-down Type 79 arrays mounted side by side, at a spacing between centres of three quarters of a wavelength. Thus there were four driven dipoles, fed in parallel, and four passive dipoles. The combined (transmitter/receiver) maximum gain was some 23 db, an important 9 db improvement over Type 79. Power was fed to (or from) the array by a twin-wire open feeder identical with that used in Type 79. A diagram of the transmitter array is shown in Figure 4.7.

The receiver array had one difference from the transmitter. The twin-line feed to the array divided close to it, so the two leads were connected off-centre but symmetrically about the centre. A motor-driven switch connected alternately to one or other of these feeds at 50 c/s, thus switching the polar diagram to one side or the other of the dead-ahead position. The scheme is shown by the diagram in Figure 4.8. The dead-ahead crossover of the two switched beams was about 6 db down from the maximum. The effect of switching on the receiver polar diagram is illustrated in Figure 4.9.

The reason for the feature was to increase the accuracy of reading an azimuth bearing significantly. Two signals were displayed on the A-scan, one from each switch position. When the antenna was on target, these signals would be of equal amplitude. As the antenna was trained off, one signal would increase, the other decrease. Because of working on the fairly steep edge of its polar diagram instead of the flat maximum, a much improved bearing accuracy of about 0.5° could be achieved, despite a combined beamwidth of some 25°.

Although this feature was excellent for obtaining an accurate bearing on an echo of reasonable amplitude, it entailed a loss of some 6 db in performance on the centre-line of the antennae, and therefore significantly affected detection ranges. This exacerbated an inherent weaknesses of Type 281, as follows. Although the higher frequency gave better low cover than Type 79Z, as intended, it also lowered the tip of the first lobe in the vertical coverage diagram. Even without beam-switching, cover was only up to about 28000-ft, compared with over 40000-ft for Type 79Z (Figure 4.5). With the performance loss from beam-switching this cover was further reduced. It was found that enemy aircraft flew over the first lobe, their detection range by the second lobe being roughly halved. Initially there was no way of stopping beam-switching, because

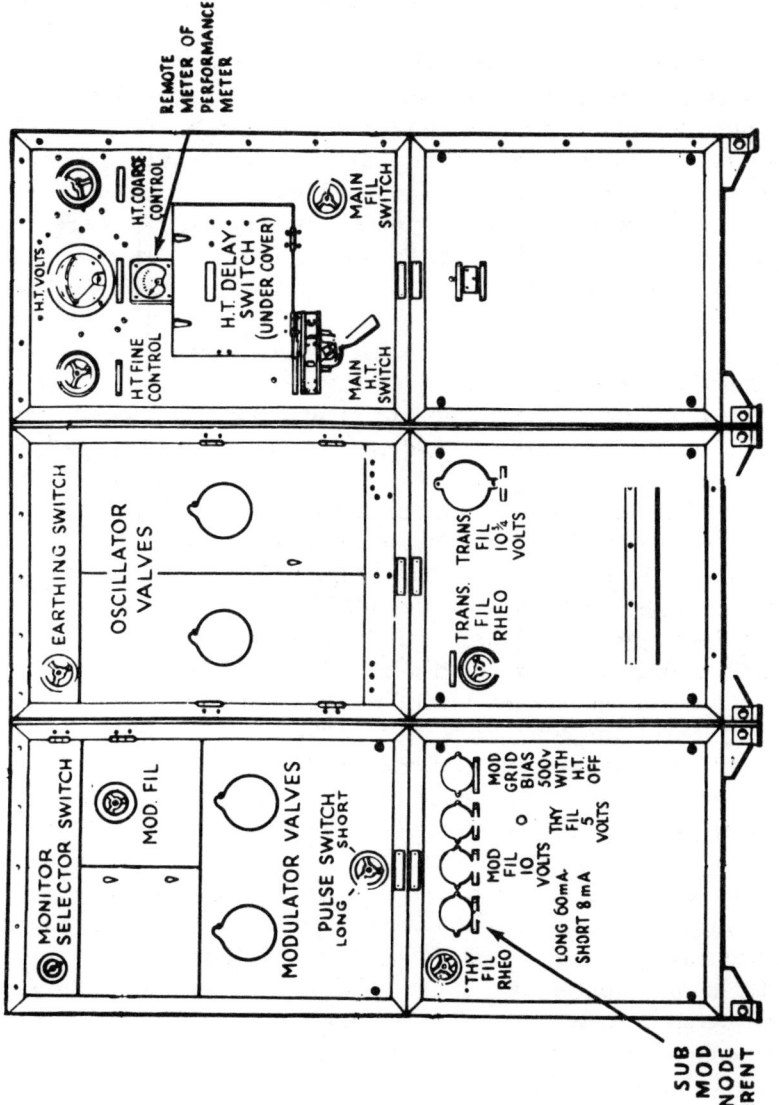

4.6 Type 281 equipment in the transmitter office (© Defence Research Agency)

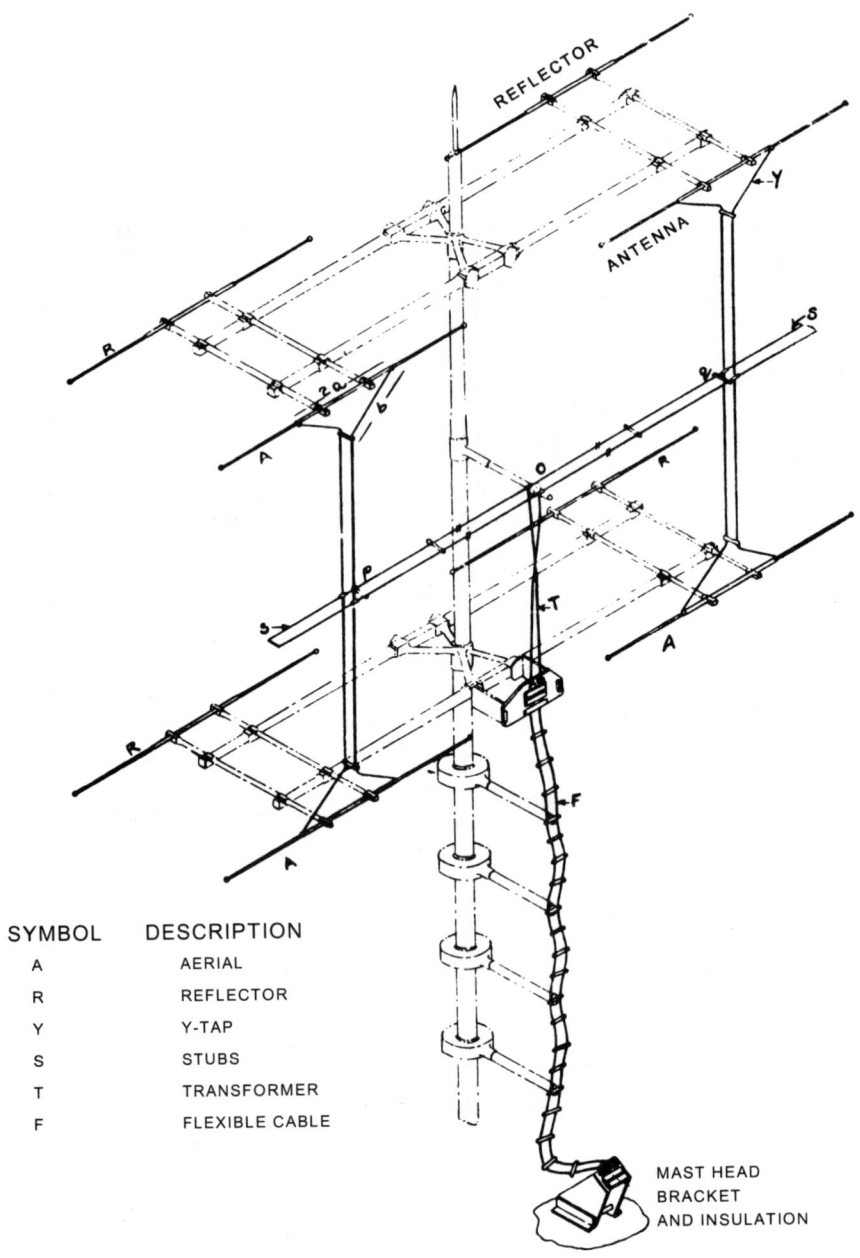

4.7 Type 281 transmitter array (scale: dipole approx. 5.5-ft long) (© Defence Research Agency)

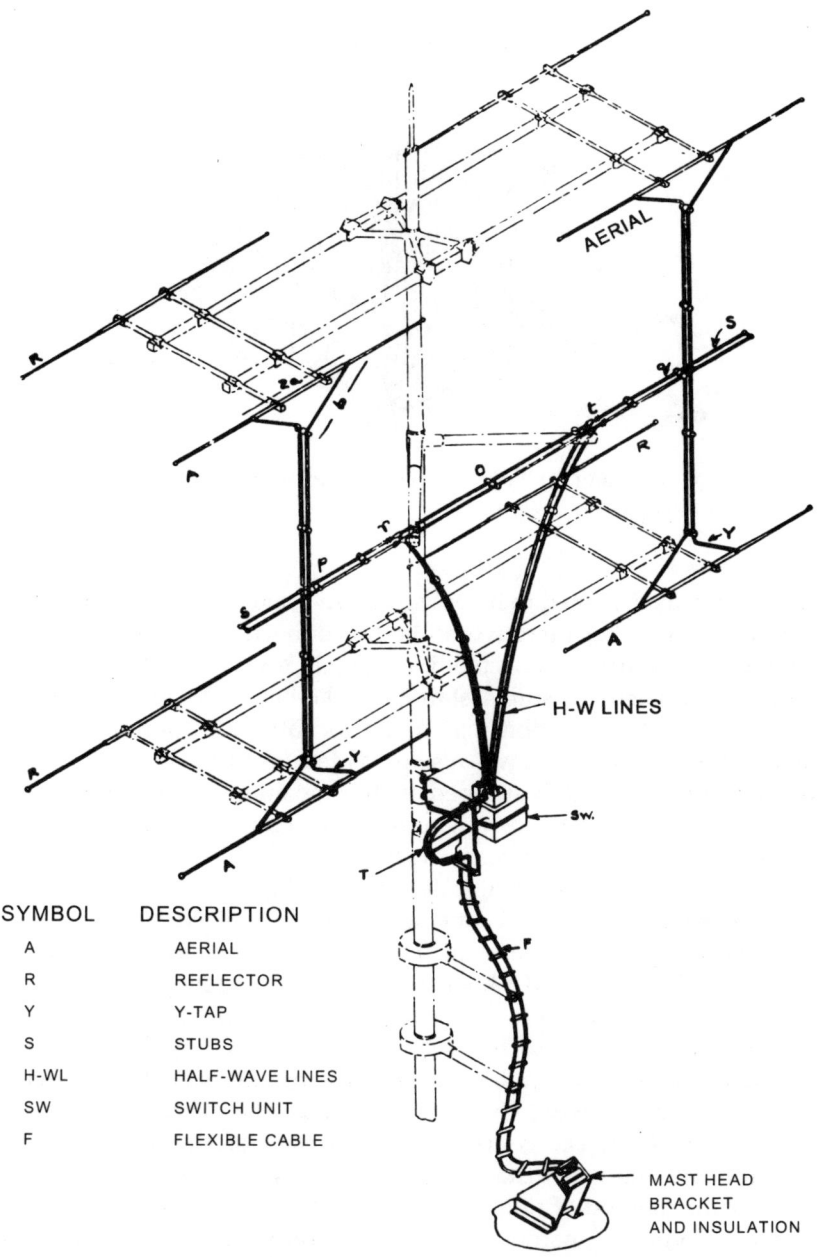

4.8 Type 281 receiver array (scale: dipole approx. 5.5-ft long) (© Defence Research Agency)

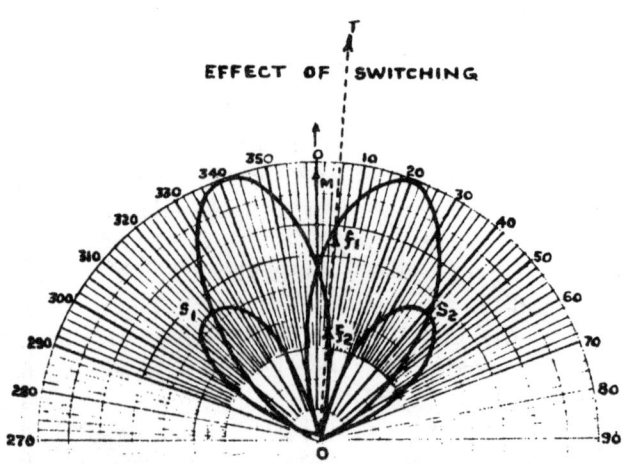

4.9 Effect of switching on Type 281 receiver array polar diagram (© Defence Research Agency)

the switch was designed to be break-before-make. Thus if the drive motor were stopped, either there would be no connection to the antenna at all, or the beam would be deflected. Fortunately it was found that the switch could be adjusted by ship's staff to make-before-break, so that at the centre of travel connection was made to both sides simultaneously. A small spring arrangement was also provided for fitting by ship's staff that ensured that the switch stopped in this position when the motor was switched off. The operator then had the choice of either maximum detection performance or beam-switching.

This problem of long-range detection of high-flying aircraft was the main weakness of Type 281, and will be further discussion later.

As with Type 79Z a band of frequencies was allocated to Type 281, namely 86, 88, 90, 92 and 94 MHz. One such frequency was allocated to each ship. The reason was to reduce mutual interference between ships and provide a spread of frequencies to make enemy jamming a little harder. While the transmitter and receiver tuning was able to cover this band, the antennae were narrow band. Thus five different designs were produced, of dimensions appropriate to each of the spot frequencies. Once the antenna was installed, the frequency for that ship was fixed.

The same method of rotating the antennae from the receiver office was used as in Type 79Z, namely Selsyn torque transmitters. However, to ease the operator's task, a motor was provided to drive them, with a mechanism for reversing at the astern bearing. Continuous rotation was not possible at that time because of the antenna feeder cable. (This facility was not introduced until 1945.)

The receiver was rack-mounted and had a separate rack-mounted power supply. The valves used for all stages were EF50 RF general-purpose pentodes. These stages consisted of an RF input stage, two RF amplification stages, and a local osciallator stage, all separately manually tuned. All four tuning controls were mounted on the front panel. These stages were followed by a mixer, and an IF stage that could be switched in or out by a switch on the front panel. This IF stage had a narrow bandwidth of 150 kHz for use with the long transmitter pulse. There followed the main IF amplifier, centred on a frequency of 12 MHz, consisting of three coupled stages with a bandwidth of 1.5 MHz for use (with the narrow bandwidth IF stage switched out) with the short transmitter pulse. Finally there was a pentode second-detector stage and an output cathode-follower. A gain control was mounted on the front panel.

An additional provision was a groundwave suppression circuit to prevent paralysis of the receiver by the very powerful groundwave. Echoes could then be followed in to short range, even when using the short pulse. An EF50 mounted close to the local oscillator took the positive DC pulse generated by the groundwave from the cathode circuit of the second RF stage and inverted it at its anode. The resulting negative DC pulse was applied to the local oscillator grid, thus making the IF circuits inoperative during the groundwave.

The display equipment in the receiver office was a repackaging of the Type 279 equipment, providing a warning A-scan, a precision ranging A-scan and a precision bearing A-scan.

The warning A-scan circuit was the same as the redesigned, triggered version used in later Types 79Z and 279. It used the same gas-focused CRT, and the circuit already described. The precision ranging and bearing displays used hard CRTs. These were repackaged versions, taking up less volume, of the Army GL1 equipment described for Type 279. The same precision wire-wound potentiometer was used for accurate ranging. A small modification on the precision bearing tube used a voltage derived from an extra contact on the receiving antenna beam switch to apply a small horizontal shift to the A-scan between the two positions of the switch. The operator could adjust this so that the echoes from the switched beams appeared side by side, so simplifying pointing the antennas to match the two amplitudes. A diagram of the equipment in Type 281 receiver offices is provided as Figure 4.10.

As with Type 79Z, interference from sea reflection produced a series of lobes in the antenna vertical-plane radiation pattern. However, because of the higher frequency the first lobe was a little lower than in Type 79Z, and there were more lobes (Figure 4.5). Resulting from experience with Type 79Z, it was always planned that height estimation of the target would be derived from range of first detection, followed up by change of echo amplitude with closing range.

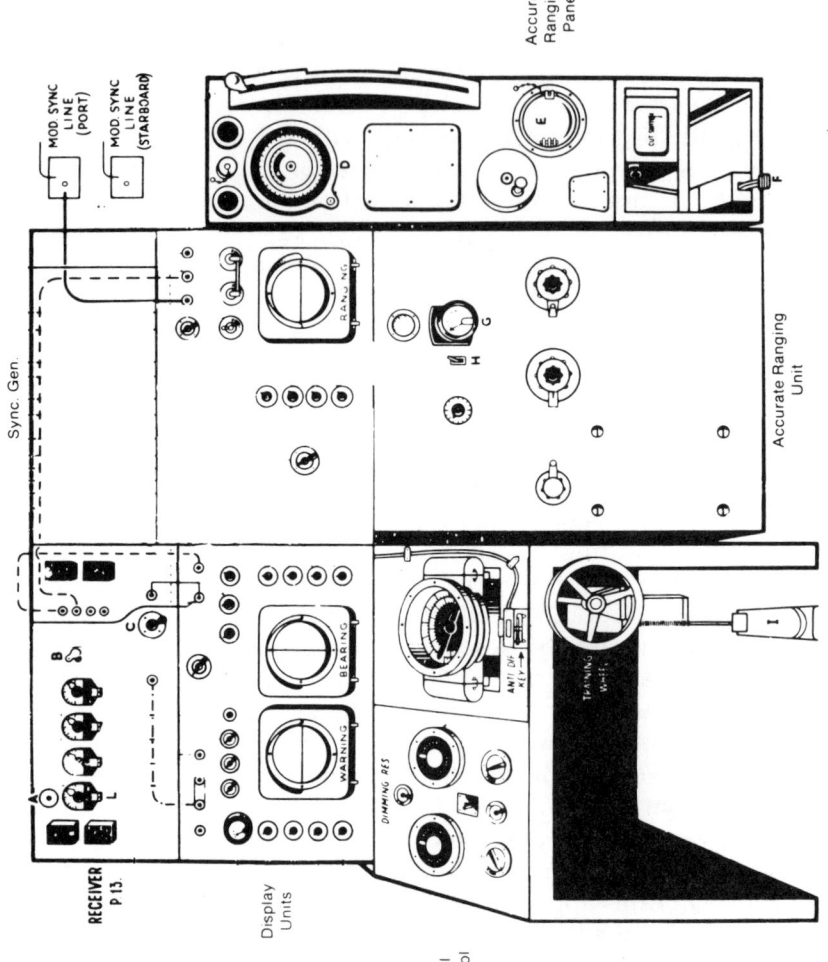

4.10 Equipment in a Type 281 Receiver Office (© Defence Research Agency)

However, as with the previous set, rather unreliable results were obtained operationally due to changes in equipment performance and variation between targets. Although some operators made valiant attempts to use the method, it is uncertain how operationally useful it was. The results were obtained from carriers, where continuous practice was available from the the use of their own aircraft as targets.

A laboratory model of all this equipment was developed at Eastney (apart from the antennae and turning equipment). As for Type 79Z, an engineered model was produced in the Experimental Workshops for trials in July 1940, some seven months after laboratory work first began (see Postscript). The trials were successful. Signal School workshops quickly produced two more equipments for operational use, one fitted in *Dido* in October 1940 and the second in *Prince of Wales* in January 1941. Aircraft-detection ranges of 60 miles at 5000-ft height and 110 miles at 20000-ft were achieved, together with detection of large surface ships at 22000-yds. This performance, together with that shown in Figure 4.5, is supported by records of trials of Type 281 in *Illustrious* and *Formidable* conducted in 1942.

In parallel with producing these two models, the Signal School Drawing Office made production drawings. Production contracts were placed with industry, where reaction was so fast that delivery began in February 1941. From this time until the end of the war about 80 ships were fitted with various versions of Type 281.

It is worth noting that the first workshop model, installed at Eastney in July 1940, was almost immediately included as part of the RAF Chain Warning System for a period. With the fall of France in June 1940, enemy aircraft were crossing the Channel much further west than envisaged in the design of the RAF Chain Home system. Although additional stations were being built they were still not operational, and much of the south coast was not covered by RDF. For several weeks the new Type 281 was manned round the clock by the Eastney staff, target information being passed by direct-line telephone to the Fighter Command plotting station. This continued until the new Chain Home stations became operational.

TYPE 281B

As for Type 79Z, pressure quickly mounted for single antenna working, so freeing one masthead. Thus a T/R switch was developed for Type 281, the first being put into service late in 1941.

It will be remembered that Type 79Z/279 generated a requirement to introduce single-antenna working in order to occupy only one masthead, and that this feature began to enter service in early 1941. Although Type

281 was also entering service at about this time, the Type 79 diode-switch development had come so late in the Type 281 programme that, in order to guarantee a dependable operational equipment, development was completed with separate transmitting and receiving antennae. Nevertheless when the Type 279 diode switch was found to operate so well, pressure quickly arose to introduce an equivalent into Type 281 and work was started at Nutbourne.[6] Virtually a scaled-down version of the Type 79 switch was developed using Pyrotenax cable, but with some engineering differences. These added more complex tuning arrangements to give a good match to the long cable up the mast and to handle the much higher transmitter power, whilst still protecting the receiver. In early models this protection was provided by having two switches in series. One was a tungsten-electrode spark gap triggered by the transmitter, which, with its circuit, gave some attenuation of power into the receiver. The other was a diode assembly using the same six diodes as Type 79. This, with its circuit, provided the major attenuation of power into the receiver. The spark gap wore with use and needed daily adjustment. To overcome this disadvantage the two-stage switch was replaced in later models by a single, high-current diode, CV 94, specially designed and manufactured by GEC. Single-antenna working was introduced into service late in 1941. From then on both metric search RDFs, Types 79B/279B and 281B, each required only a single mast.

The array used for the single antenna was similar to that for the Type 281 transmitter, that is, without beam-switching, because the beam-switch was incapable of handling the transmitter power. Thus this feature was never available in Type 281B.

TYPE 281BQ

PPI displays were introduced in quantity into ships from 1943 onwards, but the Type 281 antenna could only be rotated through 360° and back again. Continuous rotation was desirable to give a better PPI picture, and a constant time between successive illuminations of a target. Experimental work on a Type 281 antenna pedestal with RF sliprings started in 1943, but production models only became available for ship-fitting in 1945 and saw little, if any, war service.

During 1942 the Plan Position Indicator (PPI) type of display was invented and developed at TRE to aid air warning and fighter interception. Its application to Naval air warning, fighter interception, surface warning and surface plotting immediately became obvious. During the second half of 1942 a few TRE models were installed in ships. By mid-1943 a Naval version, developed and produced by EMI, became available in quantity. From an initial production order for 50, eventually over 5000 were produced and used on all RDF sets where warning or plotting was required. The scan was produced by mechanically rotating a deflection coil in synchronism with the antenna rotation, the display using a 9-in diameter long-afterglow CRT, and being North stabilised.

For best presentation the PPI required that the RDF antenna rotate continuously. Thus during the second half of 1942 the need arose to provide continuous rotation on the Type 281 antenna. This was done by designing RF sliprings, housed in the antenna pedestal. The cable from the antenna array and down the mast was of Pyrotenax, and so the sliprings had to deal with an unbalanced signal. The 'live' slipring was made of silver, with two silver-carbon brushes in contact across a diameter, the connection being made by equal lengths of wire to each. The 'earth' ring was made of phosphor bronze, with carbon brushes as the contact. In addition there was a nine-turn matching coil connected from the 'live' brush connector to earth. The whole slipring unit introduced a standing wave ratio of under 1.1 over the frequency range 80–95 MHz, and a variation with rotation of less than 0.01. The complete slipring unit also contained sliprings for a VHF antenna and the Type 941 IFF antenna, both of which were mounted on and above the Type 281 array. A diagram of the sliprings is provided as Figure 4.11.

An experimental model of this new pedestal, together with a modified control table for the receiving office, which provided continuous rotation at 2 or 4 rpm, plus reversion to manually controlled antenna pointing, started tests in June 1943 at Nutbourne. However production models did not become available for retrospective ship-fitting until mid-1945. Thus, during the whole time from 1943 to 1945, Type 281 PPIs were driven from an antenna that rotated automatically but reversed at the astern bearing.

In passing, it is worth noting that continuous rotation was never introduced to Type 279. This was mainly because its very broad azimuth beamwidth of about 60°, together with its long pulse length (minimum 8 microseconds), would have made its PPI display very cluttered and difficult to interpret if there were a number of targets present simultaneously.

By this time, 1943, the 600 MHz gunnery sets were providing accurate angular and range information for gunnery control, and 10-cm sets were providing surface warning and tracking and target-indication information for laying on the gunnery sets.

Thus the metric warning sets were needed in the single, though important role of air warning and fighter direction, a role in which the PPI and Skiatron plan displays were vital.[7] For this purpose a series of other modifications were introduced to Type 281BQ, as follows.

The normal long pulse, 15 microseconds, was reduced to 5 microseconds to improve discrimination during fighter direction without too seriously affecting detection range. An associated modification to the receiver bandwidth was introduced to handle the shorter pulse. To

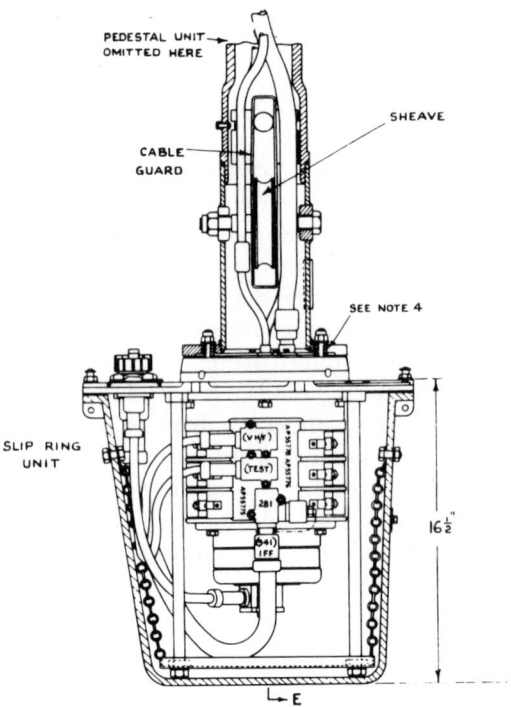

4.11 Type 281BQ RF sliprings (© Defence Research Agency)

compensate for the reduction in detection performance, a modified trigger unit was introduced to double the prf by firing the transmitter during each half cycle of the mains, that is, 100 c/s prf. In fact either 50 or 100 c/s prf could be selected.

A test set, similar to that described for Type 279, was introduced so that the overall performance of the set could be checked and kept up to standard. By this means more consistent height estimations became possible.

A special display was introduced for echo amplitude of a selected target to be plotted against range so that the target's height could be estimated without stopping the antenna rotating, and thus destroying the PPI display. The display was inappropriately called a Sector Display, although it did not display a sector of the PPI, but rather an A-scan from the sector. The display had a long-afterglow CRT. The PPI operator could select a bearing on a target of interest, and when the RDF antenna passed through this bearing, the A-scan on the Sector Display fired and displayed the echo amplitude of the target and its range. The Sector-Display operator could plot amplitude versus range and derive target height from his calibration curves.

This same display could be used to read the personal IFF code from Flight Leaders' IFF sets, a code that took several seconds to transmit. A separate IFF antenna (Type 243) was locked to, and trained with the bearing selector on the RDF's PPI. The IFF signal was displayed permanently downwards on the Sector Display A-scan, except when the RDF antenna passed through the selected bearing, when the RDF echo flashed up normally. Thus range correlation between echo and IFF signal could be established, and the IFF code read from the A-scan (see Monograph 6).

Another important addition was a preamplifier for the receiver, in its own add-on box. This consisted of a grounded-grid triode RF stage, which had a lower noise-factor than the RF pentode. It will be remembered that such a scheme was tried with Type 279, but was defeated by galactic noise. Although galactic noise is still present at 90 MHz, its significance is less than at 45 MHz (see Monograph 2). Thus although the predicted 5 db improvement was not achieved, about 3 db was obtained. This was very worthwhile, particularly as far as the vulnerable tip of the first antenna lobe was concerned. Detection at the tip of this lobe was improved from some 28000-ft to about 32000-ft, although it was still significantly less than the tip of the first lobe of Type 279, at some 45000-ft. Detection ranges at lower heights were also improved by a useful 20 per cent. This performance is confirmed by a record of the results of trials carried out on Type 281BQ installed at the test site at Tantallon in autumn 1945.

Other features introduced were swept-gain, to reduce the effects of sea clutter on the PPI, and a range of anti-jamming filters similar to those described for Type 279.

Figure 4.12 is a photograph of equipment fitted in a Type 281BQ receiver office, including some of the Types 243 and 941 IFF equipment associated with Type 281 display.

Thus Type 281BQ was the final version of Type 281 to be produced, containing all the technology for air-warning and fighter-direction RDF known at that time.

DUAL INSTALLATIONS OF TYPES 79B AND 281BQ

Both Types 79 and 281 antennae had pronounced vertical-lobe structures owing to sea reflection, so that air targets were lost in the nulls. Owing to the different operating frequencies, the lobe nulls of one set coincided with the lobe peaks of the other. Thus virtually gapless cover, badly needed for efficient aircraft direction, could be provided by installing both sets. Since each set required only a single mast, such installations were made in ships concentrating on aircraft direction, such as aircraft carriers.

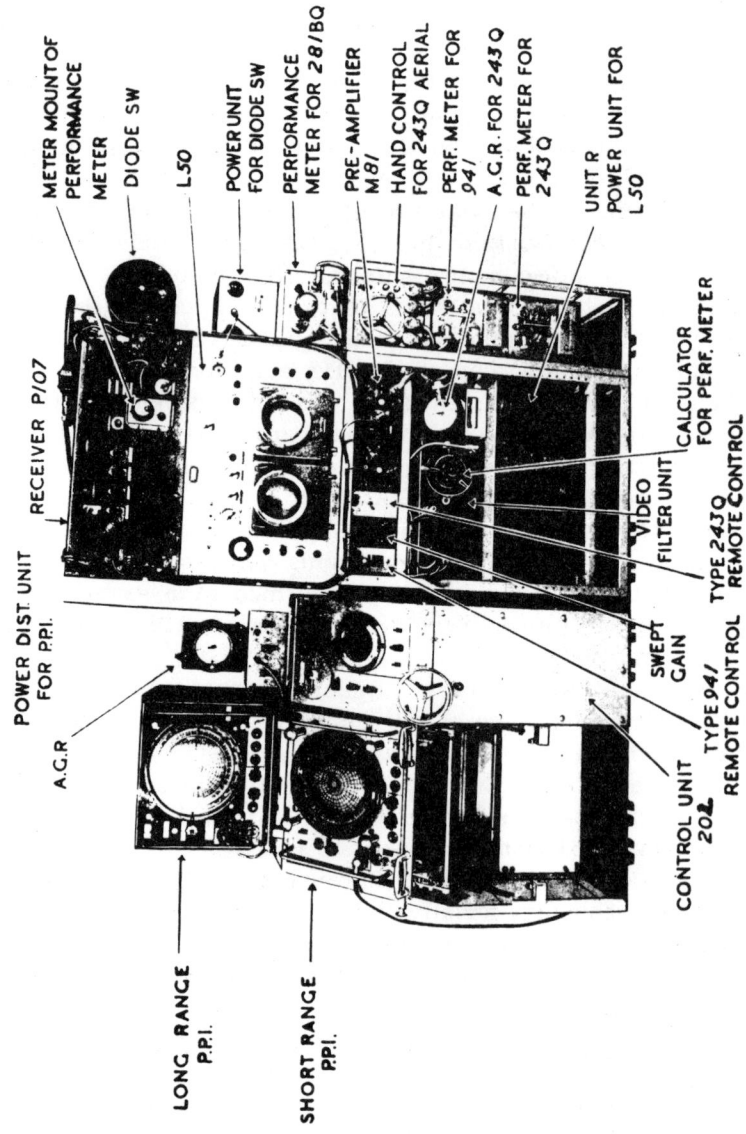

4.12 Equipment in Type 281BQ Receiver Office (© Defence Research Agency)

Although, by the standards of the time, Type 281BQ had good characteristics for aircraft detection and direction, namely long-range detection, good range discrimination and a usable azimuth paint on the PPI (some 30° echo arc), it had two weaknesses. One was for detection and the other for tracking during fighter direction. The detection weakness was the already mentioned limited altitude coverage (about 30000-ft) of the first detection lobe. Enemy aircraft could and did overfly this, reducing an expected first-detection range of 130 miles to some 70 miles, which was not enough for dependable fighter interception. The tracking weakness was the inevitable presence of the vertical lobe pattern due to interference from sea reflection. This caused echoes to disappear in the minima for 10 to 30 miles, depending on target height, often at a range where information was most wanted to complete a fighter interception.

Now the Type 79 vertical lobe structure was different from Type 281, because of the different frequency, and had two attractive features. Firstly its first lobe covered heights up to 45000-ft at ranges to well over 100 miles. Secondly, the peaks of its lobes coincided with the nulls of the Type 281 lobes (see Figure 4.5). Its main disadvantage was its very broad azimuth beamwidth (some 60°).

It followed that in ships where great importance was attached to fighter direction, such as aircraft carriers and fighter-direction ships, there would be important operational advantages from fitting both sets in a single ship. Since each only needed to occupy a single mast this was possible, and was in fact done; because of the concentration on fighter direction, the accurate ranging panel of Type 279 was unnecessary and so Type 79B rather than Type 279B was coupled with Type 281BQ.[8]

To fit both sets called for only one further modification, as follows. To eliminate cross-interference between the two RDF sets and to minimise interference with other radio equipment, it was necessary for both transmitters to fire together. This was achieved by introducing a new control panel that allowed the two 50 c/s alternators, each supplying one of the sets, to be paralleled, and for the load to be equally shared. Since the trigger units for both sets were locked to the mains, it was possible to phase them so that they coincided. It also meant that the Radio-Interference-Suppression (RIS) equipment, which suppressed the ship's radio receivers during the RDF transmission, only had to handle a single 50 c/s mains supply.

This was the last word in air-detection and fighter direction equipment, together with a 10-cm set for very low-flying-aircraft detection, from 1943 until the end of the war. A quite complex collection of remote displays was set up in the Radar Display Room (RDR), from which information on aircraft position and an estimation of height was passed on to the large manual plot in the Aircraft Direction Room (ADR).

Here there were also two Skiatrons (large, bright-screen, plan-position displays) showing raw radar information.

Part of an early RDR is shown in Figure 4.13. Equipment from different RDF sets is shown grouped together in order to achieve better coordination between operators. On the extreme left is a small portion of a Type 79 receiver rack, including display and antenna-control unit, next to which is a Sector-Display rack, on which the signal from Type 243 IFF equipment is coordinated with a particular target selected from the continuously rotating Type 281BQ equipment (shown in the centre of the picture and detailed in Figure 4.12). On the extreme right is the side of the rack holding Types 243 and 941 IFF interrogators and responders. The whole can be seen to be a collection of individual items, produced at different times as required and 'cobbled' together as a system, with a multitude of interconnecting leads. Although it may look rather amateurish, it worked surprisingly well in practice, and operators learnt to use it very efficiently.

This was the end of the development of metric warning RDF for capital ships that saw service during the war. Type 79/279 was conceived as,

4.13 Early Radar Display Room (© Defence Research Agency)

and remained, an air-warning equipment. Type 281 started as a general-purpose air and surface warning equipment, together with accurate ranging and bearing facilities for gunnery, particularly surface gunnery. However with the advent of specialist gunnery RDF sets and of centimetric warning RDF, which took over surface warning, Type 281 in the later war years reverted to a role of air warning, and very particularly to a role of fighter interception.

TYPE 286

Because of invasion fears after Dunkirk in June 1940, an immediate need arose to fit RDF in small ships, from destroyers to MTBs. There was no time to develop a Naval set from scratch, but an RAF set used for the detection of surface vessels from aircraft (ASV), had acceptable characteristics, and was just available in production. An antenna suitable for Naval service was very quickly produced, and quantity introduction into service started in mid-1940. In due course improved antennae became available and many hundreds of the various equipments were installed until well into 1942.

So far all the equipment dealt with has been aimed at capital ships since, up to mid-1940, this was where all thoughts on warning RDF in Signal School were centred. However, after the evacuation of Dunkirk in June 1940 there was a fear of an imminent German invasion. The Navy produced an urgent requirement for RDF in small craft to give some surface and air warning against such an event. In June 1940 Horton held a meeting with Nutbourne staff, and it was decided that the only realistic solution was to use as much equipment already in production as possible. Attention focused on an RAF Air-to-Surface Vessel RDF (ASV Mark 1), which was just in quantity production and was of reasonably small size.[9] It operated on a frequency of 214 MHz, ideal in principle to operate with a small antenna. The transmitter produced a pulse power of about 7 kW at a free-running prf of about 1200 c/s, and a pulse length of 1.5 microseconds. For Naval use, the equipment was mounted in a simple frame, and power was obtained from a motor generator with an output of 80 volts, 2000 c/s, as required by the airborne equipment.

The airborne antenna system consisted of a transmitter dipole on the nose of the aircraft, with a receiver dipole on each wing, angled outward at an angle of 20°. The receiver was switched between these two dipoles, the display being an A-scan vertically up the CRT screen with the echoes from the two dipoles being presented back to back. Thus when the aircraft was flying directly towards a target, the echo appeared as a 'butterfly' with equal length wings. If off target, the wings would be of different lengths.

Because of the urgency it was decided to adapt this type of antenna for ship installation, but to replace each of the three dipoles with scaled-down version of the Type 79 array, that is two driven dipoles each with a passive reflector, in order to increase gain. The first model of the array was mounted on a wooden framework made in the carpenter's shop at Nutbourne. From the first meeting between Horton and Nutbourne staff on a Friday, an equipment was fitted in a destroyer (*Ambuscade*) in Portsmouth dockyard in 10 days! Brief trials were performed, which showed worthwhile performance could be obtained.

A major problem was to provide a well-matched very long (compared with an aircraft installation) feeder from the RDF office to the antenna at masthead. A breakthrough was the choice of Pyrotenax concentric cable (its first use as an RF cable), predating its introduction into Type 79B. Operational pressure led to over 250 of these antennae being installed, despite their drawbacks. Presumably because of production and supply problems, they continued to be fitted even after a far superior rotating antenna had been developed.

The fixed antenna had several drawbacks. Firstly the need to point the ship at the target imposed tactical limitations. Secondly, side lobes and back lobes caused serious confusion in company or in convoy, and thirdly, the weight of the antenna array was such that, in high winds, the mast could break! So the pressure was on to find a lighter-weight array capable of being rotated.

Again Nutbourne found a rapid solution. They quickly produced a scaled-down version of the 90 MHz Type 281 array to 214 MHz, that is, four driven dipoles each with a passive reflector. These were fed by twin-core Pyrotenax cables threaded through steel tubes, which formed the dipole supports. The junction of these four cables was itself joined to a balance-to-unbalance stub to connect with the concentric Pyrotenax cable taken down the mast. This complex junction was housed in a bakelite moulding. The pressure exerted during the moulding process distorted some of the joints, causing shorts. The X-ray machine of a Portsmouth dentist was called into service to check the early arrays before issuing them for service!

In order to allow the antenna to rotate while using Pyrotenax feeder cable, sliprings had to be provided. A quarter-wave capacitance type with a matching stub was found to work well at this high frequency of 214 MHz.

There was no time to design an antenna pedestal, *ab initio*. Fortunately an Asdic pedestal was found that was available in production and would do the job. It could be mounted upside down (compared with its Asdic role) at the top of the mast and the Bowden cable (speedometer-type cable) mechanical drive run down the mast into the RDF office, from where the antenna was turned by hand.

Finally a T/R switch was required. Experience was being acquired at Nutbourne at this time from the initial experiments on the Type 279 and Type 281 T/R switches. A version of the Type 281 design, but using two gas switches instead of a spark gap and a diode, was produced.

Trials were performed with the above antenna in *Legion* in February 1941 and proved successful. However it is not certain how many were installed operationally. Some were used with the later Type 286s, and possibly with the short-lived Type 290 (see later). The reason for this uncertainty is that the Asdic pedestal was fairly heavy – made of steel or cast iron – and the emphasis was on minimum weight as smaller and smaller ships were required to be fitted.

Thus a new design was produced, made of cast aluminium and with a simpler gear assembly. It was still driven by the mechanical Bowden cable, which proved reliable and provided a lightweight drive. The weight of the antenna array was also reduced by persuading Pyrotenax to produce cable with a steel instead of a copper sheath. This enabled the Pyrotenax cable feeding the dipoles to be used as a physical support, as well as an electrical connection (Figure 4.14). This design, fitted with the sliprings, became available about mid-1941, and while it may have missed Types 286 and 290, it was in perfect timing for Type 291 (see later).

Type 286 with the fixed antenna was fitted in Naval vessels between June 1940 and mid-1941. This was just the period of changeover between ASV Mark I and ASV Mark II. About 200 Mark Is were produced, mostly for the RAF, during the first half of 1940. Although they provided a useful service they were unreliable, and complete re-engineering was started as

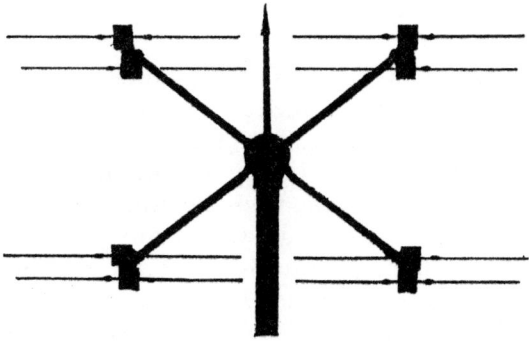

4.14 214 MHz array for later Type 286s and Types 290 and 291 (scale: dipole approx. 2-in) (© Defence Research Agency)

early as February 1940. Production models of the improved Mark II, of which eventually many hundreds were produced, started to appear in December 1940. Thus whereas the first Type 286s used ASV Mark I as described, the later ones used the more reliable Mark II. The main difference in performance was that Mark II had a pulse power of 7 kW (as did Mark I) but at a pulse length of 2.5 microseconds and a free-running prf of about 400 c/s.

Later a Naval modulator and transmitter giving 100 kW pulse power was added to Type 286P, and was designated Type 286PQ. Technical information is scant, but it is probable that the modulator/transmitter from Type 291 was used. Type 286PQ was installed only in the larger of the small ships, for example destroyers.

The designations of the various combinations described above were:

- Type 286 – ASV Mark I with fixed antenna. Mid to end of 1940.
- Type 286M – ASV Mark II with fixed antenna. January to mid-1941.
- Type 286P – ASV Mark II with rotating antenna. February 1941 to early 1942.
- Type 286PQ – ASV Mark II with higher power and rotating antenna. Late 1941 into 1942.

TYPE 290

Soon after the crash-action on Type 286 started, design began on an equipment of higher transmitter power, made to Naval requirements. Unfortunately its transmitter and modulator ran into technical problems, and production was limited. Ship-fitting started in May 1941.

At about the same time as the crash programme on Type 286 started at Nutbourne, work was started at Eastney on a more robust version, designed from scratch to meet full seagoing conditions. The aim was to produce 100 kW pulse power, a 1 microsecond pulse and a 500 cycle prf. The only suitable valves available were the 'micropup' metal-anode type used in the Type 286 transmitter, which are believed to have used thoriated-tungsten cathodes. A push–pull circuit was used with earthed (at RF) anodes, tuned-grid and tuned-cathode circuits, both of Lecher-line construction, with the antenna feed tapped off the cathode circuit. Anode modulation was used, the pulse of the transmitter anodes being derived from a delayline discharged by a thyratron, this pulse being stepped up by an air-cored pulse transformer. The thyratron was triggered by a pulse derived from the 500 cycle mains supply by distorting and differentiating it in a two-valve trigger unit. The thyratron anode was also supplied with raw mains AC, the trigger operating close to peak voltage. The later negative period of its anode voltage quenched the thyratron so that it was ready for the next trigger pulse.

The receiver (it is believed) was a re-engineered version of the Type 286 receiver, consisting of a low-noise, grounded-grid RF amplifier stage, a tuneable local oscillator and diode mixer, a four-stage IF amplifier using pentodes, a diode second detector and a pentode cathode-follower output stage. The IF frequency was 30 MHz and the bandwidth about 2 MHz.

The set never worked as planned due to problems in the transmitter and modulator. There was always difficulty in achieving the 100 kW pulse power, said to be due to limitation of cathode emission in the micropup valve. Also a high anode voltage was needed, requiring a large pulse from the modulator. Since an air-cored pulse transformer was used (suitable iron-cored pulse transformers were not available), only a small step-up could be achieved, so the delayline had to work at high voltage. The condensers used in the delayline would not stand the rapid rate of change of voltage on discharge at the start of the pulse (this was exacerbated by the elementary design of the line) and they had a very short life. Thus the transmitter always had to be run in a reduced (approximately half-power) mode.

In desperation some of these sets were fitted operationally, presumably in conjunction with the rotatable array described under Type 286 with the modified Asdic pedestal, the first fitting being in *Aurora* in May 1941. But fortunately improvements were already to hand, so the number of Type 290s was limited.

TYPE 291

Technical advances made for other RDF sets enabled a new transmitter and modulator to be designed quickly. The whole set was packaged into a remarkably small volume (for those days) and was very successful. Ship-fitting commenced late in 1942, and gradually Type 291 took over from Type 286. A version was also made for installation in submarines, with a special antenna that could be mounted outside the pressure hull.

While Type 290 was being finally engineered, technical improvements became available to solve the problem in both transmitter and modulator. Re-engineered micropup valves became available with indirectly heated oxide cathodes. This solved the emission limitation and made 100 kW a reality.

Even in the short timescale of the evolution of Type 290, the technology of delay-line and pulse-transformer design advanced significantly. This was largely due to the need for such techniques for the modulation of 10-cm magnetrons, work on which was progressing intensively at Eastney. Thus it became possible to design an iron-cored pulse transformer, using a thinly rolled Rho-metal iron strip, to feed the micropup valves. This allowed a higher step-up ratio, and thus a lower voltage on the pulse-forming line. The design of this line was also improved significantly. Thus a new modulator was designed that not only had a good life, but also extracted the required 100 kW from the new micropup valves.

The trigger unit, modulator and transmitter were engineered into a remarkably small volume by the standards of the time, and this together with the receiver units from Type 290 were packaged to form Type 291. The latter became available for operational use in 1942. It employed the rotating antenna using the newly designed aluminium pedestal, and other lightweight techniques, and provided a useful air and surface warning and tracking equipment for small ships, the surface role often being enhanced by the addition of a 10-cm set. A large number were fitted in many types of small craft, from destroyers down to MTBs.

A simple A-scan display was provided with both Type 290 and Type 291. The operator had to turn the antenna by hand, and detect and track targets on the A-scan. However in April 1943 performance tests were carried out with a Type 291 fitted in *Saltburn*. During these trials an early model of a PPI was introduced and very brief trials conducted with it. The trials report declared the results as very promising, a very useful PPI picture being obtained. The report concluded that 'With power-driven antennae, the PPI would be useful on Type 291 for aircraft observation. The watching efficiency of the operator would increase in the absence of manual training effort'. This recommendation would have been straightforward to implement (and very probably was, from late 1943, when Naval PPIs became available in quantity), although no specific evidence has been found to confirm it.

The performance results quoted in the report on the *Saltburn* trials using the A-scan for detection – illustrating the useful performance available against both air and surface targets – are given in Table 4.1.

From the start of its life in late 1942, Type 291 continued to be fitted in large numbers in small ships for the remainder of the war.

A version was also fitted in submarines for air warning, with an antenna specially designed to be mounted outside the pressure hull. Later a 3-cm (X band) transmitter was designed to be driven from the Type 291 modulator, with a 3-cm receiver front-end whose 200 MHz output fed into the Type 291 receiver. The 3-cm antenna was initially added to the Type 291 mast, and eventually was incorporated into the periscope. Thus, by switching from one RF equipment to the other, the

Table 4.1 Detection ranges on aircraft and ships in the *Saltburn* trials

Target	Height (ft)		Detection Range (yds)	
			Opening	Closing
Aircraft: Chesapeake	1000		29000	24000
	25000		59000	57000
	200	Run 1	15000	
		Run 2	15000	
		Run 3	17000	
	8500		56000	48000
	10000		62000	61000
Aircraft: Wellington	unknown		56000	
Three destroyers	Type 291	Run 1	12000	
	antenna	Run 2	12400	
	height 80-ft	Run 3	13000	
F-Class Corvette	ditto		9000	
Merchant ship	ditto		11000	

submarine could obtain a useful performance against either aircraft (214 MHz) or against surface craft (3-cm wavelength), with equipment capable of being installed in the very confined space available. The combined equipment was designated Type 267.

TYPE 960

Work started in industry in 1943 to develop a set to give a performance at least as good as a combined Type 79 and 281. The emphasis was on the engineering design to make it very 'user friendly', and this was achieved successfully. However the first sea going fitting was not until late 1945, and it saw no wartime service.

Although Type 960 did not become operational until after the war, it was developed over the last two years of the war and so it is appropriate to make brief mention of it here.

Type 960 was aimed to be the operational replacement in capital ships for both Type 79 and Type 281. These two sets had been designed at speed, and their engineering was dated. The object of Type 960 was to engineer the equipment as a complete set – rather than as a collection of individual units – to reduce and simplify setting-up and maintenance, and to make the whole equipment (in modern parlance) 'user-friendly'.

The operational frequency band was identical with that of Type 281, namely 86–94 MHz. To equal or improve the combined performance of Type 279 and 281, a far better performance than Type 281 was required, especially at long range and high altitude. Since it was planned to use an antenna identical with Type 281 in most installations, no improvement in gain could be obtained here. Neither could any improvement in receiver noise factor compared with the grounded-grid triode preamplifier version of Type 281 receiver. Thus it was necessary to increase the mean transmitted power. This was done by increasing the prf to 250 c/s and achieving a pulse power of 400–450 kW. The transmitter valves were no longer silica envelope, but were large metal-envelope triodes, the metal being the anode. Pulse lengths of 5 and 15 microseconds were provided. By this means the detection range on a medium-size aircraft was increased from about 150 miles (Type 281BQ) to 200 miles, with a first lobe height coverage to over 45000-ft. This performance is confirmed by a record of trials conducted with Type 960 installed at Tantallon in autumn 1945.

In order to ensure an overall integrated design, development of the whole equipment was entrusted to Marconi, Chelmsford. They succeeded in achieving the engineering objectives of simple setting-up and maintenance, and general 'user-friendliness'.

Technical achievements other than increased transmitter power were the provision of two complete IF chassis, one accurately tailored to each pulse-length, and with very steep bandpass cut-offs to improve anti-jamming; AFC on the receiver; and remote control of all operational parameters (pulse length, AJ features and so on) from the Radar Display Room. The transmitter could also be tuned over its band from the RDR, the receiver AFC following it, but since the wide-band antennae were never developed this anti-jamming feature was never used. Also, Marconi produced a high-power diode for the T/R switch to handle the higher mean power than Type 281B.

Another innovation was the display unit, christened the Universal Display Unit (UDU) because it was planned to use it with radars other then Type 960. This combined a PPI, still generated by a mechanically rotated magnetic coil, together with a mechanical bearing cursor on the PPI face and a Sector Display A-scan (see Type 281BQ) for plotting echo amplitude versus range to obtain height estimation, or to read personal IFF codes from selected targets. Although no new features were included that were not already provided by a collection of existing units, the ease of control and coordination by the operator was vastly improved.

The objective of providing a single metric set with sufficient range and height coverage to give air warning for the remainder of the 1940s and through the 1950s was achieved, together with a high standard of engineering and ease of operation. It is interesting that, with the demise of the requirement for rapid tuning for anti-jamming purposes, the narrow-band antenna for Type 281BQ was used throughout the life of Type 960, an antenna whose design principles could be traced back to Type 79, first produced in 1938!

One parameter that was not improved over Type 281 was the vertical-lobe structure, since the antenna was identical with Type 281, and loss of echo through the minima as the aircraft closed range was still a problem. The master plan was that Type 960 would be used mainly for first detection and early tracking, while a centimetric radar system giving 3-D target position with gapless cover from some 80 miles inwards would provide precision information to conduct fighter interception. Meeting this centimetric specification proved difficult, and was not achieved until well after the war. Nevertheless usable systems were developed during the life of Type 960, which itself gave excellent service from its first installation in late 1945 until it was superseded in the late 1950s.

POSTSCRIPT

Immediately after leaving University in 1938 the author joined HM Signal School, and after an initial period of four months spent in the silica-valve laboratory situated in the RN Barracks, Portsmouth, transferred to the Signal School Extension at Eastney Fort East in October for work on radar. The then Officer-in-charge at Eastney was A. A. Symonds; his staff were O. L. Ratsey, H. M. Bristow, W. H. Pritchard and G. Exeter, who had all been there for some time. The first Naval experimental set, Type 79X on 7-m wavelength, had recently completed satisfactory trials in *Saltburn*, and the first developmental set, Type 79Y on 7-m, had just completed installation in *Sheffield* and was giving good results. Ratsey had been responsible for developing the receiver and display, Bristow the modulator and transmitter, and Pritchard the EHT rectifier to supply the transmitter, and motor alternator and associated control gear.

Three months previously D. S. Watson had joined Signal School from BTH, the only person at Eastney having industrial experience; a few weeks before H. E. Hogben had been transferred from another section of Signal School. Watson soon became involved in the very rapid redesign of the modulator for the Type 79Y set, which had been giving serious problems, before it went into production as Type 79Z for Fleet fitting. Hogben took over receiver design, and was instrumental in producing the wider-band receiver P12 for the accurate ranging version of the 7-m set, Type 279, before being completely responsible for the receiver for the new 3.5-m set, Type 281.

The author, raw from university, first made a rig to measure the mutual characteristics of silica transmitter valves by pulsing the biased-off grid with a triangular waveform and displaying the plot of anode current against grid volts on a CRT. It was used to measure the characteristics of the improved silica transmitting valves for the new 3.5-m set, Type 281. He then went on to develop the complete transmitter for the same set. A laboratory assistant and a Naval electrical artificer completed the experimental staff at this time.

Two volunteers joined in the spring of 1939, in early anticipation of the coming war. One was S. E. A. Landale, formerly technical director at Youngers' Brewery, and the other B. Hodgson, who had already retired from a lifetime career in the valve industry. Landale immediately undertook the development of the modulator for the Type 281 3.5-m set, while Hodgson took part in specifying the characteristics required for the series modulator valves, and subsequently became adviser on all valve matters. Later in 1939 new recruits initiated the wartime expansion of effort, to be followed by a flood during 1940 and 1941.

One interesting feature, compared with today, was that during the 1930s and up to the rapid expansion at the start of 1940, all staff were known only by their surnames, Christian names being 'taboo'.

In 1938 HM Signal School Extension, Eastney Fort East embraced only two buildings. A wooden hut provided an office for the Officer-in-Charge – A. A. Symonds in 1938, succeeded by J. D. S. Rawlinson in January 1939. A single-storey brick building provided two laboratories: a smaller one for receiver and display work, and a larger one, with two offices, for transmitter and modulator work. There were also two 60-ft towers built of clamped tubular steel (as used for scaffolding) on which to mount antennae.

Amongst the equipment in the smaller laboratory was at least one commercial TV set. The RF chassis, working at around 40 MHz, and IF chassis were used in early laboratory work on what later became Type 79. The display CRT and its scanning circuits were used to provide an early A-scan.

The larger laboratory housed some half-dozen work benches. In one corner was an EHT rectifier cabinet, which provided voltages up to 30 kV, with a series of bus-bars hung from the ceiling (some 10-ft high) by insulators, so that all benches could connect to the EHT supply via a long insulated pole with a copper hook at one end.

All modulator and transmitter work was mounted in a $2 \times 2 \times 3$-ft container. This comprised a wooden framework supporting sides made of stretched, fine-mesh copper netting. This was the largest sized cabinet that could pass through ships' hatches. When the laboratory model was finally engineered by the Experimental Workshop in the RN Barracks, the wooden frames were converted to aluminium angle, supporting solid aluminium sheet sides.

All small chassis, for example for receiver, timebase, trigger circuits and so on, were made of bent sheet aluminium fixed together with brass nuts and bolts. Aluminium was used to withstand the corrosive sea atmosphere. In 1938 Naval-type resistors were still used wherever possible. These were carbon resistors in a ceramic tube with metal ends at each extremity for mounting in a clip holder, so that they could be changed without a soldering iron. They were, of course, large. Colour-coded, soldered-in resistors had just started to be used in receivers, but their general adoption was introduced slowly during 1939. Condensers (capacitors) for RF were made of mica and metal-foil sandwiches housed in bakelite cases, and were large. They were rated in Jars. Microfarad ratings were not introduced until 1940.

In the larger laboratory were also housed the general tools: a medium-sized lathe, a vertical power drill and a manual sheet-metal bender. These could be used by any of the laboratory personnel. Anything to a higher standard, or of greater intricacy, was made in the Experimental Workshops, HM Signal School. Here the only input required was a freehand sketch with the main dimensions shown, accompanied by (only if particularly complicated) a chat with the Workshop staff.

The test equipment available in the laboratories comprised a number of multirange Avo meters and Cossor double-beam oscilloscopes to share around the staff, an electronic LCR bridge and a Q-meter working up to 50 MHz. A full Standards Laboratory was available in the main HM Signal School.

Soldering irons provided from Admiralty stores had 1-in bits, which were clumsy for electronic soldering. So a ¼-in hole was drilled through the bit and a short length of ¼-in copper rod hammered through, one end being filed to a fine bit.

Total staff in autumn 1938 consisted of eight scientific officers, of various grades, one laboratory assistant and one Naval artificer.

To the modern reader, the technology of the early days of the radar era must appear very elementary. However it should be remembered that every aspect of the new radar development was breaking fresh ground at that time. The only sources of engineering expertise in the country with some knowledge of microsecond pulses, non-linear and two-state circuits, and RF technology at the limit of engineering practicability existed in the small industrial teams concerned with the development of all-electronic TV systems for the public service. These personnel were not, of course, privy to the very secret subject of radar. Nevertheless the few scientists involved at Signal School had built up a good understanding of the basics of radar technology between 1937 and 1939. At the start of the war a stream of academics was drafted in from the Universities. With the guidance of the original small team, they rapidly absorbed the new knowledge, and then initiated their own contributions to the various problems requiring solution (see Monograph 2).

The experimental workshops in Signal School (the Experimental Department of which became the independent Admiralty Signal Establishment in August, 1941) played a major role in introducing the new technology into the Fleet in the incredibly short time scales described in earlier paragraphs. They had built up a wealth of engineering expertise over many years by producing W/T equipment to meet Naval requirements, ashore and afloat. This was particularly important in relation to development of equipment capable of withstanding the severe, shock-prone environment in fighting ships. The equipments so produced reached production status in industry by direct copy of the Signal School models.

Types 79, 279 and 281 were produced in the following way. Laboratory 'breadboard' models were built by the scientists, within the overall dimensions required for ship installation. High-power equipment, such as the transmitters, was built in the largest size capable of passing through a ship's hatch – approximately a $2 \times 2 \times 3$-ft container. Receiving and display equipment was designed to fit into a standard Naval W/T rack. When the laboratory equipments had achieved a satisfactory level of performance, a senior engineer from the workshops, accompanied by a draughtsman, visited the laboratory. From his memory of what he saw, plus a circuit diagram and a few sketches from the draughtman, an engineered version suitable for shipborne use was produced by the workshops in a matter of weeks. The workshops model was tested in the laboratory, and any essential modifications introduced, up to complete assembly and test of the whole radar equipment.

Two complete copies were then made for initial ship-fitting, with the object of gaining early experience at sea. In parallel with this activity the Drawing Office prepared full production drawings, based on the workshop models. The drawings were then passed to industry for manufacture, no modifications being permitted.

In the modern sense the equipments produced were essentially ruggedised laboratory models. The exceptions were the office control equipment for the antenna and the masthead pedestal, both of which were designed by a mechanical engineer and drawn to production standards in the Drawing Office.

Whilst this procedure introduced equipments such as Types 79,281 and the later centimetric Type 271 into operational service extremely rapidly, the results were not entirely 'user-friendly'. Great credit is due to the Radar officers and men at sea, who so quickly learned to deal with the foibles of the equipments, and who were responsible for producing such valuable operational results.

APPENDIX 1: CHRONOLOGY OF DEVELOPMENT OF METRIC RADARS

Capital Ships

Sept. 1935	Directive to Signal School to start work on RDF.
Late 1937	Decision to produce an equipment for shipborne trials.
Early 1938	Type 79X (first experimental RDF set) on 43 MHz fitted in *Saltburn*.
Aug. 1938	Type 79Y (better engineered but still produced in Signal School workshops) fitted in *Sheffield*.
Aug. 1939	Type 79Z (produced in Signal School workshops, but a production prototype) fitted in *Curlew*.
Late 1940	Introduction into service of Type 279, with an accurate ranging facility.
Oct. 1940	Type 281, on 90 MHz (produced in Signal School workshops but production prototype), fitted in *Dido*.
Mar. 1941	Type 79B, with single antenna working, fitted in *Hood*.
Late 1941	Introduction of Type 281B, with single antenna working.
Mid-1943	Introduction of dual installation Type 279B/281B. Also introduction of PPI displays on Type 281B.
Mid 1945	Introduction of Type 281BQ, with addition of continuous antenna rotation.
Late 1945	Introduction of Type 960.

Small Ships

Mid 1940	Introduction of Type 286 with fixed antenna.
Early 1941	Introduction of Type 286P with rotating antenna.
Mid 1941	Introduction of Type 290 (used only briefly).
Late 1942	Introduction of Type 291.

APPENDIX 2: SUMMARY OF MAIN CHARACTERISTICS OF METRIC RADARS

	Transmitter				Receiver			Antenna	
Type	Frequency (MHz)	Pulse power	Pulse length (microseconds)	Prf (c/s)	IF frequency (MHz)	Bandwidth	Combined TR gain	Rotation	
79Z	38 to 42	70 kW	8 to 15	50 mains synchronous	5	140 kHz	14 db	±190	
279 } 279B	38 to 42	70 kW	8 to 15 / 3	50 mains synchronous	5	140 kHz / 1 MHz	14 db	±190	
281 } 281B	86 to 94	350 kW / 1 MW	15 / 1.7	50 mains synchronous	12	150 kHz / 1.5 MHz	23 db	±190	
281BQ	86 to 94	350 kW / 1 MW	5 / 1.7	50 or 100 mains synchronous	12	200 kHz / 1.5 MHz	23 db	Continuous	
286	214	7 kW	1.5	1200 free-running	45			Fixed	
286M	214	7 kW	2.5	400 free-running	45			Fixed	
286P	214	7 kW	2.5	400 free-running	45		23 db	±190	
286 PQ	214	100 kW	1	500 mains synchronous	45		23 db	±190	
290	214	50 kW (effective)	1	500 mains synchronous	30	2 MHz	23 db	±190	
291	214	100 kW	1	500 mains synchronous	30	2 MHz	23 db	±190 may have been continuous post 1943	

Acknowledgements

Thanks are due to the following wartime colleagues for provision of technical information and personal reminiscences: Messrs Allan, Bogle, Cochrane, Corner, Falloon, Fenwick, the late L. J. Fitzgerald, Foley, Reeves, the late A. W. Ross, and D. S. Watson.

Particular credit is due to Sid Wright for searching the archives at the Defence Research Establishment, Portsdown (formerly Admiralty Signal Establishment, the successor to the Experimental Department of HM Signal School), and for his provision of the relevant handbooks from *Collingwood* archives. Much of the circuit data was obtained from the latter source.

Thanks also to Janet Dudley, formerly librarian at DRE, Malvern (formerly TRE), and John Briggs, librarian at DRE, Portsdown, for the loan or photocopies of a number of valuable documents, and to Jim Hancock, chief photographer at DRE, Malvern, for assistance with the illustrations.

References

1. ADM 220/70, DSD to CSS, 13 August 1935.
2. A. W. Ross, 'Problems in Shipborne Radar', *Journal IEE*, Vol. 93, Part *IIIA*, Pg. 236 et seq.
3. E. B. Callick, *Metres to Microwaves* (Peter Peregrinus, 1990).
4. F. A. Kingsley (ed.), *The Applications of Radar and Other Electronic Systems in the Royal Navy in World War 2* (Macmillan, 1994), Monograph 1, by H. W. Pout.
5. ADM 205/2, CSS Chapter 2.
6. O. L. Ratsey, 'Radar Transmitters. A Survey of Developments' *Journal IEE*, Vol. 93, Part *IIIA* (1946), Pg. 245 et seq.
7. F. A. Kingsley op.cit, Monograph 4, by Cdr R. S. Woolrych, RN
8. F. A. Kingsley op.cit., Monograph 1 by H. W. Pout, Monograph 4 by Cdr R. S. Woolrych, RN.
9. E. G. Bowen, *Radar Days* (Adam Hilger, 1987), Pg. 98 et seq.; Professor R. A. Smith *et al.*, 'ASV. The Detection of Surface Vessels by Airborne Radar', *Proc. IEE*, Vol. 132, Part A, No 6 (1983), Pg. 359 et seq.

Monograph 5
Development of Naval Warning and Tactical Radar Operating in the 10-cm Band, 1940–45
C. A. Cochrane

SUMMARY

British inventions in 1940 of: the high-power cm-wave resonant-cavity magnetron, a tuneable receiver local-oscillator klystron, and a silicon crystal-mixer for superheterodyne receiver operation. TRE (Air Ministry) demonstrate an experimental 10-cm pulse radar against small ship targets by autumn, 1940. Urgent Naval requirement for effective radar in convoy escorts to detect surfaced U-boats at night. Resulted in first Naval *operational* 10-cm radar by March 1941.

Subsequent adaptation for larger warships, and by Army and RAF for ground applications. Rapid development of 70 kW magnetrons by mid-1942, and 500 kW version by mid-1943. Introduction of PPI display. Adoption of wave-guide techniques in antenna systems. Common T/R working, with continuous antenna rotation. Effective tactical radar for surface search, and against low-flying aircraft. Development of successful target indication (TI) radars with remote displays for use against both low and high-flying aircraft, and of fighter direction sets having accurate height-finding capability.

PREFACE

Early in the development of radar there was recognition of the desirability of using the highest possible frequency for those applications in which antenna size was limited, or good resolution was needed between multiple adjacent targets. However, in the immediate prewar period the upper limit in frequency was set by the ability to generate sufficient power from the transmitter oscillator to produce a detectable reflection from a radar target. Hence, in Naval radar development, a frequency of 600 MHz was chosen for development of radar fire-control rangefinders since, at the time, this was the highest practical frequency for which oscillators of adequate power could be manufactured in quantity (see Monograph 1). The magnetron, klystron, and even specially designed triodes could be made to oscillate at higher frequencies, but they did not provide output-power levels adequate for practical radar transmitters. By Inter-Service agreement the search for a suitable generator of microwave power was incorporated in the Admiralty-sponsored CVD Valve Research Programme, whilst responsibility for radar research on microwaves was undertaken by the Air Ministry in its own research establishments, and under contract with GEC and EMI.

The prospects for microwave radar were revolutionised by the invention of the resonant-cavity magnetron by J.T. Randall and H.A.H. Boot in the Physics Department of Birmingham University, under sponsorship of the CVD Valve Research Committee. In laboratory form, continuously vacuum-pumped, and using an electromagnet to produce the necessary magnetic field, this new form of magnetron first generated several hundred watts of CW power in the 9–10-cm waveband in February 1940. By July 1940 E.C.S. Megaw and his colleagues at the GEC Research Laboratories had developed a sealed-off version operating with a permanent magnet. With this reproducible form of the magnetron, centimetric radar for operational applications first became practicable. The question now became one of priority in relation to existing development commitments. Many considered that, given the war situation in mid-1940, all effort should be concentrated on completing developments already in hand. Fortunately this view did not prevail, and at the Telecommunications Research Establishment (TRE) of the Air Ministry a cm-wave group for airborne applications was formed in August 1940 under P.I. Dee and H.W.B. Skinner. Shortly after this a similar decision in HM Signal School led to the formation of a group under a S.E.A. Landale with the objective of establishing a capability for application of the cm-waveband in Naval radar. Undoubtedly this decision was stimulated by the urgent need to provide convoy escorts with the means to detect surfaced submarines at night, and in poor visibility.

At this date, development of metric Naval radar was located at HM Signal School Extension within the Royal Marine Barracks at Eastney Fort East, Southsea, whilst that of decimetric Naval radar was centred in requisitioned premises at Onslow Road, Southsea, with a trials site on the seafront at Southsea Castle (see Monograph 1 and 4). In September 1940 whilst development of the new 3.5-m long-range air-warning radar was just nearing completion, a major effort was in hand at Eastney to adapt TRE's 214 MHz airborne equipment to provide radar for smaller warships, coastal craft and submarines in view of the invasion threat. It was therefore an appropriate moment to make the necessary provision to initiate work on techniques for the new centimetric waveband.

Staff at Eastney at that time was organised primarily in three main divisions – transmitters, receivers and displays – with an outstation for fieldwork on antennae. Project organisation was informal, depending on the application. The group formed under Landale's direction was, in principle, a modulator and transmitter technique group. D. F. Gibbs, along with S. T. Wright and R. Durrant, were the first staff to be transferred for work on modulators and transmitters for centimetric radar. D. F. Gibbs was later to make important contributions to the design of heavy-duty transformers for very short pulses. B. W. Lythall, who also joined the group in 1940, later contributed much to understanding of the behaviour of magnetrons feeding into long waveguides. The author was also allocated to this group, only weeks after reporting for duty to Eastney as a very raw recruit to RDF development. Receivers remained the responsibility of the Receiver Division, where the early work was undertaken mainly by J. Croney and C. Shearston, whilst the early display work was contributed by C. A. Laws of the Display Division. The latter included the addition of an accurate ranging display to the first simple A-scan displays at a relatively early stage.

With the rapid development of even higher magnetron power, radar technique development expanded greatly in 1941. Waveguides were introduced for RF transmission following research work by S. Kuhn and his colleagues. Project applications also multiplied. Centimetric radar was required for all classes of ship, together with the first application of the 500 kW transmitter to mobile shore stations (G. King was the project coordinator). Larger, higher gain antennae, stabilised to hold the narrow centimetric beams in the horizontal plane, were required for longer-range surface warning for the larger warships (F. N. Scaife was the design engineer). A fast warm-up modulator was needed for a proposed low-angle centimetric rangefinder for main armament guns, later completed for blind fire as Type 274. This spark-gap modulator was developed by A. A. Symonds and a small team, including S. T. Wright. On transmitters R. V. Hughes and W. T. Davies joined the group. Finally, whilst still at Eastney and before the subsequent move to Witley and the reorganisation of the Radar Department in 1942, centimetric radar development expanded to include X-band (3-cm) application to surface-warning radar for submarines (S. R. Tanner and L. J. Fitzgerald). In October 1942, under the new staff organisation, the roles of component development and the project coordination were separated. About half the staff of the Landale group joined a new, enlarged modulator/transmitter component Division under O. L. Ratsey, the other half remaining with Landale in an enlarged Division responsible for project coordination of all Warning Radar.

This main stream of rapid evolution, and the growing importance of shipborne centimetric radar is reviewed in the present monograph. Much specific technical detail was released immediately after the war, and consequently is not repeated here. (It is readily accessible in the Proceedings of the IEE Radiolocation Convention of 1946 – see the Bibliography to this volume). Concentration is therefore on the sequence of development, function and performance, these being aspects difficult, if not impossible, to review in detail prior to the author's recent access to wartime documentation, including performance reports. Far from the least interest for the author in preparing this monograph has been the opportunity to study once again, almost 50 years later, certain of his own reports on the initial performance trials of centimetric radar, and to be able to appreciate very directly the fallibility of memory. It is a very long cry from the days when, on detached duty, one's only specialised microwave test equipment was an Admiralty Pattern neon tube, some six inches long, which could be carried in the pocket with pen and pencil. Used judiciously with a centimetric transmitter (at least the early, low-power transmitters) the length of excited column in the tube provided a useful clinical thermometer of the health of the transmitter.

INTRODUCTION

The earliest known attempt to exploit a centimetre wavelength (13.5-cm) for the detection of ships was by Dr Kühnhold of the German Nachrichtenversuchsanstalt (NVA) in 1934, using an embryo radar in Kiel harbour.[1] The transmitting valve was a split-anode magnetron of 100 mW CW output, mounted at the focus of an 80-cm diameter parabolic reflector. This attempt demonstrated the need for very much higher transmitter power, a development task in which the German valve manufacturers showed little interest at the time. Attention was therefore directed by NVA to decimetric wavelengths (48-cm), and further work in the centimetric band was abandoned. Later, work was also initiated at NVA on a wavelength of 2.4-m.[2]

Similar considerations regarding power output from available V/UHF oscillators led to adoption of the metric waveband for the initial British experiments on radar (see Monograph 4). The advantages of using the highest possible frequency for radar were nevertheless clear. Availability of a narrow beamwidth, and the associated high gain from dimensionally small antennae were particularly important considerations in both airborne and shipborne applications. In 1939 the practical upper limit of frequency for effective grid-controlled oscillators was about 600 MHz (50-cm) (see Monograph 1). However research on both magnetron and klystron oscillators continued, and was intensified in the UK in 1939 under Admiralty sponsorship, on behalf of the Committee for the Coordination of Valve Development (CVD) programme. The object was to develop both sufficient pulse-power output at wavelengths around 10-cm for radar transmitter applications, and a tuneable local-oscillator of adequate efficiency for the associated receiver[3](see Monograph 3).

By February 1940 the situation regarding radar development in the centimetric waveband had changed dramatically. The CVD-sponsored research programme under Admiralty contract at Birmingham University resulted in the invention of the resonant-cavity magnetron capable of producing a pulse-power output of the order of a kilowatt (kW). Also, successful development of the tuneable reflex-klystron oscillator, capable of up to 100 mW CW output, took place in the Valve Laboratories of HM Signal School. Availability of these devices made possible the initiation of experimental radar development work in the centimetric waveband at the Telecommunications Research Establishment (TRE) at Swanage (formerly the Air Ministry Research Establishment (AMRE), Bawdsey). H. W. B. Skinner of TRE completed the trinity of essential RF devices by proposing the use of a silicon-tungsten crystal as first detector in a superheterodyne receiver, and developing this in a form suitable for incorporation in a tuned-line RF mixer. By July 1940 both the local-oscillator klystron (NR89), and the copper-block resonant-cavity magnetron (NT98), had achieved production prototype stage. The TRE team then successfully demonstrated detection of local radar echoes using the now available 5 kW output of the magnetron in 1.5 microsecond pulses.[4] The way was now open for the development of cm-waveband airborne radar, both AI and ASV, for which compact, high-gain, narrow-beam antennae were essential.

Until this time Admiralty involvement in the development of cm-waveband radar had been limited to administration of the CVD Valve Programme, and supporting work at the Valve Research Laboratory of the Experimental Department of HM Signal School, under the direction of R. W. Sutton.

The technical aspects of the development of the resonant-cavity magnetron, the reflex klystron and the receiver mixer crystal were presented at the postwar Radiolocation Convention of the Institution of Electrical Engineers.[5]

ORIGINS OF THE FIRST NAVAL CM-RADAR

In the summer and autumn of 1940 shipping losses in the North Atlantic rose dramatically, to virtually unsustainable levels. There were many causes for this, which are analysed in detail in the official history of the

'War at Sea' by S. W. Roskill.[6] However, one factor was common to many of the successful U-boat attacks, namely attack on the surface at night. Roskill states that in September 1940 70 percent of all successful attacks followed this tactic. This exploited to the full the inability of Asdic to detect a surfaced submarine, and the near impossibility of obtaining a visual sighting of anything as small as a submarine conning tower at a useful range at night or in poor visibility. In view of the lack of an effective countermeasure to this tactic, the 'Trade Protection Meeting' in the Admiralty urged, and 'The Defence Committee' under the Prime Minister approved, inter alia,[7] that: 'An efficient radar set for anti-submarine surface and air escorts must be developed'. This directive provided authority for immediate action on any prospective improvement, such as that offered by the initial TRE experiments.

The surface escort vessels primarily envisaged at this date were the small (1000-ton) Flower-class corvettes then being built in large numbers to form the backbone of the convoy close-escort force. Antenna height in these vessels was necessarily limited. This is a most important consideration affecting range performance against surface targets, in the case of radar operating over the conducting surface of the sea. Idealising the situation, there are two paths for the radar radiation to and from the target: a direct path, and an indirect path reflected from the surface of the sea, in each direction. The difference in electrical path length between direct and reflected radiation on a given path creates an interference pattern in the vertical polar diagram of the antenna.[8] This consists of a series of upward-sloping lobes, at an angular spacing directly proportional to wavelength and inversely proportional to antenna height. The shorter the wavelength the closer and lower the lobes, and the lower the antenna above the surface, the greater their slope and spacing. Clearly one way to increase field strength near the sea surface from shipborne radar is to decrease the operating wavelength. In comparing the performance of radar operating on 3-m wavelength with one on 10-cm wavelength, there is a compression of the propagation field above the sea surface of 30:1. This still leaves a considerable advantage to the 10-cm radar in the detection of low-profile surface targets, even at as much as a third of the antenna height. In considering the other factors determining radar range performance it was estimated that, compared with performances achieved in longer-wavelength radar, the quasi-optical techniques possible for antenna design at 10-cm wavelength would result in sufficiently high antenna gain to compensate for both the lower transmitter power available, and the higher receiver noise factor.

In the history of the development of centimetric radar the immediate reaction to the availability of the resonant-cavity magnetron in a reproducible and convenient form, in August 1940, is summarised succinctly by E.G. Bowen in 'Radar Days'[9], as follows: "Following the formation of the Centimetre-Wave group at TRE, similar groups were set up at the Admiralty and at the Army Air Defence Research and Development Establishment (ADRDE) to undertake development of shipboard and gun-laying radars for Naval and Army use, respectively."

As regards the Naval Applications group, formed in the Experimental Department of HM Signal School (HMSS), the book 'Science at War'[10], records the following: 'A new group of workers of the Admiralty Signal Establishment (sic., this name for HMSS did not come into use until August 1941) was formed under Dr. S. E. A. Landale, and went to Swanage for six weeks to take advantage of this successful research.' In point of fact, it is now known that the HMSS team were at TRE Swanage for the period from 21 November until about 21 December 1940.

Valuable additional information on events at Swanage during this period of technology transfer is given by Sir Bernard Lovell in his book 'Echoes of War', under the heading 'The Naval Diversion'.[11] This includes extracts from his own, and colleagues diaries kept at the time.

Important documents[12] discovered recently in the Public Record Office files have now permitted a revised time-table to be established relating to the future successful development of Naval 10-cm radar systems. It is relevant to point out that this evidence updates that given in an associated book.[13]

The first direct exposure of the TRE's experimental work on 10-cm radar detection of small ships to Naval personnel took place on an undetermined date in October 1940. The TRE experimental equipment at that time used the GEC E1198 magnetron (9.1 cm) of nominal 5 kW pulse output, the Signal School klystron as receiver local oscillator, a crystal mixer coupled directly to the receiving antenna and feeding a 45 MHz IF stage, and a pair of 3-ft diameter dipole-fed paraboloidal reflectors for the transmitting and receiving antennae, respectively. The demonstration of this experimental equipment, which was mounted in a mobile trailer, was from a site at Leeson House, Swanage some 250-ft above sea-level. The Naval personnel concerned were Cdr. H. W. Fawcett and Lt. Cdr. E. C. Bayldon, both of the Anti-submarine Warfare Division (A/S WD) of the Admiralty. Following the demonstration of the detection of small ships in Swanage Bay by the equipment, these officers became interested in the possible use of such equipment in the detection of surfaced submarines from destroyers in charge of convoy shipping, and informed Admiralty accordingly.

The visit may well have been routine, since Cdr. Fawcett was responsible for liaison with RAF Coastal Command. However, it occurred at a time when very high priority had been given for an improved radar for air and surface convoy escorts by the Trade Protection Committee meeting in the Admiralty in September, referred to above. There was clearly no lack of authority for A/S WD immediately to alert the Signal Division of the Admiralty, and hence HMSS, of the potential opportunity for the application of the new cm-wave techniques in the field of convoy escort radar against surfaced enemy submarines.

There is still some doubt concerning the exact date of the Fawcett/Bayldon visit to Swanage. Lovell records[14] from his own diary entry for 28 October the visit of two Naval Commanders. He also quotes P. I. Dee's diary for the same date which records a recent demonstration of the experimental 10-cm equipment to Captain Willett and C. E. Horton, from HMSS. The exact dates of these preliminary visits are not of great importance. What was of consequence was the result, in the form of a request for a demonstration of the TRE equipment to Admiralty and HMSS staff. This is recorded in a letter to DCD/Ministry of Aircraft Production,[15] from which the following is an extract: 'On 7th November 1940 we were asked to show the work to a deputation of DSD (Admiralty), Captain Willett (Signal School), Dr. (sic.) Horton (Signal School), Colonel Miles (Admiralty), and Mr. Wright (Signal School).'

Lovell records[16] that 'as a result of a conference they are sending a few men to help us with a prototype in a trailer for their further tests.' This appears to be the first suggestion for sending a learning team from HMSS to TRE to expedite the acquisition of the new techniques available at TRE. Further impetus was given to the interest in the potential of 10-cm radar for A/S detection when DCD/Ministry of Aircraft Production was informed[17]: 'On Saturday, 9th November, we received a message that a submarine would pass by Swanage on its way to Portsmouth on the 11th November. A test was arranged, and Captain Willett and Dr. (sic.) Horton were present. The submarine (U-class, the *Usk*) gave very large echoes up to about 5 miles.'

A fuller account of a variety of tests against small ships and the submarine *Usk* from various sites in the Swanage vicinity is given by H. W. B. Skinner, of TRE, in a report dated 20 December 1940.[18] In the particular case of the *Usk* trial on 11 November, Skinner states that: 'Sea fairly calm. Submarine (U-class, the *Usk* was seen up to about 7.5 nautical miles. Strong signals up to 6 nautical miles. These figures apply to the stern-on view. The broadside-on view seemed to give signals about twice as strong. Horizontal polarisation of the source seemed slightly preferable to vertical polarisation. Between 4 and 4.5 miles, the submarine dived and the signal disappeared. On coming to the surface, the signal re-appeared at a time when at most the periscope and conning tower were visible. Signals at extreme range are always fluctuating, and at extreme range were liable to disappear for rather long intervals.' (Note. A nautical mile is approximately 2000 yards. All subsequent references to miles in this monograph imply nautical miles, as defined in the Glossary.)

These detection ranges were far in excess of any previous radar performance against such a small surface target. It was therefore inevitable that what had begun as an exploration of a new technique changed immediately to a concerted drive to develop a prototype 10-cm WS radar suitable for installation in convoy-escort vessels then being built in large numbers. It should, however, be borne in mind that, whilst the ability of 10-cm equipment to detect relatively small targets on the surface of the sea was amply proven, the experimental equipment would have been lethal if installed in a ship liable to severe pitch and roll, without warning. It would not be possible to judge from a superficial inspection if the equipment was suitable to be submitted for engineering prototype design. The experimental equipment was mounted in a trailer, and the antennae stood separately, mounted on a simple swivel device. Finally, the demonstrations had been made from height of 250-ft above sea level which, whilst appropriate for ASV applications in aircraft, were clearly not indicative of the performance which might be achieved from an installation in a small convoy escort vessel, with limited mast height. Nevertheless, matters moved quickly after the demonstrations. Captain Willett wrote to the Superintendent, TRE on 18 November giving the names of the party to be sent from HMSS for temporary attachment to TRE Swanage, with effect from 21 November.[19] The staff concerned were Dr S. E. A. Landale (in charge), J. Croney, C. A. Cochrane (the present author), and C. Owen. These officers arrived at Swanage on 21 November and, apart from Landale himself, were to remain at Swanage until 20 or 21 December. Landale had much work to do in transferring the acquired technology to his group at HMSS Extension, Eastney, so spent only limited time at Swanage.

The instructions given to Landale in the letter referred to above took full account of the remaining uncertainties relating to the suitability of 10-cm radar for shipborne application. Landale was instructed:

'to arrange collaboration with the staff of TRE in order to further the experiments on –
(a). Comparison of range as between a height of 40-ft above sea level and 200-ft.
(b). The practicability of finding and holding an unseen target with mirrors having a power gain of 400 times.
(c). The effect of artificial roll of the mirrors comparable to the movement of the ship.
(d). The effect of surface reflections when the sea is moderately rough and the aerial height 40-ft.'

It is clear from this remit that considerably more had been agreed at the time of the demonstrations than a simple loan of HMSS staff to TRE in order to provide an opportunity to learn the new centimetric-wave techniques. The instructions to Landale imply direct involvement of TRE staff in further necessary experimental investigations, prior to design of a shipborne prototype 10-cm radar equipment. It is also clear from various published accounts of work at TRE during this period[20] that the collaborative programme became the special responsibility of H. W. B. Skinner. The Skinner report of December 1940 provides the sole documentary evidence of the progress of the joint project at TRE. It is here recorded[21] that, on 8 December 1940 a third set of equipment, 'apparatus C', was tested in the TRE trailer installation at Leeson House, Swanage. There can be little doubt that this was the set of equipment built by the HMSS party, under instruction from experienced TRE staff. Performance in the TRE experimental set-up proving satisfactory, 'apparatus C' was brought back to the TRE Laboratory for final finishing. In the course of the next fortnight it became part of the equipment built into what became known as 'the Admiralty trailer', a rotating cabin trailer on the roof of which was built a housing to support the antenna.

At this point in the joint collaborative project, two fundamental problems remained unanswered. These related to the performance of the radar when installed at low antenna height, and the degree of holding the target under conditions of roll and pitch of the ship in which it was installed. On 15, 16 and 17 December Skinner ran three further trials of radar performance against the small ship *Titlark*. These trials used the TRE equipment 'apparatus B' fitted with free-standing paraboloid antennae for transmission and reception, respectively. Sited at Leeson House at 250-ft above sea level, *Titlark* was followed to 9 miles. Moved to a site at Peveril Point some 60-ft above sea-level demonstrated an end-on range against *Titlark* of 5 miles, whilst an abnormally low site at 20-ft above sea level still resulted in a target range of 3.5 miles. Thus, using the paraboloidal mirrors, there was clearly an effective performance obtainable even at 20-ft against a target already shown during the earlier demonstrations to be more or less equivalent to a large submarine, fully-surfaced.

As regards possible loss of target returns due to pitch and roll of the ship carrying the radar, A. C. B. (later Sir Bernard) Lovell suggested the use of a pair of cylindrical parabolic antennae in place of the original paraboloids.[22] With an aperture of 6-ft by 9-in such antennae could be expected to provide only about half the gain of the paraboloids. Nevertheless, using these 'cheese' antennae, a range of 3 miles was achieved on *Titlark* from the very low height of 20-ft. This form of antenna provides a fan beam in the vertical plane, and was destined to become the extensively-used format in Naval centimetric radar during the war whenever antenna stabilisation was not possible. It was therefore adopted for use in the prototype equipment now under active development, at HMSS Extension, Eastney.

The 'cheese' antennae used in the final Skinner trial were installed in the roof housing of the 'Admiralty trailer' on 19 December and, either on the same day or, more likely, the following day this was towed to the beach for a final general test. It was then transferred to the RAF Transport Pool to arrange towing to HMSS Extension, Eastney. Here, it was to act as a reference equipment during the final stages of design development of a sea-going prototype. Work on the latter had begun by the end of 1940, on the basis of information fed back to HMSS from the party at Swanage, via Landale.

It is clear that this final test of the 'Admiralty trailer' was not the trial conducted at 20-ft above sea level described in the Skinner report from the manuscript notes for this report which are also contained in AVIA/144.[23] In these notes it is stated that: 'Further echoes were obscured because boat was behind pier and aerial beam had to pass through the pier framework to follow course of the boat.' The site of the final check of the 'Admiralty trailer' must have been the deserted stretch of beach on the other side of Peveril Point from Swanage. There was certainly no pier in sight, or any sight of Swanage itself.

Thus was completed an outstanding example of collaboration between two large Service Organisations, each contributing their knowledge and special skills. This rendered possible the installation of a prototype Type 271 Naval radar in an escort vessel during March 1941, only six months after the desperate appeal of the Trade Protection Committee for a more effective convoy escort radar capable of detecting surfaced U-boats under conditions of poor visibility.

THE FIRST OPERATIONAL CENTIMETRIC RADAR

The proposed centimetric radar for installation in corvettes was designated Type 271, although this was later amended to 271X in order to distinguish the early models from later versions. This latter designation will be retained in order to avoid confusion. Normally the X suffix simply indicated a laboratory-built model for the purpose of early trials. In this instance, because of the urgency to provide convoy escorts with means to detect surfaced submarines at night or in conditions of poor visibility, 24 sets of equipment were produced in the Eastney Laboratories. Many of these were fitted in ships for operational service. Thus Type 271X had to meet design standards for service at sea and also, if production in quantity was to follow rapidly, all components had to be of known availability in production. By the end of February 1941 the first Type 271X of the initial batch of 12 was ready for ship installation, and the remaining 11 were well advanced in production. Also advance orders for 150 sets of components had been placed. The mechanical work on these pilot models was undertaken by the experimental workshop in Signal School, and the various electronic units were assembled and wired in the appropriate laboratory.

Up to this point the programme had involved only allocation of resources at the Eastney Extension of the Experimental Department of Signal School. Resource allocation now assumed much wider implications. A ship had to be designated for fitting and trials. Official confirmation was necessary for the manufacture of production quantities in the event of successful trials at sea. A manufacturer had to be chosen for eventual quantity production.

A meeting of all those concerned was therefore convened by the Captain Superintendent on 11 February. No minutes of this meeting have survived in archives, but it finds mention in a number of postwar interviews. Admiralty confirmation was given from the Director of A/S Warfare for the initial order for 100 sets of equipment. Allen West Ltd, whose drawing office had already been involved in the design stage of Type 271X, were appointed manufacturers. The new-construction Flower-class corvette *Orchis* was designated for the sea trials. It was also at this meeting that the number of laboratory-built pilot models to be produced was increased from 12 to 24.

Sea trials began on 25 March 1941 in The Firth of Clyde, using a small submarine target. At the antenna height of 36-ft in *Orchis*, the maximum range on the submarine, fully surfaced in calm sea conditions, was 5000-yds, echoes being strong out to 4000-yds. Later, in rough sea conditions the maximum range was reduced to 4400-yds.[24] These results, being well in excess of the visual-sighting range of a submarine at night when viewed from a ship against the dark background of the sea, were considered operationally successful.[25] Design was therefore frozen, and full production plans went ahead at high priority at Allen West. For reasons that will be discussed later, the ranges achieved on these trials were technically disappointing in comparison with: (a) observations from the trailer at Eastney (although these were not on a submarine target), and (b) the 7.5 miles of the original Swanage tests, albeit with different equipment and from a very different antenna height. The trials submarine was a much smaller boat than *Usk*, with a much lower conning-tower profile. Nevertheless this did not altogether explain the apparent shortfall in range.

Type 271X was a basically simple radar but the concepts, techniques and components were new to a world hardly yet acclimatised to the existence of television. This posed a massive training and organisational problem for the Navy (see Monograph 9), since this radar was intended for installation in a large number of small ships possessing strictly limited on-board technical facilities. The surmounting of this problem, and the establishment of adequate training in operation and maintenance, backed up by the necessary supplies ashore and afloat, was a vital contribution to achieving operational success in forcing the enemy submarines to travel underwater by night as well as day (see Monographs 7, 8 and 9).

TECHNICAL CONSTRAINTS AND CHARACTERISTICS OF TYPE 271X

The structural form of Type 271X was dictated by the performance of available coaxial cables at 3000 MHz. The most suitable cable, itself a very recent development, was a flexible coaxial with solid polythene dielectric. This cable had an attenuation of 22 db per 100-ft at 3000 MHz. The antennae feeders had therefore to be kept short; to achieve this the transmitter and the receiver crystal-mixer unit were mounted on the back of the respective parabolic reflectors (Figure 5.1). The resulting feeder length of about 4-ft limited the attenuation loss of each feeder to about 1 db.

The local oscillator, however, had to be situated in the radar office to permit viewing of the display when tuning the receiver, using either echoes or the transmitter ground wave. The required crystal-mixer current was 350–400 microamperes,[26] and to obtain this with the 50–100 mW output available from the NR89 reflection klystron ('Sutton' tube) limited the length of connecting cable to the mixer to 20-ft. To meet these requirements the antennae had to be sited immediately above the radar office, thus resulting in an integral office/antenna structure of 6-ft square base. In common with the associated transmitting and receiving

5.1 Rear view of the Type 271 Antenna: (1) magnetron with pair of filament leads on left; (2) high-voltage pulse cable from modulator; (3) IF head amplifier unit; (4) original coaxial line type crystal mixer; (5) radiation meter excited by thermocouple in back wall of cheese (© Defence Research Agency)

5.2 The office/lantern installation on HMS *Periwinkle* constructed by the Dockyard, shown before the installation of Type 271 (© Imperial War Museum)

equipment, the antennae could not be left exposed to the weather. Thus they were enclosed in an octagonal 'lantern' comprising 2-in × 3-in teak roof supports, framing perspex panels of appropriate thickness, to minimise reflection of 10-cm radiation (Figure 5.2).

The transmitting and receiving antennae were identical and constructed as a single unit, with the transmitting antenna on top (Figure 5.3). The 'cheese' form, as tested at Swanage, was retained, but the dimensions were altered.

In order to reduce the sweep of the antennae the ends of each 'cheese' were squared off, and the separation between top and bottom plates increased from the 9-in of the Swanage model to 10-in. The RF feed to the 'cheese' reflector remained a half-wave dipole, backed by a rod reflector (Figure 5.4) with the dipole supported in the centre of the aperture.

It was believed at the time that these dimensional changes would not significantly affect the gain of the antenna, though experiments indicated the existence of very large 'side lobes' in the vertical polar diagram. It was further noted that these could be reduced by replacing the rod reflector on the antenna feed by a 2-in wide section of a parabolic cylinder mounted between the top and bottom plates of the 'cheese', with the feed dipole on its focal line. However antenna design had been frozen as a result of the success of the *Orchis* trials, and therefore no change was made to the antenna design for Type 271X or subsequent versions of Type 271. Records show an early change in reference code for the antenna Outfit ANA for Type 271X. The replacement Outfit ANB marked the change to the production design of pedestal from the early models, which used a modified DF loop pedestal. Later, theoretical analysis by Dr Otto

5.3 Front view of the Type 271 antenna: aperture 4-ft × 10-in; simple dipole feed (Figure 5.4) supported by Tufnol bar in centre of aperture (in later production models dipole feed supported by rigid tube from the back of the mirror, and the dipole and rod reflector were encapsulated) (© Defence Research Agency)

Böhm[27] of the 'cheese'-type antenna made clear that the polar diagram in the plane orthogonal to the flat closing plates is critically dependent on the separation of the plates, and on the field distribution across the aperture from the antenna feed. Measurements confirmed that the change in plate separation did indeed have considerable effect both on the antenna gain and on the vertical beamwidth. In fact the gain of a single 'cheese' as fitted to Type 271X was 55 times that of a half-wave dipole, where at least 100 times had been expected. Also the overall beamwidth

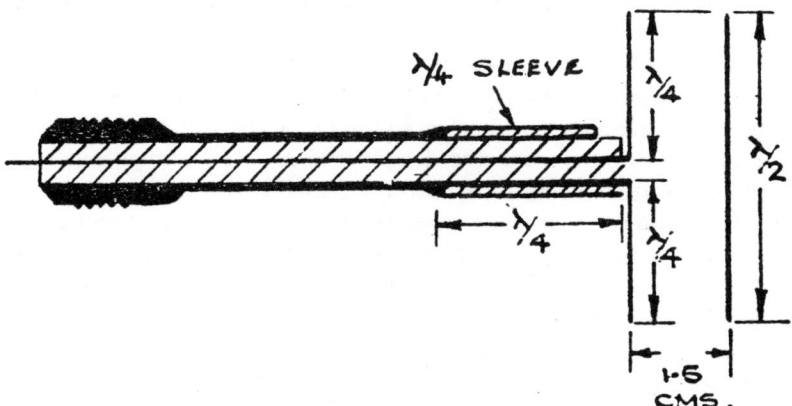

5.4 Diagram of the feed dipole and rod reflector of the Type 271 'cheese' antenna (© Defence Research Agency)

(to 6 db down) of the transmitting and receiving antennae taken together was 66°, whereas the order of 30° had been expected. It may seem surprising that no action was taken to introduce what would have been a relatively simple modification to the antennae. However it had been noted during the rough-sea trials on *Orchis* that the ship rolled to 30° without loss of target echo using the 'cheese' antenna design as fitted.

The NT98 magnetron used in the transmitter was the 10-cm version of the E1198 (9.1-cm) of the Swanage tests. Operating in a magnetic field of 1080 Oersteds at 8kV, 8 Amps, in 1.5 microsecond pulses, the peak-power output of individual valves as measured on manufacturing acceptance test varied between 4.5 kW and 10 kW,[28] the majority of valves producing between 5 and 6 kW. Modulation was by hard-valve tetrode whose cathode was held at 12 kV negative, and for which the magnetron was series anode-load.[29] The existence of a powerful, permanent magnet in the transmitter unit, rotating with the antennae, presented a particular problem aboard ship especially as Type 271X was mounted on the after-end of the bridge on corvettes, and was therefore not far from the ship's magnetic compass. The problem was solved by enclosing the transmitter in a mu-metal box of adequate thickness to ensure there was no significant magnetic effect at distances greater than 6-ft. This condition was determined and tested by the Admiralty Compass Department.

The NT98 proved to be a rugged and reliable valve in service. It was not tuneable in frequency, oscillation being set by the internal structure of the valve. However it could oscillate in several different modes, depending on the input voltage and output matching conditions. Only one of these modes was at high power. This called for experience in setting up the transmitter; a principal difficulty was the small input voltage difference separating the modes of oscillation. Operation of the NT98 at higher input voltage than the minimum needed to achieve the high-power mode was not possible since the NT100 series modulating valve was working near its limit.

At the time of the *Orchis* trials the only mixer crystals available were copies of the laboratory models made by Skinner at TRE. A more rugged version was under intensive development.[30] The aim was a crystal that was not only mechanically stronger, but also had a prepared silicon-crystal surface such that operation as a detector depended on the pressure of the tungsten whisker, without the additional need to seek a sensitive spot on the surface. The whisker was then sealed in place. However the mixer unit for Type 271X had necessarily to be designed for the laboratory-model crystal.

It took the form of a one-wavelength section of coaxial line, of which the crystal with its lead-in wires formed the inner conductor (Figure 5.5). Both the length of the line and the crystal position in the line could be altered. The production crystal was in capsule form, and an adaptor was

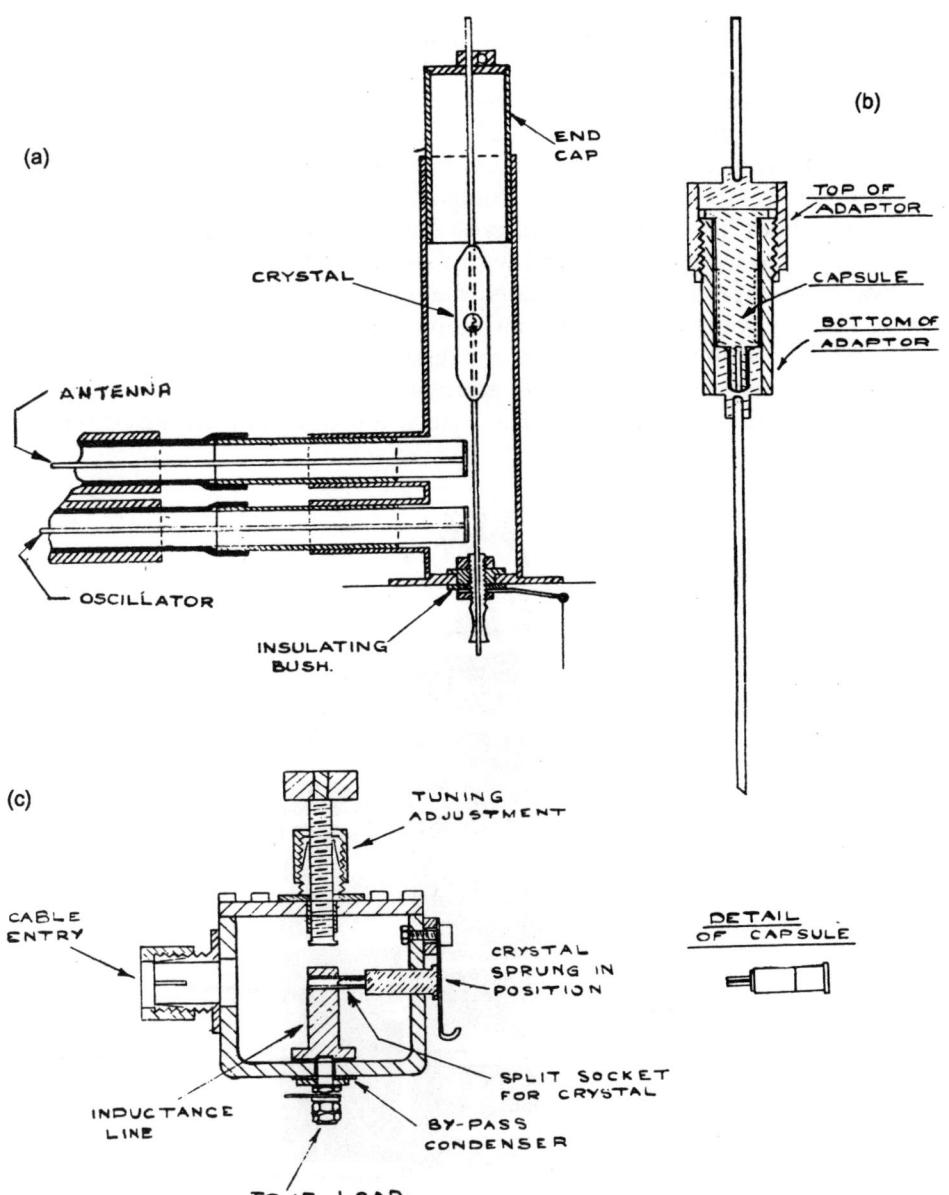

5.5 Schematic diagram of crystal holder and crystal for the mixer unit: (a) original full-wave line form for TRE laboratory crystal; (b) adaptor to US capsule-type production crystal; (c) quarter-wave inductance line mixer (© Defence Research Agency)

later supplied to permit use of this form of crystal valve in early models of the Type 271X mixer (Figure 5.5b). There is no record of how many Type 271X sets entered service using the laboratory-style crystal. However it is certain that the Type 271X on *Orchis* did so. Nor does any record exist for the overall receiver noise-factor for this trials set. It could have been as high as 20 db, since these early crystals were undoubtedly delicate, and deterioration in performance was very difficult to detect, except by comparison of performance on known fixed echoes. The cartridge-type crystal made possible the introduction of a simplified form of mixer unit with a single tuning control. This quarter-wave box-type mixer is illustrated diagrammatically in Figure 5.5(c).

Display was by simple A-scan, featuring range scales of 5000-yds and 15000-yds, with provision to use the display tube to monitor pulse shapes for maintenance purposes. The arrangement of the panels in the radar office is shown in Figure 5.6. The base of the antenna pedestal projected

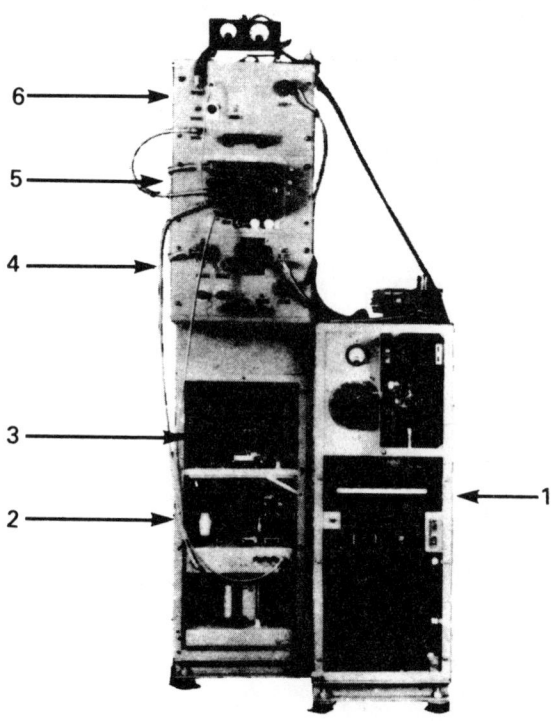

5.6 Type 271X panels in the original vertical format manufactured in the experimental workshops and by Allen West Ltd: (1) High-voltage rectifier; (2) Modulator; (3) Tetrode series-modulator valve; (4) Power-supply panel; (5) A-scan display; (6) Receiver panel (© Defence Research Agency)

directly into the office, to the right of the operator seated in front of the display. The antennae were rotated by direct hand-drive, and reporting was by voice pipe to the bridge. Initially no provision was made to aid the operator to hold an echo bearing while the ship was manoeuvring, the target bearing being reported simply by ship's head.

IN THE WAKE OF HMS *ORCHIS*

After completion of the *Orchis* sea trials the top priority at Eastney remained completion of the 24 experimental models of Type 271X, in close cooperation with Allen West Ltd, so that transfer of production tasks could be effected as soon as possible. By September 1941 the number of A/S radar-fitted corvettes was 32.[31] A few ships of other classes were also fitted with Type 271X, including the battleship *King George V*, the cruiser *Kenya* and the trawler *Avalon*. However destroyers, many of which were engaged on convoy-escort duties, could not be fitted with Type 271X as designed for corvette installation. Their multipurpose armament and limited upper-deck space could not accommodate the integral office/antennae configuration in any position that would provide a useful all-round view for the antennae. This situation led to the first major modification to Type 271X.

The obvious solution was to make possible independent siting of antennae and radar office. This became feasible with the development of the CV35 klystron, based on the NR89, to the production stage in 1941. This new reflection klystron operated at the same input power as the NR89 but, by virtue of a different electrode structure, had an efficiency of 3–5 percent compared with the 1 percent of the NR89.[32] It was also more stable in output power and frequency with respect to variations in the operating voltages. Given this trebling (or more) of local oscillator power, the cable length coupling local oscillator to mixer could be increased to 40-ft. In contrast, the lengthening of the modulator-to-transmitter cable had no significant effect on pulse shape in this radar, intended primarily for use in a warning role. The antennae were identical to the Type 271X antennae, but were mounted on a new short pedestal and rotated by a Bowden cable drive through the $\pm 200°$ sweep permitted by the cable connections. A weatherproof housing was still necessary for the antennae, and a similar form of construction was retained as for Type 271X. This was reduced as much as possible in dimensions, and therefore weight, for mast mounting. Although not possible as a top-mast antenna, the normal destroyer mounting position for this antenna was on the foremast, some 55-ft above waterline. In August 1941 the new configuration, in which the antennae were separated from the radar office, was given the designation Type 272.

The CV35 local oscillator, essential for the Type 272 development, together with the simplified mixer unit and improved receiver power pack, were also important improvements for application in Type 271X. This incorporation made tuning appreciably easier and added to reliability in service. They were therefore adopted in receiver production as early as possible, and indeed some of the 24 experimental prototypes may have been so modified. Since the unmodified Type 271X panels could not be used with Type 272, a distinction had to be made in the Type numbers of the different sets of equipment. In 1941 this was done by adding 'Mark 2' to the 271 Type number. When all radar equipments were designated in a uniform coding system in March 1942 the 'X' of 271X was simply dropped, and the radar with the CV35 modification became Type 271, thus indicating the first production version of Type 271. No records have been found of measured receiver performance of this first production model Type 271. However with its 60 MHz IF, of 2.5 MHz bandwidth, the overall-noise factor probably lay in the range 16 db to 20 db, depending on crystal performance. Despite the structural attraction of separate siting of the radar office and antennae, the preferred form operationally remained that of Type 271. The easy communication between radar office and antenna housing of Type 271 greatly facilitated tuning to maximum performance. This advantage was lost with the remote antennae of Type 272, which necessarily still carried the transmitter and mixer.

Provision of A/S radar for the larger fleet units was a simpler problem than for destroyers. Upper-deck space could be found for Type 271, and indeed the first major war vessel to fit 10-cm radar was *King George V* with a Type 271X sited on top of the tripod foremast, 100-ft above the waterline, overlooking the main armament director control tower. The cruiser *Kenya* also had an early Type 271 fitted on an upper-deck site. There was soon indication, from this latter fitting, of a problem that was to prove troublesome in cruisers. The radar lantern was too exposed to blast from the cruiser's main armament, and although the antennae and electronic equipment did not suffer damage the 'lantern' perspex windows were often cracked or broken. It was not until 1943 that an effective solution was found to this mechanical problem.

In the TRE performance tests using their experimental radar equipment, in November–December 1940, it had been demonstrated that increasing antenna height above sea level resulted in an increase in the first-detection range of the small 92-ton vessel used as radar target. Thus improved range performance against surface targets was expected from the higher siting possible for Type 271 on major war vessels. Additionally, battleships and large cruisers provided a more stable platform for the radar than the escorts for which Type 271 had been designed. It therefore seemed unnecessary to retain a form of antenna in which antenna gain had been sacrificed to provide a wide vertical beam. Replacement of the 'cheese' form by 3-ft aperture paraboloids (see Figure 5.7) as had been used in the experimental TRE equipment, raised the individual antenna gain to 250 times that of a half-wave dipole, with an overall vertical beamwidth for the radar of 7.5°, to 6 db down. Clearly, with this beamwidth, single-axis stabilisation of the reflectors to hold the beam horizontal was desirable, even for large ships. This development was not completed until the end of 1941, when the first stabilised antenna outfit was despatched for fitting in the cruiser *Nigeria*. By August 1941 six antenna pedestals had been modified to carry the paraboloid reflectors and despatched for fitting, primarily in battleships. This version of the 5 kW centimetric radar was designated Type 273 and was subjected to sea trials at the end of August 1941.

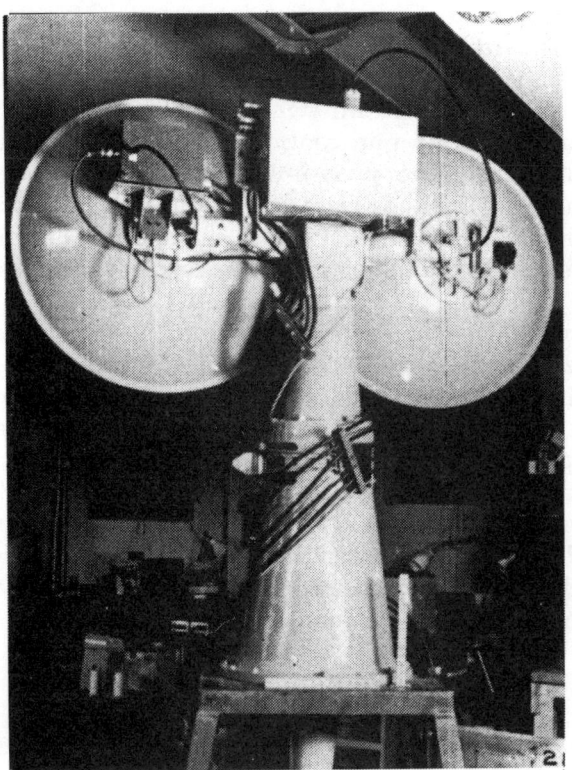

5.7 Rear view of Type 273 antenna with twin 3-ft paraboloidal mirrors and cable feed for both transmitter and receiver (© Defence Research Agency)

In a one-day exercise off Scapa Flow the results of these sea trials of Type 273 made clear the importance of 10-cm radar for general surface warning (WS radar).[33] Three ships cooperated in the exercise: the battlecruiser *Repulse*, the light cruiser *Euryalus*, and the destroyer *Lively*. Detection ranges on these targets were:

- *Repulse*:   37000–45000-yds
- *Euryalus*: 33000–36000-yds
- *Lively*:   28000–31000-yds

These ranges are well beyond hull-up visual horizon range, especially on the larger ships, and clearly indicated the capability of centimetric radar to detect targets as soon as any substantial portion of superstructure became apparent over the horizon. At 45000-yds no more than the masts, main armament director control tower and top of the bridge structure of *Repulse* would have been visible. The maximum ranges quoted were certainly for very intermittent signals, but the lower-range figures were described as reliable working ranges. This performance outranged the main armament gunnery radar, Type 284, operating in the 50-cm waveband, and led to an operational demand for an accurate ranging panel to be added to the 10-cm WS radar. Type 273 could not provide accurate bearing data for gun control, but at maximum gun range the radar ranging was more precise than the optical rangefinder and possessed the additional advantage of the capability for operating with equal accuracy at all ranges and under all conditions of visibility. This situation was exactly the same as with the corresponding German Naval radar.

By September the efficacy of 10-cm radar in ships as a detector of surface targets had been well established. The successful launch of Type 273 could have been said to mark the end of the first phase of introduction of this new waveband for shipborne radar. What had begun as a specialised A/S radar for corvettes had evolved into a general WS radar for all classes of oceangoing warship. Equipment had begun to flow from production, and a massive ship-fitting programme was then instituted. However development in the period April to September was not restricted solely to shipborne applications. Harbour and coast defence were also high-priority applications.

COAST AND HARBOUR DEFENCE APPLICATIONS

The detection and tracking of shipping in coastal waters was an Army responsibility, but it was also of great importance to the Navy: all reports of shipping movements were transmitted directly to the plotting rooms at the appropriate Naval Area HQ. Clearly the prospect of the availability of centimetric radar in production quantities was of great significance for this CD surveillance task. Thus one of the earliest Type 271X models was released to ADRDE, Christchurch, to be mounted and housed suitably for CD operations ashore. For this application the 'fan-beam' antenna was not required and a much-higher-gain paraboloid reflector antenna was used.[34] Also, by installing the radar inside a rotatable-cabin trailer, the antennae could be mounted on the side of the cabin, involving only short RF cable feeds to the transmitter and receiver mixer sited within the cabin. The use of a pair of 7-ft aperture paraboloids increased the antenna gain of this CD radar by a factor of about 25 relative to that of Type 271, and of about five relative to that of Type 273.

This first experimental conversion of Type 271X was sited at Lydden Spout (Dover) in July 1941 for trials. Officially designated CD No 1 Mark 4, it was more widely referred to as Type NT271X. By the end of August 1941 the trials team reported of this set that:

> 'Operationally NT271X was a great advance on previous sets in maximum range, discrimination, counting and accuracy. For the first time reliable cover was obtained across the Channel, so that not even E-boats could go between Calais and Boulogne undetected. Large ships could be watched at anchor in the outer harbour of Boulogne.'

The 330-ft height of the Lydden Spout site placed the French coastline more or less on the horizon; the range on the E-boats was some 18 miles. These results were so satisfactory that it was agreed to retain the experimental set at Dover, and to build 12 more at ADRDE for the most critical sites. These were to be followed by a further 50 equipments of the mobile trailer design. These formed the 'K' stations of Coastal Chain.

With the introduction of waveguide techniques in 1942, ADRDE developed a more versatile form for the CD radar. The transmitter, receiver and display were mounted in a box, large enough to be used as a radar room but of a size transportable by lorry. The antenna was single 7-ft diameter paraboloid used with common T/R working, and was rotatable. This radar, CD No 1 Mark 5, could be used as a self-contained unit on a temporary site or be dismounted and installed in a permanent building. In the latter instance it was designated as CD No 1 Mark 6. Over 200 of these CD radars were manufactured using Type 271 electronic panels. Additional demand for Type 271 equipments arose in 1942 with the development of split-beam facilities on the receiving antenna, the transmitting antenna being fed from a Type 271 transmitter. Used with appropriate display equipment this gave the necessary range and bearing accuracy for battery fire-control.

The demand for centimetric radar for specifically Naval shore-stations called for individual adaptation to particular circumstances rather than for quantity production of yet another form of Type 271.[35] The first demand recorded was from the Mediterranean Command for means to detect Italian two-man torpedoes, which were being used to attack shipping in the harbours at Gibraltar and Alexandria. The only possible response in October 1941 was Type 273, with an antenna modified to take the largest possible paraboloids side by side on a standard pedestal arm. This was a 4.5-ft diameter paraboloid. It is not known whether these sets had any success, but the lack of any information suggests not. However in due course the set at Gibraltar was moved from the mole and mounted on the Rock. In this position its antenna gain of 575 times that of a half-wave dipole permitted effective monitoring of shipping in the Straits. Other Type 273s were similarly adapted for service ashore in 1942, with designated Type 273S.

In postwar listing of Naval shore-station radar, only Type 273S is included for the 5 kW radars. However in papers in the Public Records Office[36] there is reference to a Type 271S at Ras al Tin in February 1942, and to another at Bone in July 1942. Type 273S, with its 4.5-ft mirrors, is recorded[37] as having been installed at Liverpool, Selsey Bill, Rosyth, Eide (Faroes), Trinidad and Malta specifically as watch and guard for minelaying operations. One shore installation requires special mention. A Type 273M was installed at Saebol, Iceland. This was a special modification of Type 273S, with the antennae mounted on a gun ring instead of a pedestal because of the very high winds encountered at this key site, looking across the Denmark Straits to Greenland. On trials on 29 September 1942 this set achieved a 92000-yd range on a trawler from its 1520-ft high site, that is, very nearly the radio-horizon range of 96000-yds for this height.

TYPES 271, 272, 273 IN SERVICE, 1941–42

By autumn 1941 it was clear that the demand for the 5 kW centimetric radar equipment from Naval production would far exceed the production capacity of Allen West at Brighton. In terms of Naval requirements, this was no longer an A/S radar for a limited class for escort vessels, but rather a WS radar for all oceangoing warships. In addition to the expansion of demand, the War Office requirements for the Coastal Defence Chain and the Air Ministry interest in the equipment for certain ground applications had to be added.[38] It was at this time that the initial order on Allen West was increased from 150 to 350 sets, and Metropolitan Vickers in Manchester were contracted to redesign the electronic panels (modulator, receiver and display) to facilitate production and assembly in greater quantity. This was a purely mechanical redesign (Figure 5.8), which replaced the three-panel arrangement of the Allen West production by an arrangement in two cubical panels, mounted one above the other. The prototype panels for this Mark 3 equipment were received at Eastney for acceptance tests in November 1941. The production contract for 1000 sets was completed by the end of 1942. When installed in ships with these production design panels a 'P' was added to the type number; Types 271/2/3 becoming Types 271P/ 2P/3P.

5.8 Type 271P panels: the early Type 271 panels redesigned for main production. The modulator with power supply is below; the receiver, display and associated power supplies are immediately above; the G82 tuning test set is on top (© Defence Research Agency)

In order to facilitate fitting of Type 271P on escorts, and thus avoid their withdrawal from service for longer than the minimum time necessary for their periodic dockyard boiler clean, prefabrication was introduced for the radar office, complete with antenna housing. A firm in Manchester, W. H. Smith Ltd, was contracted to manufacture the structure and install the Type 271P radar. This complete, prefabricated (PF) structure with radar installed and tested was transported by lorry to the dockside, hoisted aboard by crane and secured on an already prepared deck site. In this way the deck site could be cleared and prepared during one boiler clean, and the radar fitted during the next.

Work had also been initiated in 1941 at Eastney on a shipborne adaptation of the Plan-Position-Indicator (PPI) type of display developed by TRE. Since the antennae of the air and surface warning radars were not stabilised in azimuth against change of ship's head, stabilisation on compass bearing had to be applied. This was effected in the scanning coils of the PPI by mixing the antenna-position-signal and gyro-compass signal in a differential, whose output drove the PPI scanning coils. In February 1942 an experimental 12-in diameter PPI display was installed in the Plotting Office of *King George V*. For this trial the antennae of the Type 273 radar were temporarily modified for power drive, which was necessarily reciprocating, since the cable connections precluded continuous rotation. There was no doubt concerning the potential value of this form of display as a means of transmitting radar data to the command position. On the basis of this experience a development contract was placed with EMI for a PPI display unit for Naval use. This equipment, later designated Outfit JE, was based on the use of a 9-in CRT to reduce bulk. It was widely fitted from mid-1943 onwards.

The technical history of these first centimetric radars would be incomplete without reference to the troublesome problem of 'side echoes' experienced with shipboard fittings. As the novelty of being able to 'see' at a distance at night wore off, and more and more reliance was placed on Type 271 radar, both for the detection of surfaced U-boats and for convoy station-keeping, reports began to be received of large 'side echoes' that obscured wide arcs of the antenna sweep. These were apparently associated with the large echoes from the ships in convoy. Nothing during the development phase of Type 271 had suggested the existence in azimuth of other than the small diffraction sidelobes normally associated with this quasi-optical form of antenna. Measurements in February 1942 on the Type 271 in *Guillemot* with a dockyard-built lantern, and in March on Type 271P in *Veteran* with a factory-built PF lantern, traced the source of the radiation-scattering to the lantern roof-supporting pillars. The magnitude of the effect on the antennae performance can be seen from the two diagrams of Figure 5.9, showing echo response on the A-Scan to a strong signal as the antennae were rotated before and after replacement

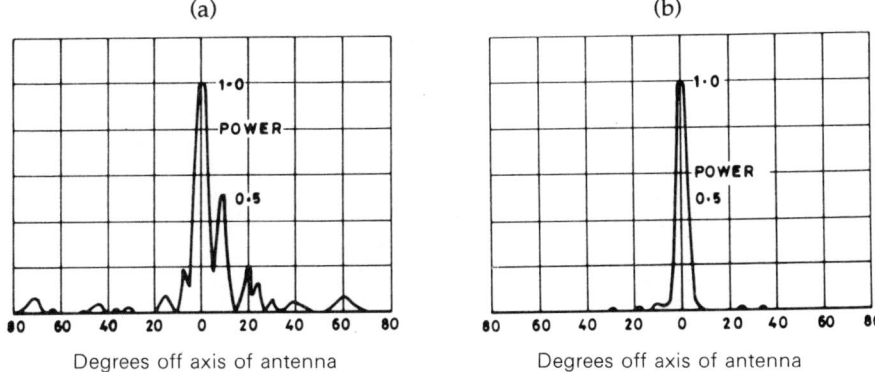

5.9 Effect of the lantern of the Type 271 radiation pattern: (a) teak lantern with plane perspex windows; (b) cylindrical perspex lantern
(© Defence Research Agency)

of the pillar and window-type lantern. The lantern was replaced by a 'radome' in the form of a perspex cylinder of uniform thickness, chosen to minimise back-reflection of radiation. The first ship fitted with this new radome was *Hesperus* in November 1942. The larger radome for Type 273 (Figures 5.10) took longer to develop, and had also to be blast-tested for cruiser installation. However all Types 271 and 273 had been fitted with the new-style radome before the end of 1943.

ANTI-SUBMARINE RANGE PERFORMANCE OF 5 KW CENTIMETRIC RADAR

Although the maximum range of 2.5 miles on a surfaced submarine achieved in the sea trials of Type 271X in *Orchis* was acceptable operationally, it was less than had been expected technically in comparison with the 7.5 mile range in the experimental tests at Swanage in November 1940. By analogy with experience at metric wavelengths this shortfall was attributed to the low antenna height, in combination with the small size of the submarine for the *Orchis* trials. As described earlier, the lower the antenna the greater the upward slope of the lobes of the interference field formed when propagating over the sea. In the limit, first detection of the target is made, not at a lobe maximum, but on the underside of the first interference lobe, closer to the radar. At the 250-ft height of the antenna on the Swanage site it was reasonable to presume that the conning tower of *Usk* had been illuminated in a maximum lobe of

5.10   HMS *Suffolk* with Type 273 raised on stalk for clear view over High-Angle Director to left and Low-Angle Direction to the right with their 50-cm Yagi radar arrays. Cylindrical perspex radome for Type 273 antenna (© Defence Research Agency)

the transmitted field. At the 37-ft height of the antenna in *Orchis* this was far from certain, especially with a smaller submarine of lower profile. In addition, judgement of range performance of Type 271X was distorted in 1941–2 by the failure to appreciate that measured gain data for the 'cheese' antenna was being quoted for an antenna with a modified feed. Estimates of possible range performance of Type 271X were therefore based on an antenna gain for the 'cheese' of approximately half that of the Swanage paraboloids, whereas in reality it was only about one fifth that gain.

For propagation in Free-Space the radar equation in a given narrow waveband takes the form:[39]

$$\tau_{max} = K \sqrt[4]{\frac{P_T.G_T.G_R.\sigma}{P_R(\min)}}$$

When propagation is over a perfectly reflecting flat surface the expression for maximum range is:

$$\tau_{max} = 2K \sin\left(\frac{2\pi hH}{\lambda \tau_{max}}\right) \sqrt[4]{\frac{P_T . G_T . G_R \sigma}{P_R(\min)}}$$

where K is sensibly constant over the waveband, h is the target height, H is the antenna height, P_T is the transmitter power, G_T, G_R are the transmitting and receiving antenna gains, σ is a coefficient measuring the reflectivity of the target, $P_R(\min)$ is the minimum detectable signal power and λ is the wavelength.

The sine function is usually referred to as the form factor for propagation over a reflecting surface, and determines the geometry of the multilobe structure. For a given antenna height H and wavelength λ, the height of the first maximum of interference field h above the surface at distance τ is given by $h = \lambda\tau/4H$. At the 7.5 mile range on *Usk* from the 250-ft site this made the height of the first field maximum above the sea 15-ft–appreciably less than the conning-tower projection of about 20-ft. However the figure for first field maximum for Type 271X was 33-ft at the 2.5 mile maximum range in *Orchis*. This situation lent credence to an 'under-the-beam' interpretation of the short range on trials.

This apparently reasonable hypothesis concerning the propagation behaviour to be expected at 10-cm wavelength was called in doubt by a further sea trial of Type 271X on 5 August 1941. This trial used the Type 271X in *Penstemmon*, and took place in the Channel off the Eddystone Lighthouse. A maximum range on the large submarine *Sea Lion* of 4.5 miles was achieved. The experience of the Army research team with the experimental CD installation at Dover was also relevant in the attempt to understand propagation behaviour in the centimetric waveband. Using the Type 271X equipment connected to high-gain, 7-ft aperture reflector antennae on a site 330-ft above sea level, E-boats were detectable at around 18-miles range.

Individual operational sighting reports were not usually a good source of data on maximum range. Too many other factors can influence the time of first detection of a small echo. However there were two operational reports of particular interest. The first of these concerned *Prince of Wales* in September 1941. The sea trials of Type 273 on *Prince of Wales* had concentrated on performance in relation to surface-ship targets, and there had been no antisubmarine trial of this new adaptation of Type 271, with its 3-ft paraboloid antennae sited at 100-ft above sea level. However in September, while escorting a fast convoy to Gibraltar, small echoes had been detected on her Type 273 at ranges between 10000-yds and 15000-yds. Two of these targets were subsequently identified as Italian submarines on the surface.[40]

The second operational report was in April 1942, when *Vetch*, with Type 271, made a first sighting of a U-boat at 7500-yds.[41] Other escorts in company in this action achieved first-sighting on their radar at ranges around 6000-yds. This excellent performance of Type 271 in *Vetch* was consistent with the experience of Type 271 already gained in radar bases. A Type 271, well-maintained and tuned to maximum performance, was expected to achieve 6000-yds to 7000-yds range on a submarine. A postwar Admiralty Operational-Research analysis of radar first-sighting ranges on U-boats during the period 1942–45 (Table 5.1) gives further confirmation. In 1942 only Types 271 and 272 would be involved in this analysis. It was not until 1943 that a higher-powered replacement for these sets entered general service. *Penstemmon's* results now had to be considered as an early experience of anomalous propagation. This was known to have been very prevalent in the Channel area in the hot summers of 1941–42. The most likely cause of the shorter range in the *Orchis* trials was the considerable reduction in target size, especially as regards freeboard.

For the four remaining measures of maximum range, namely CD Dover (18 miles), Swanage and *Prince of Wales* (7.5 miles) and *Vetch* (3 miles), the range achieved was directly proportional to the square root of antenna gain; the only other variable was the height of the antennae. This concordance over a wide range of antenna heights suggested that, in all instances, the target was illuminated near or above the first maximum of the transmitted field. In effect the combination of antenna size and its height above waterline in a corvette and a battleship, respectively, was optimum for the available power of the radar. Strictly speaking, in 1941 these two factors were more determined by the constraints imposed by ship construction than by any prescience as regards the transmitted field conditions. However the latter finds expression in the choice of sites for

Table 5.1 Radar detection ranges on U-boats, 1942–45

Period	Number of detections by range band (miles)					
	0–1	1–2	2–3	3–4	4–5	Over 5
Jan/Apr 1942	1	2	1	1	–	–
May/Aug 1942	–	8	3	3	–	–
Sep/Dec 1942	6	19	8	–	–	–
Jan/Apr 1943	18	27	13	7	6	1
May/Aug 1943	–	7	22	4	–	3
Sep/Dec 1943	3	13	10	2	9	8
Jan/Apr 1944	4	6	6	8	3	5
May/Aug 1944	1	1	–	–	–	–
Sep/Dec 1944	–	4	3	1	–	–
Jan/May 1945	–	–	1	–	–	–

Source: Admiralty Operational Research group analysis of hunts on U-boats by Surface Craft

the Army 10-cm CD stations. For these a minimum height of 200-ft above sea level was specified and adhered to, even to the extent of building 200-ft towers to carry the radar on the flat East Anglian coastline. The 'under-the-beam' hypothesis resulting from the *Orchis* trial did, however, continue to find expression in listings of performance of different radars, even postwar. Type 272 will be found ascribed a 15 per cent range advantage over Type 271 because of its higher antenna; a circumstance that does not pertain unless at both heights the target is first detected below the first lobe maximum.

THE HIGH-POWER MAGNETRONS

Progress in development of the critical RF components for 10-cm radar was rapid in 1941. The crystal, in its ceramic encapsulated form, was developed for production by April;[42] the tuneable 'gas switch' for crystal protection against the transmitter ground wave[43] followed in June, also making possible the use of a single antenna for transmit and receive; and the introduction of waveguide technology for low-loss power transmission was introduced later in the year.[44] Perhaps the least expected development was the achievement of 500 kW pulse power with the CV76 magnetron before the end of the year. As early as September 1940, the NT98 type magnetron used in Type 271 had been demonstrated to be capable of generating an RF pulse power of 100 kW, using a high magnetic field and at correspondingly high applied voltage.[45] However the efficiency of generation remained in the region of 10–15 percent.[46] The concept of linking alternate resonant cavities in the magnetron anode-block with metal straps was first proposed by J. Sayers of Birmingham University in the autumn of 1941. In its strapped condition the magnetron efficiency rose to about 40 percent using magnetic fields of 1500 Oersteds, and to the region of 50–60 percent at field strengths above 2000 Oersteds.[47] As will be seen, this early achievement of high transmitter power became a dominant factor in seeking replacement radars for Types 271 and 273 of higher performance.

In April 1941 the route to longer range performance with WS centimetric naval radar was clearly through the use of higher transmitter power. As regards the receiver, there was no immediate prospect of RF amplification without the introduction of additional noise. Also antenna size and weight were constrained by considerations other than radar performance. Furthermore the transmitter power increase had to be considerable in order to have a significant effect on surface target range. Antennae heights were already fixed for Types 271 and 273, and there could no longer be an optimisation of performance such as existed with the more powerful Type 273 by mounting the antennae higher when fed from the new, more powerful transmitters. Detection of surface targets at extreme range would be 'under the beam', and in these circumstances the radar equation can be approximated to show that the range increase to be expected from an increase in transmitter power would be proportional to the eighth root of the power.[48]

By the time of the *Orchis* trials a preproduction design of magnetron of the NT98 type had already been produced to generate 100 kW RF pulse power with 1 MW pulse-power input.[49] The series hard-valve modulator of Type 271 was operating at the limit of performance of the NT100 tetrode modulator valve, and thus the immediate development requirement was for a new, more powerful modulator. Clearly the ultimate aim in development was the '1 MW Modulator'. However, in the first instance, in order to obtain a compact, direct replacement of the existing hard-valve modulator of Types 271/273 a target was set to achieve 25 kW RF pulse-power output from the magnetron at 15 kV, 15 Amps input using a magnetic field of 1500 Oersteds.[50] By September 1941 an experimental model had been built and installed in the trailer at Eastney. This modulator was based on the use of an 'artificial discharge-line', effectively a section of transmission line composed of condensers linked by inductances.[51] This was charged to twice the required load voltage through a simple hold-off diode on one half-cycle of the mains, and discharged through the load on the other half-cycle by a triggered thyratron. With the load correctly matched to the line impedance, the duration of the resulting current pulse was determined by the number of inductance-condenser elements in the line. The line was completely discharged and quiescent after this single square-current pulse. This technique had already been used in the metric air-warning radar Type 281, but only as a pulse-shaping network in the submodulator (see Monograph 4). In this new modulator the line was the direct energy source for the magnetron. To achieve the necessary impedance match of the magnetron-to-line required design and development of a high-power transformer capable of handling the very short pulses used in centimetric radar, without distortion. Two pulse lengths were provided: a 2-microsecond pulse for search, and a 0.7-microsecond pulse for discrimination between closely spaced targets. Using this discharge-line modulator the change to short pulse was effected simply by disconnecting the necessary number of lumped condenser-inductance elements in the line.

Delivery at Eastney of the first strapped magnetron, the CV56, more or less coincided with the installation of the experimental modulator. Naturally the immediate reaction was to test this high-efficiency magnetron in the trailer installation, in the hope of seeing a dramatic improvement in the strength of the 'standard' echo from the Nab Tower.

There was indeed a dramatic demonstration of increased transmitter power, but not in the form of target echoes! Examination of the antenna revealed that all that remained of the dipole feed was a single quarter-wave rod dangling on the end of a filament of polythene. Whilst it was very disappointing to be denied an immediate test in a working radar at much higher power than the intended 25 kW, waveguide technique had by then evolved sufficiently to ensure that the antenna feed by cable and dipole could be replaced, after a short delay, by waveguide and horn components.

The CV76 magnetron followed very quickly after the CV56.[52] This strapped magnetron, using a magnetic field of 2400 Oersteds, generated 500 kW of RF pulse power from an input of 25 kV, 40 Amps. Thus the 1 MW modulator project remained an essential feature of the 10-cm radar-development programme, even though the initial aim of 100 kW transmitter output power was now assured at much lower input. The situation in late autumn 1941 can thus be summarised as follows:

- a production and fitting programme for the 5 kW A/S, WS radar was just reaching full momentum, in response to the clear Navy priority for the maximum number of ships to be fitted in the shortest possible time;
- experimental development was largely completed for a direct replacement of this radar by a medium-power (70–100 kW) radar;
- potential existed for a 500 kW radar after further modulator development; with development of the necessary waveguide components, this set could employ an antenna remote from the radar office, and the antenna could be continuously rotated.

THE MARK 4 DEVELOPMENT: TYPES 271Q AND 273Q

Experimental development of the medium-power (70 kW) Mark 4 equipment was completed by the end of 1941. Both office equipment and antennae were designed as direct replacements for the corresponding units of the Type 271P and Type 273P Mark 3 equipment (see Figures 5.11 and 5.12). Some structural alterations were necessary in the radar office when replacing the earlier preproduction Mark 2 equipment, but these did not involve any major dockyard work. Thus, despite the rapid advances in waveguide technique, Types 271Q and 273Q retained the integral office/antenna-lantern construction, with separate transmitting and receiving antennae, whose direct hand-rotation limited azimuth search to $\pm 200°$, due to the cable connections.

Three prototype equipments were built in the laboratory workshops at Eastney,[53] followed by 20 preproduction sets for delivery in 1942 (10 from Marconi and 10 from Allen West). The first prototype replaced an

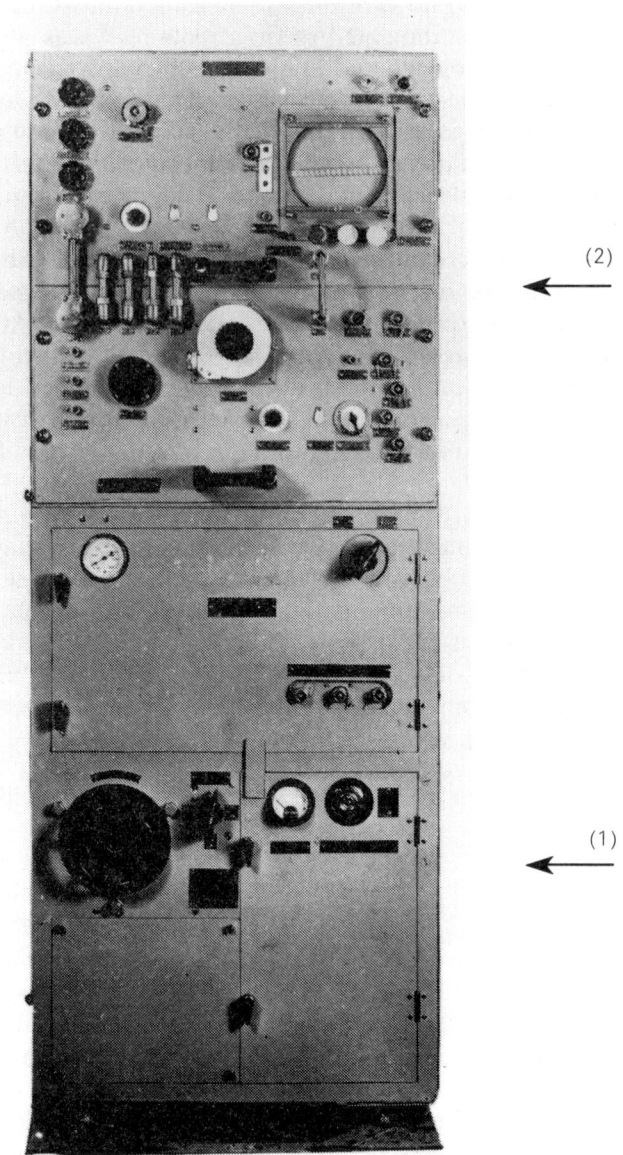

5.11 Type 271Q panels designed to fit the same space as Type 271P panels, for ease in conversion: (1) discharge-line modulator unit; (2) receiver and display unit (© Defence Research Agency)

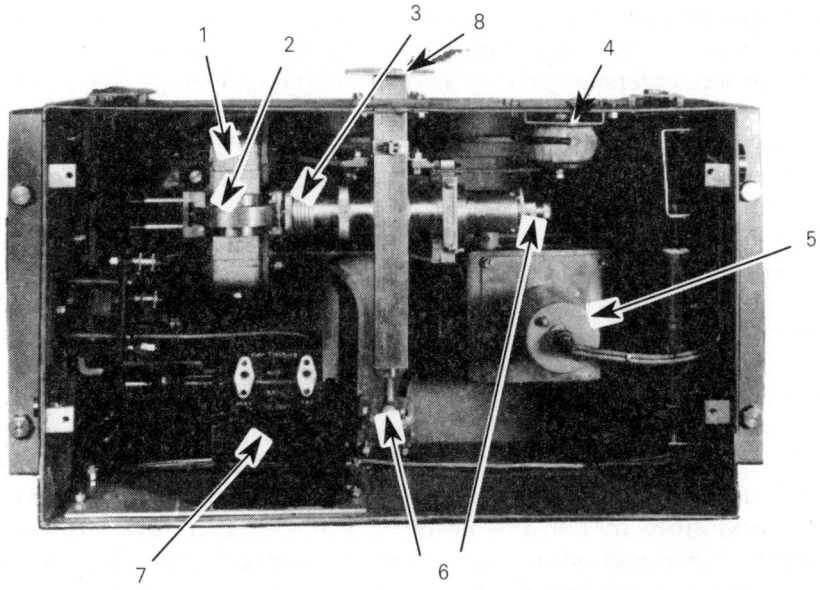

5.12 Type 271Q transmitter mounted behind the antenna: (1) magnet; (2) magnetron; (3) flexible bellows coupling of magnetron to output line; (4) current transformer to monitor magnetron pulse; (5) cooling fan; (6) matching adjustments on output line; (7) voltage step-up transformer for modulator pulse; (8) waveguide to antenna (© Defence Research Agency)

existing Type 271 (Mark 2) in the corvette *Marigold*. There were A/S and WS range performance trials in May 1942 off Tobermory, which also included assessment of the capability of 10-cm radar to spot 4-in shell splash.[54] The second prototype replaced the existing Type 273 in the battleship *King George V*, and sea-trials of WS performance were conducted off Scapa Flow in July, 1942.[55] In this instance, in addition to ship targets there were preliminary trials of the performance of Type 273Q against low-flying aircraft. The antennae of this first Type 273Q were not stabilised to hold the paraboloid reflectors in the horizontal plane; the first stabilised antenna outfit was dispatched for fitting in the cruiser *Belfast* in November 1942.[56] Main production of Mark 4 equipment for Types 271Q/273Q began in January 1943, on completion of the earlier contracts for Mark 3 equipment for Types 271P/273P. By the end of 1942 the source of excessive side echoes experienced with Types 271/273 had been traced to the pillar and window construction of the antenna lantern, and a new cylindrical perspex radome was available for general fitting.

Thus the second convoy-escort to be fitted with Type 271Q late in 1942, *Itchen* also received the new radome. This was an essential improvement in Type 271Q since even the small diffraction-pattern side lobes of an unprotected antenna could generate troublesome side echoes when operating close to large targets. The operational problem of retaining full sensitivity at maximum range while still being able to resolve targets at shorter ranges, was not finally solved until adjustable swept-gain was introduced on the receiver later in 1943.

The completed Mark 4 equipment had a nominal transmitter pulse power of 70 kW, with a choice of either 2.0 microsecond or 0.7 microsecond pulse width and corresponding receiver bandwidths of 1.25 MHz and 4.0 MHz. The receiver noise factor was about 15 db.

The modulator and transmitter, the latter still mounted behind the antenna, were designed to work with up to 100-ft of cable separation.[57] In the receiver the local oscillator was pulsed on only during transmission half-cycles, thus permitting up to 100-ft of cable feed to the crystal mixer. This ability to separate office and antennae by up to 100-ft was never in fact used, since no conversions were made of Type 272. As for the antennae, these were identical to those of Types 271 and 273, except for the change to waveguide feed for the transmitting antenna (Figures 5.13 and 5.14). This change was not, however, insignificant as regards performance. Using waveguide feed the gain of the Type 273 paraboloidal transmitting antenna was increased from 250 to 360 times that of a half-wave dipole, and that of the Type 271 transmitting 'cheese' from 55 to 180. At the time the magnitude of the contribution of higher antenna gain was not fully appreciated for Type 271Q, since the gain of the dipole-fed Type 271 'cheese' was still believed to have been around 130. It was perhaps fortunate that, in the interests of minimising change to equipment, separate antennae were retained for transmission and reception. The combination of a waveguide-fed and a dipole-fed 'cheese' retained an adequately wide vertical beamwidth to cope with the roll of the corvettes in a heavy sea.

With 70 kW pulse power from the transmitter, the receiver crystal had to be protected from the transmitter ground wave, even with separate transmitting and receiving antenna. A tuned-cavity gas switch could restrict the power loading on the crystal to less than the 1 watt maximum permissible, but the speed of ionisation was not fast enough to prevent an initial spike of energy exceeding the permissible 1 erg.[58] The problem was resolved by mid-1941 with the development of a tuneable-cavity gas switch using the type of cavity employed in the klystron valves. With the inner part of the cavity sealed within a glass envelope, a small ionising discharge could be maintained, external to the cavity, which assured extremely fast ionisation of the gas in the cavity on the rise of the transmitter pulse. The protective gas switch introduced a loss of about

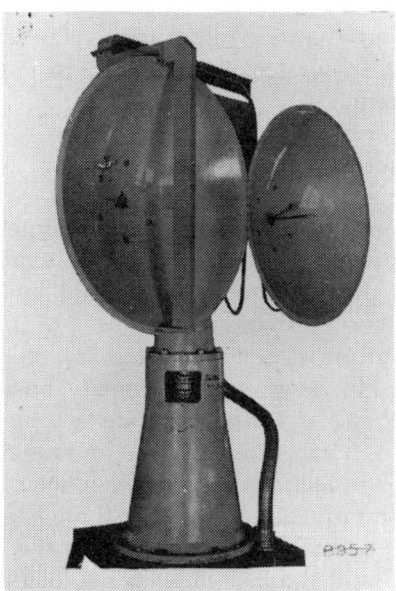

5.13 Type 273Q antenna with waveguide transmitter feed (© Defence Research Agency)

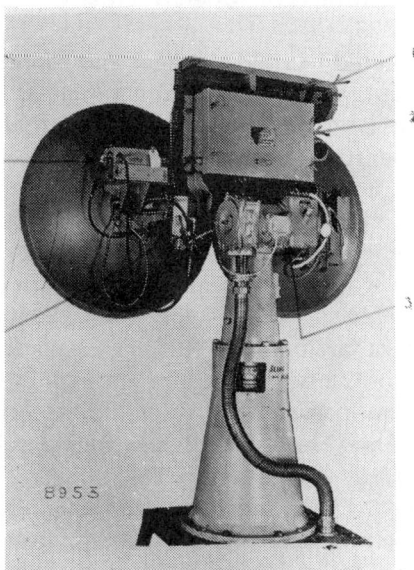

5.14 Type 273Q antenna – rear view showing (1) waveguide feed to horn; (2) transmitter unit; (3) stabiliser to hold beam horizontal; (4) gas switch to protect receiver crystal; (5) crystal mixer and IF head amplifier (© Defence Research Agency)

1 db in the receiver channel, although in practice this was fully compensated by the lower loss experienced in the waveguide feed to the transmitting antenna.

The uplift in power of Type 271Q and 273Q was considerable. In terms of equivalent transmitter power it was some 45 times that of Type 271 for the former, and some 20 times that of Type 273 for the latter. Antennae heights were unchanged on replacement of Types 271/273 by Types 271Q/273Q. Small surface targets detectable only on the underside of the radar beam at less than horizon range could be expected to show a minimum first-detection-range increase of 1.6 times and 1.45 times respectively (maximum range proportional to the eighth root of transmitter power). However, as had already been experienced in the Type 273 trials in *Prince of Wales*, large surface targets were detectable well beyond 'hull-up' horizon range (horizon range $d = 1.23\sqrt{h}$, where d is in nautical miles, h in feet, and atmospheric refraction is accounted for by assuming an earth radius of 4600 nautical miles). As already noted, all that would have been 'visible' of *Repulse* from the 100-ft high Type 273 site on *Prince of Wales* at the 22.5-miles range of first detection was the massive main armament director, the top of the tripod mast, and the pole masts. In effect, range was already limited by earth curvature with Type 273, requiring only visibility above the horizon of a sufficiently massive element of superstructure. (For superstructure of height H in feet, overall horizon range $d = 1.23[\sqrt{h} + \sqrt{H}]$ nautical miles.) In these circumstances Type 273Q was expected to yield stronger and more useful echoes, rather than any startling increase in maximum range of detection. Prediction could not usefully be carried much further than this in 1942, and determination of operational performance depended mainly on sea trials.

In the sea trials of Type 273Q in *King George V*[59] the heavy cruiser *Cumberland* was detectable at a range of 22 miles. This was not significantly different from the range achieved on *Repulse* during the Type 273 trials, but was achieved on a target with a much less massive superstructure. However, perhaps the most instructive comparison of different WS 10-cm radars on large surface targets was to be found in the operational reports of the two engagements with the *Scharnhorst*; the first in February 1942 on passage up the Channel, and the second with elements of the Home Fleet in December 1943 when *Scharnhorst* was sunk. In 1942 *Campbell*, with a newly fitted Type 271, led the destroyer attack on *Scharnhorst*. First radar detection was at a range of 9.5 miles which, from the antenna height of a bridge-mounted Type 271, implied nearly full hull exposure of *Scharnhost* to produce a detectable signal. On the other hand the Dover CD station, the original NT271X at 330-ft above sea level, with its 7-ft paraboloidal reflectors and 5 kW transmitter, reported first detection at 32.5 miles. At that range no more than the top superstructure of *Scharnhorst* would have been visible from the radar.

Type 273Q, with its smaller reflector but higher-power transmitter, had the same range performance capability as the Type NT271X, but the 100-ft site on top of the tripod mast in *Duke of York* foreshortened considerably the horizon range on surface targets. After the action, in December 1943, the Commander-in-Chief Home Fleet described the performance of Type 273Q as follows:[60]

a) The surface warning provided by Type 273Q was entirely satisfactory, giving on PPI a clear picture of the situation throughout the engagement. Blast from the ship's own broadsides so shook the office that some of the overhead supports for the panels were carried away, but the set remained functioning throughout the entire period.

b) Gyro-stabilisation of the aerial proved its worth, justifying for the first time the fitting of such gear in capital ships.

c) The successful presentation on the PPI of the tactical situation was almost entirely due to the improved performance of the set since fitting the cylindrical perspex lantern; this, by cutting down side-echoes to negligible proportions, has improved the value of the set by one hundred percent.

d) The picking-up range of *Scharnhorst* was 45,500 yards, nearly the full visual range of *Scharnhorst's* director tower from the height of the *Duke of York's* radar aerial.

e) Type 281 was able to hold the *Scharnhorst* up to 12.75 nautical miles, a reminder of the useful part this set can play as a stand-by for surface warning.

The surfaced submarine represented the other end of the scale of target size for general surface warning. For this target there was no horizon limitation effect in relation to maximum range with the earlier Types 271 and 273. The range of 3.5 miles (*Vetch*, Type 271) and 7.5 miles (*Prince of Wales*, Type 273) were well within horizon range from the height of the antennae, and therefore experienced full freeboard exposure of the targets. On sea trials of Type 271Q in *Marigold* in May 1942 a maximum range of 7.5 miles was recorded on a small H-class submarine, fully surfaced. For Type 273Q the sea trials in *King George V* gave a maximum range of 10.5 miles on *Uredd*, a large Norwegian boat. In both instances the long-range echoes were intermittent and, especially with the small boat of the Type 271Q trials, obviously dependent on the lift of the boat in the ever-present ocean swell off Tobermory. However confirmation of the marked improvement in A/S performance is to be found in the postwar analysis of first U-boat sightings quoted earlier in Table 5.1. From 1943 onwards, when fitting of Type 271Q in escorts became general, maximum first sighting range increased from the 3–5 miles range bracket to over 5 miles. In effect both Types 271Q and 273Q, given full freeboard exposure of a large submarine on the lift of the waves, had a detection capability close to horizon range from the height of the antenna. Yet higher transmitter power could not , therefore, be expected to yield significant increase in the range of first detection of surface craft with 10-cm radar at the antenna heights possible aboard ship, and indeed much of the 70 kW equipment was still in service at the end of the war.

Of more immediate consequence to the further development programme for 10-cm equipment was the urgent need for warning of very-low-level air attacks. These escaped detection on the metric WA radar until the aircraft were too close for effective counterattack. During the Type 273Q trials a number of test runs were flown by the ship's Walrus aircraft and by a Sunderland flying boat. The conclusion, as summarised in the Signal School technical report of the trials, was:

> 'It is not easy to pick up, or to follow, aircraft without the help of a PPI, in conjunction with a rapidly spinning mirror. Results achieved suggest, however, that Type 273 Mark V with a 4.5-ft spinner should be capable of picking up aircraft at a height of up to a few hundred feet at ranges at which they are just dipping beyond the visual horizon.'

THE MARK 5 DEVELOPMENT

The 500 kW CV76 magnetron became available in preproduction form only weeks after delivery of the 70 kW CV56 version used in the Mark 4 equipment. However the Mark 5 equipment was essentially a longer-term project. In the first instance, the experimental research phase on the discharge-line modulator had to be completed, to raise the output to the 1 MW level for the pulse drive needed for the higher-power magnetron. This involved much higher stress on components, and presented new problems in the design of the pulse transformer, because this component had now to handle a mean power of 1 kW. Life-testing of components was essential. Also, although separation between transmitter and modulator was still possible, even at these power levels, the rapid development of waveguide technology in 1941 eliminated the necessity for minimising the length of the RF feed to the antenna. It was therefore planned to use a long waveguide connection office-to-antenna, and thus concentrate all electronic equipment in the radar office. Common T/R working for transmit and receive functions had also become possible with the invention of the rapidly-ionising gas switch,[61] and continuous rotation of the antenna was possible using a waveguide rotating joint. This latter was already planned by ADRDE to be used in a common T/R working version of the original CD (5 kW) rotating cabin installations.

The operational situation had also changed. Before the end of 1941 radar had ceased to be a novelty, restricted to application in a few selected ships. Also the 10-cm band, in particular, was no longer regarded as only a specific A/S countermeasure. Its value for general surface warning (WS) was recognised, and as such it had become a necessary adjunct to the equipment for all classes of warship. As stated earlier, ideal sites existed for the integral office/lantern configuration of Type 271/273 on corvettes and battleships, but siting was difficult in other classes; in some, such as fleet destroyers, it was impossible. The compromise solution to this problem with Type 272, which became available from production in November 1941, was far from ideal either structurally or technically. This was because its antenna still required a protective lantern, and its transmitter and mixer were situated at a distance from the main panels. Two different antennae arrangements were therefore proposed for development for use with the Mark 5 shipborne equipment. One would be based on use of a 4.5-ft diameter paraboloidal reflector, necessarily stabilised to hold the beam horizontal. This version would replace Types 271 and 273, making full use of the opportunity presented by adoption of common T/R working to increase the gain of the antenna. The other would be a lightweight version, based on the use of a single 'cheese' reflector, which would replace Type 272. Initially designated Type 273 Mark 5 and Type 272 Mark 5, respectively, these radars became Type 277 and Type 276 in the renumbering exercise of March 1943.

A second operational development, which greatly influenced the planning of the Mark 5 equipment, was appreciation of the need to supplement the air warning (WA) capability of shipborne metric radar with means to detect the approach of low-flying aircraft. At 40-miles distance an aircraft would have to be flying at 4000-ft or above to be detectable on the Type 281 WA radar. At this range aircraft flying above 500-ft would be visible from 100-ft high antenna site, and it was hoped that these new high-power 10-cm radars would fill the 'under-the-beam' gap in air cover. As was demonstrated in the Type 273Q trials in *King George V* in July 1942, hand-trained antennae and A-scan displays were inadequate equipment in the WA role. A power-driven antenna in continuous rotation, in association with a plan position display (PPI), was needed. The PPI display itself was not restricted in application to radar using power-driven antenna rotation. The Outfit JE developed for Naval application by EMI was also fitted to earlier radar system that used hand-trained antennae. The first escort ship fitted was *Rother* a remote PPI display being installed on the bridge as well as the PPI in the radar office, in July 1943.

A further major influence on the Mark 5 development project was a Staff Requirement, formulated in December 1942, for a Target Indication

Radar (TI) to permit allocation of gun defences in the event of multiple air attack.[62] This proposed radar had to present all targets in plan display out to some 10-miles range, and up to maximum aircraft operational height. Rotation rate of the antenna had to be high enough to permit a quick estimate of target course and speed, and cover in elevation was required up to 70°. It was considered that Type 276, equipped with an antenna of much wider vertical-beamwidth than the antenna then currently under test, might form the basis for this TI radar, now designated Type 293. In larger warships Type 293 would be fitted in addition to the ship's WS radar (usually Type 277), but in destroyers, in particular, Type 293 would be expected to provide both surface and air warning (WC), as well as fulfilling the TI role.

THE MARK 5 EXPERIMENTAL SHORE TRIALS – TYPE 277T

By the spring of 1942 the feasibility of the 1 MW modulator had been established. The discharge line was oil-immersed and designed for pulse lengths of 1.8 and 0.7 microseconds.[63] The discharge was controlled by a CV12 gas triode capable of passing 160 Amps, and holding off 13 kV. The 6.5 kV, 160 Amps generated on discharge required a 4:1 transformation to match the CV76 load condition of 25 kV, 40 Amps. The pulse transformer was also oil-immersed.

At that time there was very little evidence up on which to base estimates of the range performance of high-power 10-cm radars operating against aircraft targets. Also, the behaviour of a magnetron when feeding through a long waveguide was still under laboratory investigation. Thus there were still many technical problems to resolve before initiating design of a shipborne radar equipped with an antenna remote from the radar office. Therefore it was decided to build an experimental 500 kW radar in a rotatable cabin trailer using a short waveguide connection to a single large 'cheese' antenna on the roof (Figure 5.15). The receiver was of Mark 4 design, providing alternative bandwidths of 1.25 MHz or 4.0 MHz. The A-scan display was supplemented by a TRE design of PPI. The 'cheese' antenna used in common T/R working had an aperture of 15-ft × 2.5-ft, a gain of 1300 relative to that of a half-wave dipole, and beamwidths of 2° horizontally and 8.8° vertically, to 6 db down. In 1942 this radar was designated Type 273S Mark 5. Later this became Type 277T.

The experimental radar was completed in the summer of 1942. After tests from a site on Selsey Bill, and then on an ADRDE site some 700-ft above Ventnor, the trailer was taken to the Army test site in North Wales at Great Orme Head, for range performance trials against aircraft. With the radar at 460-ft above sea level, a full series of aircraft trials was flown

5.15 Type 277T trailer installation with large 'cheese' antenna in a fixed position on the roof: (1A) horn feed; (1B) 'cheese'-type reflector; (2) interrogator antenna (a) dipole feed, (b) reflector; (3) rotatable cabin having transmitter, receiver and display (© Defence Research Agency)

between December 1942 and February 1943 using a Beaufighter and a Flying Fortress as targets.

The objective set for these trials was to obtain scientific data on the signal-to-noise ratio versus range against an aircraft target as it opened and closed the radar, at various heights within the vertical cover of the radar beam. Thus the antenna was held closely on the target bearing throughout each run. Also, laboratory-type equipment was used to tune the radar, including measurement of the magnetron-frequency spectrum; such a facility did not become available operationally in service until postwar. Under these somewhat idealised conditions a maximum range of 80 miles was recorded against the Beaufighter (which was classed as a medium bomber) flying at 8000-ft. Visual horizon range at this height would have been 110 miles. At low heights the aircraft was detected as soon as it rose above the horizon. A warning was given however, that the fluctuating nature of the aircraft echo was such as to make first sighting at the maximum possible range unlikely when sweeping the antenna in search mode at 2–4 rpm. It was considered that a reasonable estimate of first-sighting range under operational conditions would be 60 miles.

The value of this experimental radar as a prototype to improve low air cover in the Coastal Defence Chain was recognised immediately, and production design was initiated. Preproduction orders for Type 277T were placed with Allen West, Marconi and Metropolitan Vickers. By March 1943 the first preproduction set had been installed in its trailer and moved to an operational site at Capel, Dover. Before the end of March the Dover Type 277T had confirmed its value in providing warning of the approach of low-flying aircraft.[64] A flight of approaching wave-hopping FW190 fighter – bombers was detected as soon as it crossed the French coast. Reported immediately to the RAF operational filter room at Stanmore, this permitted a general air-raid warning to be sounded some two and a half minutes before the aircraft crossed the English coast. This gave just sufficient time for the school at Ashford to be evacuated before the buildings were destroyed.

In installations such as that at Dover, the mobile trailer design was considered temporary, pending a suitable design for a rotatable antenna. The first fixed-installation equipment, designated Type 277S, was despatched in mid-1943 to Saebol, Iceland, to replace the Type 273S Mark 3 installed there on a site overlooking the Denmark Straits. The second Type 277S replaced the Type 277T at Dover. During 1944–45 these radar systems were installed at various Naval Air Stations, as well as in the Coastal Defence Chain. Both the RAF and the Army were supplied with the standard panel equipment; the RAF panels were designated Type 277A to identify the fact that they contained a 45 MHz IF chain in place of the standard Naval 60 MHz IF chain. Type 277T was also used in various amphibious operations, including Operation OVERLORD, to provide air warning in the bridgeheads.

THE MARK 5 EXPERIMENTAL SHIPBORNE RADAR – TYPE 277X

There were several critical aspects of the proposed design for the shipborne version of Type 277, for which performance had to be tested under realistic conditions, before committing the final design. These included:

- Antenna: Stabilisation in elevation and azimuth.
- Waveguide: Performance achievable, as fitted by a dockyard.
- Display: Sensitivity of the PPI with a rotating antenna, compared to A-scan display.

Plans were therefore made to fit an experimental model of Type 277 (Type 277X) in *Saltburn*, the Signals Department trials ship, as early as possible in 1943.

The 4.5-ft diameter paraboloidal reflector of the antenna was fed at its focus by a waveguide flare for horizontal polarisation (Figure 5.16). The reflector was trunnion-mounted and tiltable in elevation, with a self-contained servo-system and gyro-vertical reference to hold the reflector horizontal in the normal search mode, with the antenna rotating. Drive in azimuth was by Selsyn, with a variable-speed drive motor in the control table. Rotation rate was variable from 0–16 rpm, together with hand-control. The antenna could also be swept by hand-control in elevation, by simple displacement of the gyro-vertical reference. Thus the antenna could either be rotated continuously for use in conjunction with a PPI display, held on true compass bearing under hand-control for A-scan, or swept in elevation on a given bearing for heightfinding. The PPI fitted was a laboratory model of the Outfit JE under development by EMI. The height display (HPI) was a modified TRE design of PPI with a special cursor from which height could be read directly from the range and angle-of-elevation coordinates of the display.

5.16 The Type 277X antenna – note the sheet-metal paraboloid (© Defence Research Agency)

The antenna in *Saltburn* could be mounted only at the abnormally low height of 27-ft above waterline. This was, however, of little importance since no doubts existed with respect to the performance of Type 277 against surface craft. With an antenna gain of 800 and a transmitter that (even allowing 3 db for possible waveguide losses) was radiating at least 250 kW, Type 277X was some 25 times more powerful than Type 273Q in terms of transmitter power equivalence. Type 273Q had already demonstrated adequate WS performance. The interest in these sea trials on *Saltburn* centred on the performance achievable against aircraft, and in particular on the effect of continuous antenna rotation in association with PPI display.

The waveguide run on *Saltburn* was at least 100-ft long; this was appreciably longer than was later found necessary on standard ship fittings. Installation was by the dockyard staff without any special resources other than the kit of precision-cast bends, corners and flanges, which had been developed for the standard 3-in × 1-in rectangular brass waveguide. Despite the upper-deck position of the antenna on *Saltburn*, the RF conditions of the waveguide run were not atypical of those to be found in future standard fittings with the antenna mast-mounted. A number of bends and corners near the radar office were necessary, followed by a long straight run consisting of standard lengths of waveguide tube, flanged and bolted tightly together. In *Saltburn* this run was along the upper deck; in later standard fittings it was up the mast. A further series of bends or corners was then needed to connect to the antenna pedestal, with its two pre-fitted rotating joints, and intervening bends. Thus power reflected back to the magnetron could be expected to originate either from points in the waveguide run close to the transmitter or, after appreciable transmission time, from the far end of the 100-ft run. Laboratory experiments on coupling the CV76 to its load through a waveguide had demonstrated that maximum useful power output did not necessarily coincide with tuning simply for maximum power (see Monograph 2). Widening of the frequency spectrum, dependent on the coupling, could lead to a proportion of the power output lying outside the receiver band pass. Experience with Type 277T confirmed the effect, although with short waveguide lengths the impedance changes seen by the magnetron were confined entirely within the rise-time of the pulse. On a long waveguide run, however, reflections from the far end could be expected to introduce impedance change after the pulse rise-time, which can produce a magnetron spectrum with several peaks. In *Saltburn* the delay-time in arrival at the magnetron of power reflected at the antenna was of the order of 0.25 microsecond. The maximum power output of the magnetron into the waveguide was measured as 450 kW, with some frequency spreading resulting in a loss of about 1 db outside the receiver bandpass. The voltage standing wave ratio (vswr) measured in the radar office lay between 0.55 and 0.7, but no multiple-peaking of the magnetron spectrum was detectable.

There were two periods of sea trials with Type 277X, the first from 8–20 April 1943 off Lough Foyle and the second from 2–5 May off the Isle of Man. Weather conditions and unserviceability of aircraft greatly limited flying time during the first period. However, since this period coincided with that of the equinoctial gales, the stabilisation equipment and antenna control were severely tested and valuable design changes were consequently initiated. In the second period a small number of aircraft runs were possible. These were sufficient to demonstrate that first detection of a medium bomber was possible at 40-miles range on the PPI display with the antenna rotating continuously. Heightfinding was shown to be possible at 20-miles range, but no maximum range could be established since no aircraft was available to fly high enough to be sure that there was no influence from sea-reflected radiation. With regard to the difference in performance between a PPI display associated with a continuously power-rotated antenna and following a target on an A-scan with the antenna held manually on the target bearing, it was possible to follow the target on the latter to some 15 percent further in range. This, however, appeared to be associated with the intermittency of the aircraft echo at long range, rather than any loss of sensitivity on the PPI display. There was thus no real penalty to pay for the enormous tactical advantages of using PPI display.

SEA REFLECTION AND TARGET INDICATION – TYPES 276 AND 293

Type 276 was conceived originally as a lightweight antenna version of Type 277, with a simpler antenna control giving one fixed rotation rate of 10 rpm, with the option of hand-training. Azimuth stabilisation was applied to the scan coils of the PPI, and not to the antenna itself. To avoid the need to stabilise the beam in elevation it was planned to use the Type 271Q transmitting antenna in association with common T/R working. With the antenna gain now 180 for both transmit and receive, and using the 500 kW transmitter, the WS performance of this radar was expected to be identical to that of a Type 273Q with its antenna mounted at a similar height. Therefore the range on all except small surface targets would be limited only by exposure of the target above the horizon.

In 1942 data on range performance of cm-radar against aircraft was very limited. However a prediction had been made for Type 277 of a maximum range on a medium bomber of 45 miles, with the antenna horizontal and the aircraft target above the horizon.[65] As was seen later, this estimate was slightly optimistic, but not seriously so. On the same basis Type 276 with its antenna gain of 180 compared with 800 for Type

277, was expected to provide warning of low air-approach to about 21 miles. Using the Type 271Q transmitting 'cheese' antenna for both transmission and reception restricted the vertical beamwidth for the proposed Type 276 to ± 14°, to 6 db down, leaving very little provision for pitch and roll of the ship, if the Staff Specification of 10° vertical cover was to be met. An early change was made to the Staff Requirement for low cover, and the 10° elevation cover increased to 20°.[66] Thus, by the end of 1942 an antenna of aperture 4-ft × 8-in was on test at Eastney to replace the proposed 4-ft × 10-in 'cheese' of Type 271Q.

Free of the additional complications involved with Type 277, no experimental sea trials had been planned for Type 276. The experimental 500 kW rig at Eastney, originally installed for development work on the use of long waveguides, together with test equipment for Type 277 and other ancillary equipment (such as the adjustable swept-gain unit for the receiver necessary with PPI display) was considered adequate for any overall early testing of Type 276 as a whole. It was expected that Type 276 would be in service by mid-1943. However at the end of 1942 a new Staff Requirement for a 'Target Indication' radar was formulated. This TI radar was required to provide plan display of all targets within an envelope defined by a range of 10-miles radius, maximum aircraft operational height, an angle of sight of 70°, and the horizon. A special TI display unit was to be developed to make possible a rapid initial determination of course and speed of aircraft targets, and thus permit an effective allocation of the ship's armament in the event of multiple air attack. It was thought that Type 276 could be the basis of this new radar. On cruisers, and above, this radar (designated Type 293) would be additional to their WA and WS radars. However there was special interest in fitting this TI radar in fleet destroyers to operate in a combined air and surface warning (WC) role. This would provide fleet destroyers with a single all-purpose radar. It was proposed to meet this specification through the use of a 6-ft × 4-in aperture 'cheese' antenna mounted on its pedestal and tilted upwards by 15°. No records exist of measured gain or beamwidth for this antenna, but it maximum gain along the axis cannot have exceeded 100: at 15° below the axis the gain along the horizontal would have been significantly less. Following the intention to use this radar for WS applications in destroyers, it became essential to undertake sea trials as early as possible. Since Eastney was no longer a suitable site for the necessary extensive air and surface trials, a Type 293X was fitted in the fleet destroyer *Janus* (Figure 5.17). Trials were organised for 27 August to 4 September 1943 in the Pentland Firth area. Type 293X was only experimental in respect of its Outfit AUR antenna, and in the use of brass for the second (and last) time for its 63-ft waveguide run. The straight waveguide tube in all subsequent radar fittings was high-conductivity copper, whose resistive loss was 0.025 db per metre compared with that

5.17 HMS Janus with the experimental Outfit AUR antenna of Type 293X on top of a lattice mast – Note the HF DF pole-mast installation behind (© Defence Research Agency)

of brass at 0.05 db per metre.[67] The influence of the use of brass on performance was negligibly small, being less than 1 db in the *Janus* fitting.

There was no firm basis for the prediction of performance of Type 293X prior to the sea trials. The Type 273Q sea trials gave some reference for low-angle performance, but even so there was the critical question concerning the effect of echo fluctuation at long range when using a rotating antenna. First-detection range could be expected to be variable, but there was no data on the frequency or depth of the fading pattern of the target echo at long range. Also, this was the first use of cm-radar in an 'all-height' WA role. Hitherto, in respect of low-angle performance, it had simply been assumed that the sea surface acted as a specular reflector of the incident radiation. However at 10-cm wavelength the sea represented a rough surface, more likely to scatter incident radiation at higher angles of incidence than to cause 'mirror' reflection. Some radar experiments conducted in early 1943, in which an attempt was made to measure the direct reflection from the target separately from the sea-reflected component, indicated that there was little or no specular reflection at the higher angles of incidence. It proved impossible to obtain precise data from these tests. However a rough rule-of-thumb was proposed (and adopted) for performance prediction, namely unity sea-reflection coefficient at 0° incidence, reducing linearly to zero at 5° incidence and above.[68] On this basis, and using such low-angle air-performance data as existed, a preliminary estimate was made of the expected vertical coverage-diagram of Type 293X (Figure 5.18).

The use of a PPI display in conjunction with continuous antenna rotation made possible a more precise form for recording trials data than the previous judgements of operator sensitivity to small echoes from an A-scan display. The target response was clearly a random variable that was sampled at regular intervals on successive antenna sweeps through the target bearing. With an intensity-modulated display a record could be kept of 'response/no response' for each rotation of the antenna. Presented in bar-chart form, as in Figure 5.19, this gave a very concise record for subsequent analysis. Maximum and minimum ranges of first detection were at once obvious and, depending on the number of aircraft runs, these could be given numerical meaning. For example, with 10 aircraft runs the maximum first-detection range achieved would correspond to a probability of a detectable signal on the PPI of 0.1; the minimum to a probability of 0.9 that the aircraft will have already been detected. The technique was later extended to the extraction from the experimental results of the probability of signal detection on any single antenna rotation, at a given range.

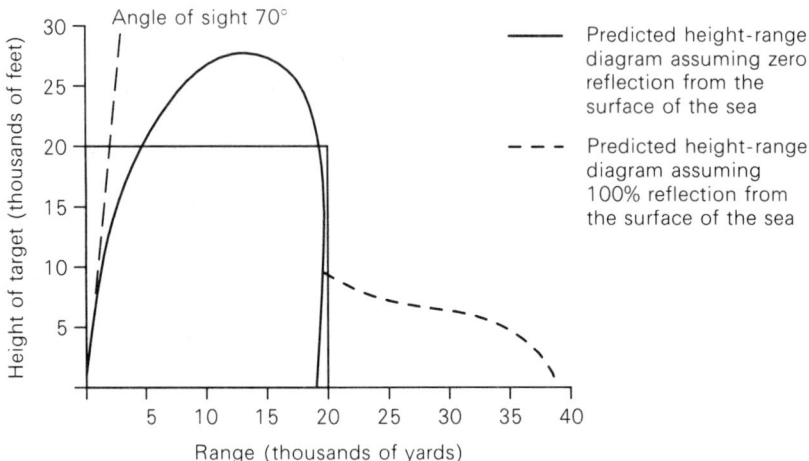

5.18 Predicted vertical coverage diagram for Type 283X (© Defence Research Agency)

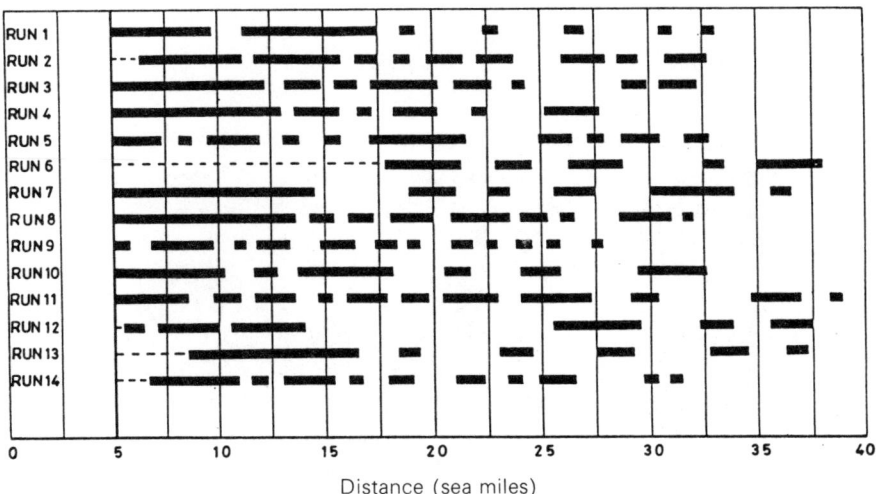

5.19 Bar-chart presentation of PPI 'signal/no signal' versus range for 14 radial flights at constant aircraft height. (Source: Type 277 trials on *Campania*) (© Defence Research Agency)

The coverage diagram of Figure 5.20 summarises the air-performance data from these trials of Type 293X.

Even though the range on an aircraft target fell short of specification, Type 293 could still have been useful in its TI role. Maximum first-detection range on the Boston target aircraft at low angle of elevation was 13 miles, reducing to 7 miles at high elevations.[69] Scientifically there was clear evidence from these trials of some range enhancement at low angles, which was lost at high angles of elevation. The total number of runs was, however, insufficient to give this effect any precise numerical value. It was nevertheless possible to give an approximate numerical estimate to the relationship between the maximum achieved range of first detection and the useful operational range. Uncertainly of the ability to make a first detection reduced the operationally reliable range to some 70 percent of maximum range at low angle of sight. At high angles of elevation there were not sufficient aircraft runs to judge whether or not the absence of sea-reflection effects made any significant difference to this estimate. However, irrespective of doubts regarding performance against aircraft, debate on the acceptability of Type 293 using this small (Outfit AUR) antenna centred on WS performance.[70] In the WS trials the destroyer *Obedient* was detectable only out to 24000-yds. At this range all but 4-ft of the hull of *Obedient* was visible from the 68-ft antenna height of Type 293X on *Janus*. Type 273Q had already demonstrated the capability to detect a destroyer when only the topmost elements of superstructure

(a)

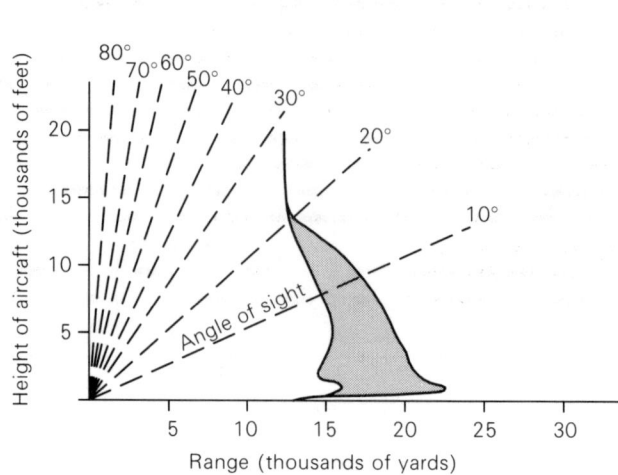

(b)

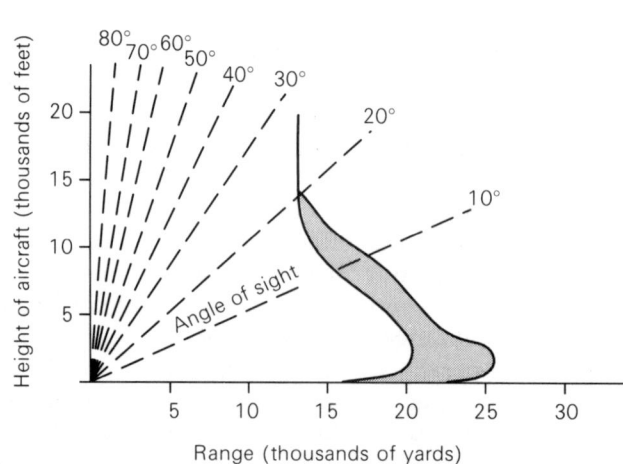

5.20 Vertical coverage of Type 293X as determined by trials against a Boston aircraft target: (a) height versus range – aircraft opening; (b) height versus range – aircraft closing (© Defence Research Agency)

were visible above the horizon from the antenna site. Type 276 had been estimated to provide a similar WS performance to that of Type 273Q, with the additional advantage that its lightweight, exposed antenna could be fitted at the masthead, so providing an extended horizon compared with that of bridge look-outs. This valuable access of surveillance range would have been lost if Type 293 had displaced Type 276 as the WS radar in light cruisers and destroyers. It remained, however, to prove that Type 276 had indeed the performance of Type 273Q.

A return to the original plan for the Type 276 antenna obviated any need for antenna development. The aperture of the Type 271Q transmitting antenna was simply sealed by a plastic sheet of thickness appropriate to minimise reflection and to prevent moisture ingress to the waveguide. This Outfit AUJ antenna could then be mounted on the same pedestal as that used for Type 293 (Figure 5.21). Type 276 was fitted in the fleet destroyer *Tuscan*, and the *Janus* trials programme was repeated in November 1943. In *Tuscan* the waveguide run was slightly longer, at 70-ft, than in *Janus*, but now utilised low-loss copper waveguide. The antenna height was slightly lower, at 63-ft. In the WS trials the destroyer *Kempenfelt* was detectable at 29000-yds.[71] This was an almost exact equivalent, as regards exposure above the horizon, of the 34000-yds range

5.21 Type 276 (Outfit AUJ) antenna identical to the Type 271Q transmitting antenna with aperture sealed to permit use in exposed sites without the need for a radome (© Defence Research Agency)

on a destroyer during the Type 273Q trials from the 100-ft antenna site on *King George V*. There was, of course, a sharp cut-off in high air-cover for aircraft flying above 8000-ft, as can be seen from the comparison of Types 293 and 276 air cover in Figure 5.22. It was now clear that design of an adequate TI/WC radar was feasible with an antenna of similar form to that used for Type 293X, but that the antenna would need to be larger, with a gain twice or thrice that of the Outfit AUR tested on Type 293X.

A number of destroyers had been fitted with Type 293 in the autumn of 1943, prior to the change of policy based on the trials data. However, as from the end of 1943 and until such time as a larger antenna had been developed for this set, Type 276 was fitted as an interim measure, not only as combined surface and low air-warning radar, but also as TI radar. The new antenna for Type 293, Outfit AQR, became available in 1945, at which time Type 276 was declared obsolescent. The larger Outfit AQR antenna (Figure 5.23) had an aperture of 8-ft × 7.5-in, was tilted 15° to the

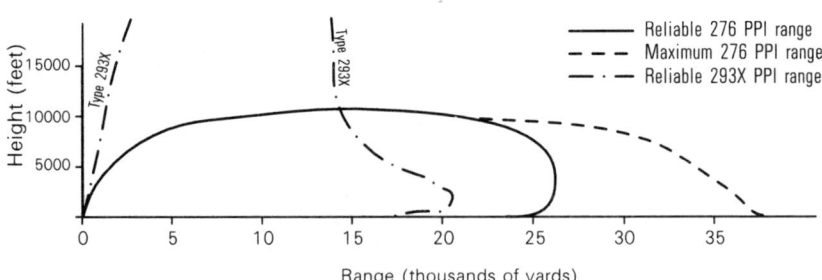

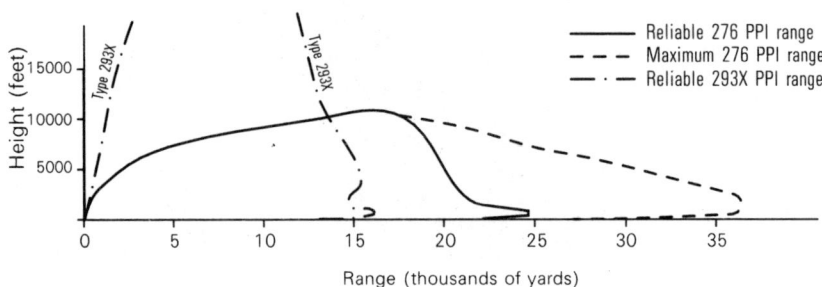

5.22 Vertical coverage of Type 276 as determined by trials (© Defence Research Agency)

5.23 Type 293M antenna replacing the Outfit AUR used in the Type 293X trials (© Defence Research Agency)

horizontal and had a gain of 220 times that of a half-wave dipole. When fitted with this antenna, Types 276 or 293 became Type 293M. Design of a yet larger antenna (Outfit ANS) for cruisers and above was in progress in 1945, but this system did not enter service until postwar. This radar, Type 293Q, gave cover against aircraft to 18-miles range at all heights above the horizon to 35000-ft.

WS TACTICAL RADAR – LOW AIR WARNING – HEIGHT-FINDING – TYPE 277

Type 277 was the most powerful British Naval 10-cm radar to enter service during the war. It became available for ship-fitting towards the end of 1943. The sheet-metal 4.5-ft diameter reflector of Type 277X had been replaced by a wire-mesh construction for the paraboloid, to reduce windage and weight (Figure 5.24). The antenna control system retained the facility of variable rotation-rate from 0 to 16 rpm, with hand-control by inching the main drive-motor in the control table. This velocity hand-control permitted fast training to a given bearing for heightfinding. For normal WS and low air search the antenna was rotated at a constant

5.24 Type 277 antenna Outfit AUK: (1) wire-mesh paraboloidal reflector; (2) waveguide rotating-joint on elevation axis; (3) gyro vertical-stabiliser (© Defence Research Agency)

speed, with the antenna stabilised in the horizontal position. During sea trials in March 1944 in the escort carrier *Campania*, sweep rates of between 2.5 rpm and 3.5 rpm were used for normal search and warning.[72] Concentration in these trials was on the air cover, and on using Type 277 in conjunction with Type 281B to determine aircraft height by sweeping the Type 277 beam in elevation. *Campania*, unlike most escort carriers, was fitted with Type 281B as the WA radar and was equipped with a Skiatron display switchable between the Types 281B and 277 for Fighter Direction. Considerable benefit was derived in these trials from use of the carrier's Fulmar aircraft. Flight planning and communications were easy, and a very complete flying programme was achieved. The Fulmar was small in comparison with the 'medium bomber' usually quoted for range performance. In size it was much the same as a Hurricane fighter, and thus an estimated 2 db down in reflection coefficient. In another respect it was a particularly useful choice since its reflection coefficient proved to be, as near as could be judged from the trials data, the same for the nose-on aspect as for tail-on. This effectively doubled the number of runs in the subsequent statistical analysis of the trials data.

Figure 5.25 summarises the range performance achieved against the Fulmar target, both as regards low cover with the antenna rotating horizontally, and in Free-Space with the antenna hand-swept in elevation for heightfinding. Within area A of this diagram there was near certainty that any aircraft approaching the ship would have been detected earlier and could have been tracked continuously on the Skiatron (or PPI) for Fighter Direction. The outer boundary of area B was determined by the maximum first-detection range achieved among flights at the same height. The performance at 2000-ft was particularly well-established since, in addition to programmed flights at this height, a test flight was flown each day. On the basis of the number of flights, the maximum range of detection determined on these trials corresponded to a probability of a detectable signal of 0.05 to 0.1 as the antenna swept through the target bearing. Similarly, the minimum range of first detection (namely, the inner boundary of area B) corresponded to a

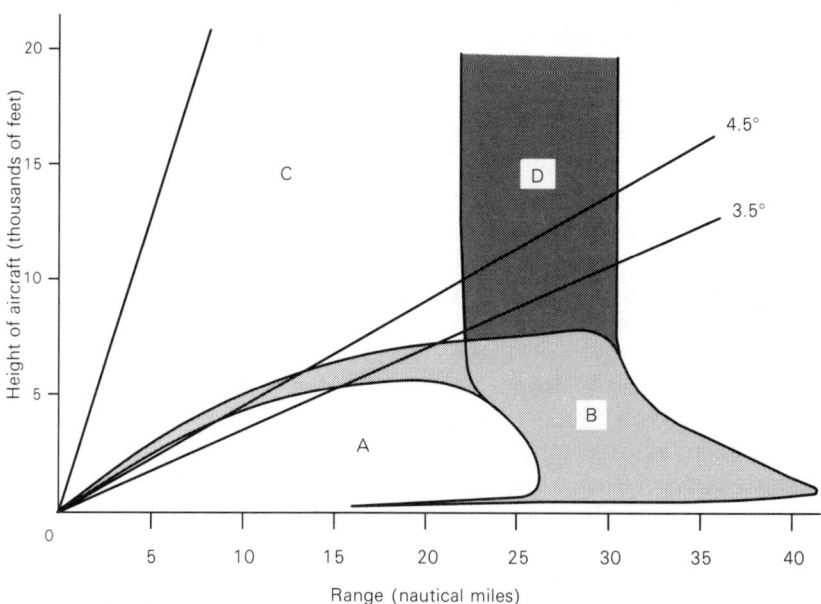

5.25 Type 277 in HMS *Campania*; height versus range diagram for a single Fulmar aircraft (© Defence Research Agency)

cumulative probability of 0.95 that all approaching aircraft had been detected.

As mentioned earlier, the bar-chart presentation of 'signal/no signal' for each rotation of the antenna made possible the extraction of the probability of a detectable signal on a single rotation of the antenna at any given range. The experimental data from the trials for flights at 2000-, 500- and 50-ft were analysed for this probability of 'painting' on the PPI. It can be seen from Figure 5.26 that, at an aircraft height of 2000-ft and at the range at which the cumulative probability of having detected the aircraft had attained 0.95, there was still only an average probability of a detectable signal on the PPI once in every three rotations of the antenna. With the antenna horizontal, low air-cover by Type 277 was limited to aircraft flying at 5000-ft and below. Figures 5.27 shows the degree to which Type 277 was able to fill in the gap in air-cover beneath the beam of Type 281. At the antenna height of 69-ft on *Campania*, range on low-flying aircraft was limited by the horizon for aircraft flying at 500-ft and below. The effect of this on the probability of 'painting' was marked. This rose steeply with closing range, as can be seen from Figure 5.26.

The heightfinding trials were primarily operational to determine how quickly height could be measured, and how accurately, with Type 277 'azicated' from the Type 281 plan display. Out to 20–22 miles, angle of elevation could be determined to an accuracy of $\pm 0.5°$, and a height reported within 40 seconds of transmission of target bearing from Type 281. Also, the target could be collected on the Type 277 HPI on the first sweep of the antenna in elevation. Thus area C of Figure 5.25 more or less corresponds to the reliable operational working area A of low-angle warning. At longer range, in area D of Figure 5.25, the intermittent nature of the echo made collection of the target on the first sweep-up of the antenna uncertain. However the Fulmar was detectable out to 32-miles range with the antenna elevated beyond the possible influence of sea-reflected radiation. Comparing this Free-Space range with that of the longest range achieved with the antenna horizontal (43 miles with the Fulmar at 1000-ft) indicated that for 10-cm radiation even at near-grazing angles of incidence on the sea surface, the sea reflection coefficient was only 0.34. Moreover, the evidence of the coverage diagram suggests that the influence of sea reflection diminished rapidly as the angle of incidence increased, and was sensibly negligible at a much lower angle of sight than the 5° that had been postulated.

One other aspect of these trials was of significance to the operation of TI radar. The rotation rate of the Types 293 and 276 antennae was fixed at 10 rpm, and therefore no test had been possible of the effect on range performance of rotation rate. A number of test flights were now made, at constant height, at rotation rates of 10 and 16 rpm using the variable speed control in the Type 277 control table. No significant effect on range

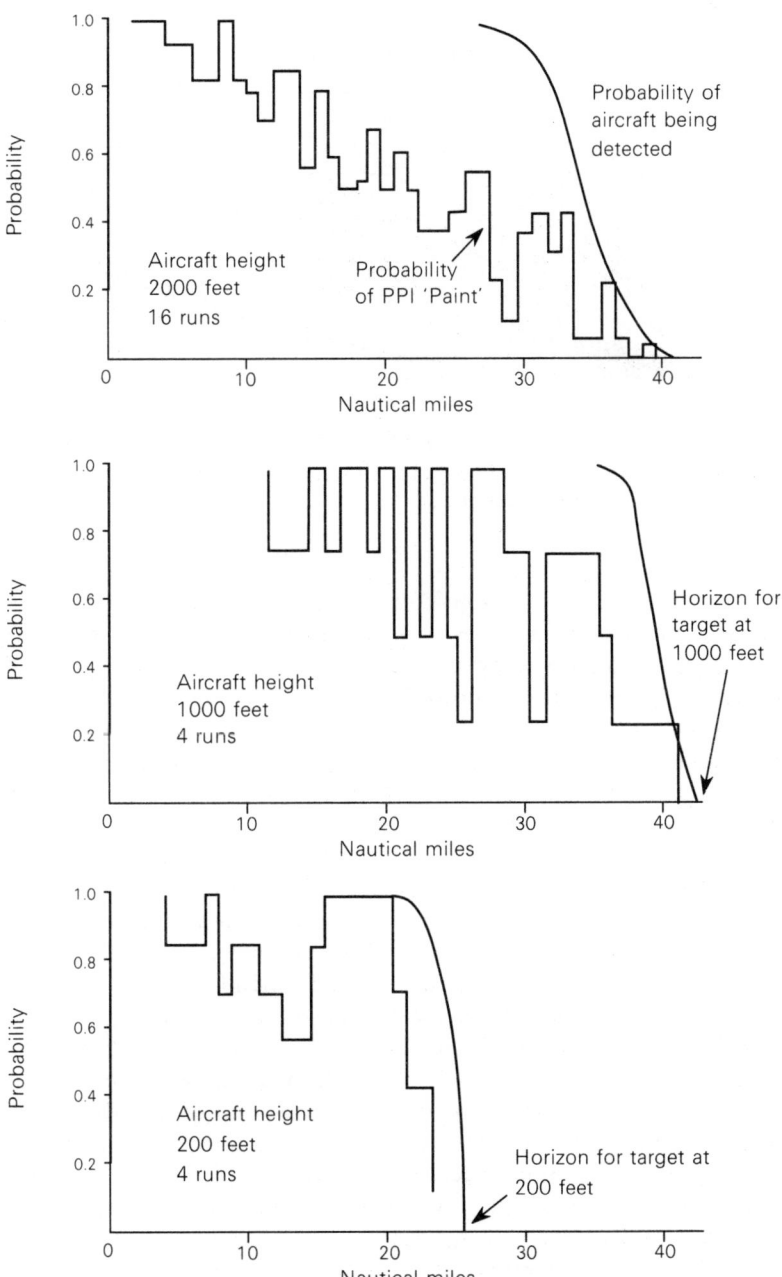

5.26 HMS *Campania*: Analysis relating probability of 'paint' to target-acquisition probability (© Defence Research Agency)

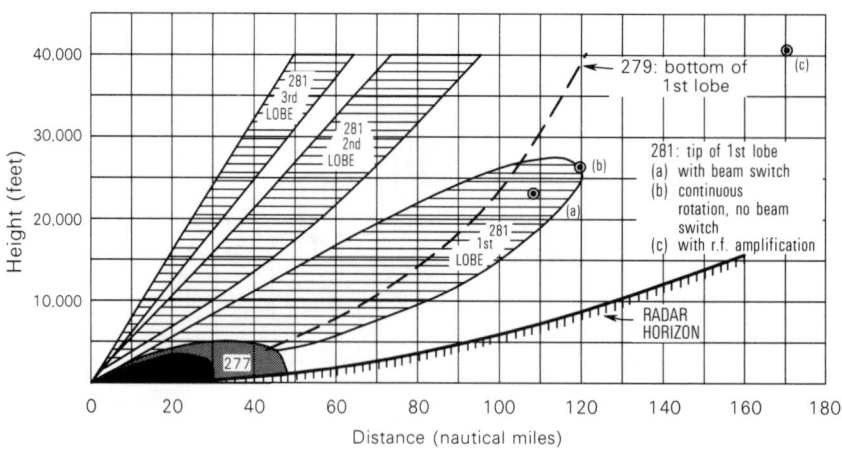

5.27 Comparison of vertical coverage diagrams of Types 281 and 277
(© Defence Research Agency)

performance was measurable compared with the 2.25 to 3.5 rpm used generally in these trials. However a count was made of the number of antenna revolutions giving detectable signals compared with the total number of antenna revolutions in the recorded section of the run. The proportion of antenna rotations with a detectable signal remained sensibly constant, the increased number of interrogations of the target apparently compensating for the reduction of total energy beamed at the target at the higher rotation speeds.

Fitting of Type 277 in escort vessels did not being until mid-1944, and there were no specific WS trials until September 1944. There was no doubt regarding the performance to be expected against ship targets. Range would be horizon-limited. However it was desirable to confirm performance achievable against a submarine target, and, of more importance, to determine detectability of the Schnorkel using 10-cm radar. Sea trials of Type 277 in the Castle-class corvette *Coppercliff* were organised for September 1944 in the Firth of Clyde and North Channel.[73] The Type 277 antenna was mounted on top of the lattice-mast in *Coppercliff*, at 65-ft above waterline. This gave a horizon range of 9.5 miles. A fully surfaced submarine was reliably detectable at 11 miles when only the conning tower would have been visible above the horizon. The Schnorkel was, however, a difficult target. Raised some 3-ft above wave-top height, it was detectable at 5.5 miles range. On the other hand, with an exposure of only 1-ft above the wave-tops the echo could not be distinguished from wave-clutter peaks. This was clearly a situation where the higher resolution of X-band (3-cm) radar with a narrower beam might have been expected to be more successful. In later comparative trials of S-band and X-band radar this did not prove to be so for minimum Schnorkel exposure in the presence of waves (see Appendix 1 to this monograph).

FURTHER DEVELOPMENT OF TYPE 277 – TYPES 277P AND 277Q

Development of 10-cm radar had been very rapid between 1941 and early 1944. In these three years 10 distinctly different radars had entered service, involving two major increases in transmitter power, a complete change in antenna-feed technology and a major change in display technique. Test equipment had been developed and added, and in place of the simple voice-pipe to the bridge the radar office had its share of repeaters and communication equipment. Maximum use had been made of existing design in successive improvements, and an add-on policy had often enabled improvements to be brought rapidly into service. As a result many pieces of equipment were external to the main panels, and the resulting plethora of plug and socket connections were a source of much minor maintenance. From all points of view, considerations of manufacture, installation and maintenance at sea required a redesign of the radar-office equipment.

In addition, during 1944 a major receiver improvement completed development. A broad-band crystal-mixer unit was available[74] that could be mounted directly on to the waveguide, and incorporated the common T/R switch unit. This development eliminated the majority of the tuning and matching adjustments on the mixer, and removed the cable connection of mixer-to-waveguide. Also, the receiver IF chain had been redesigned, the operating IF frequency being changed from 60 MHz to 13.5 MHz. Overall the receiver noise-factor was improved by about 2 db. A redesign of the office equipment was therefore undertaken, incorporating the new receiver and mixer. Anti-jamming facilities were also added, with provision of a third IF chain, of 0.5 MHz bandpass, together with various filters that could be switched into service as needed. With these new panels, Types 277 and 293M became Types 277P and 293P.

Regular production output of this improved equipment did not become available until mid-1945. However several important Fleet units were undergoing refit in the spring of 1945, preparatory to joining the Far East Fleet. Others had already left Home Waters. A limited emergency production (project Bubbly) was therefore mounted to obtain a number of conversion sets of panels to upgrade Types 277 and 293M during the spring of 1945. The battleship *Anson* was the first ship so fitted in March 1945. For ships already in the East, conversion kits were sent for installation by base and ship's staff.

A second improvement to Type 277 was also initiated in 1945. In carriers, capital ships and Fighter Direction ships, Type 277P, in conjunction with Type 281BQ (or its replacement Type 960) was expected to fulfil an FD role. For this the maximum range of Type 277P was short, both for heightfinding and for in-fill below the beam of the metric radar. Certainly the P conversion had added some 2 db to performance, and allowance for the small size of the Fulmar aircraft added a further 2 db. Even so, reliable heightfinding to some 25-miles and low cover to a maximum of 50-miles were still less than desired. It was therefore decided to design a larger antenna for Type 277P. The reflector for this antenna was an 8-ft diameter paraboloid, reduced at the sides to a 6-ft width. It was mounted, stabilised and controlled in the same manner as the Type 277 antenna. Thus there was no correction for deck-tilt or cross-roll errors with the reflector elevated. This was the first 10-cm Naval radar to employ vertical polarisation, and in consequence the reflector was constructed from parallel vertical rods to reduce windage. Engineering-design time, and the problem of installing a major antenna system without dockyard assistance, prevented this improvement being made part of the Bubbly crash programme. In fact Type 277Q, as the modified radar was designated, did not enter service until the postwar years. In sea trials in conjunction with Type 960 in *Illustrious* in March 1947, Type 277Q achieved reliable heightfinding out to 55-miles range, and low air cover to a maximum of 85 miles. Type 277Q remained in service for many years, postwar.

CENTIMETRIC FIGHTER DIRECTION (FD) RADAR – INITIAL PROPOSALS

Effective air defence of a fleet at sea depended on the timely provision of fighter cover whose efficient deployment required the individual fighter aircraft to be vectored to its target using position data supplied from ground or ship radar. This was exactly equivalent to the situation that had been demonstrated on land with ground control of interception (GCI). During 1942 in the Mediterranean Theatre, experience of the use of Type 281 metric WA radar with the Skiatron display in carriers met with limited success in this FD role. The beamwidth was too broad because of the limited size of the dipole array that could be carried at the masthead. Also, strong back-echoes from adjacent land could swamp the screen when working close to shore. The early tests of Type 277T in the autumn of 1942 gave promise of achieving ranges on aircraft with 10-cm radar that would be useful for Fighter Direction, with the advantage also of narrow beamwidth and high resolution of targets. Hence a Staff

Requirement was formulated for a 10-cm FD radar for carriers and Fighter-Direction Ships.[75] This FD radar was to provide 'fadeless' cover on all targets above the horizon up to 35000-ft altitude at ranges out to 80 miles. Presentation was to be on plan display, with accurate height measured on each sweep of a continuously rotating antenna, and without interruption of the plan display. Height was to be accurate to 500-ft out to 40-miles range, and to 1000-ft between 40 and 80 miles. It was accepted that any such radar would have to have a heavy antenna system since not only would the antenna itself be large, it would also have to be fully stabilised to a high degree of accuracy.

The term 'fadeless' in this requirement should not be confused with the random fluctuations of an aircraft echo at 10-cm wavelength. The effect that this phenomenon was to have on useful operational radar range when using continuous antenna rotation was unknown in 1942. The reference was to the deep fade in target response with metric WA radar between the first and second interference lobes in the vertical coverage pattern. As can be seen in Figure 5.27, the range at which this gap in cover occurred depended on the target height. It could easily coincide with a critical stage in the control of interception, just as the fighter was about to make its first visual sighting of the target.

Experimental data on the range performance achievable using 10-cm radar against aircraft was very limited; what there was related to maximum range at low angles of sight. However an estimate had just been made of the performance of Type 277 against low-flying aircraft of a maximum range of 45 miles. This estimate was later proved to have been remarkably accurate in the sea trials of Type 277 using the small Fulmar aircraft in 1944. A 'cheese'-type antenna of 17-in gap between top and bottom plates could provide the high cover needed for the FD radar if the axis was tilted upwards by 5°, and with a 12-ft wide horizontal aperture this antenna had a similar gain to that of the Type 277 paraboloid antenna. Thus a plan display range of 45 miles seemed readily attainable using the existing 500 kW, 2-microsecond-pulse transmitter. This was, of course, only just over halfway to the required range. However, already in 1942 a pulse power-output of 2 MW had been shown to be possible from a magnetron.[76] On the assumption of as rapid a development in the technology necessary for manufacturing design as had hitherto been possible with each previous upgrading of magnetron power, it was planned to develop a new transmitter with an output of 2 MW pulse power using 5-microsecond pulses. With this tenfold increase in pulse energy the 80-mile range appeared to be assured. The proposed radar was designated Type 295 when fitted with the 2 MW transmitter. The designation Type 294 was applied should there be need to fit the 500 kW transmitter as an interim measure. In the absence of any Free-Space range data for 10-cm radar at the time these estimates of range and vertical coverage were made for Type 295, the sea-reflection coefficient had to be assumed as unity at all angles of incidence for horizontal polarisation. The absence of specular reflection at the sea surface at more than a few degrees angle of incidence did not become apparent until the following year.

The height-measuring system proposed for Type 295 was considered more speculative, especially in relation to the requirement not to stop azimuth rotation. Moreover the proposal was based on purely theoretical study. A second reflector, with its long dimension vertical, was to be mounted on top of the plan display antenna, for reception only. Multiple close-spaced waveguide flares would provide overlapping beams in the vertical plane. Theoretically, comparison of the signals, fed through different receiver channels, could have been expected to provide a means of measuring height whilst the whole antenna system swept through the target bearing. Since the heightfinding antenna would only receive radiation directly from the target, its gain had to be greater than that of the plan display antenna to compensate for the loss of the sea-reflected component of radiation returned from the target. In addition to development of a special receiver and display for this method of operation, a four-channel waveguide rotating joint had to be developed for the antenna mount.

Much of the research and development for Type 295 was performed under contract. In particular the modulator and transmitter were contracted to the BTH Co Ltd and the antenna mounting to Howard, Grubb, Parsons Ltd. The mounting was itself, by radar standards, a major engineering development. Full stabilisation was required to 6-minutes-of-arc, and strict limits were set by DNC on height and sweep of the mounting. The several servo-control systems necessary were metadyne, and contracted to Metropolitan Vickers Ltd. Azimuth stabilisation reference was provided from the ship's gyro-compass, with correction fed in for deck-tilt from a separate unit. Roll-along and roll-across corrections were from separate gyros, with mercury level-references on the mounting. In all, each mounting (including antennae) weighed some 7 tons. It was envisaged that two such radars would be fitted on a carrier, one forward and the other aft. A full 360° clear all-round view was to be obtained by switching plan display, as necessary, from one radar to the other.

A second project was launched in 1943 with the object of extending low air cover beyond the 80 miles of Type 295. It was intended to complement the long-range warning capability of the metric radar. Relieved of the high-cover requirement of FD, an antenna using an 8-ft diameter paraboloidal reflector would have had about three times the gain of the Type 295 plan display antenna. It was planned for Type 990 to use such an antenna with the very-high-power 2 MW transmitter in prospect. This would have extended low-cover above the horizon, and below metric radar cover, to 130–40 miles.

Unfortunately, by early 1944 it was apparent that there were serious problems to be overcome to reduce the proposed height-measurement system of Type 295 to a practical operational proposition. In addition it was clear that the technological aspects of the production design of a 2 MW magnetron could not possibly be solved within the project timescale to meet first fitting in *Eagle*, then building. Proposals for an alternative FD system were therefore prepared and forwarded to Admiralty.[77] These proposals were based on use of the Type 277 transmitter, involving separation of the functions of heightfinding and plan display between two independent radars.

REVISED PROPOSALS FOR FD RADAR – TYPES 980 AND 981

The loss of 10 db in available transmitter output consequent on the need to use the Type 277 transmitter had to be compensated by larger antennae, providing higher gain. For plan display a tripling of gain was required; for heightfinding only a doubling was involved since the receive antenna of Type 295 had already been planned to have higher gain than the plan display antenna. Reversion to the Type 277 technique for heightfinding, namely antenna stopped on bearing and swept in elevation, required the provision of separate mountings for the plan display antenna and the heightfinding antenna. The resulting radars were designated Type 980 and Type 981, respectively. The stabilised antenna mounting for Type 295 was retained for both Types 980 and 981, whose antennae had to remain within weight and inertia limits dictated by the performance of the stabilisation servo-drive system of the mounting.

Type 981 provided the simpler problem.[78] The antenna reflector was a 14.5-ft diameter paraboloid, cut to 5-ft width and trunnion-mounted with the long dimension vertical (Figure 5.28). When stopped in azimuth for heightfinding, the antenna was made to nod in elevation, over an arc of - 4° to 26°, by off-set of the gyro-reference for the roll-along stabiliser. At six reciprocations per minute, the interval between interrogations fo the target was only 5 seconds. The HPI was a modified PPI upon which the signal was displayed as a vertical line in rectangular coordinates of height and range.

The gain of the Outfit AQT antenna for Type 981 was 3000 compared with that of a half-wave dipole. In the trials *Campania* Type 277 had demonstrated that a Free-Space range of at least 32 miles was attainable on the small Fulmar aircraft, using an antenna gain of 800. Allowing 2 db for the receiver improvements under development for Type 277P, and a further 2 db to equate performance with that on a medium bomber,

5.28 Types 982, 983 and 960 at the Royal Naval Air Station, Kete (© Defence Research Agency)

indicated an expected Free-Space range of Type 277P of 40 miles. On this basis Type 981 was confidently expected to achieve a maximum range of at least 77 miles. Without the need to search for its target in bearing as well as in elevation, and with the short interval between successive sweeps through the target elevation, it was expected that Type 981, would then be operational up to near the maximum range of the radar, despite the intermittency of long-range echoes. In postwar service Type 981, by then designated Type 983, was rated at 80-miles capability for heightfinding; maximum range for low air cover with the antenna in continuous rotation, at zero elevation, was over 100 miles.

The antenna for Type 980 presented a more intractable problem. The volume of sky to be scanned around the full 360° in azimuth in 10 seconds was large. The high cover required increased from the original 35000-ft up to 50000-ft as aircraft performance increased. It was not possible to obtain the necessary increase in antenna gain simply by

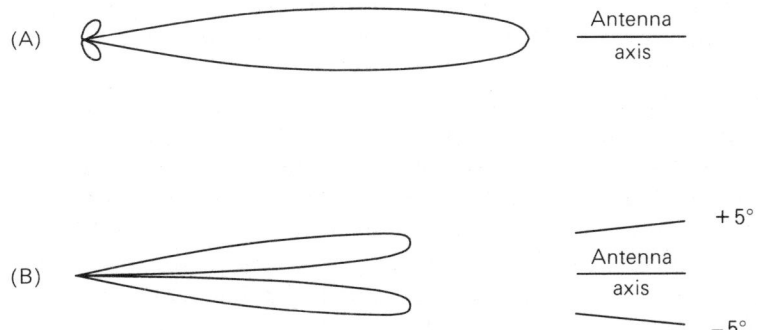

5.29 Vertical polar diagram for the Type 980 tilted 'cheese' array for different phase conditions of the feed: (a) all three elements in phase; (b) upper and lower elements in anti-phase to central element (© Defence Research Agency)

increasing the dimensions of the antenna, as had been done for Type 981. The horizontal width of the 'cheese' reflector of the Type 295 plan display antenna was limited by the performance of the antenna mounting, and the height of the aperture could not be increased without an unacceptable reduction in high cover. The solution adopted was to triple the vertical aperture and to scan the beam electronically at high speed to retain the desired vertical cover.

The method of scanning was both novel and ingenious.[79] Three 'cheese'-type reflectors, each identical to the original Type 295 plan display antenna, were stacked one above the other. The aperture dimensions of each 'cheese' were 12-ft × 17-in and the distance between aperture centres was 21-in. The stack was tilted 5° from the vertical. Transmitter power was split between the individual 'cheese' antennae in the ratio 1:4:1, with top and bottom 'cheeses' fed through rotary phase-changers. With radiation from all three elements of the antenna in-phase, the resulting interference field gave the antenna a polar diagram in the vertical plane of one main lobe and two small subsidiary lobes, one above and one below, at 10° spacing (Figure 5.29). Rotation of the phase changers in opposite senses rotated this pattern, the contour followed by the lobe tips being determined by the polar diagram of a single 'cheese' element of the antenna. The phase changers were driven at 1200 rpm. The pattern repeated twice for each rotation of the phase changers, thus effectively scanning the full required vertical cover 80 times per second. With the antenna tilt of 5° off vertical, this method of scanning implied some reduction of gain at low elevation, since with an interference lobe at 0° the pattern would have consisted of two lobes of equal size, one at 0° and the other at 10°. However it was considered that there would be some compensating range enhancement from sea-reflected radiation at low angles of elevation.

Theoretically the maximum gain of this interference scanning system on axis at 5° elevation was 2.67 times that of the single Type 295 'cheese' antenna. On the same basis of comparison with the Type 277P Free-Space range as was used with prediction of Type 981 performance, the maximum range of Type 980 was expected to be 69 miles. In this instance, however, there was adequate experience from previous trials regarding the uncertainty of first-detection range when scanning in azimuth to indicate that the reliable operational range would not exceed around 70 percent of maximum range. This would be particularly true for actual interception control, or 'azication' of the heightfinder. Thus plan display from the metric WA radar Type 960P would have to be available at the Skiatron interception control stations and at operating positions for laying Type 981 on bearing. The original plans for the FD complex of radars had limited the role of Type 960P to long-range warning and height estimation from the range of first detection.

The complete FD system comprising Types 960P/980/981 was scheduled for installation and trials at the ASE Extension, Tantallon, in 1945. The inevitable adjustments and delays to Establishment programmes in 1945 delayed the installation of Type 980, and it was September 1946 before the Outfit AQS antenna was erected and ready for trials. A standard Type 277P transmitter and receiver were used in this trial against a Meteor target aircraft. The complete records of the resulting range and vertical coverage trials have not been located in archives. However the data on seven runs, with the Meteor flying at 20000-ft have survived. This data is summarised in Figure 5.30. Far from the estimated 68-miles maximum, the best first-detection range on trials was only 52 miles, and the reliable operational range for interception control was rather less than 40 miles. There was therefore an unpredicted (and at that time unpredictable) performance loss of some 5 db. At the date of these Type 980 trials, the Type 981 Outfit AQT antenna had not been erected completely. However there had been some preliminary tests earlier using a purely experimental antenna of lower gain than that of the Outfit AQT, but of practically the same gain as the Outfit AQS antenna. In these preliminary tests, using a Mosquito aircraft as target, a 70-miles maximum range had been achieved.

It was 1951 before *Eagle* was commissioned and Type 980/981 could be subjected to operational trials at sea. One pair of Type 980/981 was fitted forward on the island and a second aft (Figure 5.31). Both radars performed marginally better than might have been predicted from the

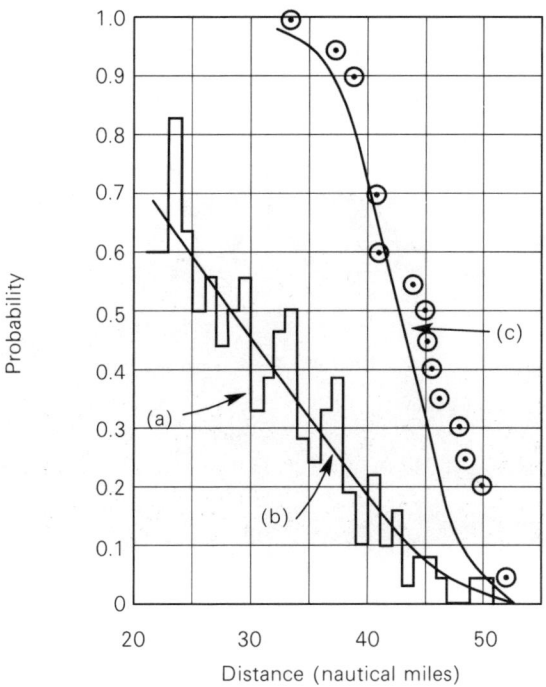

Notes:
(a) Observed probability of PPI signal each rotation of the antenna
(b) Smoothed curve of probability of visible PPI signal versus range.
(c) Probability of detection of approaching aircraft before a given range as derived from curve (b)
Cumulative probability is from the first detection ranges as observed during the trials.

5.30 Type 980: Trials against a Meteor aircraft at 20000-ft: analyses of 10 flights, opening and closing range (© Defence Research Agency)

early shore trials at Tantallon. Type 981 did prove capable of a maximum Free-Space range of at least 80 miles. Type 980 capability was rated at 40 miles for a 0.5 probability of signal detection on each rotation of the antenna. More stringent selection of components during manufacture could easily have added average performance.[80] In 1946 it was estimated that 2 db (or more) improvement in average crystal performance in centimetric radar could be obtained in this way. Types 980 and 981 were also fitted in the carriers *Ark Royal* and the Light Fleet Carriers *Bulwark, Centaur* and *Albion*.

In describing the development of Types 980/981 the original Type numbers have been used. In fact, in 1947 Type 980 became Type 982, and Type 981 became Type 983. It is these new Type numbers that will be found in postwar references to the carrier-fittings. The change merely

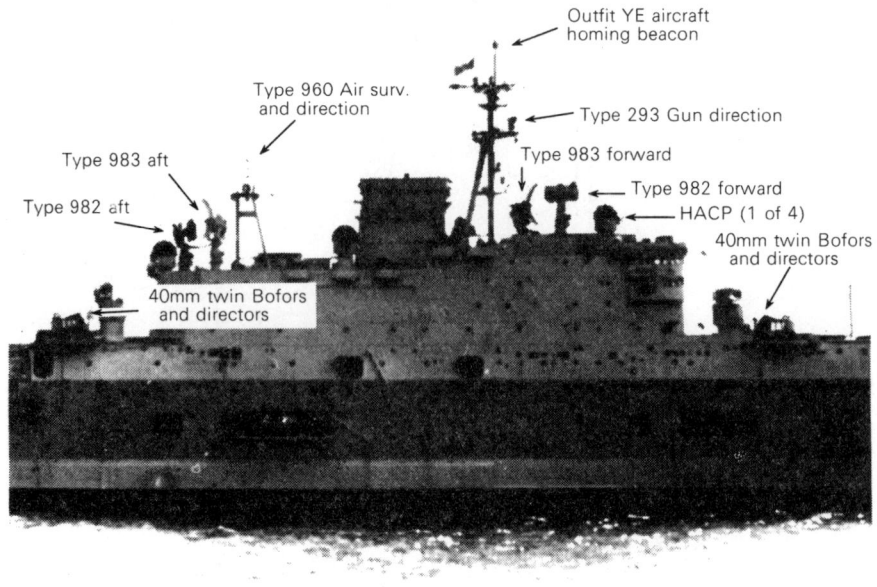

5.31 HMS *Eagle*, circa 1951 (© Imperial War Museum, Negative No FX: 10736)

reflects a decision not to complete yet one more redesign of the basic Type 277 transmitter and receiver. By 1946 the Type 277P design was complete and in production. There were no additional functional improvements affecting performance to add in a further redesign, and hence the redesign specifically for Types 980/981 was cancelled and new Type numbers issued for the radars. Indeed all the development work had used Type 277P equipment with the Outfits AQS and AQT antennae. The larger antenna for Type 960P, which would have given this metric radar a narrower beam in azimuth and a 220-miles range on aircraft at 50000-ft, was also cancelled. Thus post-1947 the FD radar complex consisted of Types 960/982/983.

Disappointment over the unexpected limitation in operational range performance of Type 982 (ex-Type 980) has tended to obscure the scope and magnitude of the technical revolution in radar operation introduced with the establishment of the Action Information Centre (AIC), with its Radar Display Room (RDR) and Air Direction Room (ADR).[81] Previous to the AIC organisation each radar had been unique with control and reporting from the radar office. Now, with the AIC, all functional control and reporting was centralised in the RDR as an Information Centre. Many plan displays were switchable between radars, or composite displays were employed to ensure a clear 360° antenna sweep in azimuth. Transmitter 'firing' times had to be synchronised, and antenna rotation coordinated. There could be over 50 displays fed from up to six radars in this complex information system.

THE UNKNOWN FACTOR IN TYPE 980 (982) PERFORMANCE

As early as 1944 it had been suggested that there could be an undetermined factor in performance assessment associated with target display. Theoretical studies of the detection of small signals in the presence of noise[82] indicated that an enhancement of signal-to-noise ratio was to be expected if there was integration following the second detector of a receiver. The long-persistence screen of a PPI could well provide this integration, and it was realised that persistence of vision in the eye could also perform the same function with an A-scan screen. Laboratory experiments[83] in 1945 using simulated radar echoes on a PPI display demonstrated that pattern recognition was also a factor in determining the minimum input signal required for recognition in the presence of the normal speckled background of noise-peaks on the display. Thus it appeared that both the total energy beamed on to the target, and the geometric form of the echo displayed, would affect the minimum signal needed for visual recognition. In both respects Type 980 differed significantly from Type 981. The azimuth sweep-rate of Type 980 over the target was faster than the vertical sweep-rate of Type 981; in addition, the number of pulses beamed at the target was further reduced by the vertical scan of the Type 980 beam. Finally, the echo displayed was broken into a series of dots instead of the normal, small, solid-line arc.

By 1946 it was clear that centimetric radar performance had to be expressed as a range versus probability of detectable signal, and not as a definitive relationship. In effect target reflection coefficient in the radar equation had to be considered a random variable, and the PPI echo as a multipulse sample of this variable. There was some evidence that target reflectivity varied greatly from pulse to pulse, in addition to the slower, deep fading that made the presence of echo on successive rotations of the antenna uncertain. However there was insufficient data to determine the frequency distribution of reflectivity of the target, and hence to incorporate integration effects numerically in range-performance prediction. Nevertheless some estimate had to be made as to whether any major improvement was possible in the performance of Type 980 and, if so, whether the required operational range of 80 miles was possible in the then current state of technological development.

In all previous plan display radar trials the 'maximum' range (or more precisely the range for 0.1 probability of detectable signal) had proved to be reasonably predictable using the established form of the radar equation. Even Type 981, under somewhat different operating conditions, had been predictable as regards maximum range. Type 980, with its double scan, was the only exception; it was therefore reasonable to consider what might be the performance of Type 980 if equipped with a single element of the Outfit AQS antenna, but having double the width in the azimuth plane. This would need no scan in elevation for high cover, and the antenna would have double the gain of the original Type 295 plan display antenna, which, as has been noted earlier, had the same gain as the Type 277P antenna. Predicting from Type 277P Free-Space range, as was done for Type 981, the maximum range of Type 980 would have been 58 miles. Although only a small improvement on the performance achieved with the Outfit AQS antenna, it indicated that longer range might well be achieved with a simpler antenna. In addition the 'cheese' form of antenna had been chosen originally to make provision for roll and pitch of a small ship. It did not generate the most efficient energy distribution for high cover from a relatively stable platform. However, even if maximum range could be extended appreciably toward the 80-miles objective, there seemed no possibility within the limits of then currently available technology to extend the achieved 40-miles operational range to the required 80 miles. This, it was estimated, would require either higher transmitter power or lower receiver noise-factor.

The opportunity arrived postwar to redesign the Type 980, now Type 982, antenna. The cumbersome and heavy, fully stabilised mounting that was necessary for Type 983 was not essential for the plan display Type 982. This needed no more than stabilisation in azimuth. The antenna for this new mounting consisted of a reflector of aperture 18-ft × 3-ft fed by a linear slotted waveguide, and shaped to produce a vertical polar diagram matched as nearly as possible to the required coverage in range and height. This both eliminated the vertical scan and achieved a more efficient distribution of energy in the vertical plane. Type 982 then became Type 982M, with an improvement in operational performance to 60 miles compared with the 40 miles of Type 980. Replacement of the Type 982 Outfit AQS antennae in carriers was effected in 1952–53. In the mid-1950s the Cathedral-class of destroyers was also fitted with Types 982M/277Q (278) for their role as Air-Defence Picket Ships (Figure 5.32), replacement of the heavy Type 983 antennae by the Type 277Q antenna being necessary from considerations of top-weight. Type 278 was a development of Type 277Q using the same antenna, but with an improved control system, and with solid-state devices introduced into the below-decks panels. Thus the FD radar system, initiated in 1944 as Types 980/981 continued in service for some 12 years postwar.

(a)

(b)

5.32 Air-defence picket *Llandaff*: (a) Type 960 – long-range air warning; Type 982M – air direction plan display; Type 277Q (278) – heightfinding; Type 293Q - target indication (guns); Type 275 – high-angle AA director (gun control); Type 974 – X-band navigation radar; (b) Type 982M. On the jackstay intership transfer: S.T. Wright, project leader for FD radar. In 1940 he was a member of the original project group under S.E.A. Landale for the Type 271 radar.

The full 1942 specification for a three dimensional radar for FD was not met until *Victorious* was fitted with Type 984, completing in 1957–58. This radar became possible with development of magnetrons of very high output power and yet further evolution of antenna technology in the 1950s. Type 984 used four magnetrons, each of 2.5 MW output in 2.5 microsecond pulses at 400 pps repetition rate. Driven by a common modulator, and in association with a 14-ft metal-plate lens antenna, the total output was divided appropriately between five scanning beams for heightfinding and one fixed-beam for surface and low-angle detection. With this radar there was no need to stop rotation for height measurement, and performance comfortably exceeded the original specified 80-mile range.

CONCLUSIONS

The pace of development of microwave radar during the war years had been remarkable. In the five years following the decision in November 1940 to capitalise on the initial, successful research at TRE by building a short-range radar in the 10-cm band (S-band), dedicated to detection of a surfaced submarine and capable of being installed in small convoy-escort vessels, Naval search and tactical radar in this waveband was in service, or on trial, with a capability of:

- equalling or bettering the capability of visual look-outs to detect surface craft as they appeared above the horizon in all visibility conditions;
- presenting on plan display the tactical situation around a ship of surface targets out to horizon range, and low air targets above the horizon out to 80–100 miles;
- by beam sweeping in elevation on a given bearing, measuring aircraft height out to 80 miles;
- plan display of all targets around a ship, including aircraft up to maximum operating height out to 40 miles.

Naval ambitions for Fighter-Direction radar were set higher than that actually achieved at 80 miles continuous tracking of aircraft on plan display, and continuous measurement of height without stopping azimuth scan. Limitation in magnetron technology placed these objectives out of reach at the time, and it was well into the postwar years before this specification could be satisfied. However, if the development of 3-cm (X-band) radar is included (see Appendix 1 to this monograph), then by 1945 all classes of Naval vessel had been equipped with surface search radar – viz. MTB/MGBs with Type 268, submarines with Type 267 – and with X-band there was a greatly enhanced capability for close in-shore navigation. In retrospect the key technological advances that made this rapid progression possible were, in the first instance, the early availability of higher transmitter power, and in the later years the development of PPI display and the rapid transition of waveguide techniques from theory to reproducible components suitable for dockyard installation.

Although range has been used as a primary measure of the progress of development of warning radar, the scope of evolutionary progress was also to be seen in the rapid extension of use of centimetric radar in all of the many functions in a ship that could benefit from surface-to-air warning, gun direction, navigation and fighter direction, and finally the welding of all tactical radar data into a single comprehensive input to the Action Information Centre.[84] The particular value of range achieved as a measure of progress lay in the ability to make numerical comparison. The constant search to increase operational radar range was a common feature in all development of warning and tactical radar in the period 1940–45, and meant that such radar was always operating at or near the upper limit of technological possibility.

APPENDIX 1: THE COMPLEMENTARY ROLE OF X-BAND FOR WS RADAR

Introduction

The first formal operational requirement for an X-band (3-cm) radar was formulated in the autumn of 1941 by the Admiralty for ASV for the Barracuda aeroplane, then under development for the Fleet Air Arm to replace the Swordfish. Work under the Admiralty-sponsored CVD valve programme had already been initiated in late 1940 on the extension of S-band (10-cm) technology to the development of oscillators for this shorter waveband.[85] By the spring of 1941 cooperation between the Clarendon Laboratory at Oxford, EMI and the HMSS Valve Development Laboratory at Bristol had resulted in the production of samples of a tuneable reflex klystron with an output of 100 mW in the waveband 3.05–3.45-cm.[86] A satisfactory design of magnetron proved more difficult to achieve. All designs tested up till late 1941 had been much less stable as regards oscillatory mode, and very variable in efficiency, compared with the original NT 98 S-band unstrapped magnetron. Strapping, when introduced in the X-band magnetrons in mid-1941, did not resolve these problems as it had done for the S-band valves.[87] Preproduction was established for the most successful of the early unstrapped designs in order that radar development could proceed, whilst the search for a more satisfactory valve continued. It was not until mid-1943 that reliable, mode-stable, strapped magnetrons for X-band (CV208, CV209, CV214) became available in production from the research programmes in UK and USA.[88]

The Type 261 Development

The significance of X-band for airborne radar lay in the prospect of very compact, high-resolution, high-gain antennae. This was to prove important also for Naval radar in the development of close-range, blind-fire control of anti-aircraft armament.[89] However, around September, 1941 the predominant interest in X-band for shipborne radar lay in the possibility of more effective illumination of small surface targets. In the absence of data on propagation factors and target reflectivity at centimetre wavelengths, judgement of the utility of X-band radar for WS application had to be determined empirically by trials at sea. An experimental shipborne radar was therefore required, and a development contract was placed with EMI for this purpose. The specification for this radar, later designated Type 261, would have made it a suitable replacement for the S-band Type 272, should X-band demonstrate any marked improvement over S-band for general surface warning. In particular the transmitter and signal mixer were required to be included in the radar-office equipment, to permit remote mounting of the antenna on the mast.[90] Several sets of equipment were made under this contract. One was delivered to the Eastney Laboratories of ASE in May 1942 and another, modified and improved, was installed in the Signal-Department trials ship *Saltburn* in October 1942.

Type 261 as fitted in *Saltburn* is shown schematically in Figure 5.33. This diagram, and the following technical data are drawn from the EMI report referred to above. The antenna unit was mounted on a bracket off the mast, at 51-ft above the waterline. It was connected by 60-ft runs of circular cross-section waveguide to the transmitter and receiver signal mixer in the radar office. Total waveguide losses were estimated at 4–5 db for the size of waveguide used. Separate transmitting and receiving antennae were installed to avoid the additional losses that would have been introduced by use of a switch for common T/R working. Connection to the rotatable antennae was by simple rotating joints in the waveguides, leakage at the break in each guide being prevented by quarter-wave flanges. One joint was placed below the antennae on the axis of rotation, and the other above. The magnetron used was the BM325, one of the many experimental designs in 1942. It operated at an input of 300 kW in a magnetic field of 2600 Oersteds, producing 1 microsecond pulses at 500 prf. The pulse-power output was estimated at

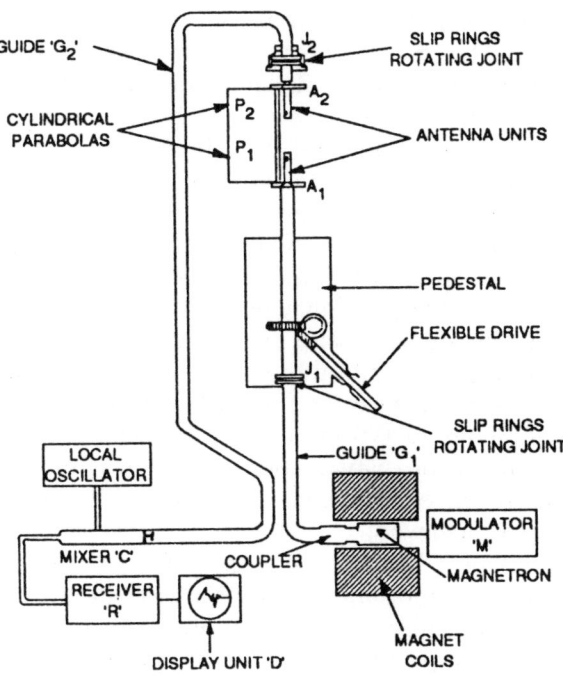

5.33 Type 261 installed in *Saltburn* for trials (© Defence Research Agency)

25 kW. Laboratory testing with a spectrometer had shown that under the load conditions in Type 261, these magnetrons could have output spectra as broad as 10 MHz, and in general not less than 4 MHz. They could also oscillate in more than one mode simultaneously. In conjunction with the Type 271 IF amplifier used in the receiver there would have been significant additional losses to those incurred simply by use of a long waveguide. The two antennae were of identical 'cheese' form, mounted one on top of the other, as in Type 271. The aperture dimensions of each 'cheese' were 48 × 12-cm. Gain, as estimated at the time, was stated to be 140 relative to that of a half-wave dipole, although this appears to be very low compared with the measured gain of other later X-band antennae of similar form. However, this particular 'cheese' antenna was designed before the theoretical solution to the effect on the vertical radiation pattern of the parallel top and bottom plates sectioning the parabolic cylinder of the 'cheese' form of antenna. Therefore, as had been the case with the original Type 271 antenna, the gain could easily have been well below optimum for the aperture dimensions chosen.

The performance during sea trials in November 1942 was good, but not outstanding. Maximum ranges of selected radar targets, using A-scan display, are shown in Table 5.2. Reliable range would have corresponded to a signal showing a distinct base-line break on the A-scan, with a maximum signal amplitude (as it varied) of about that of the rms noise level. A just-visible signal would have been detectable only as a change in the pattern of the noise trace. These results were rated in the report of the trials as comparable to the performance of Type 271 with antenna at 35-ft.[91] Perhaps a more accurate numerical assessment would have credited Type 261 using masthead antennae with a 10–20 percent range advantage over Type 271 fitted in a corvette. However, at the end of 1942 Type 271 was already obsolescent and on the point of being replaced by the more powerful Type 271Q. Hence the formal Establishment conclusion at the time was that, as far as WS radar in oceangoing ships was concerned, there seemed to be no Naval application for which Type 261 could compete with 10-cm equipments, unless a substantial improvement in performance was achieved.

This rejection of X-band for WS application in ships, from corvette upwards, still left important WS applications in situations where the possible antenna height was too low for even S-band to be effective. The need for a small, compact radar for destroyers, and below, to provide air warning and a measure of surface warning had been recognised in 1940. An existing ASV airborne radar on 214 MHz (1.4-m) had been adapted for this purpose, and with an appropriate antenna became Type 286, later Type 291 (see Monograph 4). In a destroyer, with the small dipole array of Type 286PQ or 291 mounted at the masthead 80-ft above waterline, this radar had a WS performance comparable to Type 271 in its corvette position with the antenna at 35–40-ft. With adaptation of antennae, this radar was also fitted in submarines, MTB, MGB and other small craft of Coastal Forces. In these applications the normal advantage of antenna height was lost; whilst still providing air warning, the surface performance was quite inadequate for operational requirements.

Table 5.2 Type 261 trials, November 1942

Target	Rèliable range (yds)	Just-visible range (yds)
ML466	4000–5000	7000–8000
Destroyer *Rockwood*	17000–18000	20000
Submarine *H44*		
Surfaced	6000	9000
Conning tower	3500	6000
Cage buoy	2600	

Evolution of Type 267W

It was not until late 1941/early 1942 that radar in submarines began to be considered as a possible aid in attack, as well as providing air warning when surfaced. It was clear that an effective WS radar for submarines would need to operate on the highest possible frequency, especially if used at periscope depth in attack. Even with the submarine fully surfaced the maximum antenna height could not be allowed to exceed 25–30-ft. At periscope depth the antenna might project above the surface by no more than 4-ft. A second consideration was visibility to the enemy during attack. For this, a limit was set to the maximum size of object exposed – not to exceed 25-in × 2.5-in. This is called for a very different antenna from that developed for the experimental Type 261 destroyer radar, and a further set of equipment was built for trials in the ASE workshops at Witley in mid-1942. This radar, Type 261W using the GEC E1263 magnetron operating at an input of 400 kW, was subjected to performance trials in the autumn of 1942. No detailed report has been found of these trials, but in a brief summary of the development of submarine radar written in 1945 it is stated that 'the results showed surface cover to be satisfactory, but that the probability of detecting aircraft at ranges greater than 10000-yds was very low'.[92] In view of these results the decision was taken, and subsequently confirmed by an Admiralty Staff Requirement in January 1943, to develop a radar combining the WS properties of X-band with the air-warning capability of Type 291W. This was designated as Type 267W, comprising essentially two transmitters working from a common modulator, and using a common IF and display system.

Sea trials of the first Type 267W took place in the Firth of Forth in August 1944 in the submarine *Tuna*, following shore-based trials from the ASE Extension, Tantallon. The X-band magnetrons used in the developed set yielded a peak-power output of 15–25 kW, operating in a magnetic field of 3200 Oersted at an input of 160 kW. The valves were selected for a narrow frequency range of 9610–9700 MHz (3.09–3.11-cm), tuning of the transmitter and mixer having been preplumbed to the maximum extent. (the magnetron was matched into a short section of rectangular waveguide during tests, and the receiver mixer/local oscillator unit was developed in ASE by J. Dahl and M. Surdin.) Pulse length was 1 microsecond, and the prf 500. AFC was applied to the local oscillator using a separate crystal mixer. The aperture of the 'cheese'-type antenna was 24-in × 6.75-cm, and the measured gain 230. Overall beamwidths for the set were 3.25° horizontally and 30° vertically, to 6 db down.

In general, with *Tuna* surfaced, the airguard (214 MHz) and the seaguard (3-cm) performance of the first Type 267W fitted met expectations, except for an unexpectedly low performance of 3-cm against low-flying aircraft. On 3-cm, a range of 11 miles was achieved on *Pollux*, an ex-French minelayer of around 2500 tons. At periscope depth, good performance was maintained on surface targets. Following the *Tuna* trials two further T-class boats were equipped with Type 267W, *Tapir* and *Turpin*. These entered operational service during patrols off Norway in March 1945. The X-band mast for Type 267W was raised to a fixed height. To meet the need for a periscopic mast, the mast system and associated waveguide run were modified. Type 267MW with this improvement began to be fitted in the more modern A-class boats in March 1945, beginning with *Amphion*. Neither this improvement nor a general redesign to make better use of the space allocated to radar in the more modern boats (Type 267PW with PPI display) were in operational service before the end of the war.

The Type 268 Radar Development

The X-band Type 268 WS radar was a parallel Canadian development, initiated in 1942. In view of the development of X-band magnetrons and klystrons in the United States, the Radio Branch of the National Research Council of Canada (NRC) had formed a techniques group to study the use of this waveband for radar.[93] A formal request for the development of a lightweight, high-performance WS radar for Coastal Forces craft was made to NRC by the British Admiralty in May 1942. The low antenna height possible in an MTB, MGB or ML predicated the use of X-band with a high-gain antenna if, as requested, performance was to be comparable to the longer-wave S-band radar mounted higher. A 'cheese'-type antenna was used, similar to the Type 271Q transmitting antenna, to provide a fan beam and to avoid the need for stabilisation. The ends were squared off to give an aperture of 30-in $\times$ $6\frac{1}{4}$-in, the focal line of the parabolic cylinder reflector lying outside the aperture (Figure 5.34). The gain was 840 relative to that of a half-wave dipole, and beamwidths to 6 db down for the set were 2° horizontally and 14° vertically. Transmitter power was 40 kW, and pulse length 0.75 microseconds. Display was on a 5-in PPI, with a slave PPI in the chartroom. Type 268 was entirely produced in Canada.

A prototype model of Type 268 was received by ASE in April 1943, and fitted in MGB 680 for sea trials in Spithead. A range of 7000-yds was achieved on an ML. First production sets arrived in November 1944, and general fitting began in Coastal Forces, including minesweepers, in 1945. This was later extended to Hunt-class destroyers, and as an interim navigation radar in cruisers and above. In all, some 1600 sets were produced before the end of the war.

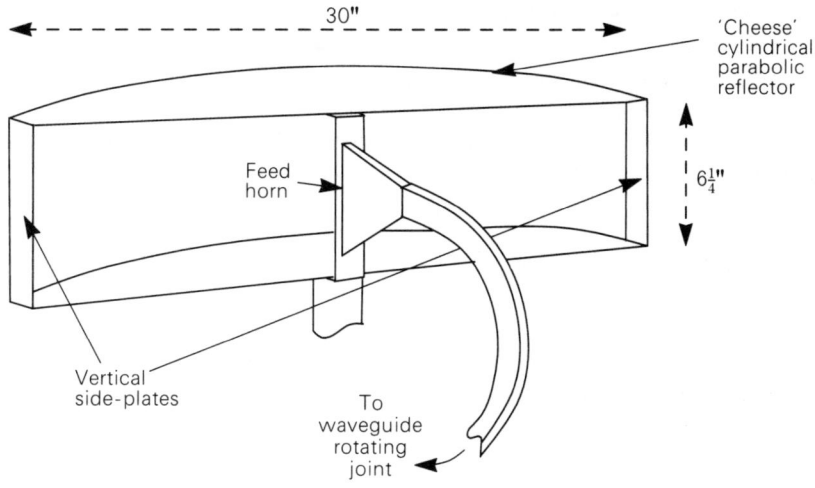

5.34 The Type 268 radar antenna

The Type 972 Development

After the sea trials of the S-band Type 277 in *Coppercliff* in September 1944, it was thought that the high resolution possible with an X-band radar might make it more effective in detecting a Schnorkel in rough sea conditions than had been the case with Type 277. In January 1945 Type 268 equipment became available. It was therefore decided to replace the lightweight Type 268 antenna by an 8-ft wide 'cheese' antenna mounted on a Type 277 pedestal for a series of comparative sea trials of this X-band radar against Type 277 with its antenna mounted at the same height as the Type 972 antenna.[94] The overall beamwidth of Type 972 in azimuth would have been about 1°, or a little under. The initial trials in *Pollux* in March 1945 were in the Solent in enclosed waters, with later trials in the Firth of Forth in rougher water conditions. Performance in the detectability of the Schnorkel proved little different between Types 277 and 972. In rough sea conditions the target became more or less impossible to detect and, if seen, impossible to follow as soon as wave height became comparable to Schnorkel exposure above the surface. However Type 972 demonstrated a clear superiority close inshore, or in enclosed waters, in the detectability of a small echo in the presence of multiple echoes from adjacent shipping, buoys and strong land echoes. The value of radar as a navigation aid had increased with each upward step in frequency. With X-band radar, the PPI presentation of the foreshore, and other shipping, was such as to make comparison with the chart relatively easy, and gave promise of the possibility of 'blind' navigation, as can be seen from the two photographs of the PPI screen of the Solent as seen by 3-cm radar (Figure 5.35). The Type 972 in *Pollux* was again used in the summer of 1945 to make a radar survey of the Thames Estuary and lower river, as part of a project to study future radio aids to navigation. With an accurate ranging panel added, Type 972 was fitted postwar to survey vessels as Type 972M.

Shipborne Radar Navigation

A considerable number of Merchant Vessels were radar-fitted as early as 1943. When the S-band Type 271 became obsolete for Naval use, following replacement by the more powerful Type 271Q, independently routed merchantmen were fitted with reconditioned Type 271 sets. However the first Naval radar specifically for accurate navigation inshore was Type 970, a conversion of the airborne H2S for shipborne use in a few LCH (Landing Craft Headquarters). These ships took part successfully in amphibious landing operations such as OVERLORD, but special charts had to be developed to aid in identification of shoreline features using the 10-cm radars available. True 'blind' navigation only became possible with

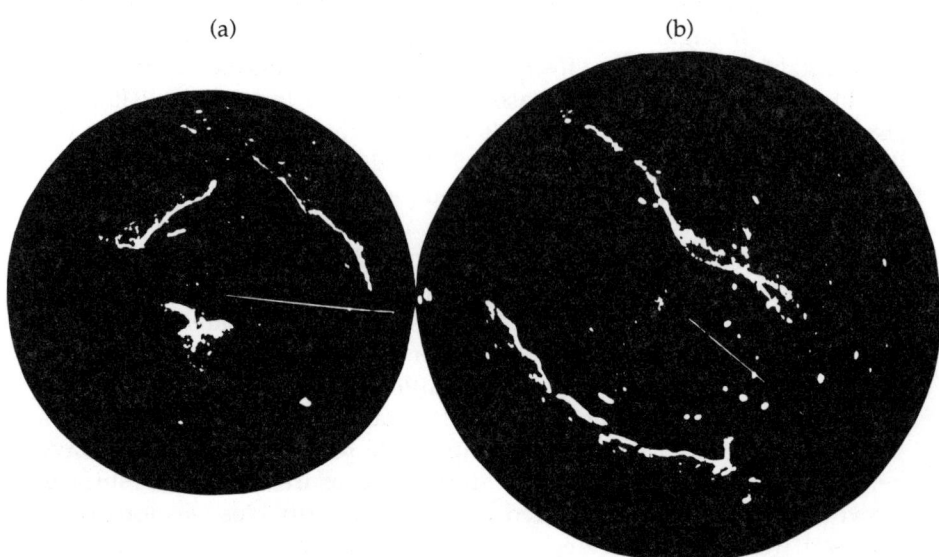

5.35 The Solent as seen on the PPI display of a 3-cm radar: (a) Southampton Water and Cowes; (b) Ryde, with pier clearly distinguishable (© Defence Research Agency)

the development of 3-cm radar. The first set in service was Type 971, a conversion of the airborne H2X, installed again in a few LCH vessels. These ships were used in the Scheldt estuary in the autumn of 1944, and early in 1945 made probably the first blind navigation of an inland waterway.[95] With this background, and the experience rapidly gained with Types 268 and 972 in 1945, it appeared obvious that X-band was the appropriate waveband for development of a commercially produced radar for the Merchant Navy.

In the summer of 1944 a Government Body known as the UK Conference on Radio for Marine Navigation was set up to consider the question of applying modern radio navigational aids for the use of the Merchant Navy. ASE was asked to collaborate in preparing a performance specification for a radar suitable for the typical merchant ship. A specification was drafted in January 1945 and issued by the Ministry of Transport to ship owners and the radio industry. At the same time ASE began construction of a demonstration model conforming closely with the major electrical features of the performance specification. This radar was fitted in *Fleetwood*, subjected to various sea trials and used for demonstrations to industry from April 1946. Production of Marine radar by six British manufacturers followed quickly. It is interesting to note that the Eastney Fort East Laboratory extension of ASE, where the first 10-cm operational radar was built in 1941, became a test facility for commercial Marine radar postwar, and so remained until 1990.

APPENDIX 2: CHRONOLOGY OF DEVELOPMENT OF NAVAL RADAR FOR AIR AND SURFACE WARNING

1939	Admiralty CVD contracts for research on transmitting and receiving valves for 10-cm wavelength.
Dec. 1939	Air Ministry contracts to GEC and EMI for research on radar at centimetric wavelengths, initially in the band 20–30-cm.
Feb. 1940	First achievement of several hundred watts CW output from a laboratory model of a resonant cavity magnetron (Boot/Randall, at Birmingham University).
July 1940	Comparative tests of sealed-off production version of resonant cavity magnetron (9.1-cm) and 'millimicropup' triode (25-cm) in GEC experimental radar at Wembley.
Aug. 1940	Formation of a Cm-Wave Applications group under P. I. Dee at TRE for S-band (9–10-cm) radar.
Oct. 1940	Formation of a Transmitter-Development group at HM Signal School Extension, Eastney under S. E. A. Landale to prepare for development work in the 10-cm waveband.

Nov. 1940   Surfaced submarine detected in tests of TRE experimental equipment at TRE, Swanage at a range of 7.5 miles. Decision to proceed immediately with the design of a 10-cm radar for convoy-escort vessels.

Type 271 Series of 5 kW Radars

Feb. 1941   Completion of Type 271X corvette A/S radar.
Mar. 1941   Sea trials of Type 271X in *Orchis*; 'cheese' antenna.
(May 1941) Improved receiver using more powerful local-oscillator. Type 271X now designated Type 271.
July 1941   Mobile trailer version NT271X at Dover. Army conversion of Type 271 for Coastal Defence using large antenna. Official designation CD No.1 Mark 4.
Aug. 1941   Sea trials Type 273 in *Prince of Wales*. Identical Type 271, except larger paraboloid antennae (3-ft diameter).
Sep. 1941   Type 272 in *Wallace, Hermione* and *Ariadne*. Identical Type 271, except for long cable connection from radar office to transmitter and receiver mixer on the antennae.
Oct. 1941   Type 273S, Naval shore-station version of Type 273, with 4.5-ft paraboloid antennae.
Dec. 1941   First Type 273 with stabilised antennae in *Nigeria*.
Jan. 1942   Types 271P/272P/273P. Redesign of radar office panels for line production; functionally identical to Types 271/272/273.
Mar. 1942   Introduction of prefabrication for Type 271P. Fully equipped radar office with antenna housing ready for hoisting aboard and securing to a prepared site on convoy escorts.
Nov. 1942   Introduction of cylindrical radome to replace 'pillar and window' style of antenna housing, to minimise side echoes.

Type 271Q Series of 70 kW Radars

Development of the 'Q' conversion equipment proceeded in parallel with diversification of application for the 5 kW Type 271. An important design objective was to make the replacement of an existing Type 271 or 273 as easy as possible, and achievable without use of dockyard facilities.

(Sep. 1941) CV56 magnetron available. Initiation of design of a suitable 'discharge-line' modulator.
May 1942   Sea trials of Type 271Q in *Marigold*.
July 1942   Sea trials of Type 273Q in *King George V*.
(Nov. 1942) Second corvette conversion to Type 271Q, but now using the new cylindrical radome.

Type 277 Series of 500 kW Radars

1941–42	Development of 1 MW pulse-power modulator, common T/R working, long waveguide operation for the CV76 magnetron, and PPI adapted for Naval use.
(July 1942)	Experimental Type 277 radar in rotating cabin trailer for low-flying aircraft trials. Designated Type 277T, this became the first version of the series in operational service. It differed from the later ship versions in having only a short waveguide connections of transmitter to antenna, and in use of a very large 'cheese' antenna fixed directly to the cabin roof.
Dec. 1942	Trials from shore at Great Orme against aircraft.
Mar. 1943	First operational Type 277T installed at Dover.
Apr. 1943	Sea trials of Type 277X in *Saltburn*. Stabilised 4.5-ft paraboloid antenna with 100-ft of waveguide run.
Aug. 1943	Sea trials of Type 293X in *Janus*. Experimental test of a 'cheese' antenna designed to give high-air cover for TI plan display, and allocation of target to individual weapons. Range proved inadequate.
Nov. 1943	Sea trials of Type 276 in *Tuscan*. WS and low-air warning for light cruisers and destroyers. Also retained as interim for TI, pending development of a larger, higher gain antenna for high-angle cover. Type 276 used the Type 271Q transmitter antenna.
Mar. 1944	Sea trials of Type 277 in *Campania*; Surface ranges to horizon, low-air to 40 miles, and heightfinding to 30 miles.
Mar. 1945	First fitting of Type 277P in *Anson*; improved receiver and broad-band mixer giving 2 db improvement in noise-factor.
Mar. 1947	Sea trials of Type 277Q in *Illustrious*. Larger antenna (8-ft × 6-ft) increased low-air cover to 85 miles and heightfinding to 55 miles.

Centimetric FD Radar

1942	Limited success in using metric radar Type 281 to vector fighters on to the target led to a Staff Requirement for Centimetric FD Radar to provide continuous tracking on plan display, and accurate height measurement out to a range of 80 miles.
1943	Type 295 proposed; two antennae on a single mount, one for plan display and the other for heightfinding. To achieve range specified, a magnetron of 2 MW pulse power was considered essential. By the end of the year it was clear that

	neither could the very-high-power magnetron be developed to production stage in time for the project, nor could the problem of height measurement using a continuously spinning antenna be resolved.
1944	Functions of Plan Display and heightfinding split between two separate radars, Types 980 and 981, respectively. By use of larger antennae the existing 500 kW transmitter could be used, and height could be measured by sweeping the beam of Type 981 in the vertical plane whilst held on the target bearing.
1945–46	Preliminary trials of the system indicated adequate performance was to be expected from Type 981, but that Type 980 would only achieve (at best) half the range required for plan display. Redesignated Types 982/983, the system was fitted in several carriers postwar.

Note: Use of brackets around the date indicates that the month quoted is only approximate.

Notes to Appendix 3 (opposite)

1. The nominal power of 500 kW for the high-power radars should be reduced by 2 to 3 db when installed with a long waveguide to allow for RF losses and irremediable frequency spread.
2. Types 271/272/276: 4-ft × 10-in aperture-'cheese', dipole fed, except for 271Q transmitter and 276. Types 293/293M: Tilted 'cheese' of apertures 6-ft × 4-in and 8-ft × 7.5-in, respectively. Types 273/273Q: 3-ft diameter paraboloids, 273Q transmitter being waveguide fed. Type 277T: large 'cheese' of aperture 15-ft × 30-in. Type 277/277P: 4-ft 6-in diameter paraboloid. Type 980: triple 'cheese', each 12-ft × 17-in aperture. Type 981: paraboloid 14-ft 6-in diameter, cut to 5-ft width.
3. Values in brackets ( ) are estimated.
4. Antenna beamwidths are measured to 6 db down, gain is referred to the gain of a half-wave dipole, and power-rating has been chosen within the acceptable spread from manufacture as an average value most likely to be met in service. Under laboratory conditions it was not difficult to ensure 100 kW from a selected CV56, or up to 750 kW from a selected CV76.

Source: This data is taken from contemporary ASE Technical Reports, with particular reference to a compilation of antenna data issued in 1944.

APPENDIX 3: CHARACTERISTIC DATA FOR NAVAL S-BAND WARNING RADAR

	271X	271/272 271P/272P	273 273P	273S	271Q	273Q	293	293M	276	277T	277	277P	277Q	980 982	981 983
Transmitter (Tx)															
Peak power (kW)	5	5	5	5	70	70	500	500	500	500	500	500	500	500	500
Pulse length (microsec)	1.5	1.5	1.5	1.5	2.0/0.7	2.0/0.7	1.8/0.7	1.8/0.7	1.8/0.7	1.8/0.7	1.8/0.7	1.8/0.7	1.8/0.7	1.8/0.7	1.8/0.7
p.r.f. (Hz)	500	500	500	500	500	500	500	500	500	500	500	500	500	500	500
Receiver (RX)															
Noise factor (db)	20	(16–18)	(16–18)	(16–18)	(14–16)	(14–16)	(14–16)	(14–16)	(14–16)	(14–16)	(14–16)	12	12	12	12
I.F. (MHz)	60	60	60	60	60	60	60	60	60	60	60	13.5	13.5	13.5	13.5
Bandpass (MHz)	2.5	2.5	2.5	2.5	1.25/ 4.0	1.25/ 4.0	1.25/ 4.0	1.25/ 4.0	1.25/ 4.0	1.25/ 4.0	1.25/ 4.0	0.5/1.0 /4.0	0.5/1.0 /4.0	0.5/1.0 /4.0	0.5/1.0 /4.0
Antenna															
Gain (Tx) × dipole	55	55	250	575	180	360	(110)	220	180	1300	800	800	1750	2320	3000
Gain (Rx) × Dipole	55	55	250	575	55	270					Common T/R				
Horiz. beam (Tx) (degrees)	8.6	8.6	14.8	9.7	6.3	9.2		3.6	6.3	2.5	6.2	6.2			
Horiz. beam (Rx) (degrees)	8.6	8.6	14.8	9.7	8.6	14.8	no data								
Horiz. beam (Set) (degrees)	6.6	6.6	10	6.8	5.6	7.6	no data	2.6	4.6	2.0	4.5	4.5			
Vert. beam (Tx) degrees	85	85		7.1	36	12		44	36	12.3	6.2	6.2		(1.5)	(4.0)
Vert. beam (Rx) degrees	85	85	11	7.1	36	10	no data								
Vert. beam (Set) degrees	66	66	7.6	5.2	27	8.8	no data	30	27	8.8	4.5	4.5		Vert. scan	(1.3)

Acknowledgements

The author wishes to express his gratitude to the Director of the Defence Research Establishment (DRE), Portsdown, for permitting access to archives of which many of the relevant reports may later be copied in the Public Records Office. Thanks are due also to many former colleagues for clarification of certain aspects of technical development not fully covered in the archive reports available; to the Librarian at DRE, John Briggs for his patient assistance, and particularly to two former colleagues, J. S. Shayler and S. T. Wright, whose assistance in the search for relevant reports and in critique of draft material have made this review possible.

References

1. E. G. Bowen, *Radar Days* (Adam Hilger, 1987), p. 141.
2. F. Trenkle, *Die Deutshcen Funkmessverfahren bis 1945* (Dr Alfred Hüthig Verlag, Heidelberg, 1986), p. 72 et seq.
3. E. B. Callick, *Metres to Microwaves* (Peter Peregrinus, 1990), p. 62.
4. A. C. B. Lovell, *Echoes of War – the Story of H2S Development* (Adam Hilger, 1991), p. 41.
5. H. A. H. Boot and J. T. Randall, 'The Cavity Magnetron', *Jour. IEE*, Vol. 93, Part IIIA, p. 928 et seq; W. E. Willshaw et al., 'The High-Power Pulsed Magnetron: Development and Design for Radar Applications', *Jour. IEE*, loc. cit., p. 985 et seq. E. C. S. Megaw, 'The High-Power Pulsed Magnetron: A Review of Early Developments', *Jour. IEE* loc. cit., p. 977 et seq. L. F. Broadway, 'Velocity-Modulated Valves', *Jour IEE* loc. cit., p. 855 et seq. A. F. Pierce and B. J. Mayo, 'the CV35 – a Velocity-Modulation Reflection Oscillator for Wavelengths of about 10-cm', *Jour IEE* op. loc. cit., p. 847 et seq. B. Bleaney, 'Crystal Valves', *Jour IEE* loc. cit., p. 847 et seq.
6. Captain S. W. Roskill, *The War at Sea* (HMSO, 1947), Vol. I, p. 350.
7. Ibid
8. C. Domb and M. H. L. Pryce, 'The Calculation of Field-Strengths Over a Spherical Earth', *Proc. IEE*, Vol. 94, Part III (1947), p. 326 et seq.
9. E. G. Bowen, op. cit., p. 148.
10. J. G. Crowther and A. Whiddington, *'Science at War'* (HMSO, 1947), p. 19.
11. A. C. B. Lovell, op. cit., p. 50 et seq.
12. PRO AVIA 7/144 (TRE File 4/4/457). The documents now located in AVIA7/144 permit a precise timetable to be drawn up for establishment of the Naval 10-cm Applications Group under S. E. A. Landale. The documents are:
    - Letter of 19th November, 1940: Superintendent TRE to The Secretary, Ministry of Aircraft Production (D. C. D.) on the subject of "Use of cm waves for Submarine Detection" (TRE 4/4/457).
    - Letter of 18th November, 1940: Captain Willett for Captain Commanding Signal School to Superintendent TRE, copied to DSD and DSR, Admiralty and circulated at TRE to Drs. Lewis and Skinner: names the HMSS party to report at Swanage on November 21st and details the instructions given Landale as the senior in charge (no formal letter reference).
    - MS notes for the 'Skinner' report as reproduced in Appendix A of *Radar at Sea* (Macmillan, 1993), containing some additional information.
13. H. D. Howse, *Radar at Sea. The Royal Navy in World War 2* (Macmillan, 1993), pp. 69–71.
14. A. C. B. Lovell, op. cit., p.51.

15. PRO AVIA 7/144. Superintendent TRE letter of 19 November, 1940, Reference TRE 4/4/457.
16. A. C. B. Lovell, op. cit., p. 52.
17. PRO AVIA 7/144. TRE letter, op. cit.
18. H. W. B. Skinner, 'Preliminary Reports on Tests on Ships Using 10-cm Waves, November/December 1940', quoted in full as Appendix A in H. D. Howse, *Radar at Sea. The Royal Navy in World War 2* (Macmillan, 1993).
19. PRO AVIA 7/144, loc. cit., Letter of 18 November, 1940 from Captain Willett, HMSS to Superintendent, TRE.
20. A. C. B. Lovell, op. cit, and E. G. Bowen, op. cit.
21. H. W. B. Skinner, op. cit.
22. A. C. B. Lovell, op. cit, p. 53.
23. PRO AVIA 7/144. Manuscript notes in file.
24. CAC NRT ND154, Appendix B, 'Type 271. Trials in HMS *Orchis*, Report dated April, 1941.
25. ADM 220/78, CSS to DSD (personal), 29 March 1941 included the following: '. . . I think that Landale and his party have done a wonderful job in taking over from the Research Department (TRE) in December, engineering and improving the technical performance of the equipment, and getting it to sea and working by mid-March. I hope one day you will have a chance to tell him so'.
26. L. F. Broadway *et al.*, 'Velocity-Modulation Valves', *Jour IEE*, loc. cit., p. 859.
27. O. Böhm, 'Cheese Aerials', *Jour. IEE*, loc. cit., p. 45.
28. W. E. Willshaw, op. cit.
29. O. L. Ratsey, 'Radar Transmitters: A Survey of Developments', *Jour. IEE*, loc. cit., p. 847 et seq.
30. B. Bleaney, 'Crystal Valves', *Jour. IEE*, loc. cit., p. 847 et seq.
31. PRO ADM 220/79, Admiralty letter SD 02053/41 of 2/8/41.
32. L. F. Broadway, op. cit., p. 183.
33. CAC NRT ND154 Appendix C, 'Report on RDF Trials with Type 271 on HMS *Prince of Wales*', August 1941.
34. Brig. A. P. Sayer, 'Army Radar – the Second World War', 1950, p. 128 et seq.
35. Ibid., p. 129.
36. PRO ADM 220/78, February 1942.
37. PRO ADM 220/80, 1942.
38. PRO ADM 220/78, Admiralty meeting of 4/8/41.
39. D. G. Fink, *Radar Engineering* (McGraw-Hill, 1947).
40. PRO ADM 220/80, Signal 11/10/1941.
41. RDF Bulletin No 1, Report No 9, April 1942, DSD.
42. J. H. E. Griffiths, 'The Development of Radio Valves', *Jour. IEE* op. cit., p. 177.
43. J. G. Crowther and A. Whiddington, op. cit., p. 47.
44. L. G. H. Huxley, *The Principles and Practice of Waveguides* (Cambridge University Press, 1947).
45. E. C. S. Megaw, op. cit., p. 977 et seq.
46. J. H. E. Griffiths, op. cit, p. 175.
47. W. E. Willshaw, *et al.*, op. cit., p. 985.
48. D. G. Fink, *Radar Engineering* (McGraw-Hill, 1947).
49. W. E. Willshaw *et al.*, op. cit., p. 986.
50. C. A. Cochrane, 'Development of Naval Warning Radar', Draft report held in CAC, NRT papers.
51. O. L. Ratsey, op. cit., p. 255.
52. J. H. E. Griffiths, 'The Development of Radio Valves', *Jour. IEE*, loc. cit., p. 175.

53. CAC NRT ND32, Extracts from SS/ASE Monthly Reports, 1941–45.
54. CAC NRT ND154, Appendix E, 'Type 271 Mark IV RDF trials in HMS *Marigold*', May 1942.
55. CAC NRT ND154, Appendix F, 'Type 273 Mark IV: Report of Trials Carried Out July 10th–13th, 1942.
56. CAC NRT ND32, Extracts from SS/ASE Monthly Reports, 1941–45.
57. O. L. Ratsey, op. cit., p. 255.
58. A. W. Cooke, 'Gas-Discharge Switches for Single Aerial Working', *Jour. IEE*, loc. cit., p. 186.
59. CAC NRT ND154, Appendix F, op. cit.
60. CAC NRT ND30 and ND34 DSD RDF Bulletins, Nos 1–6, February 1943– March 1944.
61. A. H. Cooke et al., 'Electronic Switches for Single-Aerial Working', *Jour. IEE*, loc. cit., p. 1575 et seq.
62. F. A. Kingsley (ed.), *The Application of Radar and Other Electronic Systems in the Royal Navy in World War 2* (Macmillan, 1994), Monograph 1 by H. W. Pout.
63. O. L. Ratsey, op. cit., p. 259 et seq.
64. CAC NRT 30 and ND34, op. cit.
65. PRO ADM 220/2014, 'RDF WS Types 272/273, Mark V: Expected Performance', DSD Minute No. SD 05577/42, October 1942.
66. PRO ADM 220/78, Amendment to Reference 65.
67. M. H. L. Pryce, 'Waveguides', *Jour. IEE*, loc. cit., p. 36 et seq.
68. E. C. S. Megaw, 'Experimental Studies of the Propagation of Very Short Radio Waves', *Jour. IEE*, loc. cit., p. 96 et seq.
69. CAC NRT ND154, Appendix J, 'Type 293X in HMS *Janus*: Report of Trials, Part 1, Aircraft Trials', August 1943.
70. CAC NRT ND154, Appendix K, 'Type 293X in HMS *Janus*: Report of Trials, Part 2, Surface Craft', September 1943.
71. CAC NRT ND154, Appendix L, 'Type 276 in HMS *Tuscan*: Report of Trials', November 1943.
72. CAC NRT ND154, Appendix M, 'Type 277 in HMS *Campania*, Report of Trials, 1944.
73. CAC NRT ND154, Appendix N, 'Type 277 in HMS *Coppercliff*, Interim Report of Trials on Surface Craft, Nov., 1944'.
74. H. Dahl, 'A Pre-Tuned Frequency changer for the 10-cm Region', *Jour. IEE*, loc. cit., p. 280 et seq.
75. H. D. Howse, op. cit.
76. W. E. Willshaw et al., op. cit. p. 999.
77. CAC NRT ND154, Appendix O, 'Microwave Fighter Direction Equipment', March 1944.
78. CAC NRT ND154, Appendix P, 'A Description of Types 960P, 980 and 981', January 1946.
79. L. G. H. Huxley, op. cit., p. 102 et seq.
80. L. A. Moxon, 'The Noise Characteristics of Radar Receivers', *Jour. IEE*, loc. cit., p. 1140.
81. F. A. Kingsley (ed.), op. cit., Monograph 3 by Cdr. A. E. Fanning, RN, and Monograph 4, by Cdr. R. S. Woolrych, RN.
82. R. E. Burgess, 'The Characteristics of Detected Noise and the Observation of Signals below Noise Level', Radio Division report, NPL, 1945.
83. G. Bradfield, J. G. Bartlett and D. S. Watson, 'A Survey of Cathode-Ray Tube Problems with Special Reference to Radar in Service Applications', *Jour. IEE*, loc. cit., p. 147.
84. F. A. Kingsley (ed.), op. cit., Monograph 3 by Cdr A. E. Fanning, RN.

85. E. B. Callick, op. cit., p. 69; J. H. E. Griffiths, op. cit., p. 175.
86. E. B. Callick, op. cit., p. 84; L. F. Broadway, 'Velocity-Modulated Valves', *Jour. IEE*, loc. cit., p. 860.
87. J. H. E. Griffiths, op. cit., p. 173.
88. W. E. Willshaw *et al.*, op. cit., p. 993.
89. F. A. Kingsley (ed.), op. cit., Monograph 2 by H. W. Pout.
90. CAC NRT ND154, Appendix Q, 'Report on the Hand-Controlled 3.2-cm System N3' (Extract from EMI Report RL/44).
91. CAC NRT ND154, Appendix G, 'Trials of Type 261 on HMS *Saltburn*'.
92. CAC NRT ND154, Appendix R, 'A Brief History of Submarine Radar', by S. R. Tanner.
93. W. E. Knowles-Middleton, *Radar Development in Canada* (Wilfrid Laurier University, Ontario, 1981), p. 54.
94. CAC NRT ND154, Appendix T, 'Comparative Trials of S, X, and K-band Radar for Detection of a Dummy Schnorkel'.
95. F. A. Kingsley (ed.), op. cit., Monograph 3 by Cdr. A. E. Fanning, RN.

Monograph 6
The Royal Navy and IFF – Identification Friend or Foe, 1935–45
J. S. Shayler

SUMMARY

First British IFF evolved for aircraft, which carried a unit to receive, amplify and retransmit the surveillance radar signal. Mark II version responded to RAF, Army and Naval Type 79 radars – followed by Mark IIN to cope with Naval Type 286 radar, and later Types 290 and 291 – system fitted in ships as Type 252 in early 1942. Diversification of radar frequencies required a separate low-power radar added to the Type 281 surveillance radar to trigger the responder and to receive response – but 10-cm band still not covered. Mark III IFF, devised in 1940, responded only to special interrogators, comprising low-powered secondary radars using separate antennae operating in conjunction with main radars. US agreed to adopt the British Mark III system – change-over taxed resources, but eventually resolved. Joint US – UK Mark V development almost complete at end of war, but abandoned.

INTRODUCTION

The British IFF system evolved quickly into one in which individual aircraft carried a unit to receive, amplify and retransmit the surveillance radar signal. This unit was called a Transponder. When radar frequencies became too diverse for the transponder to react to them all, a separate low-power radar was added to the primary radar to trigger and receive the response from the transponder. The transmitter was called the Interrogator and the receiver was called the Responser.

Since the earliest units were fitted solely in aircraft, the Air Ministry organisation naturally took the lead in the development and production of the system. This remained the situation throughout the war. Thus the large majority of IFF equipments used by the Navy was designed and produced by the RAF organisation, the Naval contribution being to provide suitable shipborne antennae and appropriate interfaces for Naval radars. The exception was the shipborne Type 242, forming part of IFF Mark III. This system was developed and produced entirely within the Naval organisation, and fitted in large numbers in small ships.

THE BEGINNING

Prior to mid-1938, during the early development of radar at the air Ministry Research Establishment (AMRE) at Bawdsey, and the subsequent operational use of the Chain Home (CH) stations, two systems were tried to identify a friendly aircraft:

1. the pilot sent a code by W/T which, in due course, could be correlated with the radar information, or
2. when challenged by W/T or R/T, the pilot flew in a circle.

By the summer of 1938 it was realised that a much faster method of identification was mandatory. Work began at AMRE on a system that would modify the appearance of the radar echo from a friendly aircraft on the CRT display.

Initial experiments to accomplish this used a horizontal half-wave dipole antenna on the aircraft. The dipole elements were either shorted and unshorted successively by a relay at its centre or, later, by using a tuned circuit and 'acorn' valve amplifier coupled to it. The amplifier did not oscillate, but introduced negative resistance at its tuned frequency. Neither system was satisfactory because the dipole was too directional, and the effectiveness depended on aircraft aspect. Because the system was responding directly to the CH frequency (22–30 MHz), it was impossible to overcome this problem.

By February 1939 an aircraft antenna system had been evolved that used the tailplane of the aircraft as a Y-fed dipole. This probably acted more as a loop antenna, since it produced quite a reasonable all-round polar diagram. However the antenna gain was very low, and the simple non-oscillating valve circuit mentioned above gave a range of only about 1 mile. In consequence a quench oscillator was added to improve sensitivity. In March 1939 this system produced identification at ranges greater then radar detection range.

This active device in the aircraft was christened the Transponder, and it led rapidly to the development of the first operational IFF system.

IFF MARK I

Development work on a Mark I IFF transponder started at AMRE in April 1939 to provide identification to CH stations operating in the 22–30 MHz band. The circuit was improved by adding a diode detector and pulse amplifier to drive the RF oscillator, thus increasing the output power and incorporating pulse-width modulation. This first equipment used mechanical sweep over the band 22–30 MHz in 10 seconds, and back again during the next 10 seconds. It produced selectable pulse lengths of 20 and 40 microseconds. Thirty handmade sets from AMRE were used in trials during August 1939. These were successful enough to lead to a production contract for 1000 sets being given to Ferranti in September 1939. These sets had three codes:

1. All narrow responses (15 microseconds).
2. All wide responses (40 microseconds).
3. Alternative wide and narrow responses on alternate sweeps.

A few of these production sets were modified to sweep the band 35 – 45 MHz. Towards the end of 1939 they were fitted in Naval aircraft flown at Scapa to identify them to Naval radar Type 79.

IFF MARK II

Because new radars on different frequencies were introduced by the RAF and Army, the need arose for a transponder to react to their different frequencies. Development of a Mark II transponder started in October 1939. Trials were held in February 1940, production was arranged in March 1940, and the new equipment entered service in July 1940. It covered frequency bands:

1. A(i), 22–38 MHz for identification to CH.
2. A(ii), 39–51 MHz for identification to ACH.
3. B, 54–82 MHz for identification to GL.

Times and sequences of sweeps were:

1. A(i) band in 4 seconds, B band in 2 seconds;
2. A(ii) band in 4 seconds, B in 2 seconds;

thus giving responses in each A-band every 12 seconds, and in the B-band every 6 seconds. Two pulse widths were available in each band, namely A-bands 14 and 80 microseconds, and B-band 4 and 18 microseconds. The two long pulses were used as a distress signal.

In mid-1940 the Navy introduced a crash programme for the installation of Type 286 radar, operating on 214 MHz, in small ships[1] (see Monograph 4). The equipment comprised RAF ASV Marks I and II, with a Naval design of antenna. This was superseded by Type 290 and then Type 291, both on the same frequency. Thus the need arose for a transponder to respond to this frequency. A modification of a Mark II transponder, Mark IIN, was quickly introduced for Naval aircraft. It covered the two bands 39–52 MHz and 195–220 MHz by motor-driven mechanical tuning. Responses were received by Types 79/279 and by Types 286/290/291 once every 12 seconds for a duration of about half a second. There was a separate receive/oscillator (single valve) for each frequency, their supplies being switched. An RF power of 2–3 watts was fed to the antenna for a pulse length of 10 microseconds.

This made no provision for an IFF signal for Type 281. Fortunately this radar was fitted in large ships, and there was some space available in the receiving office for a complete Type 286 radar to be installed to act as an IFF interrogator and responser. Some additional units were required to harmonise it with the Type 281. The whole interrogator was designated Type 241.

Instead of the free-running transmitter pulse in Type 286, a trigger pulse was generated to fire the Type 241 interrogator on the opposite half-cycle of the 50 cps mains from Type 281. Both series of trigger pulses were then mixed to fire the Type 281 A-scan display every half-cycle of the mains. The Type 281 video signal was fed through a buffer valve directly to the CRT display, so that radar noise appeared on both the radar and IFF A-scan traces. This did not matter because the IFF signal was large. The IFF video signal was fed through a gate valve, which was normally cut off. This was switched on by a 'flip-flop' for about 5 milliseconds by the IFF trigger pulse. This allowed the IFF video signal through, but also put it on a small pedestal, caused by the standing current in the gate valve. A negative version of this signal was fed to the CRT, so that the IFF signals appeared not only downwards under the radar signals, but also separated from them by the pedestal. There was no problem of radar signals appearing on the IFF trace since the equivalent range from the radar ground wave was too great.

Thus IFF coverage of aircraft was given to all Naval metric warning radars by the combination of Mark IIN transponder and Type 241 interrogator. In order to give IFF identification of Naval ships to Allied aircraft, an RAF transponder was modified to provide the band 173–179 MHz, which covered the frequency of 176 MHz used by the airborne ASV Mark II radar. When this transponder was fitted in ships provided with a Naval design of omnidirectional antenna it was designated Type 251.

In order to allow Allied ships to identify themselves to other ships, they were fitted with a Mark IIN transponder with a Naval design of omnidirectional antenna. With small additions to harmonise it with the ship's installation, this was designated Type 252.

The IFF Mark IIN/241/252 combination met the Naval needs of its time (mid-1940), and continued to be used until late 1942.[2] It suffered from two technical disadvantages, which contributed to a variable and unreliable operational performance. As a result the system acquired a bad reputation. The disadvantages were:

1. The aircraft antenna had to use horizontal polarisation to match the interrogating radars, and had to operate in two widely separated bands. This led inevitably to practical but inefficient aircraft antenna installations, having far from omnidirectional characteristics. Thus there was difficulty in achieving IFF range to match RDF detection range. In addition there were deep fades at some aircraft aspects.
2. The aircraft voltage supply to the transponder was not stabilised, so a set properly adjusted on the deck could be seriously out of adjustment during flight.

A fundamental limitation arose that radars were being, or were planned to be, introduced into all three Services using frequencies outside those covered by IFF Mark II.

IFF MARK III

In early 1942 a strategic decision was taken by all three Services to develop an IFF system that, apart from the final display, was completely separate from the radar. This IFF system, Mark III, would operate in its own frequency band, 157–187 MHz, and by this means overcome the problem of the proliferation of radar frequencies.[3] Of course every radar requiring identification facilities had to have an associated low-power radar in the Mark III band. However independence from the main radar allowed the major advantage of vertical polarisation to be used, to provide virtually omnidirectional airborne coverage from a simple and efficient aircraft antenna. This consisted of a quarter-wave stub mounted beneath the fuselage, or wing, giving adequate sensitivity and performance.

The completely re-engineered Mark III transponder contained a number of detailed improvements, including (perhaps of most importance) a stabilised power supply. Thus the sensitivity adjusted on deck before takeoff was retained in the air. This feature, together with the improved airborne antenna performance of Mark III, removed much of the notorious variability of operational performance of Mark II. Although not perfect, it provided an operationally useful system for the remainder of the war.

Early in 1942 approaches were made to the Americans who, even before Pearl Harbor, were seen to be contemplating the possibility of entering the war. Despite the fact that they were considerably advanced in developing their own system, commonsense prevailed. They made the vital operational decision to accept Mark III as the common system to be used by all Allies. Thus not only was complete interoperability achieved in all theatres of war, but IFF Mark III equipment came to be manufactured in vastly greater quantities than any other radar equipment in the war.[4]

The Navy decided that their needs could be met by two interrogator equipments: one of higher power to be fitted with long-range warning radars (where there was reasonable space available in the radar office); and one of lower power to be packaged as compactly as possible for use with radars in smaller ships, where space was at a premium.[5]

The Type 243 interrogator was developed for use with the Type 79/279 and 281 radars. Modulator, transmitter and receiver of Air Ministry design were used. These were mounted in an angle-iron rack with a trigger generator and video mixer developed by the Admiralty. The interrogator developed a pulse power of 1 kW during a pulse length of 10 microseconds at a prf of 50 c/s. This was radiated on a frequency of 179 MHz when a single equipment was fitted on a ship, or of 171 MHz when a second equipment was fitted. The receiver (called the responser) had a bandwidth of 4 MHz in order to allow responses of approximately one-third-of-a-second duration, since the aircraft transponder swept the band of 157–187 MHz every two and a half seconds. Early equipments incorporated a diode common-antenna switch. However experience proved that, with such low transmitter powers, enough protection could be achieved by making use of the change in impedance of the responser input circuit when saturated with RF power, in conjunction with correctly chosen lengths of connecting cable to the interrogator and responser.

A Naval design of antenna was developed, consisting of two driven vertical dipoles mounted approximately one half wave apart, each with a passive dipole reflector mounted approximately one fifth of a wavelength behind. This antenna was mounted on top of the Type 79 or Type 281 array, and rotated with it. The combined antennae could be stopped briefly to look at a particular target, any IFF response being correlated in range on the type of A-scan display described under Type 241. The Type 243 array and associated feeder system is shown diagrammatically in Figure 6.1. Figure 6.2 shows this antenna mounted above the Type 281 array.

This system sufficed until early 1943, when the PPI display was introduced on Type 281. PPIs (or Skiatrons) with a display (from Type 281) were quickly adopted in the Fighter Direction Room (FDR) to conduct fighter interceptions with carrier-borne fighters. A rapid method was needed to check whether a particular aircraft was friendly by display on the PPI, rather than on a separate A-scan. The same problem had already arisen in the RAF GCI situation. For this a Mark III G transponder had been produced, which, when the 'G' button was pressed on request by R/T from the ground, transmitted a chopped signal of 0.05 seconds on

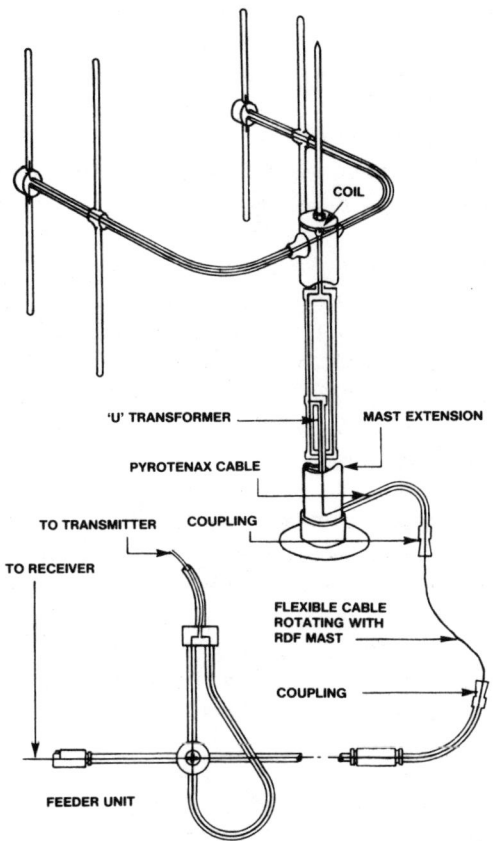

6.1 Type 243 antenna and feeder system (© Defence Research Agency)

and 0.15 seconds off on a fixed frequency for fixed time of 20 seconds. During the 0.15 seconds off period, normal Mark III operation continued, so that the transponder could give its distinctive response to the GCI without sacrificing its normal IFF function. If the IFF interrogator array was rotated with the radar antenna, then in the 'G-mode' a chopped IFF response appeared at the radar echo point on the PPI.

To meet the Naval requirement, the Mark IIIN transponder was fitted with a 'G-mode'. To trigger it, the shipborne interrogator associated with the fighter direction display had to transmit on the fixed G frequency. A Type 243 modified in this way, and with the harmonising interface required to feed PPI displays, was designated Type 941. The Type 941 antenna was mounted on top of, and rotated with, the Type 281B antenna. When continuous rotation was introduced to Type 281, the slipring unit to handle Type 281 signals also included brush-contact type

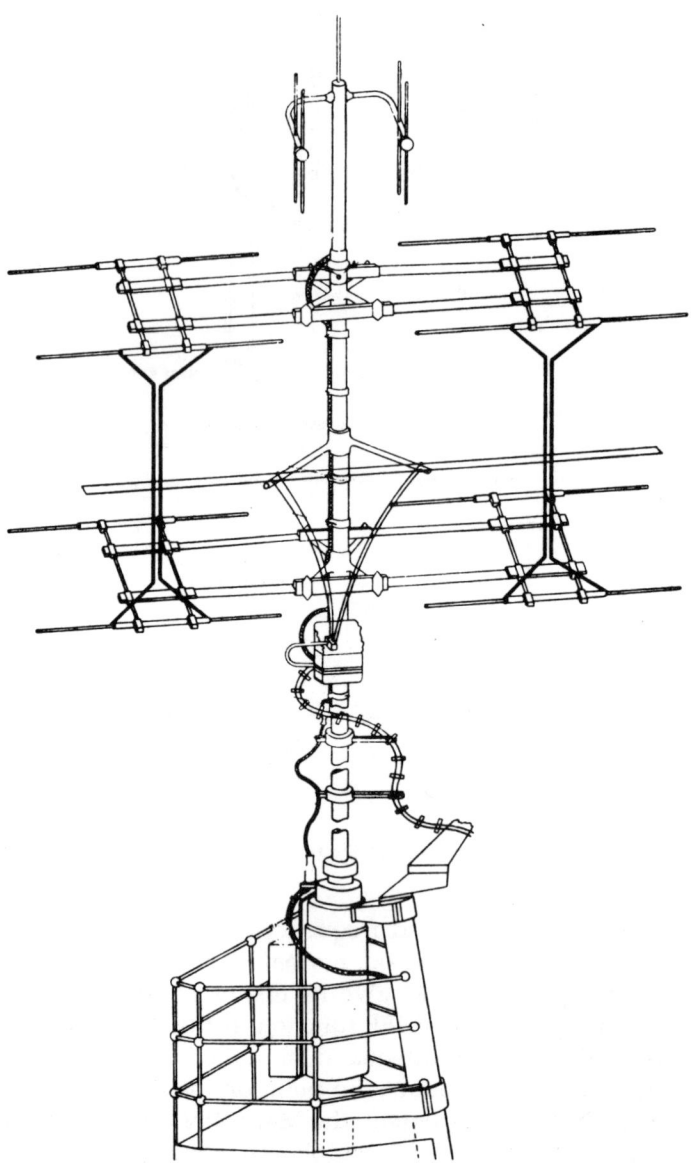

6.2 Type 243 antenna mounted above Type 281 array (scale: Type 281 dipole approx. 5.5-ft long) (© Defence Research Agency)

sliprings to handle Type 941 signals. Figure 6.3 shows a Type 941 antenna mounted above Type 281BQ array.

The sidelobes of the Type 941 antenna could not be made very small, and thus the response from an aircraft at close range appeared either as a full circle, or at least a very wide arc on the PPI display. To overcome this, and to provide a reasonably uncluttered display suitable for fighter direction, it was essential to use swept gain, a feature included in Type 941. Swept gain reduced the receiver gain when the transmitter fired and allowed it to recover to normal exponentially with time. Thus it provided reduced gain for short-range echoes, but normal gain at longer ranges.

The G-mode satisfied the need for rapid identification on the PPI required in fighter direction, but sacrificed certain features of the normal Mark IIIN mode. For instance, the transponder fitted in the Flight-Leader's aircraft could transmit an identifying Morse letter quite slowly by pulse-length modulation, when the pilot pressed a button on request by R/T from the carrier. To handle these features (and also any friendly aircraft without the G-mode) it was necessary to fit a normal Type 243 in addition to the Type 941, and arrange for its antenna to dwell on a chosen target long enough for the codes to be read. This was achieved by having the Type 243 antenna manually trainable by a power follow-up system, its position being shown by a mechanical bearing cursor over the face of a PPI displaying the Type 281 signals. Thus the Type 243 antenna could be pointed at any selected target on the type 281 display. The radar and IFF signals were correlated in range without stopping the Type 281 antenna rotating by having a special so-called Sector Scan Display (see Monograph 4).

It will be remembered that in the original combination of Type 281 and Type 243, the interrogator was triggered on the opposite half-cycle of the mains from the radar, the signals being displayed only on the radar A-scan. However the design of PPI and Skiatron available in 1943 would not accept this system. To prevent widespread modification of these displays, the triggering system was changed to fire all the transmitters, Type 281, Type 243 and Type 941, virtually simultaneously. The IFF response was large enough for the output of the IFF responser noise to be set very low so that it did not degrade the radar signal on the PPI. The only A-scan display now used, the Sector-Scan A-scan, could accept any relationship between radar and interrogator triggering.

By having both a Type 243 and a Type 941 (which were mounted in a single rack in the Type 281 receiving office, along with their appropriate interface equipment) all the needs of a warning and fighter direction organisation could be met.

So far all the systems discussed have been based on units developed and designed in the RAF organisation, the Naval organisation being responsible only for the provision of shipborne antennae, for the supply

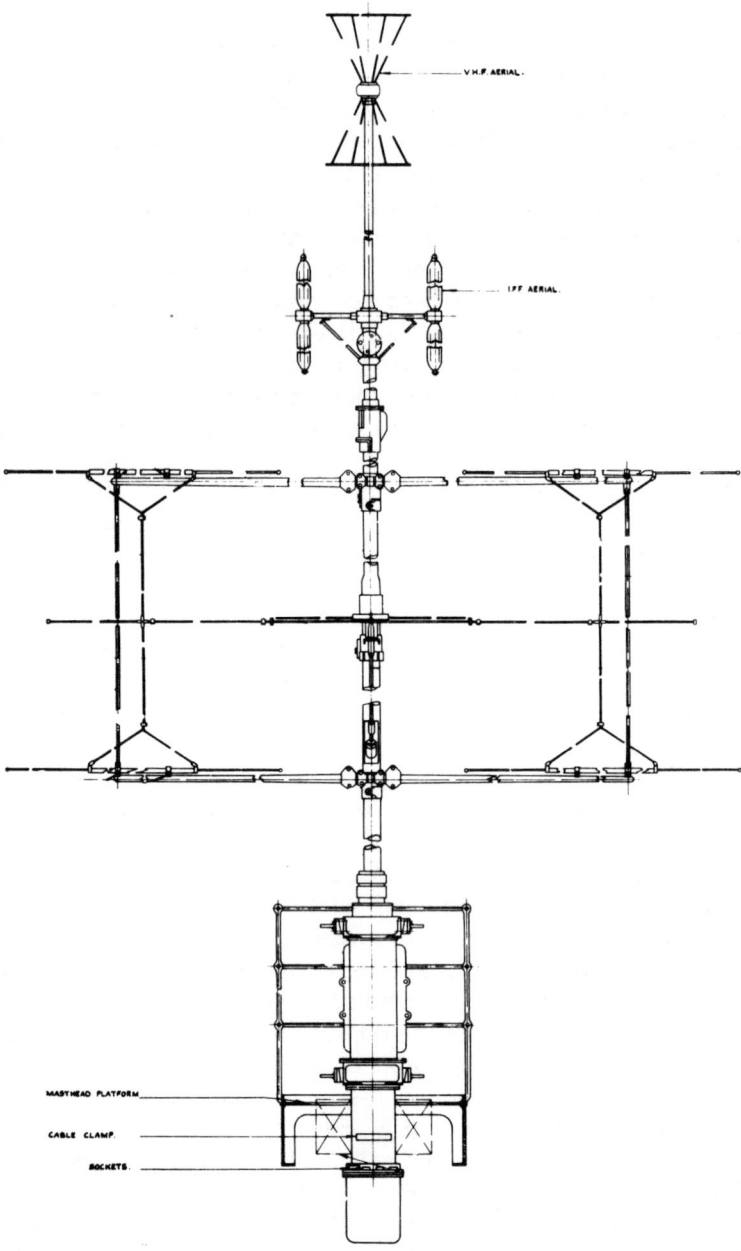

6.3 Type 941 Antenna mounted above Type 281BQ Antenna (scale: Type 281 dipole approx. 5.5-ft long) (© Defence Research Agency)

of appropriate interface equipment and for the integration of the system. The interrogator and responser equipments were designed for use in RAF ground stations, where there was little pressure on space, and these could be accommodated in the fairly spacious Type 279 and Type 281 offices. However circumstances were entirely different in small ships. The Navy was forced to design it own equipment to fit with such radars as the 10-cm Type 271 series and its successors, the 214 MHz Type 291, and later with some gunnery radar sets. This equipment was designated Type 242, and since it was the only Interrogator/Responser entirely designed by the Naval organisation, it will be described in some detail.

It consisted of three units, each a cube of about 10 x 10 x 10-in, which provided considerable flexibility in installing it amongst other equipment in small and cramped radar offices. The three units were the Interrogator (transmitter), Responser (receiver), and Mixer and Modulation (which had no connection with modulating the transmitter).

Normally the equipment was powered by the same 180V 500-cycle machine that supplied the associated radar. However Types 286, 286M and 286P radars were powered by an 80 V, 2000-cycle motor alternator. This had no spare capacity, and in consequence Type 242 could be powered by its own 80 V 2000-cycle motor alternator, which was much smaller and lighter than a 180 V 500-cycle motor alternator.

The Interrogator Unit contained the power oscillator, its series modulator and the power supplies. The push–pull oscillator used two RF triodes (CV63) producing a pulse power of about 1 kW at a pulse length of about 6 microseconds. It could be tuned between 165 and 185 MHz. The two valves had Lecher-line circuits connected to their anodes, grids and cathodes. The anode line was fixed for mid-frequency, whilst the grid and cathode lines were tuneable. The antenna output was taken from another tuneable Lecher line, closely coupled to the grid line.

Modulation was achieved by a series valve between the oscillator cathode circuit and earth. The modulator valve was driven by a cathode-follower, itself driven from the trigger pulse from the Mixer and Modulation unit. The oscillator was self-quenching, so the nominal 6 microsecond pulse length was, to some extent, affected by the oscillating conditions.

The Mixer and Modulation Unit performed the following functions:

- took the radar trigger pulse (usually 500 pps) and counted it down to 125 pps (later to 50 pps to reduce mutual interference when many transponders were in use). This reduced trigger rate fired the Interrogator;
- mixed the radar signal and responser signal in a way suitable for display on the radar CRT display; and
- provided power supplies for the responser.

Since displays differed amongst the various radars with which Type 242 had to operate, four different designs plus two minor variants of Mixer and Modulation Units were available.

The responser was a superheterodyne receiver. This had a tuned RF stage, a tuned local oscillator, a diode mixer, five IF stages at a centre frequency of 11 MHz with a bandwidth of four MHz, a diode detector and an output stage.

Owing to the range of radars with which Type 242 had to operate, a range of IFF antennae was available. These ranged from a four-dipole array – consisting of two driven elements separated by a half-wavelength and backed by two passive elements – to a single omnidirectional dipole. Vertical polarisation was used with IFF Mark III, of course. A total of six antenna types and three pedestal types were available, of which 11 combinations were used. At the radar-office terminations of all IFF antenna feeders, a passive concentric-cable network was included that isolated the Interrogator output from the responser input, so providing common transmit/receive (T/R) operation.

Type 242 was fitted in large numbers to a wide variety of small ships, from destroyers and corvettes down to MTBs. At the larger end of the scale the high-gain antenna was used, either physically fixed to the radar array or on a separate pedestal driven from the radar array by a DC motor follow-up system. At the smallest end of the scale a simple vertical dipole was used, correlation of IFF and target being obtained entirely from range coincidence. This was tolerable since the limited radar range obtained in these vessels restricted the number of simultaneous targets present.

When high-power 10-cm radars such as Types 276, 277 and 293 were introduced, giving significantly long ranges against surface targets and low-flying aircraft, the performance of Type 242 had to be upgraded. Since antenna gain could not be increased, improvements had to be made to the interrogator and responser. A new Transmitter Unit was introduced, replacing the Interrogator in Type 242, the remainder of the units remaining the same. This modified system was designated Type 242M.

The new Transmitter Unit was larger, 18-in x 20-in x 12-in high, and included:

- a new transmitter, giving a choice of 2 kw or 10 kW pulse power, and capable of tuning (to improve antijamming) over a 30 MHz band from 157 to 187 MHz;
- improved T/R switching, incorporating a gas switch, to cope with the higher transmitter power;
- a receiver grounded-grid preamplifier stage to add some 6 db to receiver sensitivity;

- an artificial load for measuring the transmitter output and facilitating transmitter tuning; and
- power supplies for all the above.

A new wide-band four-element antenna array was also introduced to cover the 30 MHz tuning range, together with a modified pedestal containing RF contact-type sliprings, uprated to handle the higher RF current generated by the increased transmitter power.

The transmitter consisted of a single RF triode (CV 199) with grounded anode (at RF) and concentric lines in grid and cathode, each tuneable by a plunger. The antenna output was taken from a variable tapping on the cathode-line inner conductor. The transmitting valve was cooled by an internal blower.

In the modulator, the trigger from the mixer and modulation unit caused a valve to discharge a pulse-forming network through the primary of a pulse transformer. The secondary of this transformer caused a power tetrode, connected between HT and the oscillator anode, to conduct for about 4 microseconds, thereby applying a pulse of about 5 kV to the oscillator anode in the high-power mode.

As with IFF Mark II, it was necessary to identify friendly ships both from aircraft and other ships. For this purpose ships were fitted with a normal IFF Mark III transponder with a wide-band omnidirectional vertical dipole antenna. This was designated Type 253.

In addition, special operational requirements arose for identifying a craft, or a point on land to allow craft to home in on it. This need was met by several different type of beacons, all based on the Mark III IFF system.

IFF MARK IV

This system, based on 400 MHz (possibly only designated Mark IV by the British), was developed to preproduction in the USA. An interconnection and interface system was developed for Type 281, and a successful trial conducted using preproduction equipment provided from the USA. However, resulting from the decision of both Allies to adopt Mark III as the common system, the commitment lapsed and Mark IV never went into production.

IFF MARK V

This was planned as a fully coordinated development to meet the joint requirements of all Allied forces, and to interface with all their radars. The USN was chosen as coordinating Service, and a Combined Research

Group was established at NRL, Anacostia. This was staffed by scientists and serving officers not only from the various US services, but also by a small group from the UK. The British Navy provided one full-time scientist, D. V. Ridgeway, to work at Anacostia and to provide liaison with the Admiralty Signal Establishment. Development work proceeded well, production was in sight, and interface design was virtually completed when the war came to a sudden end. The project was cancelled and the whole team disbanded.

The system was based on the band 950–1150 MHz, and was planned to have some coding features that, at the time, were considered to be more secure than those in Mark III.

POSTSCRIPT

The systems described above fulfilled a wartime operational need, even though they were not 100 per cent reliable. There was persistent fear, which increased as the war progressed, that the enemy would penetrate the system and turn it to their own advantage. Fortunately this never happened. Various schemes to improve the security of Mark III were worked on, such as cross-frequency working and more complex coding. However they added complication and were never introduced into operation, since the enemy threat never forced the issue.

Acknowledgements

The author is greatly indebted to A. G. Bogle, who shortly after the war wrote a comprehensive but unpublished paper[6] on the wartime development of IFF. Almost all the information in this monograph has been gleaned from that source.

Thanks are also due to Janet Dudley, formerly librarian at DRE, Malvern (formerly TRE), and John Briggs, librarian at DRE, Portsdown (formerly the Admiralty Signal Establishment, formed from the Experimental Department of HM Signal School in August 1941), for the provision of photocopies of documents and other information, and to *HMS Collingwood* for the provision of data from equipment handbooks. Also, to Jim Hancock of DRE, Malvern, chief photographer, for assistance with the illustrations.

References

1. E. G. Bowen, *Radar Days* (Adam Hilger, 1987), Pg. 98 et seq.
2. ADM CAFO 328/42 of 19 February, 1942 to all ships except submarines, operational MTBs at Home priority.
3. K. A. Wood, '200 Mc/s Radar Interrogator Beacon Systems', *Journal IEE*, Vol. 93, Part IIIA (1946), Pg. 481 et seq.
4. E. G. Bowen, op.cit., pp. 180, 181.
5. ADM 220/204, RDF Bulletin No. 5, Report No. 2, July 1943.
6. A. G. Bogle, Private communication.

PART II
Radar Ship-Fitting and Maintenance, 1939–45
(Monographs 7 to 9)

PART II

Editorial Note

In Monographs 7 to 9, the scene is shifted to the fighting ships, and the shore bases, in order to describe the unprecedented problems of supplying, installing, operating and maintaining a newly-invented weapon of war for which commanding officers and their staffs had received no training, or possessed much operational experience, in the early war years. Neither was there much in the way of any developed supporting organisation ashore. Monographs 7 to 9 present various aspects of the subject, ashore and afloat, by three individual authors who possessed different, and sometimes extremely arduous responsibilities at the 'sharp end'.

Monograph 7
Radar Ship-Fitting and Maintenance in the Royal Navy 1939–45 Experiences at Scapa Flow, May 1940 to April 1942

B. G. H. Rowley

SUMMARY

Appointed to HM Signal School in October 1939 to work on WA RDF equipment. In May 1940, seconded to the Staff of the C-in-C, Home Fleet, Scapa Flow. Enjoyed fury of the elements during sea trials of RDF equipments. From summer 1940 conducted trials of heightfinding techniques on aircraft targets. Also, supervisory support provided for local RN-manned shore-based CHL stations. Increasing activity arose as more RDF sets, and types of set, arrived at Scapa for working-up trials. Problems of equipment maintenance and supply of spares. Established informal two-way exchange of information: from Scapa to Signal School on equipment performance, trials and problems; and from Signal School to Scapa on future needs, technical possibilities, and new developments. Staff increased to cope with greater workload, including ships with multiple fire-control equipments in addition to WA sets. Appointed as Senior RDF Officer, under the aegis of the FWO. The new 10-cm wavelength RDFs for small ships. The PPI display in HMS *King George V*. Appointment to the USS *Gleaves* as an RNVR (Special Branch) Lieutenant to act as radar/sonar observer and adviser on convoy duties.

APPOINTMENT TO HM SIGNAL SCHOOL

Following my recent graduation in physics, and experience of live-ammunition firing with the Royal Artillery as a member of the University Senior OTC, I was first interviewed by the Oxford Joint Recruiting Board, and subsequently by Cecil Horton at the Eastney Fort East Extension of

HM Signal School on 25 October 1939. I then joined the group at Eastney working on the development of shipborne WA RDF, working firstly with Jack Shayler on transmitters (see Monograph 4), and subsequently with C. A. Laws on display systems.

There was considerable diversity of activity at Eastney, from laboratory to workshop, and problems of ship installation were investigated with colleagues such as Peter Morey, Michael Ellison and Bill Cox.

Two particular forays are worth mentioning, in view of what was to follow later. The first of these, in March 1940, was to Liverpool to join HMS *Illustrious* at her fitting-out berth, preparatory to going to sea for trials over a period of several days. This was the first fitting of RDF in a carrier (Type 279). Fortunately all went well due to careful supervision of the installation by Jack Shayler. The second visit was to Chatham for two days in April 1940 to examine the damage to the RDF equipment in HMS *Curacoa*, sustained in a bombing attack off the Norwegian coast during the German invasion there. It was found that much of the radar-office equipment had survived remarkably well due to the effectiveness of the shock-mounts.

APPOINTMENT TO THE STAFF OF C-IN-C, HOME FLEET

In May 1940 I was summoned for interview by Captain Basil Willett, RN, the Experimental Captain of HM Signal School. He told me that it was proposed to second me to the Staff of the C-in-C, Home Fleet, based at Scapa Flow, to work in conjunction with the Fleet Wireless Officer (FWO) 'to assist with the maintenance and use of RDF in the Fleet', In a very few days I made the first of many departures on the 'Jellicoe' express train from Euston to Thurso. From there I made the crossing in the famous *St Ninian* steamer from Scrabster to Lyness in Scapa Flow. Accommodation was provided in the SS *Dunluce Castle*, moored just off the base. At that time the base appeared to consist of a few temporary buildings and boom defence sheds, overlooked from the rear by some hill-sited oil tanks, which were being provided rather hurriedly with protective brick aprons.

Very shortly a boat arrived to take me to the flagship to meet the Fleet Wireless Officer (FWO), Commander C. M. Jacob, who speedily introduced me in rather flattering terms (as far as my RDF expertise was concerned) to his senior and junior colleagues. Commander Jacob did a great deal to convince his own seniors that RDF, at that time and in its prospective development, was a new weapon of enormous power. It became clear that one of my most important tasks was to assist in this process, and to 'spread the word' wherever appropriate!

SCAPA FLOW

In World War 2 Scapa Flow existed chiefly to provide shelter, protection, sustenance, maintenance support, training and limited recreational facilities for the Fleet. It is a clear area of open water, some 6 miles by 9 miles, deep enough for the conduct of many types of Naval exercise, enclosed by the southern islands of the Orkneys (see Figure 7.1). There is a NW entrance at Stromness and a southern one at Hoxa, both of which were then massively protected by boom defence nets, having gates operated by guardships. The shallow eastern entrance north of S. Ronaldsay was closed by a sunken blockship and causeway after its penetration by the U-boat that sank HMS *Royal Oak* in October 1939. Two days later an enemy Ju88 aircraft bombed the old battleship HMS *Iron Duke*, which was subsequently beached down the Longhope inlet south of Lyness. She then continued to provide limited but comfortable accommodation, berthing facilities and headquarters for the small craft of the Auxiliary Patrol.

Kirkwall, the capital, is at the north-east corner of the Flow, dominated by the red-sandstone Cathedral of St Magnus and overlooked by the Hatston Naval Air Station on a low hill to the east. The enclosing islands are low-lying, except to the west, where Hoy climbs up to sheer Atlantic

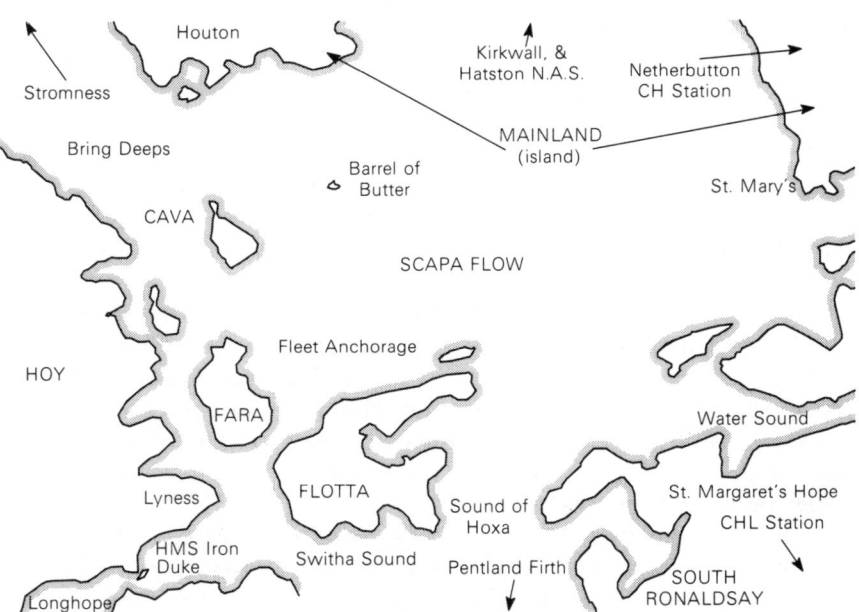

7.1   Scapa Flow: the Home Fleet anchorage

cliffs some 1200-ft high, incorporating the famous Old Man. The islands support some cattle-raising crofts, but there are very few trees – the few there are being bent almost double by the westerly winds.

At the south of Hoy is Lyness, the Fleet's 'stone-frigate' shore base, HMS *Prosperine*. The Fleet's anchorage was in the extreme south-west of the open water of the Flow, nearby. The surroundings and location of the extended Naval base at Scapa were unlike any other in the country. The rather lengthy description is necessary to provide some idea of the feeling of being stationed there in wartime. Everyone present felt part of a great operation to sustain a fighting Fleet. Some months later a military acquaintance stationed at a northern AA battery on Hoy said to me 'You Naval people always seem to act as if you were afloat, even when you are firmly on shore'.

RDF was new, very secret, and little known outside strictly limited RDF circles. Today's reader can have little idea of the extreme precautions with which the secrecy concerning the existence of RDF was then enforced. In a newly fitted ship, the RDF staff – usually in Communications and/or Gunnery Departments – found that they were responsible for a new piece of equipment for which they had not received any special training, possessed only a handbook for guidance, and were provided with very little in the way of spares. They arrived in the isolation of Scapa, where there was no dockyard or stores organisation, but were nevertheless expected to keep the equipment in effective operation throughout a testing working-up programme. In these circumstances all ranks concerned performed a remarkable job, encouraged, I hope, by the technical and moral support I was able to give them, plus the assurance that the stores and workshop back-up would follow as soon as practicable. My 'Defence of the Realm' (DR 1) pass, incorporating a photograph, demonstrated that in spite of my youth I should be afforded every assistance where necessary. I also found that the FWO, in the name of the C-in-C, had similarly prepared the way for me in Lyness.

I remained in *Dunluce Castle* until July, and succeeded in accumulating a few useful hand tools from visits to RDF-fitted ships, together with a small collection of spare valves for Types 79Y and 79Z receivers. The transport and communication channels about the base and the anchorage worked very effectively. Many of my early visits to RDF-fitted ships were made with the FWO. As a result I was able to mingle readily with all ranks, because of my status.

In July 1940 I was advised that more suitable quarters were available for me in *Iron Duke*. To provide me with independent mobility, a steam drifter from Buckie had been assigned for my use. For the next eight months it would collect me at least once a week at 0500 to join a ship slipping at 0600 prior to a day's trials of various kinds, including the RDF, outside the Flow. In mid-1940 this would often involve a Type 286 installation using a 'bedstead' antenna on a destroyer. Very often the trials took place in extremely rough seas, which can build up suddenly, at any time of year, around the Orkneys. Fortunately I have never in my life suffered from air- or seasickness and so have been able to enjoy the fury of the elements without qualm.

EXPANDING ACTIVITIES

Summer 1940

In July 1940 a Signal School report on techniques for heightfinding on aircraft targets was sent out from the Admiralty to all ships fitted with Types 79Z and 279 RDFs[1] (see Monographs 2 and 4). These were based on the range of first detection of the aircraft target, and subsequent observation of the echo maxima and minima as the aircraft progressed through the vertical-plane lobe structure of the RDF antenna array. The FWO requested that I rewrite the report in simpler language for the information of non-technical officers. At the same time I suggested that trials be conducted on a friendly aircraft flying successively in the approaching and receding modes on a given bearing, at regularly increasing height intervals. This proposal was adopted, on a bearing out to the north-west. In spite of interfering land echoes, it was considered to be a useful exercise, which was subsequently performed on a regular basis. Subsequently I organised the construction of a ranging mark for surface range-calibration on the Barrel of Butter, a rocky islet that was usually awash (see Figure 7.1 and Monograph 8).

Many RDF-fitted ships were now arriving for working-up exercises at Scapa, and some had their own RDF officers. These were young Canadian physics graduates holding Special Branch RCNVR commissions who had completed a short technical and operational course at Signal School before joining ship[2] (see Monograph 1). The secrecy of everything to do with RDF had been drummed into them, with the result that on my visits I found that they were delighted to meet someone to whom they could talk freely about the virtues, and otherwise, of the equipment that monopolised their lives. They were at this time a scattered and lonely band of brothers – but who later established a distinguished record of service.[3]

Occasionally, at this time, the FWO would ask me to give short talks about RDF, followed by question-and-answer sessions, to small groups of officers on the staff. One of these was to describe the vertical-plane lobe structure of the RDF radiation pattern, and the technique for the determination of the height of an approaching aircraft. In preparation for this talk I asked a friend on a 'run-ashore' to Kirkwall to find for me an assortment of children's balloons. During the talk I stretched and partially inflated a couple of these, tied their stalks together, and secured them one above the other to a flat surface to simulate the RDF antenna radiation lobes. I explained that the energy of the RDF transmitted pulse spread out effectively in a shape like the balloons, whose skins represented a model of a surface where equal and just enough energy was available to form a reflection from any one particular RDF pulse, so that an aircraft approaching the transmitter would be first detected at a range dependent on its height. Inside the balloon the incident, and therefore the reflected energy would increase to a maximum, then decrease to zero as the aircraft emerged through the 'skin' – and then repeat the process when it entered the second lobe. The whole process would be shown on the RDF display as the echo rose and fell in height, then rose again. This 'concrete' demonstration stimulated many excellent questions and much discussion. Afterwards a very senior officer took me aside and said 'Rowley – the FWO tried to explain all this to me with a lot of diagrams – sheer gobbledegook – but you have made it crystal clear! Come and have a drink!'

When I arrived in Scapa the Type 79Y in *Rodney* was providing some air WA, because *Curlew* had been sunk in the Norwegian campaign. The RAF CH station and associated high antenna towers situated south-east of Kirkwall at Netherbutton was now in more permanent operation, and the CHL stations on Dunnet Head, South Ronaldsay and Fair Isle were also providing the necessary low air and surface cover on a frequency of 214 MHz. These stations had been taken over by the Royal Navy and were run under continuous watches, under the command of new, young, Special Branch officers who shortly before had been zoologists, anthropologists and historians. As opportunity presented, I was sent to visit them all, bearing suitable gifts, and thus providing an opportunity to talk over problems and pass on any serious difficulties with understanding. Their names were Scott, Evans, Feachem and Carmichael, and they occupied small wardrooms and living quarters, fitted out on the Naval Rate Book scale for small craft with appropriate plate and decanters – even 'jerries' (plain, not fluted!). The stations were powered by remarkably large twin diesel sets, which were sometimes rather temperamental. As the sites had been chosen with regard to their suitability for surveillance, access for fuel and supplies was often difficult.

There were two stations on Fair Isle, defended by a detachment of Royal Marines, and the FWO said that I should spend a few days there. (The Royal Marines' presence there had already given rise to a fund of stories – many now ensconced in folklore!) I therefore set off from *Iron Duke* to Lyness to catch the 0800 drifter to Kirkwall, and then by car up to Hatston to catch the inter-island flight by DH Rapide (the windows of which were obscured) to Sumburgh at the south end of Shetland, thence to board the regular connection to Fair Isle. The latter lies midway between North Orkney and South Shetland – about 30 miles from each. From the beach at Sumburgh, Fair Isle is scarcely visible, but the tidal rip gives the sea horizon a saw-tooth appearance! The regular connection was the small motor drifter *The Good Shepherd*, which bounced about cheerfully on the heavy swell. It then started to rain, so we retreated to cosy, if crowded quarters over the screw, and proceeded to lunch off bread, tinned butter and tinned plum jam, washed down with great quantities of spoon-dissolving tea. After about two and a half hours we landed at Fair Isle's only access from the sea and I was welcomed to the island, which was still occupied by busily knitting ladies and memories of the Duchess of Bedford and her aerial visits. The island is about 3 miles north to south, and 1 mile east to west, with sheer cliffs that provide a home to many sea birds.

Soon after my return from this expedition I was in transit from one ship job to another in the Flow when there was an air raid alert, during which there was a spectacular box-barrage of AA fire for the defence of the Fleet anchorage. The sky was filled with a roughly rectangular pattern of bursts from ship and land defences – accompanied by a rain of shell fragments! On this occasion the intruders were discouraged, and soon tranquillity was restored.

From midsummer many Type 286Ms were arriving in the smaller craft, and Type 79(279)s in cruisers and above. The Type 286Ms inevitably brought with them (or later developed) a number of cases of low resistance in the pyrotenax feeders due to damp ingress. Great help was given in some cases by an RNV(W)R officer, H. W. Young, who was based there to look after radio in small ships (radio and RDF were then synonymous). Later, when the RDF workshop and spares store was set up, he joined us.

Now that nearly all newly fitted ships and new ships came to Scapa to work up, the great problem was the shortage or absence of spares. This applied particularly in the case of the silica-envelope transmitter valves used in Type 79(279), which did not travel well (see Monograph 3).

The terms of my posting were vague, and it was obvious that, as far as the C-in-C was aware, I was assigned to his Staff – whilst Signal School considered that I was one of theirs. This circumstance, handled discreetly, could be used effectively to overcome problems: keep in touch with new developments, such as Type 286P, Type 271 and the 50-cm fire-control sets; and to chase up spares (I had my own book of railway warrants!) On one expedition for this I was in London at the time of the first night-time air raids. The Lyness base was expanding rapidly, and with the backing of the FWO I managed to achieve reservation of two new Nissen huts to be built near the pier for the RDF maintenance and spares shore base.

Autumn 1940

Two very important events took place in September 1940. By an executive order dated 3 September, the USA exchanged 50 ex-World War 1 flush-deck four-stack destroyers for the use of Naval bases in the Caribbean and Newfoundland. At the end of the month successful sea trials of the new 3.5-m airwarning set, Type 281, fitted in the new Liverpool-built light cruiser *Dido*, were carried out by Landale and Watson, from Signal School – the latter then continuing on to Scapa (see Monograph 4). Further trials and operations in Fleet exercises demonstrated that the open-wire antenna feeders similar to those used in Type 279 could still carry the 1 Megawatt, 1.7 microsecond pulse without flash-over. Also, height and range performance figures against aircraft targets well fulfilled expectations. Towed-target gunnery practice outside the Flow revealed that both shell flight and shell splash could be followed out to 5000-yds range.

In November, six new-construction light and heavy cruisers joined the Fleet, whilst the battleship *Nelson* and two cruisers returned after repairs. All nine ships were fitted with Type 279, resulting in a busy period of exercises, help with maintenance and cries for spares. I organised the loading of a railway wagon in the south with a great assortment of spares, and shepherded it up to Thurso with the help of Railway Transport Officers, loaded the spares on to *St Ninian* (on a fine day), and proceeded safely to Lyness. Such expeditions allowed me to carry out a useful two-way exchange of information on an informal basis: with Signal School in the south on how sets fitted earlier were performing, results of trials, how well equipment stood up to Naval conditions, including my personal reports on what happened in the RDF office when a battleship fired a broadside, effectiveness of shock mounts, unit performance, valves and so on; with staff in the north there were discussions about future needs and technical possibilities; progress on new agreed designs and information for the FWO on Types 281, 286P, 290, 271 and so on.

Christmas 1940 aboard the *Iron Duke* was a cheerful, well-fed affair – HM Customs had allowed her to keep and consume her stocks of 'duty free', including an astonishing range of liqueurs. During explorations around the ship I found the substantial remains of the television receiver and display unit of the air-to-ship spotting system that had been installed for trials in 1938 (see Monograph 1).

Early 1941

With the opening of new accommodation in Lyness early in the new year, 1941, I moved ashore to *Prosperine* in order to be closer to our new office, store and workshop, and became the Senior RDF Officer (SRDFO), under the aegis of the FWO. I left *Iron Duke* with regret – she was a very happy ship – taking remembrances of my time aboard. 'Chippy' presented me with a boat badge – and the one-shilling I had won as a member of the PMO's crew in the Longhope Regatta. I still have both of these items.

At the beginning of 1941 three 50-cm fire-control sets had just gone to sea – two Type 284s and one Type 285 (see Monograph 1). However, by the following May about a dozen of each had joined them. Many of these arrived in Scapa, where I was joined by W. F. Drury from Signal School, who had experience of these sets in the laboratory, in ship-fitting and during trials. Spares were arriving on a more regular basis, and were followed by E. J. Toy, also from Signal School, plus three radio mechanics.

The new battleship *Prince of Wales* now arrived, equipped with metric air-warning and nine 50-cm fire-control sets, which had to be powered and triggered from multiphase alternators in order to avoid mutual interference (see Monograph 1). This provides some indication of the additional effort required, and our reinforcements were fully employed.

Summer 1941

HMS *Hood* was fitted with the prototype single-mast air-warning Type 79B (see Monograph 4), as well as the surface gunnery set Type 284 (see Monograph 1). One day in May 1941 I was in *Hood* to investigate a problem with the Type 79B because her RDF officer, Bernard Stubbs (ex BBC-Moscow) was suffering from influenza. Suddenly a signal from the C-in-C directed me to report at once to *Repulse* for urgent repair work. Shortly afterwards we saw *Hood* steam out for the last time. HMS *King George V* (KGV) was due for docking, but was delayed until HMS *Prince of Wales* (POW) returned from repairs to damage suffered in the *Bismarck* action in July. She was fitted with the first Type 273 10-cm surface warning set, developed for large ships.

The light cruiser *Aurora* arrived just before the *Bismarck* operation. She was fitted with the Type 290 high-powered development from Type 286P, operating on 214 MHz. Some difficulty was being experienced with the transmitter and modulator, and performance was not as good as hoped. The radar was in the charge of the gunnery officer, Lieutenant Michael Le Fanu, who some 20 years later became Controller of the Navy, then First Sea Lord, and finally Chief of the Defence Staff, before his premature death in November 1970.

By August new fittings of Type 271 (10-cm, see Monograph 5) in the smaller ships were arriving daily. Excitement rose at the end of August when Landale, head of the Centimetre Wave Development group at Signal School, arrived at Scapa for sea trials of Type 273 in *Prince of Wales*. Record ranges for surface detection of ships were observed.

The remainder of 1941 continued to be a very busy year in the RDF sphere.[4] Then, on 7 December came the news of the attack on Pearl Harbor, followed shortly afterwards by the tragic losses of both the *Prince of Wales* and *Repulse*.

Occasionally I would join a newly fitted ship at a southern base for a working-up passage to Scapa. One such event was in the rearmed and rearmoured *Resolution*, from Devonport via the Irish Sea, where gales were incessant. It was an extremely wet passage. The new RDF surface-warning prefabricated office was on the pole-mast aft, equipped with exterior ladder rungs for access. These made a 90° turn at the top of the mast to provide access to the hatch in the base of the office – an exciting experience in the Minches, not to mention the subsequent descent.

Early 1942

In February 1942 I received two official visitors from the USN, Commander Dow, Technical Liaison Officer from London, and Lieutenant Commander Kirkpatrick. They were to be given every facility to see RDF (soon to become radar) in operation. One remembered facility was a Burns Night dinner in the flagship!

One final technical note should be recorded. In March, an Air Ministry PPI display was fitted in *KGV*, the flagship, to operate in association with Type 273, thus opening a new era in RDF operations (see Monograph 5).

For me it had been an amazing two years – at sea in every kind of warship, new and old, British, Polish, Royal Indian Navy and so on. Then, in April 1942 I was told to hand over to Drury at Scapa. My new assignment was to join the USS *Gleaves*, a modern 2500-ton destroyer, as radar and sonar observer and adviser on North Atlantic convoy duty, as an RNVR (Special Branch) Lieutenant.

References

1. H. D. Howse, *Radar at Sea – The Royal Navy in World War 2*, Appendix C: 'Height Determination by Radar', by Basil Lythall (Macmillan, 1993).
2. E. F. Burton (ed.), *Canadian Naval Radar Officers: the Story of University Graduates for whom Preliminary Training was given in the Department of Physics, University of Toronto* (University of Toronto Press, 1946).
3. Ibid.
4. B. G. H. Rowley, unofficial report of trials of Type 271 in *KGV* on 'Walrus' aircraft and HMS *Berwick*, dated 25.10.1941. CAC NRT ND 156, Part 1, Appendix.

Monograph 8
Radar Maintenance at Sea: A Personal Story, 1940–45
R. A. Laws

SUMMARY

Introduction. Author's radar experience. Personnel. Fitting-out and commissioning. Acceptance of a new ship. At sea. Handbooks and test gear – spares. Calibration and setting-up. Preventive maintenance. Malfunctions – breakdowns – modifications. Ancillary services. Tribute to the Service.

INTRODUCTION

The introduction of RDF (radar) into the Fleet in the early years of the war brought into focus the need for an organisation adequate to keep this new scientific weapon of war working effectively and reliably at sea. Initially the radar was operated and maintained by the telegraphists. They were helped at Scapa by the appointment of Basil G. H. Rowley as assistant to the Fleet Wireless Officer (Commander C. M. Jacob) 'to assist with the maintenance and use of RDF in the Fleet' (see Monograph 7). The first RDF officers were Canadians from the University of Toronto, trained in Signal School in 1940. Thereafter many RDF officers' courses were run throughout the war. The first, starting in January 1941, lasted four weeks; the courses then gradually lengthened, with the later ones lasting some six months.

The present monograph deals with the responsibilities of the radar officer (RO) for the maintenance, calibration and repair of the ship's radar, but the reader should bear in mind that he also played an important role in its use. In most large ships the RO managed the operation of the WA, WS and WC sets in collaboration with the navigating and fighter-direction officers, and he played an active role with the gunnery officer in the operation of the GA, GS and GC radars. He also had the usual ship's divisional duties. As a result the RO was a 'day-man', working all day and on call all night.

AUTHOR'S RADAR EXPERIENCE

Any first-hand account of maintenance of radar at sea will perforce be coloured by the experience of its author and what he can remember and glean from others after a gap of about fifty years. I gained my radar experience thus: first glimpse as an RNV(W)R Telegraphist operating Type 79Z in *Suffolk* (Captain R. Durnford, RN) in the spring of 1940; on the first British Naval RDF course in January/February 1941, including a spell on Type 79 at Fort Wallington, Portsmouth (Lieutenant John Hodge, later Sir John Hodge Bt), which reported to RAF Fighter Command at Stanmore as part of the coastal chain of air defence radar; as RDF officer at Fraser Battery, Portsmouth (Lieutenant Philip E. Roche, RN, Lieutenant Jack Van der Kasteele, RN), and by making frequent visits to HM Signal School at Onslow Road (John F. Coales) while in Portsmouth; as radar officer of *Trinidad* (Captain L. S. Saunders, RN) from August 1941 in Devonport dockyard to her sinking in the Barents Sea in May 1942; as radar officer of *Bermuda* (Captain T. H. Back, RN) from July 1942 in John Brown's yard at Clydebank to January 1944 in Iceland(c) (always so written and spoken of during the war to distinguish it from Ireland(r)); as officer-in-charge at the RN Radar School, Eastney, in 1944; and as radar officer of *Anson* (Captain A. C. G. Madden, RN) from late 1944, during her first major refit in Devonport dockyard, to December 1945 in Hong Kong.

The main radar sets I encountered are listed in Appendix 1 to this Monograph.

PERSONNEL

Like any other officer, the RO had to lead his supporting staff and continue their training. He also had to ensure that the new but vitally important technology of radar took its rightful share of the time and facilities available to the ship for working up and exercising.

Great credit is due to those who set up and ran the selection and training of ROs; to officers such as Johnny Bruce (Captain the Honourable J. B. Bruce, RN) who selected many ROs. Furthermore the Admiralty very wisely decided that the RO should report directly to the Captain, and early in 1941 it admitted all sub-lieutenants over the age of 23 to the wardroom. Far-sighted commanders anticipated this requirement; *Repulse*'s Commander transfered Sub-Lieutenant G. K. Armstrong, RNVR, to the wardroom at the age of twenty on the grounds that, having direct access to the Captain as RO, that was where he should be. These arrangements gave the RO a very valuable status and independence, and the success of radar in HM ships owed much to them. It seems that German radar was less effective than it should have been through lack of officers in a position to fight for its needs during the war.

Like the engineer officers, for example, the RO viewed his technical work in a professional way, but unlike them he was directly involved in operating the ship's armament. Being in a new Branch he was able to influence the development of its ethos, unhampered by too much tradition. His combined approach to upkeep and maintenance and to operational use was in many ways similar to that of other wardroom officers with responsibilities for both operation and upkeep of complex equipment, such as the signals officer for radio equipment and the torpedo officer for electrics. Nevertheless he had fewer inhibitions about getting his hands dirty, for in the early years of the war there were no warrant officers or petty officers to help him, or to tell him how things ought to be done. Unlike many ROs, I had served on the lower deck, and I found my experience there in *Suffolk* to be invaluable during my subsequent commissioned service.

In early 1941 the RDF officer (as the RO was then called) had to do almost everything technical himself, but radar mechanics soon appeared from the training courses, thus relieving him of the more mundane aspects of maintenance and repair. Appropriate schemes of training and promotion gradually produced a stiffening of leading seamen and petty officers on the operational side, capable of taking charge of a watch and supervising operational training. Cruisers carried a second radar officer from late 1941 onwards, and at least one radar mechanic. In her second commission in 1945 *Anson* had two assistant ROs, several petty officer radar mechanics and radar petty officers, and about 120 operators.

There was great camaraderie in the radar world, and when in harbour one would invite one's opposite number from a ship in company to come aboard for dinner and for a discussion of problems of common interest. An 'RPC' (request the pleasure of your company) to him would evoke a 'WMP' (with much pleasure) or less welcome, an 'MRU' (much regret unable). Similarly, one would visit the local Port Radar Officer, and ASE itself when possible, to keep oneself up to date. AFOs and CAFOs were eagerly perused, and most ROs managed to maintain complete sets for themselves.

FITTING OUT AND COMMISSIONING

After arrival on board in the dockyard and making his number with the Captain, the RO had to establish good relationships with his fellow officers and ensure that his key ratings and junior officers were not commandeered by the Commander for general ship's duties. He then had to establish his relations with officials in the dockyard and with local Admiralty staff, particularly with the radar fitting-out party.

In familiarising himself with the plans for the installations of the radar, the RO had to look beyond those initiated by ASE, and examine the power supplies and the signal transmission and telephone circuits. He also had to turn a critical eye on the plans for ventilation and cooling of his sets. For all this, good contact with the electrical officer was essential, as well as with operational users: the gunnery, torpedo, and navigating officers.

In theory, by the time the RO joined his ship all matters of installation had already been resolved and laid down in Admiralty drawings (see Monograph 9), but in practice reasonable requests from ship's officers for change would be met wherever possible. For example, in *Anson* the space allocated for Type 281BQ was inadequate for proper operation of the set. The Warrant Officers' galley was therefore sacrificed to provide a suitable room for it, a common galley being established to serve all officers. It was unfortunate that Signal School and then ASE always referred to radar 'offices', following W/T tradition, rather than 'rooms', because the search for a real office, from which to manage an ever growing range of work, always evoked the response, 'you have plenty of offices already, why do you want another one?' In *Anson* the instrument store for the Royal Marine Band was taken over as the Radar Office. In older ships the search for spaces for radar sets was fraught with difficulty, and sometimes there was no choice.

Even matters as mundane as the transport of equipment had their troubles. During the fitting out of *Trinidad* at Devonport, a pair of six-phase motor-alternators for Types 282–5 (see Monograph 1) was despatched in its own railway truck but it failed to arrive. A second pair was therefore despatched – and this also went astray. A third pair was then despatched; it duly arrived and the machines were installed. In the ensuing weeks another pair arrived, followed shortly by yet another. We never learnt where these machines had gone astray.

The fitting-out staff in the dockyards occasionally took short cuts. In *Anson*, during its major refit between commissions, we could not rotate the antenna of Type 281BQ as the gearbox at the masthead was locked solid. A call to ASE was followed by a visit from the mechanical designer. After a short conversation I suggested that we look at the pedestal, so up the mast we went – some 70-ft of vertical ladder. The designer confessed afterwards that he had never climbed a mast before, a possibility that had never entered my mind. Each pedestal carried four lugs, which had to be offered up to a set of four brackets previously welded to the mast with the aid of a jig; the gaps then had to be carefully closed with shims and slotted wedges before the bolts were tightened. Someone had had the idea of shortening the process by using the pedestal as the jig, fastening the brackets to its lugs separately, and then offering the whole assembly up to the mast and welding the brackets to it, all in one operation. It transpired that when the welds had cooled, the consequent distortion had so strained the pedestal and gearbox as to prevent any rotation. It was necessary to loosen three of the bolts, readjust the shims and wedges, and tighten up again. This was quite tricky to do as the antenna pole had already been stepped and the array of dipoles rigged on it.

Anson was a 'four-cornered ship', with four high-angle directors each controlling four 5.25-in guns in two twin-turrets, sixteen such guns in all. During her refit at Devonport for her second commission, the four HACS directors were replaced by four Mark 6 directors; each director carried a pair of nacelles containing the transmitter and receiver of its Type 275 radar.

Each new director was lowered onto its circular mounting by a dockyard crane. Up forward all went well, but each of the two after directors was found to foul one of the two tubular struts supporting the main-mast. The remedy was to carve out a few inches from each strut, weld a curved plate over the hole thus exposed and weld strengthening material to the other side. Triangulated bracing was then welded between the two struts. The result can be seen in photographs of *Anson* including one taken in Devonport dockyard during the refit (Figure 8.1).

8.1 The modified main mast of *Anson* (© Defence Research Agency)

Anson had many remote displays: PPIs and a Skiatron, fed by a labyrinth of pyrotenax cables. After leaving Britain for Malta to work up, we went to exercise action stations but were dismayed to find that many displays were 'dead', although they had been working before we went to sea. We found that one cable tray had been badly damaged as if by a sledgehammer, with many pyrotenax cables having been severed. The damage was just below an open hatchway through the armoured deck, through which we had ammunitioned ship. It seemed too great to have been caused by a swaying case or 14-in shell. At first we thought of sabotage, then realisation dawned. The hatch, being very heavy, was raised and lowered by means of a handwheel above it turning a worm moving a very strong crank below it. It had remained open throughout the refit, and the tray and cables had been laid in the path of the crank. The hatch had been closed for the first time during exercise action stations at sea. As the men working above closed it they unwittingly drove the elbow of the crank through the cable tray, severing many cables. The hatch had been opened again before the damage was detected. We made some temporary repairs, but permanent repairs had to await our arrival at Captain Cook dock in Sydney.

Pyrotenax cables were used for RF as well as for video signals. They needed careful attention during installation. Magnesium oxide, their insulant, is very hygroscopic, so after the cable had been cut and installed the last couple of feet had to be dried out with a blowlamp and the end sealed with SeeKay wax. The seals were not always perfect, and a drop in insulation would require the work to be redone at sea.

Rats were a minor hazard during refitting and after, as they would nibble at the insulation of cables. Tradition had it that a rat could feel the presence of a lethal voltage, and I never found an electrocuted rat!

ACCEPTANCE OF A NEW SHIP

Bermuda was built at the Clydebank yard of John Brown & Company Ltd, and pride in having built the *Queen Mary* was much in evidence. Then came the day when the Admiralty was to take over *Bermuda* at a ceremony in John Brown's model room. ASCBS (Admiral Superintendent Contract-Built Ships) sat at one end of the huge oval table faced by a large book. Around the table sat heads of department from the Admiralty and the yard, the Captain of *Bermuda* (Captain Terence H. Back, RN), ship's officers, including the RO, and others. Behind most of these people sat columns of supporting staff, ready to brief their man at the table, for only he might speak at the meeting. Whispered and written messages went up and down these columns.

The questions were many and varied, designed to show up any shortcomings in the ship's equipment, from engines to guns, from cranes to galleys. They covered the testing of everything and the operation of everything as far as it could be operated with the ship in dock, and included the proper provision of stores. The Admiral worked through the book, reading out the questions and noting the answers. When he came to the question 'Have all the ship's Naval stores been received and stowed on board?', there were murmurs of 'Yes' from around the table, but I stood up to express my concern that not all the radar spares had been received on board. After some argument between the radar fitting-out officer and me, the Admiral told me to stop arguing and sit down. Later in the day the Captain caught sight of me walking in the dockyard and altered course towards me. I expected a reprimand, but no! 'Laws' he said, 'I was very pleased to hear you argue with the Admiral. I was hearing far too many yeses to the questions and I couldn't believe that things were in such perfect order!' The Captain's secretary ('Scratch') told me much later that after that episode I could do no wrong in Captain Back's eyes.

AT SEA

While at sea we had to take great care never to put anything where it could move or fall, and thus suffer or cause damage. After a spell in harbour, the importance of this rule had to be emphasised. Similarly one never took any tools aloft without securing them to one's person by a lanyard.

In general the sets coped well with own-ship's gunfire, even though they would leap in their rubber mountings quite alarmingly when the guns fired on dangerous arcs. Plug-and-socket contacts suffered most, particularly those of the back-plugged panels, but a well-placed kick would often put matters right. Nevertheless, in some ships, interlocks and contactors had to be mechanically secured to prevent their opening when nearby guns were firing.

In many cruisers Type 273 was fitted forward of the bridge and very close to B-turret. Experience showed that such a set was in danger of damage if either A- or B-turret was fired abaft the beam. Access to the office was via B gun deck and it could be perilous in rough weather or when those turrets were firing. We therefore provided the office with some essential spares such as crystals, the handbook for the set, a stock of corned beef, bread, drinking water and a bucket.

Pyrene fire extinguishers were provided for electrical fires, and I used one in the confined space of the Type 273 office in *Bermuda*. I had a persistent loose cough for many days thereafter, and only after the war did I learn that carbon tetrachloride on hot metal could produce phosgene.

Seasickness took its toll, and only those who have never known it would belittle its seriousness. After a spell in harbour, the first few days at sea could be the worst, after which all but a few had found their sea legs and could stand anything from their own ship. Even so, heavy weather produced a slight deadening of the senses and a reduction in liveliness. Immunity would not go with you to another ship, as many discovered after transhipping to take passage.

Colossus had troubles during her sea trials, as reported by Lieutenant G. K. Armstrong, RNVR, her radar officer. A Type 277 antenna was mounted on top of the office in the centre of the ADP. Its support structure vibrated so badly during the ship's turning trials that the antenna's gyro was upset, with the result that the antenna would oscillate in elevation over an arc of about 30°. The trouble was cured by welding six steel webs between the antenna pedestal's mounting and the deck, but at the cost of hampering the movements of the ADP's crew.

The mast carrying the antenna of Type 281 in *Colossus* was mounted at the after end of the island and the antenna was only 75-ft above the water line. When Type 281 was transmitting on a forward bearing at night, sparks were seen between the lockers and their lids on the flag deck. There was thus a risk to aircraft refuelling on the flight deck and a new mast was fitted in Alexandria, 15-ft higher. It was necessary to recalibrate Type 281 for heightfinding, but it proved very difficult to do, owing to the almost continuous presence of anomalous propagation off Alexandria (see Monograph 2).

Similar modifications were made in the other aircraft carriers of the *Colossus* class.

HANDBOOKS AND TEST GEAR

The quality and usefulness of handbooks for radar sets steadily improved as the war and the development of the sets progressed, and the hard work that was put into their production earned our gratitude. The general complement was one per ship for each type of set fitted. For *Anson* we sought and got further copies so that repair parties closed up at action stations would have all details available locally. ASE then adopted this provision as standard.

The RO usually armed himself with a few appropriate texts as opportunity presented itself (see Appendix 2 to this Monograph), plus a set of mathematical tables and a slide rule.

In 1941 test gear was hard to get and a multirange meter was an essential tool. I took my DC 'AvoMinor' with me, and my own soldering iron with a few hand tools. Luckily a friendly civilian on the staff of the SNSO had a Universal 'AvoMeter' at his disposal and he gave it to me just before *Trinidad* left Devonport. It was a Model 7 plus an external transformer and shunts, all in a magnificent teak box. I carried it over the side to *Matchless* when we abandoned *Trinidad* and I kept it throughout the war until I left *Anson* in Hong Kong to go home, when I had it taken on charge by the Accountant Officer. The Cossor double-beam oscilloscope was supplied to ships from late 1941 onwards and was a godsend. Its versatile time-base and trigger were exactly what the radar engineer wanted. In large ships the RO could borrow a 'Megger' (insulation tester) and sometimes even a bridge 'Megger' from the electrical officer.

Year by year, more and more specialised test gear was supplied, geared to the setting up and testing of particular sets. It included cable testers, wavemeters, oscillators, dummy loads and standing-wave detectors. A kit was supplied for plotting the polar diagrams of the 50-cm antennae and comparing them with standard. It included a Cambridge Unipivot galvanometer, which was far too good for the job. On a visit to ASE Haslemere I met a scientist bewailing the fact that the Cambridge Instrument Co. had run out of Unipivots and could no longer supply them to ASE. I told him where they had all gone, suggesting that he don a peaked cap and order a kit for his 'ship' (*Mercury II*, as ASE was known).

SPARES

Each set was allocated an establishment of spares detailed in an 'E List'. In the main, the list dealt with consumable items such as radio valves, resistors, capacitors, variable resistors and potentiometers, crystal rectifiers for S-band mixers and special cable, but in general no transformers. The RO had to make sure that he had his proper complement and do his very best to maintain it, visiting Naval Stores at Glossop as opportunity permitted, and getting his packing cases back to his ship, perhaps in remote Scapa Flow. Packages and cases of radar spares sent out by ASE were marked with a 'four-square' pattern in red to denote their contents, so that they could be recognised and given priority.

Storage of spares was a problem. Radar spares were classified as Naval Stores and the average supply assistant could not be trusted to leave delicate equipment in its protective packing. I have seen a valuable range-potentiometer wrecked by being unwrapped and dropped into a bin. The RO usually took over his stores by agreement with the Naval Stores officer, and with it the responsibility for maintaining the records and ordering replacements as necessary. A suitable man would be chosen from among the operators to do this job. Stowage had to be improvised, and spares had to be dispersed about the ship so that they would be available quickly to sets that might need them urgently. In *Bermuda* we kept spare crystal mixers for Type 273 in an old tobacco tin near the set; it vanished en route for Gibraltar, perhaps thrown out by an overzealous sweeper. On arrival there we obtained some spares from Commander F. J. Emuss, whose radar connections were disguised by his title of Port Technical Officer. One could sometimes resuscitate a defective crystal mixer by opening the wax-filled hole and moving the cat's-whisker with a pin, but I felt the loss keenly.

Some equipment panels were notorious for having a particular transformer that could suddenly fail. Type 282 had one such. The consequences of failure could well be serious for the ship, so I took the matter up with ASE. 'Well' said my civilian contact, 'we give you spare electrolytic capacitors because they have a finite life, but a transformer should not deteriorate, so we do not supply you with spare ones', 'What happens', said I, 'if such a transformer burns out while we are escorting a convoy to Murmansk and we have no spares?' 'Ah', said he, 'that's where we get back at the manufacturers for a refund for supplying a defective component'!

Oil-filled transformers and capacitors were employed in the high-tension circuits of Types 79 and 281. Failures were rare but the consequent smell of burnt insulating oil is unforgettable; so is the smell of a burst electrolytic capacitor, and the sight of the resultant mess within a densely wired panel.

Anson was issued with an American after-action radar set: the SQ. It was a portable centimetric set, packed in boxes designed to go through the hatchways of US ships, to be stowed away below for safety. When needed, it was to be unpacked on deck and assembled there to give some warning of aircraft and ships. Our hatchways were 30-in wide and the boxes would not pass through them, so we had to unpack the SQ and stow it below without its protective packing.

CALIBRATION AND SETTING UP

Metric Sets

Vertical coverage diagrams for estimation of height had to be worked out for the metric WA sets (see Monograph 4). They called for a lot of computation using trigonometrical tables, for which the instructor officer was often pressed into service. These diagrams needed verification against reality, and to provide it for the Home Fleet an aircraft flew daily in and out from the Fleet anchorage in Scapa Flow at known heights on a north-westerly route out over the waters between the islands of Hoy and Mainland. Despite these efforts I found the results to be of limited value. The vertical interference pattern obtained at sea depended on specular reflection from a large area of sea up to roughly two kilometres away from the ship, so the position of the ship was important. Berths between Cava and Flotta would have Cava in the way; berths near Mainland did not have a clear view between Hoy and Mainland; the practical coverage thus achieved showed considerable irregularity compared with simple theory (see Figure 7.1 in Monograph 7.)

As the war progressed, various specialised parties were established to help ship's staff, such as the ASE Sea Trials Party, whose members would join seagoing ships of all sizes from battleships to landing craft to help get their radar into proper working order, and they would sometimes be carried off to sea, willy-nilly. There was also the Heightfinding Working-up Team devoted to the setting up and calibration of the metric WA sets for service against Japan. It also helped with additional training for the operators. It was led by Lieutenant R. A. Jenkinson, RNVR, and it helped in many ships: *Anson, Duke of York, Sussex, Glory, Belfast, Glasgow, Jamaica, Norfolk, Liverpool, Colossus* and HMCS *Ontario*. There were other parties helping ships' staff before and during working up: one under Major G. D. Pegler, REME (later to become Commandant of the Royal Military College of Science, Shrivenham), and another under Lieutenant W. Cheeseman, RCNVR (later to become President of the Canadian Marconi Company).

Range Scales

The A-scan display of Types 282–5 employed two alternating horizontal sweeps, each at 500 per second, driven from one time-base generator but separated vertically (see Monograph 1). One sweep carried the receiver output and thus the target echoes. The operator would align his cursor with the leading edge of the echo to transmit its range. The other sweep carried ranging pips from an oscillator tuned to 163.9 kHz for 2000-yd pips for Type 285 or 819.6 kHz for 200-yd pips for Type 282. The oscillator was shut down at the end of each ranging cycle. In setting up the system one tuned the this oscillator against the continuous output of a portable crystal oscillator, bringing the beat to zero by ear. This needed care, as the random phase between the pulse trains from the range oscillator made the zero beat difficult to determine. I would leave a radar mechanic or a radar PO to make this setting, but because of its crucial importance, I would instruct him to have me listen to his final setting before he put the set back into operational use.

After switching on, the operator would align the range pips with the range-transmission system by adjusting the phase of the oscillator and the speed of the time base. With some display units one had to change a fixed resistor in the time-base circuit before one could cover the required speed with the variable control. I was discussing the matter with a fitting-out officer who was concerned mainly with small ships, and I asked him how often he met this problem. 'It is no problem' said he airily, indicating the frequency control on the range panel, 'I just turn this potentiometer with a screwdriver until I get the right spacing', I pointed out that he was altering the ruler to get the right measurement, insisting that he should never do it again.

Index Errors

Range was given by the time between the receipt of the direct pulse (the ground wave) and the echo. The leading edge of the direct pulse, because of its great amplitude, appeared relatively early and thus caused an overstatement of range. This index error was important for gunnery. It varied slightly from set to set and had to be assessed by calibration against the echo from a target of known position. For radars Types 282–5 it was one or two hundred yards in value. During the calibrations the ship would be swinging about its anchor and one had to take a series of fixes of the ship while taking ranges with the radar before the correction could be evaluated. Ships in the Fleet Anchorage in Scapa Flow regularly used the reflecting structure of metal rods on the Barrel of Butter, an islet in the Flow (see Monograph 7); I have also employed *Royal Oak*, sunk in Scapa Bay in 1939. We had a large choice of fixing points, including Flotta Signal Station, the dipole on the Calf of Flotta, Houton Signal Station, Hargaback Mark, the azimuthal extremities of Cava, Flotta and Fara, the Martello Tower at Rinnigill, Buoy A8, and the Barrel of Butter itself. South Walls W/T masts and the masts of Netherbutton CH station were occasionally used (see Figure 7.1 in Monograph 7).

Angular Alignment

The advent of Type 275 called for great precision in the alignment of the radar and optical axes, for which one used helicopters and meteorological balloons in envelopes of copper gauze. We had ordered a helicopter for *Anson* while in Scapa Flow. It arrived without warning and hovered alongside seeking instruction. All on the upper deck rushed to the side to look at it, and the Commander was furious: 'Laws', he exploded, 'What have you done? You've ruined my evening quarters!' Meteorological balloons were easier to handle, but still one had to take care to avoid explosion while filling them with hydrogen.

Anti-jamming

In anticipation of hostile jamming, staff officers in the Home Fleet tried staggering the frequencies of the 50-cm sets, allocating frequencies to ships and to the various sets that they carried (see Monograph 1). It was not a success, as each set seemed to have its preferred frequency at which it would deliver most power, and did not take kindly to being tuned off it.

Receivers

In general the receivers neither required nor received much maintenance. They could be tuned on sea returns in all but a flat calm. In the early days one did not consider noise factors. In 1945 Lieutenant R. A. Jenkinson, RNVR, made the first shipborne measurements of Galactic noise in *Glasgow* on passage to Malta in August 1945 under the scientific direction of L. A. Moxon at ASE (see Monograph 2). Jenkinson also discovered a more mundane source of interference in Malta in May 1945. Troubled by noise on Type 281BQ in *Duke of York*, he connected a pair of headphones to the receiver and heard a general background of carrier-borne noise. Eventually he heard a voice say: 'This is Warrant Officer ——— testing. This is the face that launched a thousand ships'! The source was a VHF link between Malta and Catania operating on 93.5 MHz, which was being kept active when not in use and was transmitting the clatter and chatter of a teleprinter room. Its transmission lay within the bandwidth of the ship's Type 281 centred on 94 MHz.

Transponders

Passive equipment presented few problems. The IFF and the radar beacon had to be set up and tested periodically. Many ROs took responsibility for Loran, the American LOng RAnge Navigation system supplied by the USN. It operated by the receipt and timing of pulses transmitted from base transmitters around the world, enabling the ship's position to be determined.

PREVENTIVE MAINTENANCE

State of Readiness

The RO would never put a set out of action for any purpose without first ensuring that he would not be compromising the ship's state of readiness. If in doubt he would seek authorization for doing so. Most preventive maintenance was therefore carried out in harbour. There were, however, routine examinations that were carried out during operation, and they included the examination of transmitted waveforms on small monitor CROs.

Logs

There was no firm doctrine about records. We gave every set its own file, in which we kept records of all work done on that set, with times and dates. Certain radars of a series, such as Types 282-5, contained similar panels that could be exchanged between sets. We therefore kept a separate log for each such panel, keeping it in the appropriate parent file.

Open-Wire Feeders

The earlier metric WA sets (Types 79 and 281) were connected to their antennae by 400-ohm open-wire feeders. There was a flexible section between the masthead and the rotating array, which had to be replaced from time to time. We would routinely disconnect the feeder at the set and put an Avometer across it. Incipient failure then showed up as a fluctuation in resistance during rotation. We always had spare lengths made up ready for use. One had to have permission to go aloft from the Officer of the Watch or, in harbour, from the Officer of the Day. High power transmitters, W/T and metric WA, had to be shut down and their 'safe to transmit' keys removed and held by the Officer of the Watch. Those who enjoyed working aloft did the work. It was a long climb to the little platform at the top, ending with about two yards of ladder leaning outwards from the vertical. One needed both hands free while standing on the pedestal, but there were no safety harnesses so I would borrow a tack-line from the flag deck to hold me to the pole mast carrying the array. At sea, in addition to the motion, there was the problem of funnel gases at the masthead. The heading to be avoided was agreed with the Officer of the Watch, but on one occasion he forgot about it and I was nearly asphyxiated, with no means of contacting the bridge about my plight.

Tuning the Transmitter

The RO had to watch the triode transmitter valves of his metric sets for signs of softness, given by the anodes' starting to glow red. Such valves would soon have to be replaced, and being permanent stores were returnable to ASE for repair (see Monograph 3). This change would necessitate retuning the transmitter for maximum power into a lamp-load. It was a tedious process as there was no provision for adjustment of the settings with the power on. Some ROs used to bypass the safety interlocks and move the sliding contacts with a long non-conducting rod, but I never took the risk involved, for the voltages were lethal. I would plot the power for various settings and then interpolate for the best settings. Some metric sets used by the Army had their resonant lines formed into circular arcs, with the sliders controlled by external knobs. This arrangement must have led to quick and accurate tuning. Tuning the transmitters of Types 282–5 on lamp-load was easier, as some adjustments could be made without switching off.

Common T/R

In later versions of Types 282–5M with common T/R, the receiver was protected during transmission by means of a T/R switch employing a quarter-wave line and a spark gap. These spark gaps had to be cleaned and burnished regularly, and very carefully set, for failure to spark during transmission could damage the receiver.

MALFUNCTIONS

Types 282–5

A damp and salt-laden atmosphere was the worst possible ambience for electronic equipment, especially for high-voltage circuits. *Trinidad* had two Type 282 sets in the lower part of the bridge structure. Here it was cold, damp and ill-ventilated, and during the ship's short life we never got those sets working properly.

Types 282/4/5 employed 1-in polythene coaxial cables to carry RF power to and from the antennae. The joints were weak points and it proved almost impossible to keep water out of junction boxes exposed to the weather, even though they were filled with vaseline after the connections had been soldered up (see Monograph 4). The outer sheath was the usual braided sleeve covered with thin black plastic. The sheath would crack near the box, water would penetrate the interstices of the braiding and thus pass through the glands, however much they had been tightened. We fitted canvas covers across the front of the 'pig-trough' reflectors of Type 284 in our efforts to keep out the sea water and to prevent icing. Performance seemed to be unimpaired.

A 6 kV cable tester was supplied for these systems. We found that leakage currents, in micro-amps, tended to die away during an application lasting a few tens of seconds, so the test removed the symptoms that it had detected. It might have been electrolysis of unwanted dampness.

The cables going between deck and director were subjected to regular rotation as the director was trained and elevated, and the drawings prescribed that these cables should be passed through junction boxes to isolate the rotated lengths and thus facilitate their replacement. These particular boxes were a source of weakness, especially as they had to hold the cable against the torque of rotation. Following troubles encountered with them in *Trinidad*, I arranged with the fitting-out staff in Glasgow that in *Bermuda* the boxes should be fitted but the cables should pass them by. We thus removed a weakness and I think that the boxes were never required.

Troubles were reported with the beam switch of Types 284–5, from *Emerald* for example. We met it in *Bermuda*. Remembering Nelson, we ignored an inscribed warning forbidding us to open it up, and we found that the Geneva cross mechanism was extremely stiff to turn. We made appropriate adjustments and all went well thereafter.

Antenna Control

Control Unit 20D for Type 281 gave the only example I encountered of regress in design. Unit 19D for Type 79 was well liked by the operators and was very reliable. The operator had to supply muscle power for rotation, but the steering wheel was well matched to his effort, and he could use both hands. Unit 19D was also silent. Unit 20D was designed for powered automatic rotation and reversal. There was a length of bicycle chain in the drive system and when it broke, which it did from time to time, the operator had to train by hand. He had to turn a crank handle near one knee using one and the same hand all the time; the work was punishing and exhausting, and militated against good watch-keeping. Unit 20D was also noisy. I would have replaced it by Unit 19D if I could have got hold of one.

Automatic rotation with reversal had its dangers. If the reverser failed, the rotation was arrested physically by stops on the mast, but the resulting impact could, it seems, put the upper Selsyn motor into a motoring mode at 3000 rpm. With the gear ratio of 90:1, the array could reach the other stop in about two seconds, where the torsional impact could wreck it. Nevertheless I do not remember hearing of such a disaster actually occurring.

BREAKDOWNS

There is little of interest to be said about routine breakdowns. The faults had to be diagnosed and rectified; the usual cause was a defective valve (vacuum tube), resistor or capacitor, or a bad contact in a plug and socket, or in a switch. In *Trinidad* we were faced with an unusual fault on the modulator unit of a Type 284/5. Voltage applied to the modulator was controlled by manual rotation of a 'Variac' (a variable auto-transformer) mounted on the horizontal surface of the modulator cabinet, its value being displayed on a meter above. When the voltage was being raised there was a loud tearing sound and it collapsed to zero. Starting again produced the same results. Terminals carrying the high voltage were mounted on a horizontal paxolin sheet, and we found that the underside of this sheet was covered with a fine deposit of carbon black, with tracks across it obviously made by a high-voltage breakdown. But how did the carbon get there?

We were at sea, and ship's motion has a soporific effect, especially below decks, so one's thinking is less acute. It took 24 hours to locate and clear up this fault. The main HT rectifier was mounted below this paxolin sheet, lower down in the cabinet. Its filament connections emerged from the re-entrant base of the bulb, insulated from each other with ceramic beads. Further out the leads were insulated with non-ceramic sleeving, beyond which they were secured to a pair of terminals. Each set of beads was kept in place by a lead (Pb) pinch. The leads had shifted and the two pinches had made contact with each other, putting a short circuit on the filament supply, heating the leads as far as the pinches and charring the sleeves, from which smoke had arisen and blackened the underside of the paxolin with a film of carbon.

The operator's antenna Control Unit 20D for Type 281 contained a gearbox designed by ARL that was required to be filled with Oil, Castor, Pattern 109. We had also used this oil, it seems, to fill the gearbox in the pedestal of Type 281 at the head of the mainmast. On 28 March 1942 *Trinidad* was in the Barents Sea near Bear Island, escorting convoy PQ13 to Murmansk, when this gearbox suddenly locked solid and the antenna could not be turned. We were under threat from German aircraft, with Type 281 out of action, in low visibility and freezing temperatures, though with a moderate sea. Two of us donned our thickest practicable clothing and clambered laboriously up the mast to the top platform. Removal of the cover showed that the gearbox was filled with a waxy substance of the consistency of cheese, which clearly had to be melted out.

With the help of the engineers we hauled a flexible metal hose up to the masthead and passed steam from the boiler room into the lower end. It was to no avail as the steam turned to water, and then to ice, before it was halfway up. We lowered the hose, thawed it out, had the steam passing through its entire length on the deck, and then hauled it up again. It froze up before we had got the hose to the top. One of the engineers produced a very large blowlamp with a two-gallon tank. We pumped it up to a roaring flame and hauled it up the mast, steadied by one of us below it on the ladder, and one above. We played it on the gearbox and slowly melted out the oil. Just as we were finishing an enemy aircraft approached us out of the snow and fog, so we secured the cover without refilling the gearbox and clambered down to the deck as quickly as possible. We operated Type 281 with an empty gearbox until we reached Murmansk, where we refilled it with non-freezing mineral oil from the engine room. (An account of this incident is given in its wider context in *Radar at Sea – the Royal Navy in World War 2* by H. D. Howse [1993].)

After *Trinidad* had been struck by her own (rogue) torpedo off the North Cape on 29 March 1942 she went into dry dock at Rosta, on the Kola Inlet to Murmansk. Here we had five weeks in which to effect repairs. On the radar side we had to restore cable runs damaged by the torpedo, and this involved much dragging of new cables through cable passages. The remaining radar work was not memorable in any way.

MODIFICATIONS

Equipment had to be modified from time to time, and ships were usually required to order kits of parts by notices in AFOs. In 1944–5 several modifications were introduced as Project BUBBLY. They included the introduction of swept brightening into the standard PPIs, intended to reduce the angular width of paint and clutter at close range (see Monograph 5). In an exercise immediately after the modification, we lost all aircraft indications on Type 293 as they came near to the ship (*Anson*). Captain Madden, in command, was not pleased with us! We restored our PPIs to their former state as quickly as we could, and warned ASE of the failure of that particular modification.

ANCILLARY SERVICES

Cooling

Cooling was essential to keep radar equipment from overheating. Later in the war there was air-conditioning for Radar Display Rooms and Action Information Centres, but otherwise the need for cooling and ventilation was met by ducting outside air in suitable quantities to all radar compartments. During *Anson*'s passage through the tropics to join the British Pacific Fleet, Type 293 (or 277) could not be operated because of high ambient temperatures at the transmitter. The designers of the cooling system had placed an extraction duct near the hot components rather than one blowing air directly upon them, and to obtain adequate cooling the direction of air flow had to be reversed. The work was done in the dockyard at Sydney.

Power Supplies

Motor-alternators, switch-gear and the Watford automatic-starters generally functioned without trouble in temperate climes. The services of the electrical department and of the electrical artificers were available, but were seldom called upon. Radar staff did any necessary lubrication, and care was taken not to mix greases based on soda soap with those based on lime soap, for mixing of the two types could cause stiffness and perhaps wreck a bearing. The names of greases were sometimes misleading, but our guide was a comprehensive list in an AFO.

Even so the six-phase 500 Hz motor-alternators for Types 282-5 ran extremely hot, and the DVD machine was said to have a temperature rise of 40°C. In *Bermuda*, No. 2 low power room containing a DVD had an air temperature of 130°F (54°C). Trouble with them was reported by Lieutenant R. A. Jenkinson, RNVR; *Emerald* had two radar power rooms, each containing one 500 Hz machine plus a 50 Hz machine for Type 281, which ran cool. Her after power room was over the starboard evaporators and the temperature inside the room was so hot that both machines over-speeded and neither could be used in the tropics. Even during Operation OVERLORD, her after power room could be used only occasionally.

For greater security in action, most radar sets had their AC supplies in duplicate, with one motor alternator fed from a port section of the ring main (220 V DC), the other from a starboard section. We formulated rules by which we would change over supplies in case of a failure.

Data Transmission

Most radars fed their outputs into the gunnery and air defence systems by magslip transmitters and receivers on 50 Hz, and by M-type (step-by-step) transmitters and motors on 24 volts DC. The bearings of the antennae of Types 79 and 281 were displayed to the operator by magslip. The RO had to understand how these systems worked, but he usually arranged for them to be maintained as part of the fire-control electrics. The range-tracking panel of Types 282–5 with its ball-and-disc integrator clearly belonged to the fire-control world, but it was so reliable that responsibility for its maintenance was purely academic.

TRIBUTE TO THE SERVICE

So ends my story. I have always been glad that fate put me in the Royal Navy for the duration of the war and in a Branch where I could make full use of my talents. The RN with the RNVR was a great service, magnificently supported by the Admiralty's organisation ashore, including the WRNS. I learnt a lot in the Navy, and I am grateful to all who helped me during my time there, both civilian and Naval.

APPENDIX 1: MAIN RADAR SETS ENCOUNTERED BY THE AUTHOR

Suffolk: Type 79Z, fitted in Govan on her return from the China station.

Fort Wallington: Type 79Z, reporting with the CH stations to Fighter Command, Stanmore.

Fraser Battery (a tender to *Excellent*): HACS with RDF Type 285 and an early experimental array of 50-cm dipoles with beam-switching in two axes.

Trinidad: Types 281, 284, three 285 and two 282.

Bermuda (first commission): Types 281; 284 and three 285, all with beam-switching, two 282, and 273.

RN Radar School, Eastney, in 1944: most sets in use in 1944 or destined for installation in HM ships.

Anson (second commission): Types 281BQ, 277, 293, two 274, four 275 and seven 282 or 283, appropriate interrogators, Radar Beacon Type 251M, IFF, SQ after-action radar (USN), Loran (USN).

APPENDIX 2: AUTHOR'S COLLECTION OF USEFUL TEXTS

Manual of Seamanship Vol 1 (HMSO, 1932).
Admiralty Handbook of Wireless Telegraphy (BR 230) (HMSO, 1938).
Admiralty Navigation Manual (HMSO, 1938).
Rayner, *Cathode Ray Oscillographs* (lost with *Trinidad* and not replaced).
Wireless & Electrical Trader, *Cathode Ray Oscillographs*, ditto.
M. G. Scroggie, *Radio Laboratory Handbook* (Iliffe & Sons), ditto.
The Amateur Radio Handbook, 2nd edition, and the *Radio Handbook Supplement* (Radio Society of Great Britain).
W. T. Cocking, *Wireless Servicing Manual*, 6th edition (Iliffe & Sons, 1942).
W. T. Cocking, *Television Receiving Equipment,* 1st edition, (Iliffe & Sons, 1941).
The Radio Amateur's Handbook, 20th edition, and the *ARRL Antenna Book* (American Radio Relay League, 1943, 1944).
F. E. Terman, *Radio Engineers' Handbook,* 1st edition, 4th impr. (McGraw-Hill, 1943).
M. G. Scroggie, *Foundations of Wireless*, 4th edition (Iliffe & Sons, 1943).
Emrys Williams, *Thermionic Valve Circuits*, second edition (Sir Isaac Pitman & Sons Ltd, 1944).
F. E. Terman, *Radio Engineering,* 2nd edition (McGraw-Hill, 1937).
F. Langford Smith, *Radio Designer's Handbook* (Iliffe & Sons).
Puckle, *Time Bases,* 3rd impression (Chapman & Hall, 1944).
G. Parr, *The Cathode Ray Tube*, 2nd edition (Chapman & Hall, 1944).
Sundry issues of *The Wireless World* (Iliffe & Sons).

Monograph 9
Naval Radar Fitting Policy, Matériel Procurement, Installation, Sea Trials and Shore-Based Maintenance 1939–45

A. M. Patrick

SUMMARY

Review of the manifold aspects and activities involved in the period between Admiralty approval for the production of a particular type of Naval radar, and its ultimate introduction to Fleet service. This link formed a vital lynch-pin in the overall radar programme for the Royal Navy throughout World War 2.

A review is presented of the gestation of the Admiralty's ship-fitting programme; matériel procurement; ship-fitting, in all its ramifications; sea trials; and shore-based maintenance. Also described is the impact of the development of entirely new operational systems, in which the radar sensor was the core element; on the work load under wartime pressures, at many levels within industry, the Royal Dockyards and commercial shipyards, and on Admiralty personnel, both civilian and uniformed.

INTRODUCTION

Monograph 1 of this volume described how, under the aegis of HM Signal School acting as the radar design authority for the Admiralty, whilst only two ships of the Royal Navy had been equipped by 1939 a most impressive series of radar systems was developed during the war (see Appendix 1 to this volume). Although initially at some disadvantage vis-à-vis the German Navy,[1] the dynamism of the Admiralty's radar programme quickly outstripped the opposing radar threat both technically and operationally, and sustained its lead in these respects throughout World War 2.

For each new radar, and any subsequent variants, HM Signal School's remit from the Admiralty covered responsibility for its full development, and for comprehensive sea-trials to verify its degree of compliance with the Naval Staff's Operational Requirements.

From this point onwards Signal School was responsible for producing complete manufacturing drawings and detailed installation specifications for the very substantial constructional and electrical work generally required aboard ship, prior to actual installation. Concurrently, Signal School had the task of producing supporting technical manuals and setting-up procedures for fitting-out parties.

Additionally, Signal School was responsible for carrying out post-design modifications and setting their implementation in train. These modifications were invaluable in enhancing the performance at sea of the parent design. They accrued from matching ever improving techniques to the feedback received from the Fleet regarding shortcomings in operation, recurrent component failures, and desirable improvements (see Monographs 7 and 8 of this volume).

Regarding the last item, specialist officers also reported back to the appropriate Establishments – *Dryad, Excellent* and so on – adding valuable impetus to systems development, in particular.

A further Signal School responsibility was that of liaison with the RAF and Army research authorities. This is referred to in some detail in monographs 4, 5 and 6 of this volume, in particular, in relation to cross-fertilisation of particular radar system developments.

It is perhaps worth recalling here the varied reception given to radar in its early days by the commanding officers of HM ships, of various classes. Some, almost spellbound, cherished excessively high expectations, but reality soon followed. Others, perhaps hidebound by tradition, were highly suspicious, doubting that the radar operator (usually an ordinary seaman or able seaman at that time) could outperform a visual lookout. Indeed there were isolated cases in which some commanding officers called for the keys of the locked radar offices, and used the radar operators to double as lookouts!

Fortunately there existed a substantial number of officers who looked on this revolutionary and unheralded device as a helpmate, which, if used intelligently, could aid conventional practice. They sensed that, despite its current limitations, it had the potential to become an essential component of the Royal Navy's future shipborne weapons systems.

Happily, by about mid-1941 radar had 'won its spurs'. Commanding officers then became anxious, as a refit approached, that the alterations and additions (As and As) for their ship should include the latest radar equipments. This change in attitude was particularly appreciated by Port Radar Officers and their staffs, who felt more welcome aboard certain ships, in the discharge of their duties, than hitherto.

Returning now to the *modus operandi* of the tasks covered by this monograph, two distinct objectives were fulfilled. First was the derivation by the Admiralty of a definitive radar ship-fitting programme. The second was its implementation, with the greatest possible dispatch, to which the Admiralty accorded the highest possible priority at every level.

Figures 9.1 and 9.2 show *basic* block diagrams relating to each of these phases. For clarity, many ancillary responsibilities and liaison channels (often *ad hoc*) have been omitted. The diagrams represent a somewhat polyglot but, in practice, remarkably efficient and pliant organisation, which expanded as the war progressed.

Figures 9.1 and 9.2 may present an air of substance and order, but this was not the case. The whole arrangement was initially chaotic – but there was a war on. Everyone was pointing in the same direction and, against the odds, it worked well.

Most of what follows in this monograph relates to radar matériel *per se*, but it is important to understand that, apart from small ships such as

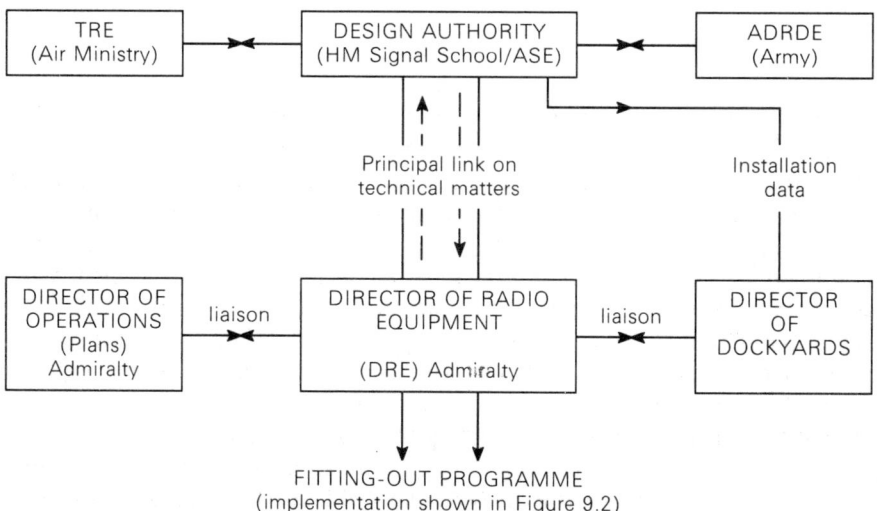

FITTING-OUT PROGRAMME
(implementation shown in Figure 9.2)

Notes
1. For simplicity, this diagram represents the status at the initial stage. As new radar systems such as action information organisations, weapons control systems, and fighter direction organisations emerged, more and more Admiralty directorates (such as those concerned with navigation, gunnery and Naval ordnance) provided an input to installation planning.
2. The later important link with the Fleet Air Arm, as fighter direction techniques evolved, is not shown.

9.1 Formulation of Admiralty radar fitting-out programme

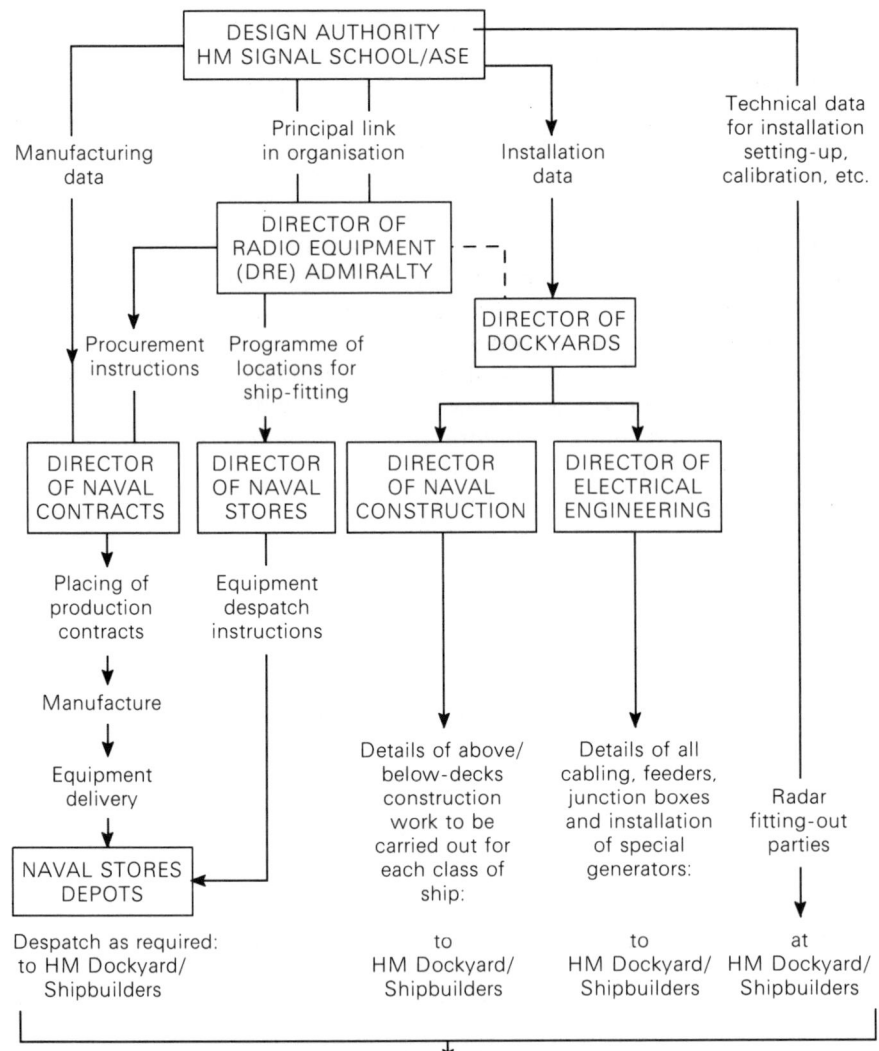

9.2 Implementation of radar ship-fitting programme

corvettes and minesweepers fitted with a single equipment, radar sets seldom operated in isolation, but were an integral part of an operational whole.[2] This had a considerable bearing on installation work. Because of the many interfaces involved, installations could not be fully tested before the total system was complete.

FORMULATION OF ADMIRALTY RADAR-FITTING POLICY AND PROGRAMMES

In the late 1930s, as the war clouds gathered, the Admiralty realised that the Royal Navy had to be placed on a wartime footing. At that time the RN had a very substantial fleet – one of the world's most powerful.

Some of the many ships serving on home and foreign stations were undergoing or were scheduled for refit, whilst the events circa 1936–7 had lent impetus to an expanded new construction programme covering for example, the *King George V*-class battleships, aircraft carriers, cruisers and destroyers.

However little provision had been made (for understandable reasons, largely concerned with security) for accommodating the calls that would be made on Naval constructors and electrical engineers in the large-scale fitting of radar equipments at short notice.

Amongst the criteria the Admiralty had to consider in creating a cogent ship-fitting programme were the following:

- Priority policy, based on operational considerations regarding the classes of ship to be fitted with extant/planned Naval radars. These ships might be in commission, undergoing refit, or at the design stage. Initially AA cruisers were awarded top priority.
- Dates from Signal School (later ASE) on the availability of manufacturing drawings for equipments, such as 50-cm equipments still at the design stage.
- The estimated production capability of industry – mainly the radio industry – to match anticipated demands, observing that the technical specifications would be of a higher order than hitherto, such as in the case of W/T equipment.
- The capacity of HM Dockyards and commercial shipyards to absorb the additional work, much of it of a highly-skilled nature, which radar ship-fitting would impose.
- The availability of HM ships for radar fitting in relation to the Admiralty's operational plans.
- Assessment of the capability of shipyards, large and small, which could be harnessed in the event of war.
- Continual review of interservice cross-fertilisation to reduce duplication of effort.

Balance of these data led to a programme, based on practicalities, being drawn up. Its updating throughout the war was an iterative process.

Responsibility for allocations of equipment rested with the Director of Radio Equipment (DRE), Admiralty. Procurement was the responsibility of the Director of Naval Contracts. Supervision of fitting-out was overseen by Section MF2 (Lieutenant-Commander J. V. Kitchingman) of the 'M' Ship-Fitting Division under Commander C. V. Robinson, RN, at ASE, Haslemere.

PROCUREMENT OF RADAR EQUIPMENT

It was the responsibility of the Director of Naval Contracts to translate DRE's directives into equipment (see Figure 9.2). This directorate had for many years handled the purchase of W/T equipment (see Monograph 1), and so a ready-made channel existed that could be expanded to service the radar requirement. The scale of operations increased dramatically in terms of the numbers of radar types, the numbers ordered, and the number of component units per equipment and their complexity.

For example, a whole series of new items, relative to W/T practice, appeared. These were much larger antennae, with their pedestals and turning gears, new feeders such as polythene coaxials, Pyrotenax and precision-machined waveguides; CRT displays; special electrical generators; and smaller special items.

The search for frequency stability was the Holy Grail of design engineers, and this included not only valve improvements, but also the development of high-tolerance components.

Not only did the number of main contractors grow enormously but, in turn, so too did the number of sub-contractors: 'Big fleas have little fleas upon their backs to bite 'em, and little fleas have lesser fleas, and so ad infinitum'. Deliveries were therefore according to a very irregular sawtooth profile.

One Admiralty stipulation was that bulk production contracts should be shared by two, preferably three, contractors, sited in well-separated locations, to minimise disruption in the event of air attack.

Another important stipulation was that, for security reasons, no one manufacturer should, in general, make a complete radar equipment. The latter would be assembled as a complete entity only on-board ship, under RN supervision.

Appendices 3 and 4 to this volume give some idea of the pressure of the task undertaken. Much of it was carried out by a workforce recruited from amongst ordinary people who had responded to the war-effort call. These were predominantly women who, prewar, had never seen a soldering iron or a lathe, but whose nimble fingers worked with a will. Without their efforts the incredible output of radar equipment produced throughout heavy air attacks would have been much diminished.

As shown in Figure 9.2, equipment deliveries were made to Naval Stores depots at Glossop, Oldham and (to a lesser extent) Haslemere.

This brings us to the distribution of the equipment to fitting-out bases. For a time, this was an area of distinct difficulty.

FROM NAVAL STORES TO DOCKYARDS/SHIPYARDS

Naval Stores depots were the depositories for a plethora of radar items from suppliers that had to be identified and systematically stored, ready for despatch when called for. Many items were already required for urgent despatch on receipt, but the warehousemen were not trained to identify these. The depot staffs were largely newcomers – strong men who humped and dumped (and, on occasion, dropped their valuable cargoes), supported by an army of clerical personnel who struggled with a cumbersome Kardex system, there being no such thing as a computer in those days.

The principle was that no piece of equipment existed until entered in the Kardex system. Then, and only then, could the equipment be despatched, assuming someone knew its destination.

In the early days distribution did not present a problem. For example, initially Type 79 equipments were coming forward at the rate of only three or four a month, and these were destined mainly for the Royal Dockyards and major shipyards.

However, from about 1941 onwards equipment was being delivered in large quantities of many different types, and the number of shipyards involved had multiplied, with the result that there was a period of muddle, involving serious delay. This was overcome by a system known as plan-packing, illustrated in its simplest form in Figure 9.3.

The starting point was to imagine the ideal situation in which a complete radar equipment was available, and then to decide on the best way to package it. This would provide the number and size of each packing case, and its contents. Figure 9.3 shows the configuration for a hypothetical radar.

A printed pro-forma packing note was produced showing the specified contents, with columns relating to the actual contents for any one consignment. An illustration of this is shown in Figure 9.4. In this example, only three consignments were needed to clear the parent pro-forma's total contents.

The concept meant that a set number of standardised packing cases could be produced in quantity. Many of these could be used (often with the aid of wedges) for items from different radars.

The system worked well and brought order out of chaos. Most crates were despatched by goods train, but last-minute vital elements went by passenger train under the care of a WRNS courier.

One frequent delay, not the fault of plan-packing itself, was the lag on the part of stores depots in responding to change in a ship's location. Battleships and cruisers for the Royal Navy built on the Clyde, on Tyneside and elsewhere came to Rosyth to have their armament fitted (guns were never fitted in commercial yards). These ships invariably

	TYPICAL CONTENTS	TYPICAL NO. OF CASES
Case 'A'	Antenna Pedestal/turning gear. Antenna framework (unassembled). Antenna dipoles, fittings etc.	1–2
Case 'B'	Transmitter and receiver panels. Other large-sized panels	2–3
Case 'C'	PPIs, and units of similar size.	4–6
Case 'D'	Valves, carbon-pile regulators, waveguide driers, spares	5–7

Note: Only four packing cases are shown. There were many more such standard-size cases. Many were common to several radars.

9.3 Plan-packing scheme for a hypothetical radar

arrived with some radars incompletely fitted, due to shortages. The latter were seldom sent direct to Rosyth, and locating and transferring them was a problem.

Shortages were always sent in the nominated case, even if only one item was involved. The situation would then arise where a sizeable packing case would arrive with the final, often very small, item.

In HM Dockyards, Naval Stores personnel persevered at their rather mundane jobs, and responded nobly to encouragements. A big problem was that they naturally could not identify incoming items, which were often desperately needed. Originally their depots were inviolate, but most Port Radar Officers eventually succeeded in infiltrating them by various means, in the interests of progressing work on board.

Throughout the war, spares were at a premium on-board ship and at shore bases. More than once a ship coming in for refit would have to surrender its spares to a refitted ship sailing the same day.

333

Item	Detail	Lots					
		2	3	4	5	6	
			Estab-lishment	Already sent	Enclosed	Balance	
1	CV1		4	–	4	–	
2	CV2		2	–	–	2	
3	CV3		6	–	3	3	
4	CV4		3	–	3	–	
5	CV5		1	–	–	1	
- - -							
10	CV10		10	–	4	6	

1st Consignment

	Lots				
	3	4	5	6	
	Estab-lishment	Already sent	Enclosed	Balance	
	4	4	–	–	
	2	–	–	2	
	6	3	–	3	
	3	3	–	–	
	1	–	–	1	
- - -					
	10	4	–	6	

2nd Consignment

	Lots					Cols 4+5= Col. 3
	3	4	5	6		
	Estab-lishment	Already sent	Enclosed	Balance		
	4	4	–	–		(4)
	2	–	2	–		(2)
	6	3	3	–		(6)
	3	3	–	–		(3)
	1	–	1	–		(1)
- - -						
	10	4	6	–		(10)

3rd Consignment

9.4 Relevant sections of plan-packing notes covering complete delivery in three consignments

THE SHIP FITTING-OUT TASK, 1939–45: AN OVERVIEW

As already stated, in September 1939 the Royal Navy had but one design of radar equipment, the Type 79 (39–42 MHz), for long-range aircraft warning. Only two ships – *Rodney* and *Sheffield* – were fitted, their equipments having been 'hand-built' in HM Signal School (see Monographs 1 and 4). The sets were installed by members of its scientific staff, who also had to supervise the immediately following installations.

Non-existent were radar fitting-out or maintenance bases, radar officers or radar mechanics. In particular, HM Dockyard and commercial shipyards had no appreciation of the special problems that lay ahead in fitting radar.

The need to redress this situation was all the more vital because equipment production was beginning to roll, and the Signal School scientists were pressing ahead with new designs. In July 1939 the Admiralty had placed contracts for 40 Type 79s, by far the largest order in the world for radar at that time. Work was well advanced on Type 281 (86–94 MHz), the planned successor for Type 79 and its Type 79B variant, with lighter antennae, higher power and working at twice the earlier frequency. In early 1940 an order was placed for 200 50-cm equipments, the first of the gunnery sets, which became designated Type 282, with Types 283, 284 and variants following on.

In the absence of an RN surface warning radar until Type 290 (quickly replaced by Type 291) could be designed, a number of RAF air-to-surface vessel (ASV) equipments were obtained from the Ministry of Aircraft Production and fitted in HM ships, with a fixed 'bedstead' array, designed in double-quick time by Signal School (see Monograph 4). Some 30 of these equipments (Type 286) were fitted by the end of 1940.

Indeed, by the end of 1941 a substantial number of HM ships had radar of one kind or another, and radar bases had been established in the main ports. The installation programme was being implemented with real drive, giving it an impetus that never slackened.

Now, so long after the event, it is needless to chronicle achievements in detail. Suffice it to say that after the small beginnings in 1939, by 1943 many types of Naval radar equipments (including variants) had been developed by the Admiralty (see Appendix 1 to this volume) and were brought rapidly into service. These were mainly shipborne, but some were land-based. Almost every ship in the Royal Navy, except for those under construction or undergoing refit, had its approved complement of radars. Many received a second fit before the end of the war. In addition to this, the numerous designs of radar beacons, jammers and IFF/interrogators involved will provide some idea of the scale of events leading from hardware manufacture to its becoming operational.

PRE-FITTING WORK ON BOARD SHIP

By 'pre-fitting' is meant the preparatory work necessary before radar units could be installed and wired up.

Prime responsibility for specifying the work to be done lay with the Director of Naval Construction and the Director of Electrical Engineering, both at Admiralty (Bath). The work was implemented under the direction and supervision of their representatives appointed to cover dockyards/shipyards in specific regions. For security reasons dockyards had been kept in the dark (like nearly everyone else!) and had to start from scratch. They were fortunate, however, that initially they were handling only one equipment – Type 79. Installation was relatively simple, it being a single 'set', relative to the much more complex radar systems that subsequently were to evolve quite rapidly.

As stated earlier, Admiralty policy was to fit existing ships undergoing refit so that radar could 'go to sea' as soon as possible. Most of the early fittings were in C-class AA cruisers. While Naval constructors designing new ships could make ample provision for fitting radar, the problems of fitting the equipment in existing ships was a classical case of 'putting a quart in a pint pot', especially as 'Topsy' kept growing as time progressed. There were always ships becoming due for refit, and the problem was never really resolved.

Generally, constructional work posed more problems above-deck than below deck. Above deck the main task was fitting the antennae. Two masts were required for Type 79 – one for the transmitter antenna and one of the receiver antenna (see Monograph 4). It was most important to maximise antenna height in the interest of achieving maximum radar range; on some ships an extra mast was required. In other cases masts had to be strengthened and/or an extra mast had to be designed. In some ships masts had to be heightened and/or strengthened. These matters involved calculations 'back at the drawing board'.

In every case the additional weight at the masthead increased the ship's turning-moment, and means of top-weight compensation had to be devised, often at some inconvenience. The tilt-test calculation had to be revised to verify that the ship's stability was not adversely affected.

Additionally, fitting of radar antennae had to mate with the installation of other radio antennae for communications, jammers, HF DF, radio beacons and so on. This could be difficult in some aircraft carriers, where the Type 79 had to be made retractable to 'time share' with the radio beacon.

Early 1941 saw the introduction of common transmit/receive working, which permitted single-mast operation for Types 79 and 79B, plus a ranging panel (see Monograph 4). Later in the war a number of ships were fitted with both Types 79 and 281, operating in synchronism.

Blending of their polar diagrams yielded valuable operational advantage in the resolution of the heightfinding problem.

Preparatory work for installation of the 50-cm gunnery sets (Types 282, 283, 284 and variants) offered not much difference in constructional terms – the work was largely electrical.

The first of the 10-cm equipments, Type 271 had its antenna encased in a 'lantern' of teak and perspex (see Monograph 5). Siting the lantern to reduce 'blank arcs' was difficult, and sometimes called for structural changes. The teak in the lantern was found to give rise to unacceptable side lobes, with the result that the lantern was replaced by a prefabricated 6-ft perspex cylindrical dome in 1942. Later, a larger cylinder was produced for the Type 271's bigger (but younger) brother, Type 273.

The early versions of 10-cm radars used coaxial feeders, but waveguide feeders later became standard. Fitting these fell to the shipwrights and it was some time before the technique was mastered. Also, it was not always appreciated that the cleanliness of the inside of the waveguide had to be scrupulously checked before fitting. Standing waves presented considerable problems – and their eradication was not simple.

Reverting to Types 286 and 290/291, the antenna feeders were of Pyrotenax, which again was not easy to install due to the problems of eradicating moisture ingress during installation (see Monograph 4).

Below deck the constructional work in existing ships was largely concerned with revamping compartments to accommodate the radar units. The smaller the ship, the greater the problem. Inevitably space had to be surrendered, greatly to the discomfiture of the occupants of many a 'glory hole'. It was all a matter of expediency, and various solutions had to be found for different classes of ship.

One typical example was the Hunt-class destroyer *Cottesmore*, whose coal-fired galley was supplied from an adjacent coal bunker. To fit the Type 286 the galley was converted to oil-burning, and the coal bunker became the radar office! This applied to a number of Hunt-class destroyers, though perhaps not all.

One difficulty was that, due to design alterations during the building of a complete class of ship, the 'last-of-class' might be structurally different internally from the 'first-of-class', although looking the same externally.

Fitting radar in existing ships added to the size of the ship's company, involving extra accommodation and victualling stores. A prewar-built County-class cruiser, such as *Cumberland*, was fitted around 1943 with Types 281, 273, one 284, two 285s and four 282s. The radar personnel complement was two radar officers, four radar mechanics and some 20-plus radar operators of various rates. Probably they were squeezed in without too much shipyard work.

It was when one came to the small ships – submarines, coastal-forces ships (MLs, MTBs and so on) and trawlers being converted for minesweeping – that the problems became acute, both in finding room for the radar and for the radar operators. Horror stories abounded concerning the privations suffered by trawler skippers, seldom men of modest stature (and vocabulary), who returned from refit leave to find a severely truncated bunk!

As the operational applications of radar grew, starting with radar plots and a number of ingenious 'home-grown' operational aids, there appeared fully-developed Action Information Organisations (AIOs)[3] and Fighter Direction Rooms (FDRs)[4] involving much more complex constructional alterations.

Electrical Work

Electrical preparatory work, although on a large scale, presented fewer difficulties. Much of it consisted of running of multicored, lead-covered cables linking the radar generators (sometimes multiphased), the radar offices and operational compartments (often widely spaced), and runs to antenna turning-gears at the masthead.

Again, the task was much more difficult in existing ships than in new construction types. In the former, routing the channelling that carried the cables often presented problems, involving burrowing around existing compartments and bulkheads. Demarcation practices often introduced delays. Only shipwrights could bore holes in metal and insert screws. Only carpenters could put wood screws into wood. Often, cable runs traversed both wood and metal surfaces at intervals. Shipwrights and carpenters were at the ready, eyeing each other suspiciously, but never the twain would meet. They seemed to have different 'break' times, which were inviolate, and this held things up.

Despite the miles of lead-covered cable and thousands of connections to junction boxes and radar units, there were relatively few errors. One rather splendid exception happened in a *King George V*-class battleship nearing completion. The radar fitting-out officer was told that the Type 281 was ready for testing. He pressed the power 'On' button, without result. Pressing the 'On' and 'Off' buttons in succession brought no success, and he was blissfully unaware that he had caused consternation amongst fitters in a distant galley when the (unattended) electric potato peeler, fitted the previous day, sprang into sporadic operation. The opinion was freely expressed that the b— ship was already haunted before it had even sailed! The respective electrical foremen were summoned, and were completely nonplussed. Fortunately these two met over their sandwiches at lunch and, exchanging details of the morning's travails, recognised the convergent nature of their mutual problems. Within two hours the Type 281 had its power, and the galley fitters were (reasonably) satisfied that exorcism was complete!

The introduction of lead-covered polythene coaxial-cables for inter-compartmental wiring and antenna feeders required added skills in fitting. An extra complication was the partiality of rats for polythene. Overnight, several lengths of 6–9 inches of this type of cable would be completely stripped by rats, baring the central copper conductor as clean as a whistle (see monograph 8 of this volume). How the rats divined the presence of their delicacy – encased, as it was, in lead – was a mystery; not to mention their seeming awareness that lead was a poison. The deck would be littered with tiny shards of lead spat out by the rapacious rodents.

In 1939 the Royal Navy was fortunate in that the workforce throughout the Royal Dockyards had undergone apprenticeships in these establishments, giving them the benefit of a training recognised as being *par excellence*. As time went on, as the radar-fitting programme expanded, so too did the workforce. However new recruits to all trades were of an increasingly low calibre, leading to what was known euphemistically as 'diluted labour'. This laid a heavier than usual load on chargehands and inspectors as regards supervision.

In practice the electrical pre-fitting task was one of increasing scale, rather than of complexity.

This brings us to the point near the bottom of Figure 9.2 where the radar hardware is on site, the constructional and electrical pre-fitting complete, and the radar fitting-out party available. The main hurdles are astern!

RADAR INSTALLATION, TESTING AND TRIALS

Installation – the fixing in place of the radar equipment and making the electrical connections, was essentially straightforward. Only when this was complete could power be switched on, and the correctness of the installation verified. In general matters were satisfactory – nothing normally on the scale of the potato-peeler fiasco referred to above. Nevertheless tracing wiring faults, bad connections, missing fuses and so on could take time and prove irritating to fitting-out officers anxious to get going.

The fitting-out officers were part of the local Port Radar Officer's staff. The term 'fitting-out' officer was a misnomer. Apart from valves, they didn't fit anything – and if they tried to, they were in trouble. On occasion, newly qualified radar officers, recently down from university and unfamiliar with the *modus operandi* of artisans, would, in their zeal, try to speed things up by making ad hoc connections where work was incomplete. More than once, conciliatory words from the Port Radar Officer in a shop steward's ear narrowly forestalled a cry of 'Everybody out!'

Shortages of a few vital items could delay setting-up procedures, but one could often borrow the necessary items from another ship, under pain of death if they were not returned quickly.

At this stage, in ships carrying their own radar officers and radar mechanics (generally cruisers and above) the ship's radar staff would have joined the ship (see Monograph 8 of this volume). Always keen to be familiar with the 'radar fit', especially in the case of new equipments, they worked in harmony alongside the fitting-out officers, but observing that it was the fitting-out officer's responsibility to prepare everything for handover to the ship.

Equipment Trials

In the case of cruisers and above, even in the dockyards, experienced fitting-out officers could use permanent echoes to satisfy themselves that the radars were working reasonably well – recalling that, in those days, there was little in the way of built-in test equipment or diagnostic fault detection.

With new-construction ships, acceptance trials, carried out in waters accessible from the builders' yard, gave the first opportunity to appraise radar behaviour. Similarly with the less exhaustive trials carried out after major refits. It was only when such ships reached Scapa Flow for 'working-up' that everything could be made 'tickety-boo' on such essential exercises as establishing the polar diagrams of the air-warning sets (see Monograph 7).

Smaller ships, destroyers, corvettes and so on worked-up at Tobermory on the island of Mull, on the West Coast of Scotland, under the sharp-eyed and indomitable Commodore G. ('Puggy') Stephenson, RN, whose strict regimen is well described in *The Tiger of Tobermory* by Richard Baker. In spite of his acerbity and being the scourge of commanding officers found wanting in his eyes, he could be generous in his praise when he deemed it was merited.

Throughout working-up periods the ASE Radar Sea Trials Party, a peripatetic band of Naval radar and civilian officers led by Peter S. B. Morey, lent their support as 'trouble-shooters', especially where early fittings of new radars were involved. In the latter capacity they rendered great service in combing out the faults, and cheerfully accepted being continually driven from 'pillar to post'.

Shore Maintenance

Shore maintenance resources were disposed according to operational needs. Obviously the greatest concentration was in the Western Approaches Command, centred at Derby House, Liverpool, throughout the Battle of the Atlantic. Other important maintenance bases were at Belfast, Londonderry, the Clyde (Sherbrooke House, Glasgow), Scapa Flow, the Harwich area (Coastal Forces) and Devonport. Due to bomb vulnerability, operations at Chatham and Portsmouth were on a reduced scale. Apart from the Rosyth Escort Force, which shepherded convoys from Methil (Fife) to Southend, there were virtually no operations off the North of England coast or the east coast of Scotland.

Initially radars were operated by selected wireless operators who, apart from being able to change valves (if they had any spares!), had no technical knowledge (see Monograph 7 of this volume). However, things improved with the introduction of qualified radar mechanics.

In general, large ships visiting port were self-sufficient as regards maintenance, provided that they could draw spares from the port store depot.

Ships from destroyer and other escort flotillas depended on the local Port Radar Organisation for maintenance, and during 'boiler cleans' there was always a thorough check.

One maintenance procedure carried out automatically was putting a 'Megger' insulation tester on the Pyrotenex feeders of Types 286, 290 and 291, and their variants, each time a ship reached port. Invariably it was necessary to repack the junction-box at the masthead with molten pitch.

Even more than running repairs and routine checking, carrying out modifications was perhaps the most important role in shore maintenance. The modifications, particularly in destroyers and below, were usually carried out in the local radar base-workshops. Devised by Signal School, these modifications arrived in kit form with a very precise set of step-by-step instructions.

Some modifications were straightforward, such as substituting an improved component or two. Others were more complicated, such as taking out several components, keeping some for reinsertion, whilst scrapping others; taking out wiring and then rewiring as necessary; and perhaps fitting and wiring a new valve holder. The writer noticed that his WRNS radar mechanics never had any problems, and worked quite quickly and purposefully alongside their male counterparts. They were a conscientious and companionable crowd, who enlivened many a drab radar workshop.

Abroad, radar-maintenance bases were scattered and, apart from Gibraltar, operated on a fairly small scale, continually frustrated by lack of spares. The Gibraltar base was set up by Lieutenant F. J. (Jack) Emuss, later Commander (L) RN, who was effectively the first Naval radar fitting-out officer, and paved the way for setting up the Port Radar Officer organisation. Whilst on the staff of VACNA Gibraltar, although in effect the Port Radar Officer, he had (for security reasons) the title of Port Technical Officer. Radar-maintenance bases operated at Malta, Alexandria, Durban, Mombassa, Colombo and Trincomalee.

Although technically 'afloat', depot ships such as *Tyne, Hecla* and *Maidstone*, strategically disposed according to Fleet requirements, were a valuable source of base logistic support for radar-fitted ships.

Replenishment

In 1945 two radar-replenishment projects were carried out, codenamed 'Bubbly' and 'Brandy'.[5] At this point ASE had designed a range of equipments to update existing radars to give greatly improved operational performance. These changes were on a far larger scale than the modifications referred to earlier, including completely new equipment panels and, often, antennae.

The sets concerned were mainly Types 281, 277 and 293. After the update was implemented these were designated Types 281BQ, 277P and 293P, respectively.

As stated earlier, electrical labour had become 'diluted', and also became depleted in numbers due to 'call-ups' into HM Forces. It was agreed that the 'Bubbly' installation work should be carried out by ships' personnel with help from the local Port Radar Officer's staff. The first phase was a 'crash' fitting-out of fifteen ships completing refits prior to joining the British Pacific Fleet (BPF). Later, ships were fitted-out as practicable, but the work was on more of an *ad hoc* basis, depending on the individual ship's operational programme.

Execution of the 'Bubbly' programme was a great achievement by all those who saw it through at such speed. Perhaps the most noteworthy achievement was hoisting and fitting a new Type 281BQ antenna assembly at the masthead.

Early in 1945 the headquarters of the BPF, which had formed up in Ceylon in November 1944 under its C-in-C, Admiral Sir Bruce Fraser, had moved to Sydney, Australia. An RN Fleet repair base had been set up there by Captain (E) John H. Illingworth, RN, occupying wharves and warehouses belonging to the P & O Shipping Company, at Wooloomooloo, to service the BPF. A most efficient base was set up in a remarkably short time.

Early in 1945 a further tranche of 'updated hardware' became available for a wide range of radars. The big problem was, against the odds, how to get all this gear out to Sydney and fitted in the BPF ships due for a replenishment visit to Sydney before leaving for the final assault on the Japanese mainland as Task Force 37 of the USN 3rd Fleet.

No airlift was available, and merchant ships were far too slow. The solution was to use *Abdiel*, a high-speed minelayer of the *Manxman* class, no longer required for her original purpose. Stripped of her minelaying gear and virtually gutted, she became the first radar transport ship the world has ever known, carrying her vital load of radar equipment from the UK to Sydney, easily breaking the record for this journey by sea. As the war was still in progress, this could not become public knowledge.

Abdiel arrived just over a day before the Fleet, giving time to segregate the equipment for delivery to each ship immediately on berthing. All the equipment was installed by ship's and base staff before sailing – only days before the dropping of the first atomic bomb signalled the end of the war.

This was operation 'Brandy', named by the C-in-C himself, who said 'Well, we've had "Bubbly", and brandy comes after bubbly. As the party may soon be over, let's call it "Brandy" '. And 'Brandy' it became.

Another replenishment venture was 'Knobbly', a high-priority programme fitting-out fast, long-range fighter-direction ships (LSF) for the Pacific War.[6] None of these reached the Pacific before the War had ended, but had hostilities continued 'Knobbly' would have added considerable support in what had become virtually an air war.

The 'Fleet Train' was another indispensable source of logistic support for the British Pacific Fleet, which, once its headquarters had been established in Sydney, Australia, could be operating at a distance of 3500 miles or so from base. The Fleet Train, consisting of a number of ships providing comprehensive maintenance and repair facilities 'across the board', included a radar/radio maintenance ship – HMNZS *Arbutus*, a corvette modified for this function.

THE PORT RADAR OFFICER ORGANISATION

General

Little needs to be said on this topic, its functions being clear from earlier parts of this monograph.

Port Radar Organisations were set up quite rapidly in Portsmouth, Plymouth, Chatham, Liverpool and, but not until mid-1941, Rosyth. Belfast, Londonderry and Newcastle followed. The arrangements at Scapa are described in Monograph 7. Those in the Clyde follow.

To some extent the title 'Area Radar Officer' would have been more apposite, because the Port Radar Officer's parish lay far beyond the named port and covered many small ports – indeed every port where radar fitting was afoot. For example, Port Radar Officer, Newcastle, Lieutenant Commander G. W. B. Mills, RNVR (Sp), covered Blyth, Sunderland, Hartlepool and so on, while Rosyth covered Leith, Granton, Dundee, Aberdeen, Fraserburgh and, in the months preceding D-day, Invergordon.

Suffice to say that wherever there was a fitting-out 'parish', a radar officer would appear. Several Port Radar Officers were awarded the MBE.

Sherbrooke House, Glasgow

This monograph would be incomplete without mention of Sherbrooke House, unique amongst shore radar bases, and remembered vividly (and with affection) by all who served or underwent training there (see Figure 9.5).

It was a law unto itself because its progenitor, Commander Oswald Stuart Neill, RNVR, was a law unto himself. A teammate at HMS *Osborne* of Prince George, later King George VI, Neill came out of the RN after World War 1 to pursue an engineering career, but still serving in the RNVR. He had an ebullient nature, was a born entrepreneur, and was irrevocably set on getting his own way. There are those who say that in Elizabethan times he would have been a prince amongst buccaneers.

At the end of 1940, when the Clyde area was at its busiest with major new construction work and refits, the fitting-out party consisted of Neill and three sub-lieutenants – Bob Goudie, Roger Laurence and the writer (who also had a watching brief on Rosyth until mid-1941, when he was appointed there). The cramped accommodation of the Clyde Division RNVR Gunnery School was completely inadequate.

9.5 From left to right, Commander C. V. Robinson, in charge of radar ship-fitting organisation at Signal School and ASE; Captain B. R. Willett, Experimental Commander, HM Signal School 1937 and Experimental Captain 1939, and Commander O. S. Neill, RNVR, Commanding Officer, Sherbrooke House, Glasgow, 1941–44.

Neill secured the requisition of Sherbrooke House Hotel and its spacious grounds as a new base in the select district of Pollockshields, overlooking the Clyde. He hoisted the White Ensign with great ceremony, with five Naval officers and his wife, Carol, in attendance!

Neill reported to Flag Officer-in-Charge, Glasgow, Vice-Admiral J. A. G. ('Jaggers') Troup, RN, and although technically Port Radar Officer, Glasgow, eschewed the title and designated himself Commanding Officer, Sherbrooke House!

Once installed in Sherbrooke House, Neill caused Nissen huts to appear all over the place. As well as being the radar fitting- out base for the Clyde, Sherbrooke House suddenly became a training establishment for radar operators and, later, radar mechanics. One-day 'acquaintance' courses for non-radar officers from ships in the Clyde were a feature. Wherever he saw a need, Neill would organise a solution.

Furniture, transport, equipment and staff kept flowing. 'Where did it all come from?' the writer once asked Neill. 'Whatever you need, "plus-up" your request. If you need a desk, ask for five, and you may get one. If you need three, ask for ten, and you will get three. Never forget that.'

The site being at a good height, permanent echoes were not a great problem and aircraft could be detected on Types 286, 291 and so on for training purposes.

Where did the equipment come from? Some, certainly, by official allocation – but perhaps not all. It is said that there was a close correlation between superseded or faulty radar sets, equipment and antennae, removed from ships in the Clyde for return to Naval stores, but which never arrived there. Much hardware, some in need of repair, somehow became integrated into Sherbrooke's radar fit!!.

Sherbrooke House never had any other title. It was never an 'establishment' or a 'stone-frigate'. It was Neill's 'ship'. It served a unique purpose and much of O.S. Neill's drive and sense of purpose rubbed off on to those who had the privilege of serving under him.

THE RADAR OFFICER ORGANISATION, 1939–45

As already stated, there was no corps of trained radar personnel available early in World War 2. Similarly, a deficiency existed in respect of Asdic, 'de-gaussing' and other new 'high-tech' devices in RN service.

An urgent programme was instituted to recruit personnel, largely graduates from the universities, the radio industry and elsewhere, whose talents fitted them for training in these new scientific devices. (For example, a six-month crash course in radio-physics was provided at certain British Universities for students in physics, electrical engineering and mathematics to provide an expanded source of radar officers for each of the three Services. Some of these students were appointed as civilian officers at the three Services' Radar Research and Development Establishments). Those selected for RN service were commissioned into the newly formed Special Branch (Sp) of the RNVR, and were immediately trained in one of those specialisations. (They received, too, minimum training in how to salute a senior officer, and how to conceal their surprise when saluted by a passing sailor or WRNS!) Special Branch officers wore a pale-green stripe next to the gold braid on their uniforms, as an identification.

'Long' radar courses for embryo radar officers began in HM Signal School in 1940 and continued for several years, transferring to the new Radar School in HMS *Collingwood* in 1944. During this period some several hundred radar officers qualified. The radar officer appointments were made by Commander the Hon. J. B. (Johnny) Bruce, RN (a direct descendant of Robert the Bruce), and a former Experimental Commander at Signal School, from 1940-4 (see Monograph 1). He had an incredible memory, and could say without reference to files where any particular radar officer was serving.[7]

Until qualified radar officers were available for ship and shore duty, momentum was maintained by Signal School staff and a number of RN signal officers, such as Captain A. V. S. Yates, RN, Captain G. C. P. Whitaker and Captain C. M. Jacob, who were unfailingly helpful to radar officers later on.

A welcome breakthrough in the radar officer shortage came in early 1941 with the arrival of some twenty young RCNVR (Sp) officers, all of whom held a degree in radio physics from McGill University, Toronto – a qualification not then available at UK universities. They quickly became radar officers, and were all posted to important ships in the Home Fleet. A succession of fellow RCNVR (Sp) officers followed, giving a total of in excess of 40 from this source. They came largely at the instigation of Canadian-born Dr C. S. Wright, Director of Scientific Research at the Admiralty (1934-46), and were to carve out a great reputation for themselves (see Monograph 1). They were always in the thick of things. Sub-Lieutenant Stuart ('Stu') Paddon (later Rear Admiral RCN) served with distinction in *Duke of York* in the *Bismarck* action, and the later action in the Gulf of Siam where he was happily one of the few survivors when the *Prince of Wales* was sunk in that action. Later, Paddon was in *Warspite* when she was flagship to C-in-C, Eastern Fleet.

These Canadians, efficient and warm-hearted, were held in the highest regard by officers and men of the ranks wherever they served. To this day, when British ex-radar officers meet, the names of their Canadian counterparts crop up. They are still remembered with great affection.

Though in smaller numbers, there were valuable contributions by radar officers from the Royal Australian Navy, the Royal New Zealand

Navy and the South African Naval Force, in addition to contributions from the Norwegian, Netherlands and 'Free-French' forces. The bulk of radar officers, however, were British.

The present monograph has been concerned mainly with matters ashore, where radar officers slogged away for long hours to achieve desired ends. Nevertheless, what they accomplished was as nothing compared with the achievements of ship's radar officers. They were at the 'sharp' end, and it is only fitting that somewhere there should be written a testament to them, and to their radar mechanics, several of whom qualified as radar officers towards the end of the war. They spent long periods at sea, often in foul weather, and served in every theatre of war. They saw much action, apart from the major battles such as Cape Matapan and the sinking of the *Bismarck*, where radar played such an important part. They contrived through thick and thin, and with success, to keep often recalcitrant radars working, and showed great ingenuity and flair in devising ad hoc 'action-damage' repairs and modifications aboard ship. These were sometimes copied by ships in company, but in the heat of war never 'went through the system' for formal Admiralty adoption.

Achievements in action did not go unrecognised. Lieutenant (Sp) H. R. Kingsmill ('King') Bates, RNVR, (later Captain (L), RN) received an immediate DSC in the *Scharnhörst* action. In other actions, radar officers, one of whom was Lieutenant (Sp) Michael Law, RNVR, were mentioned in despatches, as were a number of radar mechanics, and several radar operators were awarded the DSM and BEM.

Not all radar officers served throughout the war. Some, whose exceptional talents fitted them for higher appointments, were transferred to civilian duties. A particular example was Lieutenant (Sp) Michael Carey (son of the novelist Joyce Carey), who left *Illustrious* in 1941 for a very distinguished career in the Civil Service. He eventually became Sir Michael Carey, Secretary to the Admiralty. Sadly, he died at a relatively early age.

Another was Lieutenant (Sp) Roger F. Laurence, RNVR, whose gift for innovation and development led to his being transferred as a development engineer at ASE, circa 1942.

Inevitably, with the loss of HM ships in action, some radar officers (though remarkably few overall) lost their lives. The most notable loss was Bernard Stubbs (see Monograph 7) who went down with the *Hood*. He was the only radar officer who had been known to the public prewar when, as a highly regarded BBC overseas correspondent, his voice was familiar to radio listeners. Fortunately there were many radar officers who survived the sinking of their ships including Alan Laws, author of Monograph 8 of this volume.

By the end of the war in 1945 the radar officer community had become, in all but name, a 'Branch' of the Royal Navy. It had earned respect within the Naval establishment, despite early and sometimes hurtful disparagement as 'civvies in uniform',

After the war, while serving in a personnel role in *Collingwood* 'demobbing' radar officers, the writer had a rare opportunity to see all the confidential reports on them, written by their commanding officers. It was most gratifying to find that the merit of his colleagues had been recognised, the great majority of reports being 'above average', and many in the highest terms, written by officers of up to Flag level.

In 1945 the RN's total radar experience lay in the hands of the few who had joined the RNVR in the 1930s and the host of 'Hostilities Only' radar officers, radar mechanics and operators due for automatic demobilisation, taking with them collectively the Navy's technical experience of radar at work.

Fortunately the formation of the new Electrical Branch (L-Branch), merging the existing Torpedo Branch with that of radar in implementation of the Middleton Report, gave the opportunity for selected radar officers to take up permanent commissions in that branch.

Some 35 to 40 radar officers transferred to permanent commissions, providing a valuable carry-forward of wartime experience, well qualified to handle the new generation of radars already at an advanced state of development. Many were from the early batches of radar officers who, on-board ship, worked closely with the ship's officers in developing the operational use of radar – perhaps the most important role radar officers ever filled. Thus they brought to the peacetime Navy a keen perception of both technical and operational aspects.

It is pleasing to record that these officers integrated rapidly into their new role, as their promotion record showed. As well as many reaching the rank of Commander, several were made Captain, and some four or five Rear-Admiral.

A similar situation arose in the Royal Canadian Navy in implementation of the Mainguy Report, with similar beneficial results. Again, there were many promotions – Terry Burchill, 'Stu' Paddon and 'Mac' Lynch became Rear-Admirals, and Bob McCormick (amongst others) a Captain. All well-remembered names in the radar-officer community.

CONCLUSION

The topic of this monograph is a wide one, and has been treated only partially. Even at the time, probably no one knew the full picture. It is of little consequence that, in 1994, the story is necessarily incomplete and, though trivial and mundane as it may seem now, its pursuit at the time was far from trivial.

What has been written is largely conditioned by the writer's personal experience – in Signal School, Sherbrooke House (Glasgow), as Port Radar Officer (Rosyth), and as Base Radio Officer (Sydney), the first time the posts of Port Radar Officer and Port W/T Officer were combined, enhanced by what he gleaned of Naval radar activities elsewhere.

His fear is that readers will almost certainly be those who played a significant role and have been overlooked. He apologises in advance.

If this monograph has any merit it is this. Between the brilliant scientists at HMSS/ASE and those at sea who extracted maximum operational value from radar as it then was, there was a multitude of people, some highly skilled, some relatively unskilled, who performed prodigies without ever, for security reasons, knowing how vital were their tasks.

One must not be chauvinistic or over-preen. Radar was but one aspect of the RN's multipronged technical attack, which contributed significantly to victory in 1945.

POSTSCRIPT

Early in this century capital ships of the Royal Navy were fitted with very high masts, early warning then being provided by visual lookouts. When Rosyth Dockyard was built in 1914–15, about one mile up river from the Forth Bridge (completed 1894), a new parameter for Naval constructors was set by the height of 150-ft from river level, at the highest tide, to the underside of the structure linking the main cantilevers.

In World War 2, mast height again became paramount because of the need to maximise WA radar antenna height. Accordingly, air-warning antennae were fitted on retractable telescopic masts at the masthead. Ships came up the Forth at high speed to counter possible submarine attack, maintaining an air-warning radar watch until past the 'boom', east of the Isle of May, when the ship's speed slackened. It was only then that the Type 281 antenna could be lowered. Because the Permali nuts were often heavily encrusted with salt, it was a tricky operation. However, despite some 'close shaves', no radar antenna ever struck the Forth Bridge.

Ironically, one day the Type 281 antenna of a ship that had only just 'made it' the previous day was knocked clean off its mast by a casual operator of the 300-ton crane, the dockyard's most distinguished feature!

References

1. D. Pritchard, *The Radar War* (Patrick Stephens, 1989), pp. 190, 191.
2. F. A. Kingsley (ed.), *The Application of Radar and Other Electronic Systems in the Royal Navy in World War 2* (Macmillan, 1994), Monographs 1 and 2 by H. W. Pout, Monograph 3 by Cdr A. E. Fanning and Monograph 4 by Cdr R. S. Woolrych.
3. F. A. Kingsley, op.cit., Monograph 3 by Cdr A. E. Fanning, RN.
4. F. A. Kingsley, op.cit., Monograph 4 by Cdr R. S. Woolrych, RN.
5. H. D. Howse, *Radar at Sea. The Royal Navy in WW2* (Macmillan, 1993).
6. PRO ADM 220/91, 'Report of Fitting, Operation and Sea Trials in HMS *Boxer*' (originally ASE Monograph M761, dated 1946).
7. CAC NRT ND 156, Part 3, Appendix, 'List of Special Branch Radar Officers serving in the RNVR, 1941–1945', compiled by Commander J. Emuss.
8. E. F. Burton (ed.), *Canadian Naval Radar Officers: the Story of University Graduates for whom Preliminary Training was given in the Department of Physics, University of Toronto* (University of Toronto Press, 1946).

Appendices

Appendix 1: Type Numbers of Naval Radar Sets, Operational or Designed, 1935–45

Derek Howse

AI	– Air interception	IFF	– Identification Friend or Foe
ASV	– Air to surface vessel	INT	– Interrogator
BCN	– Beacon	LA	– Low angle
CCA	– Carrier Controlled Approach	Rg.	– In the ranging mode (Types 279 and 281)
FD	– Fighter direction	Rx	– Receiver
GA	– Gunnery fire control, aircraft, high angle (or combined low angle and high angle)	Sw.	– sweeps
		TI	– Target indication
		Tx	– Transmitter
		WA	– Warning of aircraft
GB	– Gunnery fire control, barrage	WC	– Warning (combined aircraft and surface)
GC	– Gunnery fire control, close range, high angle	Wg.	– In the warning mode (Types 279 and 281)
GS	– Gunnery fire control, surface (i.e. low angle)	WS	– Warning of surface craft
HA	– High angle	≈	– Variable around this frequency
Ht-fndr	– Height-finder		

In the numbering of Naval radar sets, the first, second and third major modifications to the basic set were indicated by the suffixes M, P and Q respectively, eg. Type 286P. The suffix B indicated adaptation to single-mast working.

Information not available is generally marked with a dash.

Type number	Classification	Wavelength nominal	Freq. (MHz)	Power (kW)	To sea (abandoned)	Description
79	WA	7.5m	39-42	70	1938	Long range air warning for large ships. See 279.
79B	WA	7.5m	39-42	70	1941	Single-masted version of 79 (originally 79M).
91	Jammer	50cm-3m	90-600	10-25w[1]	1941	Jamming of German metric and decimetric radar. Initially sine-wave modulation, ultimately noise.
241	INT	1.5m	214	—	1941	For use with 281 & IFF Mk 2N.
242	Int	1.5m	182 or 179	1	1943	For use with WS and WC sets and IFF Mk 3.
242M	INT	1.5m	182 or 179	2-10	1943	Ditto
243	INT	1.5m	179 or 171	1	1943	For use with 281 & IFF Mk 3.
244	INT	1.5m	—	—	1943	For use with US Type SL & IFF Mk 3.
245	INT	1.5m	—	—	1944	For FD ships.
251/M/P	BCN	1.5m	176 or 177	7	1942	Modified RAF transponder coded to give ships' identity.
252	IFF	1.5m	38-52 & 195-220	—	1942	IFF Mk 2 in ships.
253/P	IFF	1.5m	Sw.157-187	10	1943	IFF Mk 3 in ships. Sweeps frequency.
253S	IFF	1.5m	Sw.157-187	10	1943	253 when fitted ashore.
255	BCN	1.5m	214	—	1944	Marker buoy for use only with 291.
256	BCN	1.5m	214	—	1944	Shore radar beacon for use only with 291.
257	CCA	3cm	—	—	1945?	Carrier controlled approach (BABS).
258	BCN	1.5m	179/182	—	1943	Mk 3 shore radar beacon responding to 242.
259	BCN	1.5m	—	—	1944	Mk 3 beacon for carriers responding to AI Mk 10.
261	WS	50cm	—	—	(1941)	Based on 282.
261W	WS	3cm	—	—	(1942)	Early 3cm development, leading to 267W.
262	GC	3cm	≈9650	20	1945	STAAG, CRBFD
263	GB	3cm	—	—	(1945)	Auto-barrage for main or secondary armament, replacing 283.
267W/MW/PW	WS/WC	3cm & 1.5m	≈9670 & 214	15-25 & 100	1945	Submarines. Hybrid WS/WC with common display.
268	WS	3cm	9400	—	1945	For coastal forces, replacing 291U.
269	GS	3cm	10000	—	(1943)	Modified 3-cm AI set for coastal forces gunnery.

Type	Cat	λ	Freq	Power	Year	Notes
271/M/P	WS	10cm	≈3000	5–10	1941	Small ships.
271Q	WS	10cm	≈3000	70	1943	Small ships.
272/M/P	WS	10cm	≈3000	5–10	1941	Small cruisers, carriers, sloops, etc.
273/M/P	WS	10cm	≈3000	5–10	1941	Large ships.
273Q	WS	10cm	≈3000	70	1943	Large ships.
274	GS	10cm	≈3300	400	1944	Main armament directors, replacing 284.
275	GA	10cm	3530	400	1945	HA directors (HA/LA directors in destroyers), replacing 285.
276	WS	10cm	3000	500	1944	Small ships, replacing 271/2.
277/P/Q	WS	10cm	3,000	500	1944	Replaced 271/2/3. Could measure approximate elevation.
277S	WS/low air	10cm	3000	500	1943	277 permanent shore installation for surface and low air.
277T	WS/low air	10cm	3000	500	1943	Trailer-mounted 277. 'Monrads'.
279	WA	7.5m	39–42	Wg.70 Rg.60	1940	Long-range air warning for large ships. Type 79 with gunnery ranging.
280	WA/GA	3.6m	82	25	1940	Based on Army GL1. In *Carlisle* and Bank-class ships only.
281	WA	3.5m	86–94	Wg.600 Rg.1,000	1940	Long-range air warning for large ships.
281B	WA	3.5m	86–94	600	1943	Single-masted version of 281.
281BM/BP/BQ	WA	3.5m	86–94	350	1945	Continuous rotation.
282	GC	50cm	≈600	15	1941	Pom-pom directors, etc.
282M1/M2/M3	GC	50cm	≈600	60 or 80	1942	Increased power.
282M4	GC	50cm	≈600	60 or 80	1942	Beam-switching.
282Q	GC	50cm	≈600	150	—	Beam-switching and increased power.
283/M	GB	50cm	≈600	150	1943	Auto-barrage for main or secondary armaments.
284/M/P	GS	50cm	≈600	As 282	1940	Main armament directors.
285/M/P/Q	GA	50cm	≈600	As 282	1940	HA directors (HA/LA directors in destroyers).
286M/P	WC	1.5m	214	7	1940	Small ships. 286M fixed aerial, 286P revolving aerial
286PQ	WC	1.5m	214	100	1943	Small ships. Higher power.
286U	WC	1.5m	214	7	1941	Coastal forces.
286W	WC	1.5m	214	7	1941	Submarines.
287	Minewatch	50cm	≈600	15	1941	284 adapted for minewatching ashore.
288(1)	GC	50cm	≈600	15	(1941)	284 adapted for Armed Merchant Cruisers.

Type number	Classification	Wavelength nominal	Freq. (MHz)	Power (kW)	To sea (abandoned)	Description
288(2)	GC	50cm	≈600	15	—	284 adapted for training ashore.
289	GA	70cm	≈430	—	1940	Dutch. Fitted in *Isaac Sweers* and *Heemskerck* only.
290	WC	1.5m	214	100	1941	Small ships, replacing 286 but abandoned in favour of 291.
291/M	WC	1.5m	214	100	1941	Small ships, replacing 286/290.
291U	WC	1.5m	214	100	1943	Coastal forces.
291W	WC	1.5m	214	100	1943	Submarines.
293/M	WC/TI	10cm	3000	500	1944	Destroyers and above. Replaced 271/2/3.
294	WC/FD	10cm	3000	—	(1944)	Combined plan-display and heightfinding, replacing 277.
295	WC/FD	10cm	3000	—	(1944)	Higher-powered 294.
650	Jammer	≈6m	≈50	10–20w	1944	Jamming of air-launched Fx.1400 and Hs.293 anti-ship guided weapons. Sine-wave modulation.
651	Jammer	≈6m	≈50	2 × 1kw	1944	As Type 650 but capable of handling multi- missile attack. CW modulation.
930	GS/splash	10cm	3000	7	1945	Splash-spotting; Naval version of Army CA No.1, Mk.5 ('William').
931	GS/splash	1.25cm	≈24000	—	1945	Splash-spotting, Canadian.
940/1	INT	1.5m	209	1	1944	G-band interrogator with 281BP/BQ.
951	BCN	—	—	—	—	Marker beacon for use with 10cm WS and WC sets.
952	BCN	Rx 3cm Tx 1.5m	Rx ≈9400 Tx 182	—	1945	Portable combined ops. navigational. Triggered by X-band, response on Type 242.
960	WA	3.4m	≈88	450	1946	Long-range air warning for large ships, replacing 281/79/279.
961	CCA	3cm	≈9320	—	—	Carrier controlled approach. Modified ASV II.
970	WS	10cm	≈3300	—	1943	Combined operations. Modified RAF H$_2$S II.
971/M	WS	3cm	≈9320	—	1945	As 970 but based on H$_2$S III.
972	WS	3cm	≈9375	—	1946	Surveying.
980	WC/FD	10cm	3000	500	(1949)	Fighter direction plan display, replacing 294/5.
981	Ht-fndr	10cm	3000	500	(1949)	Fighter direction heightfinder, replacing 294/5.
990	WC	10cm	3000	—	(1944)	Low cover, to go with 960, 294/5.
992	TI	10cm	3000	—	1959	Target indication, replacing 293.

American sets fitted in British ships

SA	WA	1.5m	—	—	1943	Captain and Colony-class frigates.
SG	WS	10cm	≈3195	—	1943	Escort carriers, *Indomitable*, *Victorious*.
SJ	WS	10cm	—	—	1945	Submarines *Tiptoe* and *Trump*.
SK	WA	1.5m	—	—	1943	Escort carriers.
SL	WS	10cm	—	—	1943	'Captain' and 'Colony' class frigates.
SM-1	FD	10.7cm	≈2800	—	1944	Carriers *Indomitable*, *Ocean*; FD ships *Boxer*, *Palomares*.
SO	WS	10cm	—	—	1944	Coastal forces.
SQ	WS	12cm	—	—	1945	Big ships' portable 'after-action' set.

Naval Airborne Radar

ASV Mk.II	ASV	1.5m	176	—	1940	Standard RAF version.
ASV Mk.IIN	ASV	1.5m	214	7–22	1941	Naval version of ASV II. Swordfish, Walrus, Albacore, Barracuda
ASV Mk.XI	ASV	3cm	—	35–50	1943	Swordfish III, Barracuda III.
ASB	ASV	1.5m	214	—	1944	US copy of ASV IIN. Avenger.
ASH	ASV/AI	3cm	≈9375	35	1944	US AN/APS-4. Avenger, Firefly. Barracuda V.
AI Mk.IV	AI	1.5m	—	10	1944	Fulmar.
AIA	AI	3cm	—	—	1945	US AN/APS-6. Hellcat.

Principal source: Admiralty, CB 4497, *Simple Guide to Naval Radar* (1949).

---

[1] Depending upon frequency.

Appendix 2: Tabulations of Radar System Data, 1935–45

R. A. Laws

INTRODUCTION

The tables in this appendix have been compiled to present, in a detailed but compact form, technical descriptions of the principal radar systems developed for the Royal Navy from 1935 to 1945 and used during the war. Each table includes a short history of the development of the set or system and a broad indication of the ships fitted with it; its primary characteristics, from which its performance may be assessed; its associated IFF interrogators; followed by expanded technical details, its interface with its operators and the ship's armament, and its power supplies.

Notation

Variants of a set or system are indicated by suffixes as follows:

X – Experimental model
Y – Prototype model
Z – Production model
B – Single mast working (79/279 and 281 with common T/R)
M – First major modification
P – Second major modification
Q – Third major modification
R – With an accurate ranging panel
S – Adapted for use in shore-stations
U – Adapted for fitting in coastal craft
W – Adapted for fitting in submarines

The solidus (/) was not used in the designation of a set. It is used in these tables to mean 'and of the same set-number'; for example Type 281B/BQ means: Type 281B and Type 281BQ. Type 290/1 means Types 290 and 291.

Numerical Values

Antenna gains, unless otherwise stated, are one-way power gains on axis over an isotropic radiator, with a receiving antenna considered as a radiator. At the time of development, Admiralty scientists usually gave antenna gains in dBs over the peak gain of a half-wave dipole, but the later standard has been adopted in these tables for consistency with the radar equation. The error in the figures given is probably within ± 12 per cent (± 0.5 dB). The half-wave dipole may be taken as having a power gain of 1.63 (2.12 dB) over that of an isotropic radiator.

Unless otherwise stated, beam-widths are the full width of the one-way beam at the half-power points, laterally and vertically.

Ranges are expressed either in yards (yds) or in nautical miles (nm) of 2000 yds (see Glossary). For brevity, the abbreviations Tx and Rx are used to mean 'transmitter' and 'receiver' respectively.

Each gunnery radar had an index error, being a fixed amount by which the range of a target was overstated when measured from the start of the range scan. It was therefore necessary to calibrate each gunnery set against a point target at a known range. The index error varied between sets, even between sets of the same kind, but there are no statistical parameters available, only a few contemporary notes and recollections, which place the correction approximately in the region of 100–200-yds.

The receiver and displays were saturated during transmission, even though protective circuits were operational, and they were paralysed for a short time thereafter, showing a maximum deflection on the range displays, which was known to operators as the 'ground wave'. Echoes could not be received within the ground wave, which therefore determined the minimum range of the set. The only values of the ground wave available today are those given in CB 4497 of 14 January 1949: 'A Simple Guide to Naval Radar'. In default of anything better, these values are given in the tables as 'Minimum range'. The number of zeroes in the figures is an indication of the accuracy assumed by the authors of CB 4497.

Physical Realization

The designers of these radar sets were breaking new ground in electronics and radio, and as they made use of ever shorter wavelengths, so they had to develop the test gear to measure their sets' performance at RF. The values of RF performance quoted are the best estimates that could be made at the time of development but their accuracy was undoubtedly much lower than can be achieved half a century later. The figures should therefore be read with this fact in mind.

The construction of the electronic part of these sets would seem primitive to the electronic engineer of 1994. The transistor had yet to be invented. The radio valve (electron tube) was 'king' and its working voltages could give the unwary a nasty shock. Valves were plugged in for easy replacement. Integrated circuits and printed circuit boards lay in the future, components being mounted on vertical panels and their associated horizontal baseboards, with interconnections being made with tinned copper wire, soldered by hand. Currents were high, resistors and capacitors were large and unreliable by today's standards, and a lot of heat was generated, calling for supplies of cooling air.

A-scans used in the RN displayed echoes upwards, as opposed to the ground-based sets of the RAF, which displayed them downwards.

Power Supplies

All primary power was supplied by motor alternators driven by DC at 220 volts in large ships and 110 volts in small ships. These power units are identified in CB 3090 'Instructions for installation and fitting of RDF equipment and associated communications', 1943 with later amendments. CB 3090 also contains extensive tables of ships and the sets allotted to them, including the nomenclature of the many different Aerial Outfits that were fitted.

Tabulations of Radar System Data 361

Sources and Acknowledgements

Each set was described in detail in one or more handbooks, which may be identified through lists in CB 3090. These books may be consulted on application to the Captain, HMS *Collingwood*, Fareham, Hants, marked 'For the attention of the Curator of the Museum'.

These tables combine information from many sources, but they would not exist without the active help of Alec Cochrane, Harry Pout and Jack Shayler. Not only had these designers previously prepared some written descriptions of sets, but they have answered my questions with patience and care. They have supplied me unstintingly with helpful comments on my drafts over a long period of preparation and have cleared up many points of difficulty. I thank them for all their help. I also thank Derek Howse for his early drafts of the tables, Ken Armstrong and Bob Jenkinson for operational data about Types 79 and 281, Sid Wright for data on Type 268 and for general help with sources. There are gaps in the information presented, but some questions cannot be answered after a lapse of half a century. I apologise for any errors that might have occurred.

LIST OF TABLES

Metric Radars — *Page*

	Table A2.1	Types 79 and 79B (WA) Type 279 (WA and GS)	362
	Table A2.2	Types 281, 281B, 281BQ (WA)	364
	Table A2.3	Type 286; Types 290 and 291 (WC)	367

Decimetric Radars

(L-band)	Table A2.4	Types 282 and 283 (GC) Type 284 (GS); Type 285 (GA)	369

Centimetric Radars

(S-band)	Table A2.5	Types 271, 272, 273, X to S (WS)	372
	Table A2.6	Types 271Q and 273Q (WS)	374
	Table A2.7	Types 277, 277P (WS and WCH), 276 Types 293, 293M, 293P (WC)	376
	Table A2.8	Type 274 (GS)	378
	Table A2.9	Type 275 (GA)	380
(X-band)	Table A2.10	Type 262 (GC)	382
	Table A2.11	Type 268 (WS)	384

TABLE A2.1   RADAR TYPES 79 AND 79B (WA); 279 (WA AND GS & GA)

Chronology

Type 79 was the first Naval RDF set: 79X in *Saltburn* early 1938. First fittings: 79Y August and October 1938; 79Z (pre-production) August 1939; 79Z (first production model) April 1940; 279 September 1940; 79B April 1941.

Primary Characteristics

- Frequency: 39, 40, 41 or 42 MHz, according to model of antenna outfit fitted.
- Pulse length: 8 to 30 microsecond in warning (WA) mode, or 3 microseconds in ranging (GS) mode (279 only).
- Prf: 50 per second.
- Peak power: 79X and 79Y, 20 kW; 79Z et seq, 70 kW (all modes).
- Losses in feeders: very low, see below.
- Antenna gain: 8 (9.2dB).
- Polarisation: horizontal.
- Beamwidth: (lat) ca 84°.
- Receiver bandwidth: 140 kHz with long pulse, 1 MHz with short pulse (279 only).
- Noise factor of receiver: ca 5 dB, effectively increased to at least 10dB by galactic noise.
- Ground wave: 2–6 nm; 279 1 nm in ranging.
- Antennae heights above waterline: (ca.) 95–130-ft.

Interrogator

- Type 243 for IFF Mk III. IFF Mk II responded directly to Type 79, etc.

Technical Details

Transmitter

- Push–pull triode oscillator, using NT57 for 79X/Y and NT57T for 79Z et seq. Tuned Lecher-line anode, fixed-coil grid, no cathode RF circuit; 79X and 79Y were grid modulated; 79Z and after were cathode modulated. Supply at 20 kV for all modes.

Receiver

- Superhet. IF: 5MHz. Bandwidths: Receiver P11 140 kHz (for Type 79 and for Type 279 in WA mode); P12 1 MHz (for Type 279 in GS mode).
- Effective noise factor: ca 5 dB in lab.; on antenna, increased by cosmic and galactic noise to at least 10 dB. Hence a preamplifier would have been ineffective.

Antennae

Arrays Each 'Aerial Outfit ATD' was supplied in one of four versions, depending on the frequency allotted to the ship. It consisted of an array of two horizontal delta-fed half-wave dipoles, separated vertically by half a wavelength, each with a parasitic reflecting dipole one fifth of a wavelength behind it. The supporting structure was made of 'Permali', a form of resin-impregnated wood, and was carried by a vertical rotatable steel mast held in a pedestal at the top of the ship's mast. There was a small platform giving access to the pedestal. Types 79 and 279 had two such antennae, one at each masthead (for Tx and Rx); Type 79B had one antenna with a diode switch for common T/R. 279B was not made, since common T/R came after the 279's use for gunnery was discontinued. Their height above sea level varied between ships, for example 102-ft in *Liverpool*, 130-ft in *Glory*, each on 39 MHz.

Feeders 79 and 279: 400-ohm twin open-wire feeders to mastheads (separation 2.5-in) with a flexible open-wire section to the array, thus limiting rotation to about 10° beyond the stern from either side, 380° in all. 79B: coaxial Pyrotenax cable to masthead, followed by the flexible section, with common T/R.

Losses Open wire: less than 0.3 dB/100-ft; Pyrotenax: negligible.

Control of rotation Using Control Unit 19D, the operator provided the physical effort for turning the array(s) in azimuth via electro-mechanical (Selsyn) links. For lining up, the bearing of each array was presented to the operator by magslip repeater. The bearing of the set was given by a cursor over a relative bearing ring with a concentric gyro-repeater. The operator estimated the bearing of a target by means of an angle-divider, in effect using very slow beam-switching with a large angle of split. The target's true bearing was understated by half the change of its relative bearing during the estimating. The system gave good results (ca. $\pm 2°$, ship on a steady course). Targets astern presented difficulties owing to the limits on rotation across the stern.

Gunnery Applications

Type 279 was fitted with ranging unit RBL 10, which would transmit range of target to the TS and log range to the HACS within the scale 2000–14 000-yds. in steps of 50-yds. It was used with the short pulse of 3 microseconds. With the introduction of 282/4/5, 279 reverted to 79 and its variants.

Scales and Accuracy

- Warning: 120 nm ± 5 nm and 24 nm ± 1 nm
- Gunnery (279 only): scale between 2000 and 14 000-yds ± 25 yds.

Primary Power Supplies

230 V, 50 Hz.

Appendix 2

TABLE A2.2 RADAR TYPES 281, 281B, 281BQ (WA & WS)

Type 281 was the second major metric WA radar used in cruisers and above during the war. Other metric sets were used later (for example 960). It employed a shorter wavelength than that of Type 79 (qv) so the gap between its main and second vertical lobes was covered by 79's main lobe. The coverages were thus complementary when both sets were employed in company (or in an aircraft carrier).

Chronology

Into service: 281 in *Dido*, October 1940; 281B (common T/R) late 1941; 281BQ (continuous rotation) *Anson*, etc., early 1945.

Primary Characteristics

- Frequency: 86, 88, 90, 92, 94 MHz according to the Aerial Outfit ATE used.
- Pulse length: 281: 15 and (ranging) 1.7 microseconds; 281BQ: 5 or 1.7 microseconds.
- Prf: 50 per second; 281BQ: 50 or 100 per second.
- Peak power: 350 kW at 15 microsecond and 1000 at 1.7; 281BQ: 350 kW at 15 and 5 microseconds, 800 kW at 1.7 microseconds.
- Losses in feeders: negligible.
- Antenna gain 23 (13.5 dB).
- Polarisation: horizontal.
- Beamwidth: (lat) 27°.
- Beam-switching: (281 only).
- Combined beamwidth at half power: without beam-switching, ca 20°; with beam-switching, ca 34°.
- Sweep-rate in searching mode: 281BQ: 2 or (later) 4 rpm, according to prf.
- Receiver bandwidth: 150 kHz for long pulse; 1.5 MHz for short pulse.
- Noise factor (limited by galactic noise): 281 9.3 dB; others 8.3 dB.
- Antennae heights above waterline: ca 97–136-ft.
- Minimum range: 281/B 2–3 nm, ranging 1000–2000 yds; BQ 1 nm or 1000–2000 yds.

Interrogators

- Type 241 for IFF Mk II on 214 MHz; Type 243 for IFF Mk III on 179 MHz (each used a modified ASV MkII Tx and Rx).

Technical Details

Transmitter

- Two NT86 in push–pull; tuned Lecher-line grid, tuned Lecher-line cathode, cathode-modulated.
- EHT: 15 kV for 15 microsecond, 20 kV for 5 microseconds, 27 KV for 1.7 microseconds.

Receiver

- Superheterodyne IF 12 MHz.
- Noise factor: in lab 9.3 dB, reduced to 4.3 dB with pre-amplifier. On antenna, galactic noise increased these figures to 9.8 and 6.8 dB respectively; with the diode switch (common T/R), pre-amp and galactic noise the noise factor was about 8.3 dB.
- Bandwidth: see primary characteristics.

Antennae

- Aerial Outfit ATE was supplied in one of five versions, depending on the frequency allotted to the ship. It consisted essentially of two Type 79 arrays (qv) scaled down for wavelength and mounted side by side at a spacing, centre to centre, of 0.75 of a wavelength. The supporting structure was of Permali as for Type 79 (qv) and was mounted on a rotating mast similar to that used for Type 79.
- 281 had two such arrays, one on each mast (for T and R). 281B had one array with common T/R, 281BQ also had continuous rotation.
- Height above waterline and frequency: *Sussex*, 116-ft, 94 MHz; *Duke of York*, 136-ft, 94 MHz; *Belfast*, 91.5-ft, 90 MHz; *Glasgow*, 97-ft, 86 MHz; *Norfolk*, 114-ft, 86 MHz; *Anson*, 131-ft; *Colossus*, 90-ft; HMCS *Ontario*, 97-ft, 88 MHz.
- Peak power gain: see above.
- Beam-switching (when selected) on the receiving array with crossover at -3 dB (not 281B or BQ).

Colossus reported that 281 induced sparking at the locker lids on the flag deck, and might therefore put aircraft at risk. The height of the array was therefore raised by 15-ft from 35-ft to 50-ft above the flight deck (90-ft above the waterline) for *Colossus* and its class.

Feeders

- 281: twin open-wire feeder to masthead with flexible twin-feeder to array, thus limiting rotation to about 400°.
- 281B: coaxial cable to masthead with flexible section as in 281.
- 281BQ: as 281B but with slip-rings at RF to allow continuous rotation.
- Losses in feeders: open wire, 0.5 dB/100-ft; coax and slip-rings 0.25 dB/100-ft + 0.2 dB.

Control of Rotation

- The array(s) were rotated by Selsyn motors in gearboxes as for Types 79 and 279 (qv) but Type 281 employed control unit 20D by which the Selsyn transmitters were turned automatically by an electric motor, giving a fixed angular rate of rotation of the arrays of 2 or 4 rpm with automatic reversal after each complete circle. For lining up the Selsyn system, the bearing of each array was displayed to the operator by magslip. The bearing of the set was presented to the operator by a cursor over a relative bearing ring and a concentric gyro-repeater. When examining a selected target, the operator had to turn the system with one hand alone, using a small cranked handwheel near his knee (see

Monograph 4). *Colossus* reported that the noise of the 20D hindered communications. There was an improved unit for 281BQ, with continuous rotation at 2 or 4 rpm for use with PPIs.

Displays and Accuracy

- Warning: A-scans 0 to 120 nm ±5 nm, 0 to 20 nm ±1 nm; 281B/Q: PPIs and Skiatrons. *Colossus* reported that near the equator, when she was making large alterations of course to operate her aircraft, the changes in relative bearing of the earth's magnetic field could move the start of the time-base ca 3/8-in off centre. Ship's staff installed mechanisms to permit rapid correction by the operators. The Skiatron was unaffected.
- Gunnery: 281 only, with short pulse: range transmission to TS and log range to HACS from ranging unit RBL 11, 4000–28 000-yds in steps of 100-yds, 2000–14 000 in steps of 50-yds. Type 281's use in fire-control was short-lived as the 50-cm gunnery sets 282–5 quickly took over the task.
- Bearing: ±2°

Power supplies

230 V, 50 Hz.

Tabulations of Radar System Data 367

TABLE A2.3 RADAR TYPES 286 AND 290/291 (WC)

Fitted in destroyers and below:

Chronology

- 286: ASV Mk I with fixed antenna, ca 200 from mid-1940.
- 286M: ASV Mk II with improved fixed antenna, many hundreds from January 1941.
- 286P: ASV Mk II with rotating antenna and common T/R, from February 1941.
- 286PQ: as 286P but with higher power, late 1941.
- 286PW: as 286P, for fitting in submarines.
- 286PU: as 286P, for fitting in coastal-forces craft and trawlers, with a forward cover of $\pm 70°$.
- 290: fitted from mid-1941, 50 kW transmitter.
- 291: with re-engineered receiver and 100 kW transmitter, fitted from late 1942.

Primary Characteristics

- Frequency: 214 MHz.
- Pulse length: 286, 1.5 microseconds; 286M/P, 2.5 microseconds.
- Prf: 286, 1200 per second; 286M and P, 400 per second, all free-running; 286PQ, 290 and 291, 500 synchronous with power supply.
- Peak power: 286/M/P, 7 kW; 290, 50 kW, 286PQ and 291, 100 kW.
- Losses in pyrotenax feeders: negligible.
- Antenna gain: 286P/PQ, 23 (13.5 dB); 286/M, not known.
- Polarisation: horizontal.
- Beamwidths: rotating array: 27° (as 281).
- Receiver noise factor: not known, but estimated at ca 9 dB.
- Minimum range: M and P, 800–1500-yds; PQ, 800-yds.

Interrogator

- IFF Mk 2N responded directly; IFF Mk 3 on 179 MHz responded to interrogator Type 242.

Technical Details

Transmitter

- 286/M/P: push–pull squegger, HT 8 kV, 7 kW.
- 286PQ: anode modulated 12 kV, 286PQ probably used the modulator and transmitter of 291. Larger ships only.
- 290, 50 kW achieved.
- 291 with its new delay-line and pulse transformer and indirectly heated valves achieved 100 kW.

Receiver

- IF: 286/M/P, believed to have been 45 MHz; 290/1, 30 MHz.
- 286P, protected by a T/R switch with gas tube.

- Re-engineered for 291 with grounded-grid RF amp., diode mixer, four-stage IF 30 MHz, diode 2nd detector.
- Bandwidth: 290/1, 2 MHz.

Antennae

- 286/M employed Aerial Outfit ATQ: three separate fixed arrays of two driven and two parasitic dipoles (scaled down from Type 79 qv). The central array transmitted, the others looked slightly to port or to starboard and were used alternately, pulse to pulse, which with the display (see below) gave beam-switching. The ship could therefore be steered towards a target. The effective angle of detection lay between red and green 50°, through the bow.
- 286P/PQ employed (as did Type 290/1) a scaled-down version of the 281 array, mounted on an inverted Asdic (Sonar) pedestal and rotated in azimuth by the operator via Bowden cable, though without beam-switching but with common T/R. The weight of the pedestal confined fitting to the larger ships. The array was redesigned using steel-sheathed Pyrotenax, which was also used as the supporting structure, and a new pedestal was designed in aluminium to reduce top-weight, permitting its fitting in smaller ships, still manually rotated by Bowden cable. This redesigned array was also used by Type 291.

Displays

- 286/M: vertical A-scan upwards with port and starboard echoes deflected left and right for the fixed arrays.
- 286/P/PQ and 291: horizontal A-scan. PPIs were employed with 291.
- Scales: 0 to 9, to 18, and to 36 nm, accuracy ± 5 per cent of full scale.
- Accuracy of bearing: 286/P/PQ, 3° to 5°.

Primary power supplies

- 286/M/P/Q,1 80 V, 2000 Hz.
- 286PQ, 180 V, 2000 Hz for Rx and display, and 180 V, 500 Hz for modulator and Tx.
- 290/91, 180 V, 500 Hz.

Tabulations of Radar System Data

TABLE A2.4   RADAR TYPES 282 AND 283 (GC), 284 (GS), 285 (GA)

These were the first Naval gunnery radar sets. They were on L-band (ca 50 cm). The transmitter and receiver were common to the series, as were some of the displays.

Their original function was rangefinding, holding in bearing by beam-switching was added later. These sets went through many variants, the set number denoted its main purpose and the type of antenna used, the suffixes denoted changes and improvements.

General Description

- 284: ranging for main armament with 'pig-trough' array(s) fixed at zero elevation on DCT, in battleships and cruisers. Fitted in *King George V* in 1940.
- 285 HA/LA: ranging for secondary armament in large ships and main armament in small ships. Five or six Yagis ('fishbones') on the director, elevating with the sight. Fitted from 1940.
- 282: ranging for close-range guns in frigates and above. Two Yagis on the director, elevating with the sight. Fitted from 1941.
- 283: anti-aircraft barrage control of main and secondary armament. Two Yagi arrays on barrage director, elevating with the sight. Fitted from 1942.

Chronology

- 1938: work started at Signal School in February.
- 1939: sea trials in *Sardonyx* in June and October.
- 1939: Staff Requirement for ranging against dive-bombers to control main armament barrage.
- 1940: trials at Eastney of two-Yagi array on pom-pom director. 200 sets ordered in April.
- 1940: May – design of arrays started for GS and GA sets using same Tx and Rx, etc.; June – trials of 284X (GS) in *Nelson*; orders increased to 900 L-band sets; December – 284 in *King George V*, 285 in *Southdown*.
- 1941: July – Staff Requirement for longer ranges and more accurate bearing.
- 1942: August – modifications Types 282/4/5P promulgated to Fleet.
- 1942: first fitting of 283; trials in *Charybdis*, four sets to *Berwick*.
- 1943: April – modifications Types 282/4/5P promulgated to the Fleet.

Common Primary Characteristics

- Frequency: first model 600 MHz, thereafter ca 580 to 610 MHz. Tunings were meant to be staggered to mitigate the effects of interference between ships and of jamming.
- Pulse length: 2 microseconds; suffixes M/P, 1 microsecond.
- Pulse shape: triangular, with very rapid rise.
- Prf: 500 per second synchronised to AC supply (500 Hz) from six-phase alternator or from two such in tandem.
- Peak power: 25 kW, M/P by stages to 150.
- Feeder losses: 3 dB (est).
- Antenna gains: see below.
- Polarisation: horizontal.
- Receiver bandwidth: 4 MHz.
- Noise factor: 282–5, 25 dB; M and P, with grounded-grid triode pre-amp, 15 dB.
- Minimum range: 284/5, 700–1000-yds; M and 282, 600-yds.

Technical Details

Transmitter

- Anode modulation: thyratron-driven modulator, supplying 200 kW for 2 microseconds, 400 kW for 1 microsecond.
- Pair of balanced triodes in resonant cylinder, coupled via dipole to coaxial feeder. The RF power was 25 kW, 60 kW, 80 kW or 150 kW according to the particular variant of set.

Antennae

With horizontal polarisation from half-wave centre-fed dipoles in assemblies of the following units:
- Yagi(8) with semicircular cylindrical reflector and eight directors, or
- Yagi(7), similar but in lightweight form with seven directors. Yagi(7)s were always assembled in pairs with a common semi-cylindrical reflector.
- Collinear array of dipoles in parabolic reflector 'pig-trough' 21-ft × 2.5–ft.
- M and P introduced Common T/R by spark gap in all sets. M3 and P3 introduced lateral beam-switching in 284 and 285.

Use of antennae

For close-range weapons, for example pom-poms:

- 282 and 283: array of two Yagi(8)s, one for Tx – gain 24, one for Rec; M/P: two for each – gain 48.

With large HA directors:

- 285: array of six Yagi(8)s, three for Tx – gain 80, three for rec.
- 285M/P: six Yagi(8)s, gain 160.

With smaller rangefinder directors:

- 285: array of five Yagi(8)s, three for Tx – gain 80, and two for Rx with gain of 48.
- 285M/P: three pairs of Yagi(7)s, gain 160. For HA gunnery in smaller ships, often with the fuze-keeping clock.

With director control tower (DCT) for main low-angle armament, cruisers and above:

- 284: two 'pig-troughs', one for Tx – gain 195, and one for Rx; or one for common T/R.

Losses:

- Feeders alone: 3 dB each way (est.)
- With beam-switching: plus 1.5 dB each way.

Receiver

- IF: 6 MHz. Standard IF amplifier using Mullard EF50 pentodes (VR91). M and P and 283 with preamplifier with grounded-grid triode for reduced noise factor (see above).
- Diode mixer.

Displays

A-scan on 12-in NC7 CRT in panel L12:
- 282: 200-yd pips, 6000-yd scale.
- 284: 1000-yd pips, 24 000-yd scale.
- 285: 1000-yd pips, 15 000-yd scale.

The calibration pips were generated by an oscillator, which had to be set up by the ship's radar officer against a portable crystal-controlled standard. Later variants employed accurate ranging panels L22, L24, or L34 for use with fire-control system, including ABU and CRP Mk 2.

Connection to fire-control system

Each array trained with its director. All but those of Type 284 elevated with the sights. The vertical polar-diagram of 284 was adequate to keep surface targets in its beam at all realistic angles of roll. Under M/P, the director trainer had a supplementary display of the error in bearing as given by the beam-switching, and he could train the director to minimise that error.

Using the Range Transmission Unit (RTU) with panel L12, the range operator of 285 kept his cursor aligned with the leading edge of the target's echo, using a rate-aided handwheel. While he was depressing his foot-pedal (cut-push), the setting of his cursor, as log range, was transmitted automatically by M-type motor to the range plot of the HACS. Future range (for fuze-setting) was generated there by the alignment of a cursor along the range plot.

For the 282 series the operator worked similarly, his output going to the pom-pom director Mk 2 or the Bofors director; 282 provided range only, 282M/P provided range, bearing control by beam-switching and a degree of elevation control.

With 284 the ranges were transmitted by telephone to the TS. Type 284M/P transmitted the range automatically to the TS by panel L24 when the operator pressed his foot pedal (cut-push). Type 285M/P employed the L24 in a similar way.

Panel L24 generated accurate range by means of a crystal-controlled calibration marker on a separate trace, with both traces expanded in the region of the marker for greater accuracy. Transmission of range to the HACS was as described above for the RTU. Panel L22 performed a similar function for Type 282 and 283.

Variants of 282/3/5 were also fitted with the Auto Barrage Unit (ABU) whereby the shells (4-in and above) were fuzed for a future range of 2000-yds and the guns were fired automatically. Greater ranges could be set into the ABU for the main armament of cruisers and battleships.

285Q worked with the Close-Range Predictor Mk 2 (CRP 2). 283 worked with the Continuous Prediction Unit Mk 5 (CPU 5, part of the HACS).

Primary power supplies

- 180 V, 500 Hz, from six-phase or twelve-phase motor alternators according to the number of sets in the ship.

TABLE A2.5 RADAR TYPES 271, 272, 273 TO VARIANT S (WS)

Chronology

The first centimetric sets; designed for surface warning.
- 271X: experimental design 1941.
- 271: early change to a more powerful local oscillator permitted use of a remote antenna for 272 and a larger antenna for 273. Production by Allen West, Brighton; hundreds fitted.
- 271P/2P/3P: mechanical design of panels changed to facilitate line production of a further 1000 sets. Also used in prefabricated offices from 1942. 1000 made.
- Suffix PR (M before 1943): with range transmission to the fire-control system.

About 1300 sets of all variants were made. Several hundred were used ashore by all three Services as part of the coastal defence chain (271S and 273S) but with higher-gain antennae and different displays. See Table A2.6 for later and major variant: Q.

Primary Characteristics

- Frequency: 2940–3060 MHz.
- Pulse length: 1.5 microseconds.
- Prf: 500 per second.
- Nominal peak power: X to P, 5 kW.
- Antenna gains: see below.
- Polarisation: horizontal.
- Receiver bandwidth: 2.5 MHz possibly later reduced to 2.0 MHz.
- Receiver noise factor: ca 15 dB.
- Minimum range: 500-yds.

Technical Details

Transmitter

- Unstrapped resonant-cavity magnetron (NT98) in a magnetic field of 1080 Oersted on the back of the antenna.
- Modulator: hard-valve, grid-controlled series-modulator (NT100), passing 8 amp with 8 kV developed across the magnetron.

Antennae

- Polarisation: horizontal.
- 271/P and 272/P: two 10-inch sections of parabolic cylinder of focal length 18-in, cut from 6-ft to 48-in wide, with the ends of the cylinders closed by flat metal plates. These two 'cheeses' were stacked one above the other with their axes vertical, the Tx 'cheese' uppermost. Each was fed by or fed a dipole backed by a reflecting dipole, gains: 90 (20 dB).
- 273/P: two paraboloids, 'dishes', of 3-ft diameter mounted side by side, each was fed by or fed a dipole backed by a reflecting dipole, gains: 410 (26 dB).
- Combined beamwidths (-6 dB): cheeses, lateral 6.5°, vertical 65°; dishes, lateral; 10°, vertical 7.5°.

Feeders

- Losses in the LO feeder cables: 0.22 dB per foot; hence the short feeders to antennae in 271/3 and the need for a more powerful LO for the 40-ft run to the antenna of 272.

Receiver

- Silicon/tungsten crystal-mixer on the back of the antenna. Local oscillator in office. 271X: NR89 klystron, input 10 W, output 50–100 mW. 271/P: CV35 klystron (a development of the NR89), 300–500 mW.
- IF: 60 MHz.

Displays

- A-scan, scales in yards: 5000 and 25 000, with an optional modification to 25 000 and 75 000 (instructions in handbook).
- Fire control. Variant R: the operator had a pedal ('cut-push') and either an RTU transmitting range to the bridge and plots, or panel L17 on which he aligned a strobe on the echo in an enlarged portion of scan and thus transmitted range to the fire-control system.

Housing

In 271 and 273, the set and operator were immediately below the antennae, which were in a 'lantern' with 5/16-in plane perspex windows set in a teak frame (3-in × 2-in) (271, eight-sided; 273, sixteen–sided). This discontinuous construction gave trouble with side-echoes and frequency pulling; it was replaced by a perspex cylinder from November 1942. The operator trained the antenna by turning a crank. In 272 the antennae plus magnetron and local oscillator were up to 40-ft away from the rest of the set; the operator turned the array by a crank connected by Bowden cable to the antenna mounting. In all these sets the rotation was limited to ca $\pm 200°$.

Primary power supplies

180 V, 500 Hz.

Appendix 2

TABLE A2.6   RADAR TYPES 271Q AND 273Q (WS)

Chronology

These sets were a complete redesign of 271P and 273P (Table A2.5) to accommodate the first strapped-cavity magnetron. Mechanical design was aligned with 271P/273P to facilitate exchange. They were direct replacements for 271/P and 273/P.

Suffix R (M before 1943): with range-transmission to the fire-control system. Some 20 prototypes in service in 1942. Main production and fitting in 1943. 1000 were made; several hundred went to upgrade the coastal chain.

Primary Characteristics

As for 271P/3P except for:

- Pulse length: 2 microseconds, or 0.7 microseconds for discrimination.
- Peak power: 70 kW nominal.
- Antenna gains: see below.
- Receiver bandwidth: 1.25 MHz and 4 MHz.

Technical Details

Transmitter

- Strapped-cavity magnetron (CV56) in magnetic field of 1500 Oersted; power 70 kW at input of 15 A, 15 kV.
- Modulator: discharge line switched by CV22 thyratron.

Antennae

- Polarisation: horizontal.
- 271Q: two 'cheeses' as for 271/P, Tx 'cheese' fed by waveguide and horn; gain, 300. Receiver 'cheese' as 271/P, gain 90.
- 273Q: two paraboloidal dishes as for 273/P. Tx dish fed by waveguide and horn, gain 590. Receiver dish as for 273, gain 410.
- full beamwidths (two way, -6dB).
  271Q: lateral 5.6°; vertical (estimated), 45°
  273Q: lateral 7.6° vertical; 8.8°.
- Stabilisation in elevation was introduced from November 1942, in *Belfast*.

Receiver

- as for 271 and 271P (Table 5). Protection of crystal by a gas-switch during transmission.

Displays

- A-scan, scales: 5000 and 25 000-yds, or optional modification to 15 000 and 75 000-yds (instructions in handbook).
- M or R: with separate display panel L17, on which an operator aligned a strobe on the echo in an enlarged portion of scan and thus transmitted range to the fire-control system and the Asdic (Sonar).
- Remote PPIs fitted from 1943.

Housing

The housing was prefabricated. It enclosed the antennae in sixteen-sided 'lantern' with a frame of 3-in × 2-in teak and with 5/16-in plane perspex windows. The office was immediately below and the operator rotated the antennae by turning a crank. The lantern gave trouble with side-echoes and was replaced by a perspex cylinder from November 1942. Only one 271Q went out in the old lantern but the cylinder for 273Q was not available before mid-1943.

Primary power supplies

180 V, 500 Hz.

TABLE A2.7 RADAR TYPES 277, 277P (1945) (WS AND WCH), FOR HEAVY CRUISERS AND ABOVE; RADAR TYPES 293, 293M, 293P (WC), FOR FLEET DESTROYERS AND ABOVE; RADAR TYPE 276 (WC), FOR SMALL SHIPS

General Note

This second generation of 3000 MHz search radars was designed around the use of the 500 kW magnetron and the separation of antenna and transmitter by use of waveguides. The sets employed continuously rotating antennae and associated PPI displays. They were in development from mid-1942, with 277 to replace 273 and 271, and 276 with a much lighter antenna was intended to replace 272. Each set was to provide PPI displays of surface and low-flying targets. The antennae were mounted on the mast 60–130-ft above the surface; the added height and ease of siting outweighed the small losses in the longer waveguides.

By early 1943, 293 with its wider vertical air cover (ca 70°) was seen as displacing 276 out to 15 miles. With a special target-indicating display (TIU), weapons would be allocated to targets and in a small ship 293 would provide both WA and WC cover. The original 293 antenna proved to be inadequate and 276 was reinstated as a temporary substitute. By the end of 1944 the larger antenna of 293M began to replace each of them.

All these sets used the same modulator, transmitter and receiver, designed for the mobile shore radar 277T. They were eventually redesigned to ease maintenance at sea and to incorporate other improvements. Suffix P denoted the use of the redesigned versions.

Common Primary Characteristics

- Frequency: 2940–3060 MHz.
- Pulse length: 0.7 or 1.8 microseconds.
- Prf: 500/second.
- Peak power: 500 kW nominal.
- Feeder: up to 100-ft of waveguide, losses small, see below.
- Polarisation: horizontal (except 277Q, vertical).
- Common T/R.
- Beam shapes: 277, narrow circular; 276/293, vertical fan.
- Receiver bandwidths: 1.25 MHz or 4.0 MHz; P, 0.5 MHz, 1.25 MHz or 4.0 MHz.
- Receiver noise factor: 15 dB; P, 12 dB.
- Minimum range: 150–500 yds.

Interrogator

242M or P.

Common Technical Details

Transmitter

- CV76 magnetron in magnetic field of 2300 Oersted, taking 40 A at 25 kV.
- Modulator: discharge line switched by a CV12 gas triode at 13 kV, 160 A.
- Feeder (waveguide) losses: 0.025 dB/metre with high-grade copper, and an SWR of 0.6 to 0.7.

Tabulations of Radar System Data

Receiver

- Crystal mixer, P: broadband (Dahl) mixer.
- IF: 60 MHz; P, 13.5 MHz.

Differing Features

Antennae

For 277/P Aerial Outfit AUK (the Great Auk), mounted high on foremast, with expanded metal paraboloid 54-in diameter, waveguide sealed at horn feed. Stabilized in elevation. Elevating to 40°, 277P to 70°. Continuous rotation in either direction in azimuth at up to 16 rpm.

- One-way gain: 277/P, 1300 (31 dB).
- Two-way beamwidth: circular 6.2° diameter at −6 dB.

For 277Q In 1947 Aerial Outfit ANU was brought into service, giving 277Q. Polarisation was vertical. Its 96-in paraboloid was cut to 72-in wide in azimuth.

- Gain: 3400 (35 dB). Beamwidths: lateral, 3.5°; vertical 2.5°.

For 276 Type 271Q Tx 'cheese' of aperture 48-in × 10-in, unstabilised, sealed across the front by plastic sheet and horn- fed.

- Gain: 300 (25 dB)
- Beamwidths (two-way, −6 dB): lateral 4.6°; vertical, 27°.

For 293 Aerial Outfit AUR: horn-fed 'cheese', usually mounted on top of a mast to give all-round cover. No stabilisation against ship's roll and pitch, but wide vertical beam with antenna tilted up 15° maintained cover.

- 6ft wide × 4-in; gain, 160 (22 dB). Beamwidths: lateral 3.5°; vertical, 65° to 70°.
- M and P: 8-ft × 7.5-in; gain 360. Beamwidths (two-way, −6 dB): lateral 2.6°; vertical, 30°.

Azimuthal Scans

Continuous rotation in rpm: 10; M and P: 7.5 or 15.

Displays

- PPI: range scales: 15 000 ±50-yds, 30 000 ±100-yds, 75 000 ±250-yds, 150 000 ±500-yds. Discrimination in bearing ±0.5°.
- HPI: PPI-type polar plot in the vertical plane for use with 277, horizontal sweep stopped and set bracketing in elevation; cursor for direct reading of height.
- 293 was used in cruisers and above for target indication (TIU Mk 2) to 275 and 262.

Primary Power Supplies

250 V, 50 Hz three-phase; 150 V, 500 Hz single phase.

378 *Appendix 2*

TABLE A2.8   RADAR TYPE 274 (GS)

Chronology

- 1942: main features established.
- Late 1943: first fitting, in *Swiftsure*.
- 1944 and 1945: fitted in several large ships.
- 1945: *Norfolk* sank a German convoy at 11 nm with 8-in blind-fire controlled by 274.

Primary Characteristics

- Frequency: 3230–3380 MHz.
- Pulse length: 0.5 microsecond, rectangular pulse.
- Prf: 500 per second.
- Peak power: 500 kW.
- Feeder losses: 1.5 dB each way.
- Antenna gains: 1350 peak (31 dB); Rx, 535 (27 dB) at crossover of beam-switching.
- Combined beamwidth: lateral 1.2°; vertical, 12.5°.
- Polarisation: horizontal.
- Receiver bandwidth: 4 MHz.
- Noise-factor: 13 dB.
- Minimum range: 200-yds.

Interrogator

Type 243.

Technical Details

Transmitter

- Strapped-anode magnetron sited behind the Tx antenna, 12 kV modulator, peak power 500 kW.

Antennae

- Horizontal parabolic 'cheeses', horn-fed with horizontal polarization, 14-ft wide × 14.5-in high, one for Tx, one for Rx, mounted on the main armament director (DCT), each training and elevating with the sights.
- Lateral beam-switching on Rx beam only, ±1° at 25 Hz giving ±0.5° combined. Combined gain at peaks: 1350 × 865, at crossover, 1350 × 535.
- One-way gain: see Primary Characteristics.

Receiver

- Crystal mixer; RF head and frequency changer with AFC sited behind the Rx antenna. IF, 60 MHz; bandwidth, 4 MHz; noise factor, 13 dB; AGC on target echo.

Displays

Displays were sited in the TS (Transmitting Station), which housed the fire-control table for the main armament. Panel L31(RS) transmitted target range in a scale up to 50 000-yds. Panel L31(B) controlled the bearing of the DCT. Fall-of-shot errors were measured. The director would be put on target by the TIU. There were two operators, tracking in range and bearing each with his own display. Maximum range 40 000-yds.

Calibration

Index errors in range were removed and radar and optical bearings were aligned by calibration against radar reflectors such as those on the Barrel of Butter in Scapa Flow (see Monograph 7, Figure 7.1).

Primary power supplies

Believed to have been 180 V, 500 Hz.

TABLE A2.9 RADAR TYPE 275 (GA)

Chronology

- April 1942: preliminary work began.
- Development model at Eastney.
- March 1944: first and second development models.
- March 1945: second development model in *Barfleur*.
- 9 July 1945: *Anson* 'crosses the line' en route for the British Pacific Fleet with four preproduction models of Type 275 on four Mk 6 directors.
- September 1945 onwards: fitted in *Camperdown* and other ships

Primary Characteristics

- Frequency: 3448–3700 MHz.
- Pulse length: 0.5 microsecond, rectangular.
- Prf: 500 per second.
- Peak power: 400 kW.
- Feeder losses: 1.5 dB each.
- Antenna gains: ca 900 (30 dB).
- Polarisation: horizontal.
- Conical scan: at 25 Hz on receiving dish only.
- Receiver bandwidth: 4 MHz.
- Noise-factor: 13 dB.
- Minimum range: 200-yds.

Interrogator

Type 243

Technical Details

Transmitter

- Strapped-anode magnetron, 12 kV modulator, sited behind the Tx antenna.

Antennae

- Antennae were sited on either side of the new Mk 6 director, elevating and training with it. They used reflecting paraboloids of 4-ft diameter behind perspex domes, horn-fed with horizontal polarisation. One was for Rx with nutating horn to give a conical scan at 25 Hz, one was for Tx with centralised beam. Combined scan offset: ca 1.75°.
- Loss on axis due to scanning: ca 2 dB.
- Beamwidths: Rx, ca 5°. Tx, ca 4° lateral, ca 6.5° vertical.

Receiver

- Crystal mixer; RF head and frequency changer with AFC sited behind the Rx paraboloid. IF: 60 MHz. AGC on target echo.

Displays

Displays were in the HACP and the Mk 6 director was laid and trained from there for blind-firing. The director would be put on target by the TIU. There were three operators, tracking in range, bearing and elevation, each with his own display. Maximum range 40 000 yds, maximum range rate 300 knots (500-ft/s).

Calibration

Index errors in range were removed by calibration against radar reflectors such as those on the Barrel of Butter in Scapa Flow (see Monograph 7, Figure 7.1). Radar and optical directions were aligned using meteorological balloons or helicopters.

Primary power supplies

Believed to have been 180 V, 500 Hz.

TABLE A2.10   RADAR TYPE 262 (GC)

Chronology

- With CRBFD: first fittings 1944.
- With STAAG: first fully realistic firing trials 4 July 1945; first fitting August 1945.

Primary Characteristics

- Polarisation: vertical (to reduce forward reflection from sea surface).
- Peak power: 30 kW.
- Losses: feeder, 0.4 dB; conical scan, 1.1 dB each way.
- Pulse length: 0.5 microseconds, square.
- Prf: 1500/sec.
- Receiver bandwidth: 4 MHz.
- Antenna gain: 2250 (33.5 dB).
- Receiver noise factor: 18 dB.
- Ground wave: 500-yds.

Technical Details

Transmitter

- Modulator power: 30 kW from spark-discharged delay line.
- Magnetron: BM 313.

Antenna

- 2-ft diameter balanced paraboloid, offset 1.3° from line of sight and spinning about it at 30 Hz to give conical scan, driven by a governed DC motor.
- Beam width: 3.8°.
- Feed: stationary horn.

Receiver

- IF: 30 MHz.

Displays

- A-scan: 0–7500-yds. Angles displayed by a CRT simulating an open sight.
- Rate-aided tracking and automatic following.

Fire Control

STAAG

- 262 was mounted directly on the stabilised two-axis twin Bofors-gun mounting STAAG. It was put onto bearing and range of target by radar Type 293, then it searched automatically for the target and locked on. Search: ±750-yds about the indicated range at 30 sweeps/sec, ±12° about the indicated bearing at 0.5 sweeps/sec while the gun-mounting was increasing its elevation at 2°/s from 0° to 60°. Lock-on was followed by automatic tracking in range and direction, the computations being made in the fly-plane.

- Scan-modulated output at 30 Hz was fed to resolvers on the motor shaft to generate the two error signals.
- The mounting was stabilised by a single gyro, precessed by springs from the operator's joy-stick, or by the radar error signals. The mounting followed the gyro via magslips and hydraulic servos. The manual phase was rate-aided to give a smooth transfer to automatic control.

CRBFD

- an independent barrage-fire director carrying 262 and controlling the barrage fire of separate guns. It was mechanically simpler than STAAG.

Primary Power Supplies

Believed to have been 180 V, 500 Hz.

Appendix 2

TABLE A2.11   RADAR TYPE 268 (WS)

Fitted in coastal forces, Hunt-class destroyers, minesweepers, and as an interim navigational set in some cruisers.

Chronology

- Type 268 was a Canadian development, which came into production early in 1945. In the RN it was fitted in coastal-forces craft, including minesweepers and Hunt-class destroyers, and as an interim measure in some cruisers. (Several hundred equipments were also allocated to the Merchant Navy as an interim precise radar navigational aid, postwar.)
- Its design and production followed a request from the British Admiralty to the Canadian Government in May 1942 for the development of a high-precision warning radar for use in coastal craft. Beginning in January 1945, production reached 300 equipments per month by the following September. A total of about 1500 sets were produced by the end of the war.

Primary Characteristics

- Frequency: 9290–9460 MHz.
- Peak power: 40 kW.
- Feeder losses: estimated at 0.5 dB.
- Pulse length: 0.75 microseconds.
- Prf: 500/second.
- Antenna gain: estimated 1370 (31 dB).
- Polarisation: horizontal.
- Beamwidth: lateral 2.0°; vertical, 14°.
- Antenna rotation: 22 rpm.
- Receiver bandwidth: 8 MHz.
- Receiver noise factor: no details available.
- Minimum range: 200-yds.

Technical Details

Transmitter

- Details as above.

Antenna

- Parabolic 'cheese' at masthead, 30-in. wide × 6.25-in high, fed by waveguide and horn, the focal line lying outside the aperture, rotating about vertical axis at 22 rpm.
- Gain: 31 dB.
- Beamwidth: 2.0° lateral, 14° vertical.

Receiver

- IF: 31 MHz.

Displays

- 5-in PPI, 7500 yds ±50, 30 000 yd ±500 and 60 000 yds ±1000.
- One remote PPI in chart room.
- A-scan for monitoring.

Primary Power Supplies

- No details available.

Appendix 3: Ship-Fitting Tables, 1938-45
Derek Howse

Surprisingly, one of the more difficult aspects of this project has been finding out what radar was fitted in what ships and when, as ASE's ship-fitting records seem not to have survived. Radar sets fitted in HM Ships were listed in the following documents:

25 Sept. 1939, Somerville to Controller (ADM 205/2)
25 Sept. 1940, CSS to DSD (ADM 220/72)
6 Mar. 1941, CAFO 509/41 (ADM 182 series)
19 June 1941, CAFO 1250/41
1 Oct. 1941, Admiralty letter SDO.2053/41 (ADM 220/79)
27 Nov. 1941, CAFO 2295(F)/41
19 Mar. 1942, CAFO 503(F)/41
Aug. 1943, CB 3090 (NHB and CAC NRT, not in PRO)
June 1946, ASE Gunnery Department Turnover Notes (ADM 220/221)

From February 1941 to June 1944 the weekly Admiralty Pink Lists (ADM 187 series, which gave the location of all major warships week by week) listed the sets fitted in each ship, but they became hopelessly out of date from 1943 onwards. After 1943 – and particularly for the British Pacific Fleet – no systematic records have been found and we have had to rely upon secondary sources such as Raven & Roberts, *Battleships* (1976) and *Cruisers* (1980), Whitley, *Destroyers* (1988), Elliott, *Allied Escort Ships* (1977), dated documents and photographs, and personal recollections. The tables below should therefore be used with caution: if a set is shown as fitted, that is probably correct, but a blank does not necessarily mean the set was never fitted.

For large ships, Table A3.1 gives details of radar fitted in each individual ship, the dates being those on which the ship became operational after fitting. For destroyers and below, sheer numbers make it impracticable to publish similar data for individual ships, so Table A3.2 lists only the typical radar fitting in each *class* of ship.

ABBREVIATIONS USED IN TABLES

AA or A/A	Anti-aircraft	GS	Gunnery, Surface: low angle
AMC	Armed Merchant Cruiser	l	launched
A/S	Anti-submarine	LST	Landing Ship Tank
Aust	Royal Australian Navy	Neth	Royal Netherlands Navy
c	circa	NZ	Royal New Zealand Navy
Can	Royal Canadian Navy	OBV	Ocean Boarding Vessel
cv	Converted	p	Purchased
C&M	Care and maintenance	RN	Royal Navy
DD	Destroyer (US)	US	US-built
DE	Destroyer escort (US)	WA	Warning, Aircraft
GA	Gunnery, Air: HA or combined HA/LA	WC	Warning, Combined
		WS	Warning, Surface
GB	Gunnery, Barrage: main armament AA	<	By date given
		>	After date given
GC	Gunnery, Close-range weapons	x	Set probably fitted, but direct evidence lacking

A3.1 LARGE SHIPS, WARTIME RADAR

Ship	Completed –lost	WA 79	79B	279	279B	281	281B	WC 286	290
CAPITAL SHIPS									
Anson	6/42	—	—	—	—	6/42–3/45	3/45	—	—
Barham	10/15–11/41	None when sunk						—	—
Duke of York	11/41	—	—	—	—	11/41–4/45	4/45	—	—
Hood	3/20–5/41	—	3/41	—	—	—	—	—	—
Howe	8/42	—	—	—	—	8/42–5/44	5/44	—	—
King George V	12/40	—	—	12/40–7/44	7/44	—	—	—	—
Malaya	2/16	—	—	—	—	7/41–3/44	3/44	—	—
Nelson	6/27	—	—	7/40	—	—	—	—	—
Prince of Wales	3/41–12/41	—	—	—	—	3/41	—	—	—
Queen Elizabeth	1/15	—	—	2/41	—	—	—	—	—
Ramillies	10/17	<11/41	—	—	—	—	—	—	—
Repulse	8/16–12/41	—	—	—	—	—	—	8/41	—
Renown	9/16	—	—	—	—	10/41	—	—	—
Resolution	12/16	<3/41	—	—	—	—	—	—	—
Revenge	3/16	—	—	3/41	—	—	—	—	—
Rodney	8/27	11/38(79Y)–9/41	—	—	—	9/41	—	—	—
Royal Oak	5/16–10/39	None when sunk				—	—	—	—
Royal Sovereign (USSR 45)	5/16	<8/43	—	—	—	—	—	<11/41	—
Valiant	2/15	11/39(79Z)–Mid-40	—	Mid-40–7/42	—	7/42	—	—	—
Warspite	3/15	—	—	—	—	12/41–6/44	—	6/44	—
MONITORS									
Abercrombie	5/43	—	—	—	—	5/43	—	—	—
Erebus	6/16	—	—	—	<8/43	—	—	<11/41–<7/43	—
Roberts	10/41	—	—	<8/43	—	10/41–<43	—	—	—
Terror	5/16–2/41	None when sunk				—	—	—	—
FLEET AIRCRAFT CARRIERS (L=Light Fleet Carrier)									
Argus	9/18	—	—	—	—	—	—	<2/43–<7/43	—
Ark Royal	11/38–11/41	None when sunk			—	—	—	—	—
Colossus (L)	12/44	—	12/44	—	—	—	12/44	—	—
Courageous	cv.5/28–9/39	None when sunk			—	—	—	—	—
Eagle	2/24–8/42	—	—	—	<8/42??	—	—	—	<11/41
Formidable	10/40	—	3/44	10/40–<1/42	—	1/42	3/44	—	—
Furious	cv.25	—	??<3/44	—	—	—	—	10/42–6/43	—
Glorious	cv.3/30–6/40	None when sunk			—	—	—	—	—
Glory (L)	4/45	—	4/45	—	—	—	4/45	—	—
Hermes	2/24–4/42	—	—	—	—	—	—	10/41	—

				WS			GC	GB	GS		GA	
291	293	271	272	273	276	277	282	283	284	274	285	275
—	3/45	—	—	6/42–3/45	—	3/45	6/42(6) 3/45(1)	3/45(4)	6/42–3/45	3/45(2)	6/42(4)–3/45	3/45(4)
—	4/45	11/41–?	—	?–4/45	—	4/45	11/41(4) 4/45(7)	—	11/41–4/45	4/45(2)	11/41(4)	—
—	—	—	—	—	—	—	—	—	3/41	—	—	—
—	5/44	—	—	8/42–5/44	—	5/44	8/42(6) 5/44(1)	5/44(3)	8/42–4/44	4/44	8/42(4)–5/44	—
—	7/44	6/41–<11/41	—	>11/41–7/44	—	7/44	7/44(7)	—	12/40–7/44	7/44	<8/43(3) 7/44(4)	—
6/41 (SW-1)	—	—	—	Early 42	—	3/44(SQ)	—	—	7/41	—	7/41(1) <8/43(2)	—
—	—	—	—	3/42	—	—	3/42(4)	—	<1945(4)	<11/41	3/42(2)	—
—	—	—	—	8/41	—	—	5/41(4)	—	3/41	—	3/41(4)	—
—	—	—	—	6/43	—	—	6/43(4)	6/43	2/41	—	2/41(4)	—
—	—	—	—	<11/41	—	—	1/44(2)	<11/41	<11/41	—	<11/41(2)	—
—	—	—	—	—	—	—	—	—	7/41	—	—	—
—	—	?6/41–<11/41	—	<11/41	—	—	10/41(2)	<8/43(3)	10/41–2/45	2/45	10/41(2)	—
—	—	—	—	11/41	—	—	—	—	<3/41	—	<6/41(2)	—
—	—	—	—	<11/41	—	—	—	—	<11/41	—	<8/43(2)	—
—	—	?6/41–<11/41	-	<11/41	—	—	<3/42(3)	<1945(6)	9/41	—	<3/41(1)	—
—	—	—	—	<11/41	—	—	<8/43(2)	—	<11/41	—	<11/41(2)	—
—	<45	—	—	<2/43	—	—	<8/43(4)	—	7/42	—	<3/43(4)	—
—	—	?6/41–<11/41	—	<11/41	—	—	—	—	12/41–>6/44	>6/44	12/41(2)	—
—	<45	—	5/43	—	—	—	5/43(3)	—	—	—	5/43(2)	—
—	—	—	<7/43	—	—	—	—	—	—	—	<11/41(2)	—
—	—	—	<8/43	—	—	—	10/41(3)	—	—	—	10/41(2)	—
—	—	—	—	—	—	—	—	—	—	—	—	—
<7/43	—	—	—	—	—	—	—	—	—	—	—	—
—	12/44	—	—	—	—	12/44	12/44(6)	—	—	—	—	—
—	—	—	—	—	—	—	—	—	—	—	1/42(3)	—
—	—	—	—	—	3/44	3/44	—	—	—	—	11/41(1)	—
6/43	—	—	—	—	—	—	—	—	—	—	<3/42(2)	—
—	4/45	—	—	—	—	4/45	4/45(6)	—	—	—	—	—
—	—	—	—	—	—	—	—	—	—	—	<2/43(1)	—

Appendix 3

Ship	Completed –lost	WA						WC	
		79	79B	279	279B	281	281B	286	290
FLEET AIRCRAFT CARRIERS (cont.)									
Illustrious	4/40	5/40–4/42	6/43	—	—	4/42–6/43	6/43	—	—
Implacable	8/44	—	8/44	—	—	—	8/44	—	—
Indefatigable	5/44	—	5/44	—	—	—	5/44	—	—
Indomitable	10/41	—	5/43	—	10/41	—	5/43	—	—
Ocean (L)	6/45	—	6/45	—	—	—	6/45	—	—
Unicorn (Repair)	3/43	—	—	—	—	—	3/43	—	—
Venerable (L)	1/45	—	1/45	—	—	—	1/45	—	—
Vengeance (L)	1/45	—	1/45	—	—	—	1/45	—	—
Victorious	3/41	—	<5/45	5/41–5/45	—	—	3/44	—	—

Ship	Completed –lost	WA						WC	
		79	79B	279	279B	SK (US)	281B	286	290
ESCORT AIRCRAFT CARRIERS									
(A) =Assault; (B) =British-built; (C) =Convoy; (F) =Ferry; (T) =Training.									
Activity (C) (B)	9/42	—	10/42	—	—	—	—	—	—
Ameer (A)	7/43	—	—	—	—	7/43	—	—	—
Arbiter (F)	12/43	—	—	—	—	12/43	—	—	—
Archer (C)	11/41–10/43(C&M)	—	<2/43	—	—	—	—	—	—
Atheling (C)	11/43	—	—	—	—	7/43	—	—	—
Attacker (A)	9/42	—	<7/43	—	—	—	—	—	—
Audacity (C) (B)	cv.6/41–12/41	—	9/41	—	—	—	—	—	—
Avenger (C)	3/42–11/42	—	<9/42	—	—	—	—	—	—
Battler (C)	10/43	—	—	—	—	—	—	—	—
Begum (C)	8/43	—	—	—	—	8/43	—	—	—
Biter (C)	5/42	—	5/42	—	—	—	—	—	—
Campania (C) (B)	3/44	—	—	—	—	—	3/44	—	—
Chaser (C)	4/43	—	4/43	—	—	—	—	—	—
Dasher (C)	7/42–3/43	—	7/42	—	—	—	—	—	—
Emperor (A)	8/43	—	—	—	—	8/43	—	—	—
Empress (C/A)	8/43	—	—	—	—	8/43	—	—	—
Fencer (C)	2/43	—	3/43	—	—	—	—	—	—
Hunter (A)	1/43	—	1/43	—	—	—	—	—	—
Khedive (A)	8/43	—	—	—	—	8/43	—	—	—
Nabob (C)	9/43–8/44	—	—	—	—	9/43	—	—	—
Nairana (C) (B)	12/43	—	—	—	—	—	12/43	—	—
Patroller (F)	10/43	—	—	—	—	10/43	—	—	—
Premier (C)	11/43	—	—	—	—	11/43	—	—	—
Pretoria Castle (T) (B)	4/43	—	—	—	—	—	4/43	—	—
Puncher (C) (Can)	2/44	—	—	—	—	2/44	—	—	—
Pursuer (C)	4/43	—	4/43	—	—	—	—	—	—
Queen (C)	12/43	—	—	—	—	12/43	—	—	—
Rajah (F)	1/44	—	—	—	—	1/44	—	—	—
Ranee (F)	11/43	—	—	—	—	11/43	—	—	—
Ravager (T)	4/43	—	4/43	—	—	—	—	—	—
Reaper (F)	2/44	—	—	—	—	2/44	—	—	—
Ruler (A)	12/43	—	—	—	—	12/43	—	—	—
Searcher (A)	4/43	—	4/43	—	—	—	—	—	—
Shah (C)	11/43	—	—	—	—	11/43	—	—	—
Slinger (F)	8/43	—	—	—	—	8/43	—	—	—
Smiter (C)	1/44	—	—	—	—	1/44	—	—	—
Speaker (A)	11/43	—	—	—	—	11/43	—	—	—
Stalker (A)	12/42	—	12/42	—	—	—	—	—	—
Striker (C)	4/43	—	4/43	—	—	—	—	—	—

Ship-Fitting Tables by Class of Ship

291	293	271	272	273	WS 276	277	GC 282	GB 283	GS 284	274	GA 285	275
—	<46	—	6/43?	—	—	?6/43	6/43	—	—	—	3/41(1) 6/43(1)	—
—	8/44	—	—	—	—	8/44	8/44(6)	—	—	—	8/44(3)	—
—	5/44	—	—	—	—	5/44	5/44(6)	—	—	—	5/44(3)	—
—	—	—	<8/43	—	2/44(SG)	2/44 (SM-1)	<3/42(3) <8/43(6)	—	—	—	11/41(1)	—
—	6/45	—	—	—	—	6/45 (SM-1)	6/45(2) (262)	—	—	—	—	—
—	—	—	—	—	<45	—	3/43(4)	—	—	—	3/43(3)	—
—	1/45	—	—	—	—	1/45	1/45(6)	—	—	—	—	—
—	1/45	—	—	—	—	1/45	1/45(6)	—	—	—	—	—
—	<5/45	—	<3/43	—	3/44(SG)	3/44	<9/41(2) ×(4)	—	—	—	<3/41(1) <8/43(1)	—

291	293	SG (US)	272	273	WS 276	277	GC 282	US set	GS 284	274	GA 285	275
—	—	—	10/42	—	—	—	—	—	—	—	—	—
—	—	7/43	—	—	—	—	—	7/43(2)	—	—	—	—
—	—	x	—	—	—	—	—	x	—	—	—	—
—	—	—	<2/43	—	—	—	—	—	—	—	—	—
—	—	7/43	—	—	—	—	—	7/43(2)	—	—	—	—
—	—	—	<7/43	—	—	—	—	—	—	—	—	—
—	—	—	—	—	—	—	—	—	—	—	—	—
—	—	—	—	—	—	—	—	—	—	—	—	—
—	—	10/43	—	—	—	—	—	—	—	—	—	—
—	—	8/43	—	—	—	—	—	8/43(2)	—	—	—	—
—	—	—	5/42	—	—	—	—	—	—	—	—	—
—	—	—	—	—	3/44	3/44	—	—	—	—	—	—
—	—	—	4/43	—	—	—	—	—	—	—	—	—
—	—	—	7/42	—	—	—	—	—	—	—	—	—
—	—	8/43	—	—	—	—	—	8/43(2)	—	—	—	—
—	—	8/43	—	—	—	—	—	8/43(2)	—	—	—	—
—	—	—	3/43	—	—	—	—	—	—	—	—	—
—	—	—	1/43	—	—	—	—	—	—	—	—	—
—	—	8/43	—	—	—	—	—	8/43(2)	—	—	—	—
—	—	9/43	—	—	—	—	—	9/43(2)	—	—	—	—
—	<45	—	12/43	—	—	?1944	—	—	—	—	—	—
—	—	—	—	—	—	—	—	—	—	—	—	—
—	—	11/43	—	—	—	—	—	11/43(2)	—	—	—	—
—	—	—	4/43	—	—	—	—	—	—	—	—	—
—	—	x	—	—	—	—	—	x	—	—	—	—
—	—	—	6/43	—	—	—	—	—	—	—	—	—
—	—	12/43	—	—	—	—	—	12/43(2)	—	—	—	—
—	—	1/44	—	—	—	—	—	1/44(2)	—	—	—	—
—	—	11/43	—	—	—	—	—	11/43(2)	—	—	—	—
—	—	—	4/43	—	—	—	—	—	—	—	—	—
—	—	—	—	—	—	—	—	—	—	—	—	—
—	—	12/43	—	—	—	—	—	12/43(2)	—	—	—	—
—	—	—	—	—	—	—	—	—	—	—	—	—
—	—	11/43	—	—	—	—	—	11/43(2)	—	—	—	—
—	—	8/43	—	—	—	—	—	8/43(2)	—	—	—	—
—	—	1/44	—	—	—	—	—	1/44(2)	—	—	—	—
—	—	11/43	—	—	—	—	—	11/43(2)	—	—	—	—
—	—	—	12/42	—	—	—	—	—	—	—	—	—
—	—	—	4/43	—	—	—	—	—	—	—	—	—

Appendix 3

Ship	Completed –lost	WA 79	79B	279	279B	SK (US)	281B	WC 286	290
ESCORT AIRCRAFT CARRIERS (cont.)									
Thane (A)	11/43–1/45	—	—	—	—	11/43	—	—	—
Tracker (C)	1/43	—	1/43	—	—	—	—	—	—
Trouncer (F)	1/44	—	—	—	—	1/44	—	—	—
Trumpeter (C)	8/43	—	—	—	—	8/43	—	—	—
Vindex (C) (B)	12/43	—	—	—	—	—	12/43	—	—
AIRCRAFT TRANSPORTS AND CATAPULT SHIPS									
Albatross	—	—	—	—	—	—	—	<2/43	—
Ariguani	cv.41	–	–	<3/41–<2/43(280)	—	—	—	—	—
Athene	<7/43–<1/44	—	—	—	<1/44	—	—	—	—
Engadine	8/14	3/41	—	—	—	—	—	—	—
Maplin	cv.41	<3/41	—	—	—	—	—	—	—
Patia	cv.1/41–4/41	—	<3/41	—	—	—	—	—	—
Pegasus	cv.4/41	—	—	—	>5/42	—	—	<3/41–>5/42	—

Ship	Completed –lost	WA 79	79B	279	279B	281	281B	WC 286	290
CRUISERS (See also AA CRUISERS)									
Achilles (NZ 36, RN 43)	10/33	—	—	—	—	—	5/44	<2/43–5/44	—
Adelaide (Aust)	1922	—	—	—	—	<6/44	—	—	<2/43–<6/44
Ajax	4/35	—	—	7/40	—	—	—	—	—
Arethusa	5/35	—	—	—	—	4/42	—	<7/41–4/42	—
Argonaut	8/42	—	—	—	—	8/42–11/44	11/44	—	—
Aurora	11/37	—	—	—	—	6/42	—	—	5/41(290X)–6/42
Australia (Aust)	4/28	—	—	—	—	<6/44	—	12/40–	<3/43–<8/43
Belfast	8/39	—	—	—	—	<3/41–5/45	5/45	—	—
Bellona	10/43	—	—	—	—	10/43	—	—	—
Bermuda	8/42	—	—	—	—	8/42–4/45	4/45	—	—
Berwick	2/28	—	—	—	—	6/41–12/43	12/43	<2/41–6/41	—
Birmingham	11/37	—	—	—	—	—	8/43	4/42–8/43	—
Black Prince	11/43	—	—	—	—	11/43	—	—	—
Bonaventure	5/40–3/41	—	—	5/40	—	—	—	—	—
Calypso	6/17–6/40	None when sunk?							
Canberra (Aust)	7/28–8/42	—	—	—	—	—	—	—	—
Capetown	4/22	—	—	—	—	—	—	<3/42	—
Caradoc	6/17	—	—	—	—	—	—	<11/41–<3/43	<10/42
Cardiff	6/17	—	—	—	—	—	—	—	<2/43
Ceres	6/17	—	—	—	—	—	—	<8/43–<4/44	<4/44
Ceylon	7/43	—	—	—	—	7/43	—	—	—
Charybdis	12/41–10/43	—	—	—	—	12/41	—	—	—
Cleopatra	12/41	—	—	—	—	12/41–11/44	—	—	—
Cornwall	1928–4/42	—	—	—	—	11/44	—	—	—
Cumberland	2/28	—	—	—	—	10/41–<8/43	<8/43	—	—
Danae	6/18	—	—	—	—	—	—	—	—
Dauntless	12/18	—	—	—	—	—	—	<3/42	—
Despatch	6/22	—	—	—	—	—	—	7/43	—

Ship-Fitting Tables by Class of Ship

			WS				GC		GS		GA	
291	293	SG (US)	272	273	276	277	282	US set	284	274	285	275
—	—	11/43	—	—	—	—	—	11/43(2)	—	—	—	—
—	—	—	1/43	—	—	—	—	—	—	—	—	—
—	—	1/44	—	—	—	—	—	1/44(2)	—	—	—	—
—	—	8/43	—	—	—	—	—	8/43(2)	—	—	—	—
—	<45	—	12/43	—	—	—	12/43(4)	—	—	—	—	—
—	—	—	—	—	—	—	—	—	—	—	—	—
—	—	—	—	—	—	—	—	—	—	—	—	—
—	—	—	—	—	—	—	—	—	—	—	—	—
—	—	—	—	—	—	—	—	—	—	—	—	—
—	—	—	—	—	—	—	—	—	—	—	—	—
—	—	—	—	—	—	—	—	—	—	—	—	—
—	—	—	—	—	—	—	—	—	—	—	—	—
—	—	—	—	—	—	—	—	—	—	—	—	—
—	—	—	—	—	—	—	—	—	—	—	—	—

			WS				GC	GB	GS		GA	
291	293	271	272	273	276	277	282	283	284	274	285	275
—	5/44	—	—	—	—	5/44	5/44(4)	(2)	<2/43	—	<3/42(1) 5/44(3)	—
—	—	—	<8/43	—	—	—	—	—	—	—	<8/43(1)	—
—	—	—	10/42	—	—	—	—	10/42(2)	10/42	—	<8/43(2)	—
—	—	—	10/41	—	—	—	—	12/43(3)	<11/41	—	<8/43(2)	—
—	<46	—	8/42–11/44	—	<6/44	<6/44	8/42(2)	—	8/42	—	8/42(2)	—
—	—	<12/42	—	—	—	—	—	—	5/41	—	6/42(2)	—
<8/43	<45	—	—	<8/43	—	—	—	—	—	—	—	—
—	5/45	—	—	12/42–5/45	—	5/45	12/42(2)	12/42(4)	12/42–5/45	5/45(2)	12/42(3)	—
—	—	—	10/43	—	—	—	10/43(3)	—	10/43	—	10/43(2)	—
—	<46	—	?4/43	8/42–?43	—	—	8/42(2)	—	8/42	—	8/42(3)	—
—	—	—	—	8/42	—	—	<8/43(2)	<8/42(4)	6/41	—	<7/43(2)	—
—	—	—	—	8/43	—	—	<8/43(2)	<8/43(4)	12/40	—	<3/41(3)	—
—	—	—	11/43	—	1945(SG)	—	11/43(3)	—	11/43	—	11/43(2)	—
—	—	—	—	—	—	—	—	—	—	—	—	—
—	—	<8/42	—	—	—	—	—	—	—	—	11/41(1)	—
—	—	<8/43	—	—	—	—	—	—	—	—	—	—
—	—	<2/43	—	—	—	—	—	—	—	—	—	—
<7/43	—	<7/43	—	—	—	—	—	—	—	—	—	—
—	—	<2/43	—	—	—	—	—	—	—	—	—	—
—	—	—	7/43	—	—	—	7/43(3)	7/43(1)	7/43	—	7/43(3)	—
—	—	—	<2/42	—	—	—	12/41(2)	(283x)7/43	—	—	12/41(2)	—
—	11/44	—	<8/43	—	—	—	12/41(2)	—	12/41	—	12/41(2)	—
—	—	—	—	<3/42	—	—	—	—	—	—	—	—
—	—	—	—	<3/43	—	—	<8/43(2)	<8/43(4)	10/41	—	10/41(2)	—
<4/43	—	<8/43	—	—	—	—	<8/43(2)	<8/43(1)	—	—	—	—
—	—	<3/42	—	—	—	—	—	—	—	—	—	—
—	—	7/42	—	—	—	—	—	—	—	—	—	—

Appendix 3

Ship	Completed –lost	WA						WC	
		79	79B	279	279B	281	281B	286	290
CRUISERS (cont.)									
Devonshire	3/29	—	—	—	—	5/41–3/44	3/44	—	—
Diadem	1/44	—	—	—	—	1/44	—	—	—
Dido	9/40	—	—	—	—	10/40(281x)	—	<5/41	—
Diomede NZ	10/22	—	—	—	—	—	—	—	—
Dorsetshire	4/30–4/42	—	—	—	—	—	—	—	5/41(290x)
Dragon (Poland 41)	8/18	—	—	—	—	—	—	—	—
Dunedin NZ	10/19–11/41	—	—	—	—	—	—	12/40	—
Durban	10/21–6/44	—	—	—	—	—	—	8/42	—
Edinburgh	7/38–5/42	—	—	10/40	—	—	—	—	—
Effingham	7/25–5/40	None when sunk				—	—	—	—
Emerald	1/26	—	—	—	—	4/43	—	<2/41–4/43	—
Enterprise	3/26	—	-	—	—	10/43	—	—	—
Euryalus	6/41	—	—	6/41–7/44	7/44	—	—	—	—
Exeter	7/31–3/42	—	—	3/41	—	—	—	—	—
Fiji	5/40–5/41	—	—	5/40	—	—	—	—	—
Frobisher	9/24	—	—	—	—	3/42	—	—	—
Galatea	8/35–12/41	—	—	10/40	—	—	—	—	—
Gambia (NZ 43)	2/42	—	—	—	—	2/42	—	—	—
Glasgow	9/37	—	—	—	—	8/42–5/45	5/45	7/40–8/42	—
Gloucester	1/39–5/41	—	—	??3/41	—	—	—	—	—
Hawkins	7/19	—	—	—	—	5/42	—	—	—
Hermione	3/41–6/42	—	—	3/41	—	—	—	—	—
Hobart (Aust)	1/36	—	—	10/42–1/45	—	—	1/45	—	<2/42–10/42
Jamaica	6/42	—	—	—	—	6/42–<9/45	<6/45	—	—
Kent	6/28	—	—	—	—	9/41	—	—	—
Kenya	9/40	—	—	9/40	—	—	—	—	—
Leander NZ	3/33	—	<8/43	—	—	—	—	<11/41	Mid 42
Liverpool	11/38	—	7/45	—	—	<3/41–7/45	—	—	—
London	1/29	—	—	<3/41	—	—	—	—	—
Manchester	8/38–8/42	—	—	2/42	—	—	—	11/40(2)–2/47	—
Mauritius	1/41	—	—	<3/41	—	—	—	—	—
Naiad	7/40–3/42	—	—	9/40	—	—	—	—	—
Neptune	2/34–12/41	—	—	—	—	5/41	—	—	—
Newcastle	3/37	—	—	—	—	11/42	—	<3/41–11/41	11/41–11/42
Newfoundland	1/43	—	—	—	—	1/43–11/44	11/44	—	—
Nigeria	9/40	—	—	9/40	—	—	—	—	—
Norfolk	9/30	—	—	—	—	9/41–11/44	11/44	<5/41	—
Ontario (Can, ex Minotaur)	5/45	—	—	—	—	—	5/45	—	—
Orion	1/34	—	—	2/42	—	—	—	<5/41	—
Penelope	11/36–2/44	—	—	—	—	7/41	—	—	—
Perth (Aust)	7/37–2/42	—	—	—	—	—	—	<5/41	—
Phoebe	9/40	—	—	9/40–4/42	—	4/42	—	—	—
Royalist	8/43	—	—	—	—	8/43	—	—	—
Scylla	6/42	—	—	—	—	6/42	—	—	—
Sheffield	8/37	—	—	10/38(79Y)–7/42	—	7/42	—	—	—
Shropshire (Aust 6/43)	9/29	—	—	—	—	2/42	—	—	—

Ship-Fitting Tables by Class of Ship

				WS			GC	GB	GS		GA	
291	293	271	272	273	276	277	282	283	284	274	285	275
—	—	—	—	3/42	<45	—	3/44(2)	3/44(3)	—	—	3/41(1) 3/44(2)	—
—	—	—	1/44	—	—	—	1/44(2)	—	1/44	—	1/44(2)	—
—	—	—	6/43	—	—	—	—	—	<1/44	—	<3/41(1)x(2)	—
6/43	—	—	—	6/43	—	—	—	—	—	—	—	—
—	—	—	—	—	—	—	—	—	<6/44	—	<9/41(2)	—
<8/43	—	—	<8/43	—	—	—	—	<8/43(1)	—	—	—	—
—	—	—	—	<8/43	—	—	—	—	—	—	—	—
—	—	—	—	3/42	—	—	—	—	3/42	—	—	—
—	—	—	—	4/43	—	—	4/43(2)	—	—	—	4/43	—
—	—	—	10/43	—	—	—	10/43(2)	—	10/43	—	10/43(1)	—
—	7/44	—	Mid 43–7/44	—	—	7/44	—	—	6/41	—	6/41(1) x(2)	—
—	—	—	—	—	—	—	—	—	<3/41	—	—	—
—	—	—	—	—	—	—	—	—	3/41	—	—	—
—	—	—	—	3/42	—	—	—	—	—	—	3/42(2)	—
—	—	—	—	—	—	—	—	—	<11/41	—	<11/41(1)	—
—	—	—	x	2/42	—	—	2/42(2)	<8/43(4)	2/42	—	2/42(2)	—
—	5/45	—	—	8/42–5/45	—	—	8/42(2)	<7/43(1)	8/42–5/45	5/45	8/42(3)	—
—	—	—	—	5/42	—	—	—	—	—	—	5/42(1)	—
—	—	—	—	—	—	—	—	—	<5/41	—	—	—
—	1/45 (SG-1)	—	—	10/42–1/45	1/45	1/45	10/42	10/42(3)	10/42–1/45	1/45 (FC 1)(2)	10/42	—
—	?<6/45	—	<12/43	6/42–<12/93	—	<6/45	6/42(2)	—	6/42	<6/45	6/42(2) <8/43(3)	—
—	—	—	—	6/43	—	—	<2/43(2)	<7/43(4)	9/41	—	9/41(2)	—
—	<46	8/41	—	—	—	—	7/43(2)	12/42(4)	<11/41	—	12/42(3)	—
—	—	—	—	<7/43	—	—	—	—	<2/43	—	<3/42(1)	—
—	7/45	—	—	1/42–7/45	—	7/45	1/42(2)	7/45(4)	3/41–7/45	7/45	—	—
—	—	—	—	1/42	—	—	<3/41(2)	<8/43(4)	<3/41	—	<3/41(2)	—
—	—	—	—	2/42	—	—	2/42(2)	—	<5/41	—	2/42(2)	—
—	<46	—	x	3/42	—	—	—	<1946(3)	3/42	—	3/42(2)	—
—	—	—	—	—	—	—	—	—	5/41	—	1/41	—
—	—	—	—	11/41	—	—	11/42(2)	<8/43(4)	11/42	—	3/42(1) 11/42(2)	—
—	11/44	—	1/43–11/44	—	—	11/44	<1/44(3)	<8/43(1)	<8/43–11/44	11/44	1/44(3)	—
—	—	—	<8/43	<11/41–<8/43	—	—	6/43(2)	<8/43(4)	<11/41	—	<11/41(1) <8/43(3)	—
—	11/44	—	—	<3/42–11/44	—	11/44	—	11/44(4)	9/41–11/44	11/44	5/41(2)	—
—	5/45	—	—	—	—	—	5/45(3)	5/45(2)	—	5/45	—	5/45(3)
—	—	—	—	4/42	—	—	—	<1945(3)	2/42	—	2/42(1)	—
—	—	9/42	—	—	—	—	—	<8/43(3)	7/41	—	7/41(2)	—
—	—	—	6/43	—	—	—	—	—	4/42	—	4/42(2)	—
—	<46	—	—	—	<3/44	—	8/43(2)	8/43(3)	8/43	—	8/43(2)	—
—	<46	—	6/42	—	—	—	<1/44(2)	—	—	—	6/42(2)	—
—	<46	—	—	7/42–5/45	—	5/45	(3)	7/42(1)	<9/41	—	8/41(1) <8/43(3)	—
—	6/43(SGA)	—	—	2/42–<late 44	<45	<late 44	2/42(2)	<1/44(4)	—	—	2/42(2)	—

Appendix 3

Ship	Completed –lost	WA 79	79B	279	279B	281	281B	WC 286	290
CRUISERS (cont.)									
Sirius	5/42	—	—	—	—	5/42	—	—	—
Southampton	3/37–1/41	None when sunk			—	—	—	—	—
Spartan	8/43–1/44	—	—	—	—	8/43	—	—	—
Suffolk	5/28	<9/39(79Z)–8/40	—	—	8/40–7/42	7/42	—	—	—
Superb	11/45	—	11/45	—	—	—	—	—	—
Sussex	3/29	—	—	—	—	8/42–3/45	3/45	—	—
Swiftsure	6/44	—	—	—	—	—	6/44	—	—
Sydney (Aust)	9/35–11/41	None when sunk			—	—	—	—	—
Trinidad	10/41–5/42	—	—	—	—	10/41	—	—	—
Uganda (Can 44, renamed Quebec)	1/43	—	—	—	—	1/43–6/95	6/45	—	—
York	5/30–3/41	—	—	—	—	—	—	<5/41	—
ANTI-AIRCRAFT CRUISERS									
Cairo (AA 39)	10/19–8/42	—	—	8/40	—	—	—	—	—
Calcutta (AA 39)	8/19–6/41	—	None when sunk		—	—	—	—	—
Caledon (AA 12/43)	3/17	—	—	—	12/43	—	—	—	—
Carlisle (AA 39)	11/18–10/43	1/40(280)–<8/43	—	—	—	<8/43	—	—	—
Colombo (AA 11/43)	10/19	—	—	—	11/43	—	—	—	—
Coventry (AA 37)	2/18–9/42	—	—	5/40	—	—	—	—	—
Curacoa (AA 12/39)	2/18–10/42	—	—	1/40	—	—	—	—	—
Curlew (AA 38)	12/17–5/40	8/39(79Z)	—	—	—	—	—	—	—
Delhi (AA >9/42)	6/19	—	—	—	—	4/42–<8/43	<8/43	—	—
Heemskerck (Neth)	—	—	<3/41	—	—	—	—	—	—
AUXILIARY A.A. SHIPS									
Alynbank	cv.39–6/44	—	—	8/40(280)	—	—	—	—	—
Foylebank	cv.39–7/40	—	—	×(280)	—	—	—	—	—
Pozarica	cv.39–1/43	—	—	<3/41	—	—	—	—	—
Spring Bank	cv.39–9/41	—	—	<3/41(280)	—	—	—	—	—
Tynwald	cv.39–11/42	—	—	<3/41	—	—	—	—	—
HEADQUARTERS SHIPS (LSH(L))									
Bulolo (ex AMC)	cv.42	—	—	—	—	—	—	—	<2/42
Hilary (ex OBV)	cv.43	—	—	—	—	—	—	—	—
Largs (ex OBV)	cv.42	—	—	—	—	—	—	<2/43	—
Lothian	cv.44	—	—	—	—	—	—	—	—
MINELAYERS									
Abdiel	4/41–9/43	—	—	—	—	—	—	<5/41–<2/43	<2/43
Adventure	5/27	—	x	—	—	—	—	<3/42	—
Apollo	2/44	—	—	—	—	—	—	—	—
Ariadne	11/41	—	—	—	—	—	—	11/41–<8/43	—
Latona	5/41–10/41	—	—	—	—	—	—	5/41	—
Manxman	6/41	—	—	—	—	—	—	6/41	—
Welshman	8/41–2/43	—	—	—	—	—	—	8/41–<2/43	2/43

Ship-Fitting Tables by Class of Ship

				WS			GC	GB	GS		GA	
291	293	271	272	273	276	277	282	283	284	274	285	275
—	—	—	5/42	—	—	—	5/42(2)	—	5/42	—	5/42(2)	—
—	—	—	—	—	—	—	—	—	—	—	—	—
—	—	—	8/43	—	—	—	8/43(3)	—	8/43	—	8/43(2)	—
—	—	—	—	7/42	—	—	1/44(2)	1/44(4)	2/41	—	2/41(2)	—
—	11/45	—	—	—	—	—	11/45(3)	11/45(3)	—	—	—	11/45(3)
—	3/45	—	—	8/42–3/45	—	3/45	8/42(2)	3/45(4)	<8/43	—	8/42(2)	—
—	6/44	—	—	—	—	<4/45	6/44(3)	6/44(2)	—	6/44	6/44(3)	—
—	—	—	—	—	—	—	—	—	—	—	—	—
—	—	—	—	—	—	—	10/41(2)	—	10/41	—	10/41(3)	—
—	6/45	—	1/43–6/45	—	—	6/45	1/43(3)	<1946(1)	1/43–6/45	6/95	1/43(3)	—
—	—	—	—	—	—	—	—	—	—	—	—	—
—	—	<3/42	—	—	—	—	—	—	—	—	<11/41(1) <3/42(2)	—
—	—	—	—	—	—	—	—	—	—	—	—	—
—	—	—	—	12/43	—	—	12/43(4)	12/43(1)	—	—	12/43(1)	—
—	—	<8/43	—	—	—	—	7/41(1)	—	—	—	<11/41(1) –7/41(2)	—
—	—	—	—	11/43	—	—	11/43(2)	11/43(1)	—	—	11/43(1)	—
—	—	—	—	—	—	—	—	—	—	—	—	—
—	—	<9/42	—	—	—	—	<6/41(1)	—	—	—	<6/41(2)	—
—	—	—	—	—	—	—	—	—	—	—	—	—
—	—	4/42	—	—	—	—	4/42(2)	—	—	—	4/42(2)	—
—	—	—	<8/43	—	—	—	—	—	—	—	11/41(2)	—
—	—	—	—	—	—	—	—	—	—	—	<2/43	—
—	—	—	—	—	—	—	—	—	—	—	—	—
—	—	—	—	—	—	—	—	—	—	—	<6/41(2)	—
—	—	—	—	—	—	—	—	—	—	—	—	—
—	—	—	<11/41	—	—	—	—	—	—	—	<3/41(2)	—
<6/44	—	<2/43	—	—	—	—	—	—	—	—	—	—
<1/44	—	<1/44	—	—	—	—	—	—	—	—	—	—
<6/44	—	—	—	2/42	—	—	—	—	—	—	—	—
?	—	—	—	—	—	—	—	—	—	—	—	—
—	—	—	—	—	—	—	<2/43(1)	—	—	—	4/41(1)	—
—	—	—	<3/42	—	—	—	—	—	—	—	<6/41(1)	—
—	2/44	—	—	—	—	—	2/44(2)	—	—	—	2/44	—
<8/43	—	—	11/41	—	—	—	<11/41(1)	—	—	—	11/41(1)	—
—	—	—	—	—	—	—	—	—	—	—	5/41	—
—	—	—	—	—	—	—	—	—	—	—	6/41(1)	—
—	—	—	—	—	—	—	—	—	—	—	8/41	—

Appendix 3

Ship	Completed –lost	79	79B	WA 279	279B	281	281B	WC 286	290
FIGHTER DIRECTION SHIPS (LSF)									
Antwerp	cv.44	—	5/44	—	—	—	—	—	—
Boxer (ex LST 1)	1.12/42, cv.44	—	6/45	—	—	—	6/45	—	—
FDT 13 (ex LST 2)	cv.44	GCI Type 11 (1.5m)	2×GCI Type 15 (50 cm)	—	—	—	—	—	—
FDT 216 (ex LST 2)	cv.44–6/44			—	—	—	—	—	—
FDT 217 (ex LST 2)	cv.44			—	—	—	—	—	—
Palomares (ex Aux.AA)	cv.43	—	<6/43	<3/41–<6/43	—	—	<6/43	—	—
Stuart Prince (ex Aux.AA)	cv.43	—	<8/43	—	—	—	<8/43	—	—
Ulster Queen (ex Aux.AA)	cv.43	—	<6/43	<3/41–<6/43	—	—	<6/43	9/41–3/42	—
DEPOT AND REPAIR SHIPS									
(des) =destroyer depot ship; (rep) =repair ship; (S/M) =submarine depot ship									
Adamant (S/M)	2/42	<2/42–<1/44	<1/44	—	—	—	—	<2/43	—
Artifex (ex Aurania)	cv.42	<2/42–<8/43	—	—	—	—	—	—	<2/43 (CSC)
Ausonia (ex AMC) (rep)	c.39	—	—	—	—	—	—	<9/41–<8/43	—
Blenheim	p.40	3/41	—	—	—	—	—	<11/41	6/41
Cyclops (rep)	1905	—	—	—	—	—	—	—	—
Forth (S/M)	5/39	—	—	—	—	—	—	—	—
Hecla (des)	3/40–11/42	—	—	3/40	—	—	—	—	—
Maidstone (S/M)	5/38	—	—	—	—	—	—	—	—
Medway (S/M)	1.7/28–6/42	None when sunk		—	—	—	—	—	—
Montclare	c.42	<8/43	—	—	—	—	—	—	—
Philoctetes	1/42	—	—	—	—	—	—	1/42	6/41
Ranpura (ex AMC)	c.42	—	—	—	—	—	—	<2/43	—
Resource (rep)	1.28	—	—	—	—	—	—	<1/44	—
Tyne (des)	2/41	—	—	2/41	—	—	—	—	—
Sandhurst (rep)	cv.16	—	—	—	—	—	—	—	6/41
Vindictive (rep) ex cruiser	18	—	—	—	—	—	—	<8/43	—
Wayland (ex Antonia)	cv.41	<11/41	—	—	—	—	—	—	—
Wolfe (ex Montcalm)	cv.4/41	<8/43	—	—	—	—	—	—	—
Woolwich (des)	6/35	—	—	—	—	—	—	—	—

		WS					GC	GB	GS		GA	
291	293	271	272	273	276	277	282	283	284	274	285	275
—	—	—	—	—	—	—	—	—	—	—	—	—
6/45(2) (SMM)	6/45(SK)	—	—	—	6/45(2) (SM–1)	6/45(2)	—	—	—	—	—	—
—	—	—	—	—	—	—	—	—	—	—	—	—
—	—	—	—	—	—	—	—	—	—	—	—	—
—	—	—	<6/43	—	—	3/45 (SM–1)	—	—	—	—	<6/41(2)	—
—	—	—	12/43	—	—	—	—	—	—	—	—	—
—	<9/44 (GCI)	2/43	2/43– <8/43	<8/43	—	—	—	—	—	—	<3/41(1)	—
—	—	—	—	—	—	—	—	—	—	—	2/42(2)	—
<8/43	—	—	—	—	—	—	—	—	—	—	—	—
<8/43	—	—	—	<3/42 <8/43	—	—	<8/43(4)	—	—	—	—	—
—	—	<3/42	—	—	<45	—	—	—	—	—	<8/43	—
—	—	—	—	—	—	—	—	—	—	—	—	—
<8/43	—	—	—	—	—	—	—	—	—	—	<3/42(1)	—
—	—	—	—	—	—	—	—	—	—	—	—	—
<8/43	<45	—	—	—	—	—	—	—	—	—	<8/43(1)	—
—	—	—	—	—	—	—	—	—	—	—	1/42(1)	—
—	—	<2/43	—	—	—	—	—	—	—	—	—	—
—	—	—	—	—	—	—	—	—	—	—	<8/43(1)	—
<1/44	—	—	—	—	—	—	—	—	—	—	<8/43	—
—	—	<1/44	—	—	—	—	—	—	—	—	<3/42(1)	—
<8/43	—	—	—	—	—	—	—	—	—	—	<8/43(1)	—

A3.2 DESTROYERS AND BELOW, WARTIME RADAR

Class of ship	Date completed	WS 271	WS 272	WS 277	WC 286 291	WC 276 293	GA 275	GA 285	GC 282
FLEET DESTROYERS									
A, B	1927–31	s	—	—	1	—	—	—	—
D, E	1932–34	s	—	—	1	—	—	s	—
F, G	1935–36	1	—	—	1	—	—	s	—
H, I	1936–38	s	—	—	1	—	—	s	—
Tribal	1938–39	s	—	—	1	1*	—	1	—
J, K	1938–39	—	s	—	1	1*	—	1	—
L	1940–41	—	—	—	1	1	—	1	—
M, N, O	1941–42	s	s	—	1	s*	—	1	—
P, Q	1941–43	—	—	—	1	s	—	1	—
R	1942–43	—	s	—	1	—	—	1	—
S, T, U, V	1943–44	s	s	—	1	1*	—	1	1
W, Z, Ca	1944	—	—	—	1	1	—	1	1
Battle	1944–45	—	—	—	1	1	1	—	4
Ch, Co, Cr	1945–47	—	—	—	1	1	1	—	1
ESCORT DESTROYERS									
V & W (A/A)	1917–24	—	—	—	1	—	—	1	—
V & W (A/S)	1917–24	1	—	—	1	—	—	—	—
Town (US DD)	1919–20	1	—	—	s	—	—	—	—
S (A/S)	1919–20	—	—	—	1	—	—	—	—
Hunt (A/A)	1940–42	s	—	—	1	—	—	1	—
SLOOPS									
Hastings, etc.	1929–36	1	—	—	1	—	—	—	—
Bird	1935–46	—	1	—	1	1*	—	1	2
FRIGATES									
River (A/S)	1942–44	1	—	1*	1	—	—	—	—
Loch (A/S)	1943–45	—	1	1*	—	—	—	—	—
Bay (A/A)	1945–49	—	—	—	1	1	—	1	—
Captain (US DE)	1943–44	—	—	SL	sSA	—	—	—	—
Colony (US)	1943–44	—	—	SL	SA	—	—	—	—
CORVETTES									
Kingfisher	1935–39	1	—	—	—	—	—	—	—
Duck	1936–39	s	—	—	s	—	—	—	—
Flower (A/S)	1940–42	1	—	—	—	—	—	—	—
Castle (A/S)	1943–44	s	—	1*	—	—	—	—	—
COASTGUARD CUTTERS									
ex-USN	1920s	1	—	—	—	—	—	—	—
FLEET MINESWEEPERS									
Halcyon	1934–39	1	—	—	—	—	—	—	—
Bangor	1940–42	—	—	—	—	1	—	—	—
Algerine	1942–45	1	—	—	—	s	—	—	—
Catherine (US)	1943–44	—	—	SL	—	—	—	—	—

Notes: * Replacing 271/2; s=Some ships; SA, SL=US radar.

Appendix 4: Table of Manufacturers Employed on Naval Radar

The following manufacturers collaborated closely with HM Signal School (later ASE) during World War 2 in the development and production of radar, countermeasures and direction-finding equipments for the Royal Navy.

A.C.I. Ltd	Cooling equipment (Messrs Moule and Clifford)
Joseph Adamson & Co Ltd (crane-builders)	Type 277S housing and antenna control (G. Adamson, A. V. Flinn)
Advanced Components Ltd	Voltage regulators (Mr Armstrong)
Aeronautical & General Inst. Ltd	Initial production of improved Type 79 (Mr Joseph – chairman and managing director)
Allen West Ltd	Manufacture of Types 277T, 279 (Messrs J. Bunyan, Paddick, James, Robb); design and manufacture of Type 91 (Messrs Sherrin and Strong) (seconded to ASE); design and manufacture of Type 271; manufacture of S25 DF frame-coil
Aluminium Plant & Vessel Co	Manufacture of antennae reflectors
Barr & Stroud Ltd	Antennae and mountings. CRSI (Dr French)
Bentley & Co of Leicester	Antenna mounting for Type 262 (A family firm)
Boulton and Paul Construction Ltd	Hydraulic control (Dr J. D. North)
British Insulated Callender Cables	Cables (Dr Welbourn, Dr D. T. Hollingsworth)
British Thomson Houston Co	H. N. Sporberg, Man.Dir., F. Clough, Ch.Eng.) Thyratrons, modulators, TR cells (L. J. Davies, H. de B. Knight, L. Herbert); 6-phase inductor alternators and power systems (H. Jack, J. Walker, Ian Yorke) (seconded to HM Signal School/ASE); design assistance with 600 MHz sets (F. A. Fossey) (seconded to HM Signal School/ASE); servo systems and voltage regulators (A. L. Whiteley, E. W. Forster, K. J. R. Wilkinson); magnetron production (A. King, D. J. Mynall, T. H. Kinman); silicon-crystal diodes (R. Latham, L. Rushforth); receivers types P19 and P29; lamps for skiatron displays (H. Bourne)
Bruce, Peebles Ltd	Motor and resolvers for Type 262 (Dr Bruce)
Brush Electric Co Ltd	
Bull Motors Ltd, Ipswich	
Clarke Chapman Ltd	Antennae, stabilised mountings
E. K. Cole Ltd	Indicators and display units for Type 271; Type 242 – receiving valves (A. R. Chance)
Cooke & Ferguson Ltd	

Appendix 4

A. C. Cossor Ltd	Cathode-ray tubes, ranging panels (L. H. Bedford, O. S. Puckle, L. Jofeh, H. Moss); receiving valves (Messrs Cotton and Brooker, F. M. Walker)
Coventry Gauge & Tool Co	(A. Harley)
Crabtree Ltd (Leeds)	Barrage Director for Type 283
Crompton Parkinson Ltd	Switchgear
Croydon Engineering Ltd	
Decca Radio Ltd	Navigational and marine radar (Mr Schwarz); Decca navigator
Dubilier Condenser Co Ltd	Precision components (Mr Philip Coursey); modulator pulse-forming networks
Edison Swan Co	Receiving Valves (E. Yeoman Robinson)
Electrical & Musical Inds. Ltd	(Dr Schoenberg, G. E. Condliffe, C. S. Agate, A. D. Blumlein). Cathode-ray tubes, klystrons (L. F. Broadway, Mr McGee, J. F. Cairns, N. C. Barford, A. F. Pearce, B. J. Mayo); design of target indication display (E. L. C. White, R. Puleston, S. T. Henderson, H. A. M. Clarke, Mr Wootton); design and production of Type 262 (10 000 MHz) M. B. Manifold, Messrs Cork, Turnbull, C. Metcalf, E. A. Nind, Ora, W.H. Connell, Pemberton); PPI outfit JE (Messrs Partridge, Bowman)
Elliott Brothers (London) Ltd	Fire-control equipment (Messrs Attwood, Harben)
English Electric Co Ltd	Antennae mountings (A. J. Young, Dr Sargent, F. H. Thompson); special valves (Messrs A. Aisenstein, Bratby)
Ferguson Radio Ltd	Naval adaptation of Army transmitter panel GL 1(T) (Messrs Adams, Sayer)
Ferranti Ltd	Ranging panels (Messrs E. Grundy, N. Searby, J. S. Toothill, Best); small-valve production (A. Chilcott, R. Cantelo); high-power X-band magnetrons; travelling-wave tubes; Type 251M, also submarine beacon (Messrs Norman Bell, Humber)
Foster Engineering Co	Transformers
General Electric Company	Development and production of transmitting valves and magnetrons (C. C. Patterson [Director], R. le Rossignol, M. Gavin, B. S. Gossling, W. H. Aldous, J. Bell, W. E. Willshaw, H. R. L. Lamont, E. Kettlewell, A. G. Stainsby, E. C. S. Megaw, H. Archer-Thompson, C. E. M. Hickin, R. J. Clayton); development and production of TR cells (N. L. Harris, E. G. James, D. G. Espley); design and production of receivers (G. W. Warren, N. Bligh, C. R. Dunham, H. T. Ramsay); design and production of modulators (Messrs Hickey, A. G. Stainsby); cathode-ray tubes (L. C. Jesty, J. Sharpe, R. G. Hopkinson, W. H. Aldous); design and production of display

	equipment; design and production of crystal diodes (J. W. Ryde, B. J. O'Kane, G. C. Edwards); radar propagation research (E. C. S. Megaw, J. W. Ryde)
G.E.C. Bradford	Receiver M68 (D. A. Drybrough)
G.E.C. Telephone Works, Coventry	Type 74 rotary capacity switches for Types 282/4/5
Gestetner Ltd	Consultants (Messrs D. Gestetner, (Snr) and S. Gestetner, (Jnr))
Grubb Parsons & Co Ltd	Antenna reflectors, mountings and servo-drives (G. M. Sisson)
J. & E. Hall & Co	Cooling units for Type 262
W. T. Henley Cables Ltd, Johnson & Phillips Ltd	Polythene cables
Laurence Scott Electro-Motors	Design and production of auto-radar plot (K. Worship, J. Platt, Dr Strang)
The Marconi Company	Antenna components and arrays (T. L. Eckersley, S. B. Smith, W. H. Picken, seconded to Admiralty as secretary CVD); reproduction of Type 271 Mark 4, Type 273Q and Type 277T (also on loan to Signal School: Messrs S. Falloon, N. Levin, Brett, G. Coop); monitoring instruments (N. Lea); countermeasures (including Types 650 and 651 jammers); HF DF (on loan to Signal School: Messrs N. E. Davis, W. Hall, O. E. Keall, G. Coop, Robb, R. F. Armstrong)
Marconi-Osram Valve Co	Small-valve production (Dr S. A. W. Jolliffe)
Mellor Bromley	Type 273 Mk IV, Type 277T
Metropolitan Vickers Ltd	Ranging-units, receiver for Type 79, modulators, receivers and displays, and remote power controls (Messrs B. A. G. Churcher, J. M. Dodds, Alexander, Nutter, Ghalib, A. E. Tustin, Davis); nacelles for Type 275 (E. G. Rowe, J. Davies); Master and slave PP1 for submarines (Messrs Scholes, Whalley); trigatrons (J. D. Cragg, M. E. Haine, J. M. Meek)
Mullard Radio Valve Co	Silica-envelope valve production (T. E. Goldup, D. S. G. Lewis); production of EF50 pentode valves (J. D. Stephenson)
Nash and Thompson	Hydraulically controlled antenna mountings (Messrs Morley, Usher)
Painton Instruments Ltd	Development and production of precision potentiometers (C. M. Benham)
Parkinson, Cowan	Ranging panels (Dr Lunt)
Parmeko Ltd	Small transformers
Permali Ltd	Antenna components and dipole arrays (C. Heath)
Pirelli General Cable Co	Development and production of polythene insulated cable (Messrs R. E. Soper, Cottrell)
Plessey Radio Ltd	HF and VHF direction-finding equipment (Messrs Rhys-Jones, J. V. G. Barrett, W. J. Pitts, R. F. Howard, J. Howell, J. B. Smith)
Precision Engineering (Haslemere)	Manufacture of antenna systems for shorebased Type 91 (T. Puttick, proprietor)

Pye Ltd	Ranging units development and production (Messrs C. O. Stanley, Dalgleish, Hawke, Stevenson, Dr Edwards)
Pyrotenax Cables Ltd	VHF cables (Messrs Davies, Tomlinson, Lewis, Hart)
Rediffusion Ltd	Various receiving and display equipment (Robert Renwick)
Research Associates Ltd, Reyrolle Electric Ltd	Control gear for antenna mountings (Messrs Goodhart, Peter Gladstone)
Rose Brothers of Gainsborough Ltd	Mounting of antennae on rangefinder directors, CRBFD and STAAG Bofors mounting with Type 262 (Mr Rose)
W. H. Smith (Engineers) Ltd	Prefabricated Type 271 P (W. H. Smith)
Sperry Ltd	Servo controls and stabilised mounts (R. Broadbent)
Standard Telephones & Cables Ltd	Development and production of amplifiers for 200 MHz and 600 MHz (Messrs C. E. Strong, C. N. Smyth, E. H. Ulrich, D. C. Rogers, Tomlyn, J. H. Fremlin, W. J. Gibson); development and production of microwave signal generators (C. W. Earp, P. H. Spagnoletti, L. J. I. Nickels); test equipment (L. S. Heaton-Armstrong); HF and VHF direction-finding equipment (R. H. Godfrey, R. F. Cleaver)
Telegraph Condenser Co Ltd	Modulator components (Mr Conyers)
Telegraph Construction & Maintenance Co. Ltd	600 MHz dipoles, polythene cable (J. N. Dean, Dr Randall, Dr W. T. Smith, Dr Schofield)
Teneplas (Pangbourne)	Antenna components
Vickers-Armstrong Ltd (Crayford)	Mountings for Types 285 and 275 (Messrs Winder, Watson, Henderson)
Vickers-Armstrong Ltd (Elswick)	Mountings for Types 284 and 274

Glossary of Technical Terms and Abbreviations Used in This Work

AA	Anti-aircraft
AB	Able Seaman
ABU	Auto-barrage unit for radar-controlled blind-fire
ACNS	Assistant Chief of Naval Staff
Acorn valve	A small high-frequency valve about 1-in in diameter shaped like an acorn, with the leads arranged radially around the middle of the envelope
ADEE	Air Defence Experimental Establishment (Army)
Admiralty organization	*Divisions of the Naval Staff* were responsible for operational matters, such as the Operations, Plans, Training and Staff Duties, and Naval Intelligence Divisions, reported to the First Sea Lord through the Deputy and Assistant Chiefs of Naval Staff. *Admiralty Departments* generally responsible for *matériel* matters, such as the Naval Ordnance, Naval Construction, Scientific Research, and Signals Departments, reported to the Controller. Operational sections of the Signal Department and the Naval Ordnance Department were in London, *matériel* sections in Bath from the outbreak of war
ADP	Air Defence Position
ADR	Aircraft Direction Room
ADRDE	Air Defence Research and Development Establishment (Army) (formerly ADEE, qv)
AFC	Automatic Frequency Control
AFCC	Admiralty Fire Control Clock
AFCT	Admiralty Fire Control Table
AFO	Admiralty Fleet Order
AGC	Automatic Gain Control, in a receiver
AGE	Admiralty Gunnery Establishment
AI	Air interception (airborne radar)
AIC	Action Information Centre
AIO	Action Information Organization
AITC	Action Information Training Centre
A/J	Anti-jamming (techniques)
AMRE	Air Ministry Research Establishment, later the Telecommunications Research Establishment
ANCXF	Allied Naval Commander-in-Chief Expeditionary Force
AOP	Angle of presentation (of an aircraft target)
AOR	Air Operations Room
ARL	Admiralty Research Laboratory
ARM	Availability, Reliability, Maintainability (of electronic equipment)
ARP	Auto-radar plot
A/S	Anti-submarine
A-scan	See *Display*
ASCBS	Admiral Superintendent, Contract-Built Ships

Glossary

Asdic	Underwater detection equipment using acoustic waves. An acronym based on Allied Submarine Detection Investigational Committee (c.1920). US – *Sonar* (q.v.)
ASE	Admiralty Signal Establishment (formed in August 1941 from the Experimental Department of HM Signal School)
ASV	Air to surface vessel (airborne radar)
Azicate	'Azimuth indicate' – to direct a radar of narrow antenna beamwidth onto a target, using an associated radar of wider search beamwidth
Azimuth	Horizontal angle measured about a nominally vertical axis
Back echoes	The main beam of a directional antenna is accompanied by some radiation in other, unwanted, directions, known as side lobes or back lobes. Such lobes can give rise to echoes from large targets at short ranges in those directions, which can be wrongly attributed to the direction of the main beam. Such unwanted responses are known as side echoes or back echoes
Beam-riding	A method of controlling a missile in flight whereby a radar beam held on the target is used by the missile to derive its own guidance signals
Beam-switching	A method of obtaining accurate bearing and/or elevation of a target by rapid switching of the radar beam between two adjacent directions, thus producing two echoes. When the echoes are equally matched, the bearing is correct
BFO	Beat frequency oscillator, in a receiver, to produce an audible output from an unmodulated carrier input
Blind fire	Using radar bearing, range, and if necessary elevation to fire at an unseen target
BPF	British Pacific Fleet
BPR	Bridge Plotting Room
BR	Book of Reference – an Admiralty publication with a security classification of Restricted or lower
BTH	The British Thomson-Houston Co Ltd
BTL	Bell Telephone Laboratories
BWO	Bridge Wireless Office
CA	Coast Artillery (Army radar)
CAAIC	Computer-Assisted Action Information Centre
CAFO	Confidential Admiralty Fleet Order
CAP	Combat air patrol
CB	Confidential Book – an Admiralty publication with a security classification above Restricted
CCU	Chart comparison unit, for optical matching to PPI display
CD	Coast Defence (army radar)
CDS	Comprehensive display system
CG	Centre of gravity
CH	Chain Home (RAF radar)
CHL	Chain Home Low (RAF radar)
C-in-C	Commander-in-Chief
CO	Commanding Officer, *or* Chain Overseas (RAF radar) see context
Common T/R working	The use of a single antenna array for both transmitting and receiving
CPO	Chief Petty Officer
CPU	Continuous Prediction Unit

Glossary

CRBFD	Close-range blind-fire director
CRDF	Cathode-ray direction finder (passive receiving equipment)
CRO	Cathode-ray oscillograph
CRP	Close-range predictor
CRT	Cathode-ray tube
CSWS	Civilian Shore-Wireless Station
CW	Continuous-wave (transmission)
CVD	Coordination of Valve Development (Committee) (sometimes assumed to be Communications Valve Development)
CVE	Escort aircraft carrier (US)
DC	Direct current
DCNS	Deputy Chief of Naval Staff
DCT	Director control tower, on which gunnery radar antennae were mounted, placed high in the ship to get the best possible view of the target, visually and by radar. Its laying and training on the target was followed (via the *TS* (qv)) by individual turrets and guns when in 'director firing', the normal method. Often abbreviated to *Director*. See also *HA Director*
DE	Destroyer escort (US)
DF	Direction finding
DGD	Director of the Gunnery and AA Warfare Division
Diplexer	Device to enable common transmit/receive functions
Dipole	Antenna element of two equal collinear rods, centre-fed, of total length equal to about one half the wavelength
Director	Common term for *DCT* (director control tower) (qv).
Discrimination	In radar, the ability to distinguish between (and if necessary range on) two targets close together. *Range discrimination* is achieved by very short pulse length, *bearing discrimination*, by narrow beamwidth. Also used to define the ability of radio receivers to distinguish between signals of different frequency
Display, radar	The method of presenting radar echoes to the observer. The most common naval types were:

'A' Display (formerly *A-scan*, the term generally used in this book), which shows the target's range (but not bearing) when the radar beam is trained on it.

PPI (plan position indicator), which shows simultaneously both range and bearing of targets as bright spots on a CRT with a long afterglow. It thus gives a complete picture of the surroundings as detected by radar. PPI display normally requires the antenna to be kept spinning or sweeping.

Skiatron; a display involving the optical projection of a form of PPI onto a ground-glass screen, to facilitate plotting. Echoes appear as dark paints, as opposed to PPI where they are normally bright echoes on a dark background.

Sector display, which shows on a type of 'A' Display all echoes within one or more selected small sectors of bearing. Used in height estimation and interrogation positions.

Meters are sometimes used to match echoes produced by beam-switching.

HPI (height position indicator), which, for a given bearing, shows elevation, range, and height on a display like that of a PPI

DNC	Director of Naval Construction
DNI	Director of Naval Intelligence
DNO	Director of Naval Ordnance
DRE	Director of Radio Equipment
DSD	Director of the Signal Department/Division
DSIR	Department of Scientific and Industrial Research
DSR	Director of Scientific Research and Experiment (Admiralty)
DTM	Director of Torpedoes and Mining
DTSD	Director of Training and Staff Duties
E-boat	German *MTB* (qv)
ECM	Electronic countermeasures – anti-radio/radar techniques and tactics, particularly the jamming of transmissions, or the production of artificial target echoes. Formerly *RCM* (qv)
EHT	Extra high tension (high voltage)
Elint	Electronic intelligence – the gathering of data by *ECM* (qv) techniques. Formerly known as 'Y' (qv)
EM	Electro-magnetic (waves)
EMF	Electro-motive force (voltage)
EMI	Electrical & Musical Industries Ltd
ERA	Electrical Research Association
EW	Electronic warfare, or Early Warning (radar) (see context)
FAA	Fleet Air Arm
FC	Fire control
FD	Fighter direction
FDO	Fighter Direction Officer
FKC	Fuze-keeping clock
Free-Space	A hypothetical concept used in simplified calculations of radar performance, in which the presence of the earth's surface in giving rise to a reflection component is ignored.
FWO	Fleet Wireless Officer
GA	Gunnery Fire Control, **A**ircraft (RN radar from 1943)
Gain	Factor by which power is increased
GB	Gunnery Fire Control, **B**arrage (RN radar)
GC	Gunnery Fire Control, **C**lose-range (RN radar)
GCI	Ground Control of Interception (RAF radar)
GDR	Gun direction room
GEC	The General Electric Company Limited (of England)
GEMA	Gesellschaft für Electroakustische und Mechanische Aufbau (for manufacture of German Naval radars)
GL	Gun Laying (Army anti-aircraft radar)
GMS	Guided missile system
GOP	General Operations Plot
GRU	Gyro-rate unit
GRUB	Gyro-rate unit box
GS	Gunnery Fire Control, **S**urface (RN radar)
HA	High angle (gunnery)
HACP	High Angle Control Position. Analogous to the *TS* (qv)
HACS	High angle control system
HA Director	A *DCT* (qv) when applied to long-range AA systems
HE	Height estimation of an aircraft target
HF	High frequency
HF DF	High-frequency direction-finding

Glossary

HFP	Height Filtering Position, where all available radar heightfinding information is received, filtered, and passed to where it is required
HMSS	His Majesty's Signal School, until 1941 housed in the Royal Naval Barracks, Portsmouth. HMSS's Experimental Department, responsible for radar development, became *ASE* in 1941
HPI	Height **p**osition **i**ndicator. See *Display*
HT	Height **t**ransmission (of target data)
IF	Intermediate frequency of a radar receiver
IFF	Identification Friend or Foe – an ultra-high-frequency radio *interrogator/responser* (qv) and *transponder* (qv) system used in association with warning radars to differentiate between friendly and hostile or unidentified contacts
Interrogator	A secondary radar transmitter that could activate an IFF *transponder*. See *IFF*
K-band	Electromagnetic wavelengths of about 1.5-cm (20000 MHz).
Klystron	Radio valve for amplifying or generating centimetric microwaves by forming electrons into bunches as they cross a gap, in a resonator
kyd	Kiloyard (one thousand yards)
LA	Low **a**ngle (gunnery)
LO	Local oscillator, in a superheterodyne receiver
Loran	American **LO**ng **RA**nge **N**avigational aid operating in the low-frequency band.
LOP	Local operations plot
LOS	Line of sight
LRS	Long-range system
LSF	Landing ship, fighter direction (a landing ship adapted for fighter direction)
MADP	Main air display plot
Magnetron	Microwave generator employing an external magnetic field. In its resonant-cavity development capable of producing extremely high power-output
ManP	Manoeuvre **p**redictor
MGB	Motor gunboat.
MHz	Unit of operating frequency (MegaHertz); frequency in MHz × wavelength in metres = 300
Mile	Where 'mile' is used in this book as a unit for distance at sea, the nautical or sea mile should be assumed – one minute of latitude, equivalent to 6080 ft (usually rounded off to 2000 yds in Naval practice, including wartime British Naval radar) and 1.8532 kilometres. The English land mile is 1760-yds
MIT	Massachusetts Institute of Technology
ML	Motor launch
MN	Merchant Navy
MPI	Mean point of impact (gunnery)
MPP	Most probable position, from cross-bearings of shore-based DF.
MRU	'Much regret, unable'; signal response to an *RPC* invitation.
MTB	Motor Torpedo Boat
NIB	Noise Investigation Bureau, for classifying and analysing intercepted enemy transmissions.

Glossary

NID	Naval Intelligence Division
nm	Nautical mile – see *Mile*.
NOD	Naval Ordnance Department
NPL	National Physical Laboratory
NRC	National Research Council of Canada
NRL	Naval Research Laboratory (USA)
NVA	Nachrichtenversuchsanstalt (German equivalent of HM Signal School)
NZDSIR	New Zealand Department of Scientific and Industrial Research
OR	Operations Room
OTC	(University) Officers Training Corps
PDF	Probability distribution function
PMO	Principal Medical Officer
PO	Petty Officer
PPI	Plan position indicator. See *Display*
Prf	Pulse repetition frequency
PTT	German Post, Telegraph and Telephone organisation
Radar	An acronym from **RA**dio **D**etection **A**nd **R**anging
Radar beacon	A type of IFF transponder that could be used for navigational or homing purposes by any suitably equipped ship, aircraft or land vehicle.
Radiolocation	The early name for radar
Radome	Radar dome, protecting the antenna array from wind, weather and spray
RAE	Royal Aircraft Establishment (RAF)
Rate-aiding	A mechanical or sometimes electrical means of establishing the rate of change of a target's range, bearing or elevation by following its position as smoothly as possible. The fire-control director or radar antenna was then driven at these rates by remote power control (*RPC*)/*servo-mechanism* (qv), which helped the operators follow the target more smoothly, which in turn resulted in more accurate determination of the target rates
RC	Radar control (ratings)
RCM	Radar countermeasures – earlier name for *ECM* (qv).
RCNVR	Royal Canadian Naval Volunteer Reserve
RDF	Radar
RDR	Radar Display Room. The compartment of the Action Information Centre where displays from all warning sets, and the Height Filtering Position, were situated
Responser	A secondary radar receiver to accept the response from a *transponder*
RF	Radio frequency
rms	Root-mean-square – square root of the arithmetic mean of the squares of a set of numbers (statistics)
RN	Royal Navy
RNR	Royal Naval Reserve
RNVR	Royal Naval Volunteer Reserve
RNV(W) R	Royal Naval Volunteer (Wireless) Reserve
RNZNVR	Royal New Zealand Naval Volunteer Reserve
RO	Radar officer
RP	Radar plot (ratings)

Glossary

RPC	Remote power control. See *Servo-mechanism*. Also, *see context*, 'Request the pleasure of your company', an invitation by signal
RRDE	Radar Research and Development Establishment (Army)
R/T	Radio telephony
RTU	Range transmission unit
SA	Ship-to-Air (RN radar up to 1943)
S-band	Electromagnetic wavelengths of about 10 cm (3000 MHz)
Scanning	The technical process whereby some radar sets automatically search in azimuth or in elevation, or in both simultaneously
Servo-mechanism	A closed-loop control system in which a small input power controls a large output power in a strictly proportionate manner. In this book, it is usually synonymous with remote power control
Side echoes	See *Back echoes*
Skiatron	See *Display*
SNSO	Superintending Naval Store Officer
Sonar	Underwater detection equipment. Acronym for **SO**und **N**avigation **A**nd **R**anging. The US version of British *Asdic* (qv)
Sp.	The Special Branch of the RNVR, to which many wartime radar officers belonged, for example, Lieutenant (Sp.) A. B. Jones, RNVR.
Special intelligence	from a particularly sensitive and absolutely reliable source, available as a result of the solution of high-grade codes and cyphers, distributed to specially selected and severely restricted numbers of recipients by means of one-time pad cyphers. Code name: *Ultra* (qv)
SRDE	Signals Research and Development Establishment (Army)
SRDFO	Senior Radio Direction Finding Officer
SS	Ship-to-Ship (RN radar up to 1943). Or Signal School (see context)
S/T	Sonic Telegraphy, for communications with or between submerged submarines
STAAG	Stabilized tachymetric anti-aircraft gun
Staff Requirements	When the development of a new Naval system or equipment was proposed, the operational requirements – in terms of range (maximum and minimum), accuracy, tactical deployment and, for radar, operating frequency – would be prepared by the Naval Staff with help and discussions with Admiralty Departments. These would be formalized as 'Staff Requirements' and sanctioned or rejected by the Admiralty Board. They then became the agreed basis for technical development
STC	Standard Telephones & Cables Ltd
SWR	Standing-wave ratio, a measure of the mismatch between an antenna and its associated feeder.
TB cell	Unit for blocking transmissions from entering a radar receiver
Thyratron	A gas-filled valve with heated cathode, able to carry very high currents, which operated as a switch
TI	Target indication
TIR	Target indication room
TIU	Target indication unit
Transponder	A radio or radar device that, on receiving a signal, transmits a signal of its own, as with *IFF* (qv)

Glossary

TR cell	Unit to permit Transmit/Receive operation on a single radar antenna
TRE	Telecommunications Research Establishment (RAF). Formerly AMRE (qv)
TS	Transmitting Station – a compartment between decks which housed the fire-control predictors. They were fed with target information (range, bearing, elevation) from optical instruments (sights and rangefinders) and/or radar equipment trained on the target; they calculated the future position of the target; and they transmitted to the gun mountings the predicted ranges, bearings and elevations for the guns to hit the target. In larger ships, the TS was concerned with LA (surface) fire control only, the HA equivalent being the *HACP* (qv)
TVA	Torpedoversuchsanstalt (German torpedo research Establishment)
U-boat	German submarine
UDU	Universal display unit, containing a PPI and sector display (see *Display*), which could be used with a range of Naval radar systems
UHF	Ultra high frequency, 300–3000 MHz
Ultra	Code name and message prefix for *Special Intelligence* (qv), and for messages containing special intelligence
USCGC	United States Coast Guard Cutter
USN	United States Navy
VACNA	Vice-Admiral Commanding, North Atlantic
VHF	Very high frequency, 30–300 MHz
VLF	Very low frequency, 3–30 kHz
V/S	Visual signalling
VST	Variable smoothing time
Vswr	Voltage standing-wave ratio (maximum-to-minimum) in a mismatched transmission system.
VT fuze	Velocity trigger fuze – a proximity fuze that uses radar principles to initiate the detonation of a shell or bomb at a supposedly lethal distance from an air target or at a set height above a surface target.
WA	Warning of Aircraft (RN radar from 1943)
Wavelength	Wavelength of operation in metres × frequency in MHz = 300
WC	Warning, Combined aircraft and surface (RN radar)
WD	Weapon direction
Window	Metal-foil strips, dropped in quantity from aircraft, which gave echoes capable of screening the aircraft from air defence radars. Also used on occasion to simulate large convoys of Allied ships in the deception role during invasion operations
WMP	'With much pleasure', signal response to an *RPC* invitation.
Working up	A period spent in exercises working up the efficiency of the ship's company of ships newly commissioned after building or refit
WRNS	Women's Royal Naval Service
WS	Warning of Surface craft (RN radar)
W/T	Wireless telegraphy
WW1	World War 1, 1914–18
WW2	World War 2, 1939–45
X-band	Electromagnetic wavelengths of about 3-cm (10 000 MHz)

Y equipment	Receiving and DF equipment employed for obtaining and analysing radio intelligence. Later known as *Elint* (qv)
Y service	Organization for the interception and DF of enemy radio/radar signals

Select Bibliography

General

By the very nature of this volume, the principal sources of information on the various topics presented, from historical background through innovative concepts, experiment, development, production, installation, maintenance and operational application lie in the Ministry of Defence sphere. The major source is the archive collection of formal documents, technical reports and appreciations contained in the surviving records of the Defence Research Establishment, Portsdown (originally the Experimental Department of HM Signal School, and, from 1941, the Admiralty Signal Establishment). Some of these are already held at the Public Record Office, Kew, as the ADM 220 series of documents. Others, still held in the Establishment, but not yet accessible to the general public, await review before submission to the PRO archives. Nevertheless access to these documents has been permitted by the Director for the various contributory authors to this work, to supplement their own recollections of wartime activities. This is acknowledged with gratitude. Thanks are also due to the Director of the Defence Research Establishment, Malvern (originally the Air Ministry Research Establishment, Bawdsey, and later the Telecommunications Research Establishment), for granting similar facilities.

Other important sources of information exist at HMS *Collingwood* (such as the collection of handbooks for wartime naval radar), and at the Ministry of Defence's Naval Historical Branch in London. The Admiralty's *Radar Manual (Use of Radar)* of 1945 (CB 4182/45, ADM 239/307) contains useful contemporary information on user matters. Both these sources have contributed handsomely in the preparation of this work.

One major published source of technical information on the whole range of Service radar developments during the war exists in the form of the volumes of the 'Proceedings of the Radiolocation Convention' held by the Institution of Electrical Engineers in London in 1946, published in a special edition of the Institution's *Journal*: volume 93, part IIIA, parts 1 to 10. Considerable reference to individual papers in these Proceedings has been made, and identified, by various authors contributing to this work.

In 1986 the Naval Radar Trust initiated the collection and classification of recollections, notebooks, diaries and so on, as the basis for the preparation of the technical monographs ultimately presented herein. The original material is to be preserved in the Churchill Archive Centre, Churchill College, Cambridge (CAC) as the 'NRT' series. Ultimately, these will be available for inspection on application.

Finally this bibliography includes reference to several published works that provide further technical background of a supporting nature, or as accounts of operational events in which wartime radar played a significant part.

Select Bibliography

Printed sources

ADMIRALTY, BR 1736 – *Naval Staff Histories* (select list)
- (6) *Mediterranean, Selected Operations 1940* (1943) – ADM 234/325.
- (7) *The Passage of the* Scharnhorst, Gneisenau *and* Prinz Eugen *Through the English Channel* (1948) – ADM 186/803
- (11) *Selected Convoys (Mediterranean), 1941—1942* (1957) – ADM 234/336.
- (17) *Sinking of the* Scharnhorst (1950) – ADM 234/343.
- (42) *Operation 'Neptune'*, 2 vols (1947) – ADM 234/366–7.
- (43) *Naval Operations in the Assault and Capture of Okinawa, March–June 1945 (Operation 'Iceberg')* (1950) – ADM 234/368.
- (44) *Arctic Convoys, 1941–1945* (1954) – ADM 234/369.
- (48/2) *Home Waters and Atlantic, April 1940 – December 1941* (1961) – ADM 234/372.
- (50) *War with Japan – Vol. VI, The Advance on Japan* (1959) – ADM 234/379.
- (51) *Defeat of the Enemy Attack on Shipping, 1939–1945*, 2 vols (1957) – ADM 234/578.
- (52) *Submarines Vol. 1, Home & Atlantic; vol.2, Mediterranean; Vol. 3, Far East* (1953, 1955, 1956) – ADM 234/380–2.
- (53) *The Development of British Naval Aviation, 1919–1945*, 2 vols (1954, 1956) – ADM 234/383–4.

———, Admiralty Fleet Orders (AFOs) and Confidential Admiralty Fleet Orders (CAFOs) – ADM 182 series.

———, CB 04050 series, *Monthly Anti-submarine Reports* – ADM 199/2057–62.

———, CB 04110 series, *HMSS/ASE monthly reports, 1941–6* – copies DRE.

———, CB 04272 series, *Coastal Forces Periodic Review* – copies NHB.

———, CB 4224 (42), *Heightfinding by R.D.F.* (1942) – copy HMS *Collingwood*

———, CB 04092/42, *Instructions for the Use of IFF Sets and RDF Beacons* (1942) – ADM 239/293.

———, CB 04092A/42, *Summary of RDF Identification (IFF, RDF Beacons and Interrogators)* (1942) – ADM 239/294.

———, CB 04262, *Notes on the Direction of Fighters by HM Ships* (1942 and 1944–5) – ADM 239/352 and photocopy CAC.

———, CB 3090, *Instructions for Installation and Fitting of R.D.F. Equipment and Associated Communications* (1943) – copy NHB and photocopy CAC; not found in PRO.

———, CB 04092/44, *Instructions for the Use of IFF Transponders and Radar Beacons by Allied Forces* (1944) – ADM 239/295.

———, CB 4224(44), *Height-finding by Radar* (1944) – copy HMS *Collingwood*.

———, CB 004385 A, B, C, *Report by the Allied Naval Commander-in-Chief Expeditionary Force on Operation 'Neptune'. (The Assault Phase of the Invasion of NW Europe, Operation 'Overlord')*, 3 vols (Oct. 1944) – ADM 239/367.

———, CB 03143, *Instructions for Coastal Force Warfare* (1944), with appendix on Control Ship technique (1945) – ADM 239/220.

———, CB 4182/45, *Radar Manual (Use of Radar)* (1945) – ADM 239/307 and photocopy CAC.

———, CB 3180, *Height Determination by Radar* (1949) – copy HMS *Collingwood*.

———, CB 4497, *Simple Guide to Naval Radar* (1949) – copy NHB and photocopy CAC; not found in PRO.

———, BR 2435, ex-CB 3213, *Technical Staff Monograph: Radar 1939–45* (1954) – ADM 234/539 and photocopy CAC.

Select Bibliography

AIR MINISTRY (Air Historical Branch), Second World War, RAF Signals – Vol. IV (CD 1063), *Radar in Raid Reporting* (1950) – AIR 10/5519; Vol. VI (SD736), *Radio in Maritime Warfare* (1954) – AIR 10/5555.

ALLISON, D. K., *New Eye for the Navy: the Origin of Radar in the Naval Research Laboratory*, NRL report 8466 (Washington DC, 1981).

BALDWIN, Ralph B., *The Deadly Fuze: Secret Weapon of World War 2* (Janes, London, 1980).

BARNETT, Correlli, *Engage the Enemy More Closely: The Royal Navy in the Second World War* (Hodder & Stoughton, London, 1991).

BASSETT, Ronald, *H.M.S. Sheffield: the Life and Times of 'Old Shiny'*, (Arms & Armour Press, London, New York and Sydney, 1988).

BEESLY, Patrick, *Very Special Intelligence: the Story of the Admiralty's Operational Intelligence Centre* (Hamish Hamilton, London, 1977).

BEKKER, C., *Augen durch Nacht und Nebel* (Gerhard Stalling Verlag, Oldenburg and Hamburg, 1958).

———, *The Luftwaffe War Diaries* (Macdonald, London, 1967).

BELCHEM, Major-General D., *Victory in Normandy* (Chatto & Windus, London, 1981).

BOWEN, E. G., *Radar Days* (Adam Hilger, Bristol, 1987).

BRICKHILL, P., *The Dam Busters* (Evans Bros., London, 1950).

BROWN, David, *Carrier Operations in World War II: Vol. 1, the Royal Navy* (Ian Allan, London, 1968, revised edition 1974).

———, *The Royal Navy and the Falklands War* (Leo Cooper, London, 1987).

BURNS, Russell (ed.), *Radar Development to 1945* (Peter Perigrinus/ IEE, London, 1988).

BURTON, E. F. (ed.), *Canadian Naval Radar Officers: the Story of University Graduates for whom Preliminary Training was given in the Department of Physics, University of Toronto* (University of Toronto Press, 1946).

CALLICK, E. B., *Metres to Microwaves: British development of active components of radar systems 1937 to 1944* (Peter Perigrinus/IEE, London, 1990).

CHURCHILL, Winston S., *The Second World War* (6 vols) (Cassell, London, 1948–54).

———, see also GILBERT.

CLARK, Ronald W., *The Rise of the Boffins* (Phoenix House, London, 1962).

———, *Tizard* (Methuen, London, 1965).

CLAYTON, Robert, and ALGAR, Joan, *The GEC Research Laboratories, 1919–1984* (Peter Peregrinus/Science Museum, London, 1989).

CONNELL, G. G., *Valiant Quartet: His Majesty's Anti-aircraft Cruisers Curlew, Cairo, Calcutta and Coventry* (William Kimber, London, 1979).

CONWAY, *Conway's All the World's Fighting Ships, Part I, the Western Powers* (Conway Maritime Press, London, 1983).

COSTELLO, John, and HUGHES, Terry, *The Battle of the Atlantic* (Collins, London, 1977).

CROWTHER, J. G. and WHIDDINGTON, R., *Science at War* (HMSO, London, 1945).

CUNNINGHAM, Admiral of the Fleet Lord, *A Sailor's Odyssey* (Hutchinson, London, 1951).

CUNNINGHAME GRAHAM, Angus, *Random Naval Recollections 1905–1951, Admiral Sir Angus Cunninghame Graham, KBE, CB, JP, of Gartmore* (Famedram Publishers, Gartochan, 1979).

ELLIOTT, Peter, *Allied Escort Ships of World War II: A Complete Survey* (Macdonald & Janes, London, 1977).

ERSKIN, R., *U-boats, Homing Signals and HF DF*, Intelligence and National Security (Frank Cass, London, 1987).
FRIEDMAN, Norman, *Naval Radar* (Conway Maritime Press, Greenwich, 1981).
———, *The Postwar Naval Revolution* (Conway Maritime Press, London, 1986).
———, *British Carrier Aviation* (Conway Maritime Press, London, 1989).
GILBERT, Martin, *Winston S. Churchill*, Vol. V, *1922–1939* (Heinemann, London, 1976); Vol. VI, *Finest Hour, 1939–1941* (Book Club Associates, London, 1983); Vol. VII, *Road to Victory, 1941–1945* (Heinemann, London, 1986).
GRETTON, Vice-Admiral Sir Peter, *Crisis Convoy: The Story of HX.231* (Peter Davies, London, 1974).
GUERLAC, Henry E., *Radar in World War II* (2 vols) (Tomash/American Institute of Physics, New York, 1987).
HANBURY-BROWN, R., *Boffin* (Adam Hilger, Bristol, 1991).
HARTCUP, Guy, *The Challenge of War: Scientific and Engineering Contributions to World War Two* (David & Charles, Newton Abbot, 1970).
— and ALLIBONE, T. E., *Cockcroft and the Atom* (Adam Hilger, Bristol, 1981).
HENNEY, K., *Radio Engineering Handbook*, 5th Edition (McGraw-Hill, New York, 1959).
HESSLER, G., *The U-boat War in the Atlantic* (HMSO, London, 1989).
HEZLET, Sir Arthur, *The Submarine and Sea Power* (Peter Davies, London, 1967).
———, *The Electron and Sea Power* (Peter Davies, London, 1975).
HINSLEY, F. H. et al., *British Intelligence in the Second World War* (3 vols) (HMSO, London, 1979, 1981, 1984).
HOWSE, Derek, *Radar at Sea: The Royal Navy in World War 2* (Macmillan, Basingstoke and London, 1993).
JOHNSON, B., *The Secret War* (BBC Books, London, 1979).
JONES, R. V., *Most Secret War* (Hamish Hamilton, London, 1978).
KNOWLES-MIDDLETON, W. E., *Radar Development in Canada: the Radio Branch of the National Research Council of Canada, 1939–1946* (Wilfred Laurier University Press, Ontario, 1981).
LEWIN, R., *Ultra Goes to War* (Hutchinson, London, 1978).
LOVELL, Bernard, *P.M.S.Blackett: a Biographical Memoir* (Royal Society, London, 1976).
———, *Echoes of War: The Story of H_2S Radar* (Adam Hilger, Bristol, 1991).
———, see also SAWARD.
MACINTYRE, Donald G. F. W., *U-boat Killer* (Weidenfeld & Nicolson, London, 1956).
———, *The Battle of the Atlantic* (Batsford, London, 1961).
———, *Fighting Admiral: the Life of Admiral of the Fleet Sir James Somerville, GCB, GBE, DSO* (Evans Bros., London, 1961).
MILLINGTON DRAKE, Sir Eugene, *The Drama of the Graf Spee and the Battle of the Plate* (Peter Davies, London, 1964).
MONTGOMERY, Field Marshal the Viscount, *Normandy to the Baltic* (Hutchinson & Co., London, 1958).
MONTGOMERY HYDE, H., *British Air Policy Between the Wars, 1918–1939* (Heinemann, London, 1976).
MORISON, S.E., *History of USN Operations in World War II*, Vol. I, *The Battle of the Atlantic, 1939–1943* (1953); Vol. X, *The Atlantic Battle Won, May 1943–May 1945* (1956) (Atlantic-Little, Brown Books/Oxford University Press).
MÜLLENHEIM-RECHBERG, Baron Burkard von (SWEETMAN, J. tr.), *Battleship Bismarck: a Survivor's Story* (The Bodley Head, London, 1980).
OTTAWA, Naval Officers Association of Canada, *Salty Dips*, Vol. 1 (Ottawa, 1983).

PACK, S.W.C., *Night Action off Matapan* (Ian Allan, London, 1972).
———, *The Battle of Sirte* (Ian Allan, London, 1975).
PAGE, Robert Morris, *The Origin of Radar* (Doubleday/Anchor, New York, 1962).
PARSONS, I. (ed.), *The Encyclopaedia of Sea Warfare* (Salamander Books, London, 1975).
POOLMAN, Kenneth, *Allied Escort Carriers in World War Two in Action* (Blandford, London, 1988).
POSTAN, M.M., HAY, D., and SCOTT, J.D., *History of the Second World War – Design and Development of Weapons*, Chapter XV – 'The Development of Radar' (HMSO/Longman Green, London, 1964).
POTTER, John Deane, *Fiasco: the Break-out of the German Battleships* (Heinemann, London, 1970).
PRICE, Alfred, *Instruments of Darkness: the History of Electronic Warfare* (William Kimber, London, 1967) (2nd ed., London, 1977).
PRITCHARD, David, *The Radar War: Germany's Pioneering Achievements 1905–1945* (Patrick Stephens, Wellingborough, 1989).
RAVEN, A., and ROBERTS, J., *British Battleships in World War II* (Arms and Armour Press, London, 1976).
———, *British Cruisers in World War II* (Arms & Armour Press, London, 1980).
ROBERTSON, Terence, *Walker, R.N.* (Evans Bros., London, 1958).
ROHWER, Jürgen, *The Critical Convoy Battles of March 1943: the Battle for HX.229/SC.122* (Ian Allan, London, 1977).
ROSKILL, S.W., *The War at Sea 1939–1945* (3 vols) (HMSO, London, 1954, 1956, 1960–1).
———, *Hankey* (3 vols) (Collins, London, 1970–4).
———, *Churchill and the Admirals* (Collins, London, 1977).
ROWE, A.P., *One Story of Radar* (Cambridge University Press, 1948).
SAWARD, Dudley, *Bernard Lovell: a Biography* (Hall, London, 1984).
SAYER, Brig. A.P., *Second World War, Army Radar* (War Office, London, 1958).
SCHOFIELD, Vice Admiral B.B., *The Loss of the Bismarck* (Ian Allan, London, 1972).
———, *Navigation and Direction: the Story of HMS Dryad* (Kenneth Mason, Havant, 1977).
SCOTT, Peter, *The Battle of the Narrow Seas: a History of the Light Coastal Forces in the Channel and North Sea, 1939–1945* (Country Life, London, 1945).
SMITH, Peter C., *Task Force 57: The British Pacific Fleet 1944–1945* (William Kimber, London, 1969).
SOMERVILLE, Sir James, see MACINTYRE.
SWORDS, S.S., *Technical History of the Beginnings of Radar* (Peter Perigrinus/IEE, London, 1986).
TAYLOR, Denis, and WESTCOTT, C.H., *Principles of Radar* (Cambridge University Press, 1948).
TERMAN, F.E., *Radio Engineer's Handbook* (McGraw-Hill, New York).
TIZARD, Sir Henry, see CLARK.
TRENKLE, F., *Die deutschen Funkpeil-und-Horchverfahren bis 1945* (AEG/Dr Hüthig-Verlag, Heidelberg, 1981).
———, *Die deutschen Funkstörverfahren bis 1945* (AEG/Dr Hüthig-Verlag, Heidelberg, 1981).
———, *Die deutschen Funkmeßverfahren bis 1945* (AEG/Dr Hüthig-Verlag, Heidelberg, 1986).
VIAN, Sir Philip, *Action This Day: a War Memoir* (Frederick Muller, London, 1960).
WAR OFFICE, see SAYER.

WATSON, D.W., and WRIGHT, H.E., *Radio Direction Finding* (Van Nostrand Rheinhold [Marconi Series], 1971).
WATSON-WATT, Sir Robert, *Three Steps to Victory* (Odhams Press, London, 1957).
WHITLEY, M.J., *Destroyers of World War II: an International Encyclopedia* (Arms & Armour Press, London, Sydney, 1988).
WILMOT, C., *The Struggle for Europe* (Collins, London, 1952).
WINTON, John, *The Forgotten Fleet* (Michael Joseph, London, 1969).
———, *Sink the Haguro* (Seeley, London, 1978).
———, *Find, Fix, and Strike* (Batsford, London, 1980).
———, *The Death of the Scharnhorst* (A. Bird Publications, Chichester and New York, 1983).
———, *Ultra at Sea* (Leo Cooper, London, 1988).

General Index

Primary references to radio and radar equipments are gathered together in the 'Equipment Index' on page 467 et seq.

Civilian staff of HM Signal School and Admiralty Signal Establishment are indicated by the terms HMSS, HMSS/ASE or ASE according as their period of service was before, through or after the change from HMSS to ASE (August 1941).

Illustrations and Tables in the main text are indicated by page numbers in **bold** type.

A

Abdiel, HMS (minelayer), 342, 396
Abercrombie, HMS (monitor), 388
Aberdeen, 343
Aberporth, propagation trials unit, 78
acceptance trials, 339
A. C. Cossor Ltd., 37, 64, 313, 402
Accountant Officer, 313
accuracy/precision of radar:
   bearing accuracy: early results, 15; beam swinging/switching, 32, 51, 142, 152, 154; split-lobe method, 133; gunnery precision, 51; Type 280 v aircraft, 33; Type 281, 152, 154; Type 271 and shore battery FC, 207; Type 284, 51
   elevation accuracy, 51
   height estimation precision, 55
   range accuracy, 32, 33, **59**, 145, 206, 207
Achilles, HMNZS (cruiser), 392
ACI Ltd, 401
A-class submarines, 264
acoustic detection, *see* Asdic
Action Information Centre, *xxxv*, 254, 259, 322
Action Information Organisation, 56, 254, 337
Activity, HMS (escort carrier), 390
Adamant, HMS (submarine depot ship), 398
Adams, Mr (Ferguson Radio Ltd), 402
Adamson, G. (Joseph Adamson & Co. Ltd), 401
Adamson (Joseph) & Co. Ltd, 401

Adelaide, HMAS (light cruiser), 392
A-Display (A-Scan), *see* Displays
Admiral's Plot, *see* Plots & Plotting
Admiral Superintendent Contract-Built Ships, 310
Admiralty, Board of:
   call for A/S radar, 190
   endorsed frequency choice, 21
   FD, 249
   First Lord of, 30, 31
   First Sea Lord, 303
   HMSS staffing and Leave, 16, 46
   Naval radar authorised, *xxxiii*, 6
   priorities, 15
   propagation, 78
   Report on Heightfinding, 299
   request to Canada, 264, 384
   ship fitting, 10, 308, 324, 325, 329
   sponsorship of CVD, 186, 189
   Third Sea Lord, *see* Controller of Navy
   under pressure to centralise research, 16
   and Type 284, 41
Admiralty:
   Fleet Order (AFO), 307, 322, 323
   Pattern (AP), 188
   Stores, 179
Admiralty Departments, 405
   see also Compass; Electrical Engineering; Naval Construction; Naval Contracts; Naval Ordnance; Radio Equipment; Scientific Research and Experiment;

General Index

Admiralty Departments (*cont.*)
   Signal (up to 1944); Torpedoes and Mining
Admiralty Gunnery Establishment (AGE), *xxxi*
Admiralty Mining Establishment, 17
Admiralty, Naval Staff Divisions:
   see Anti-submarine Warfare; Signal (from 1944); Tactical Training and Staff Duties
Admiralty Research Laboratory (ARL), 12, 30, 321
Admiralty Staff Requirements, *see* Naval Staff Requirements
Admiralty Signal Establishment (ASE)
– (formerly HMSS):
   Application Officers, *xxviii*
   auto-following, 58
   Chief Scientist, 61
   cosmic noise, 80, 81
   crystals, 122
   Departments/Divisions/Sections/Groups, **50**, **62**, **63**; Antenna Division, 70, 90; Anti-Jamming Division, 70, 84; Anti-submarine Warfare Division (A/S WD), 45, 191; Communications Department, **63**; Direction Finding Division, 60, **63**; Display Division, 73; Electronic Warfare, 61; Landale's Group, 187; M Division, 330; MF2 Section, 329; Radar Countermeasures Division, 84; Radar Department, **50**, 69, 70; Radio Warfare Division, **63**; 'R' Department, **50**; RE2, **50**, 58; XRC4, **62**, 73; XRC8, **62**, 70, 72, 73; XRC9, **62**, 70, 83, 85
   echo studies, 73
   ECM, 60, 61
   FD trials, 252
   founding of ASE, 53, 70, 122, 180, 406
   guided missiles, 58, 60
   handbooks, **50**, 312
   IFF, 60
   information handling, 56
   liaison, **50**, 60, **62**, 290, 307, 313, 322, 326
   locations (main premises and extensions); Bristol, 79, 259; Eastney, **50**, **62**, 260, 267; Haslemere, 61, 330; Nutbourne, 62, 70, 79, 80, 90, 163; Pinewood, **62**; Tantallon, **62**, 80, 165, 176, 252–3, 263; Witley, *xxvii*, 53, 61, **62**, 70, 187
   Mercury II, HMS (ASE) 64, 313
   PPI for ship use, 56
   propagation studies, 78, 79
   post-WW2, *xxix*
   radar camouflage, 85–6
   radar for Merchant Navy, 267
   re-organisations, *xxviii*, **50**, **62**, **63**, 70
   sea-clutter, 78, 82–3
   Sea-Trials Party, 315, 339
   secrecy, 73, 80, 84
   ship-fitting, **328**, 329
   Skiatron, 56, 125
   spares, 313
   staffing, 61
   Type 901 development, 60
   Ultra-Short-Wave Propagation Panel, 78
   valve development laboratory, 259
   waveguides, 86, 91
   'Window', 73, 79
   for earlier references, *see* HMSS
'Admiralty Trailer', 46, 194, 195
Advanced Components Ltd 64, 401
Adventure, HMS (minelayer), 396
aerials, *see* Antennae
Aeronautical & General Instruments Ltd (AGI), 36, 401
Africa, 53
'after action' radar sets, 314, 324, 357
Agamemnon, HMS (battleship), 7
Agate, C. S. (EMI), 402
AI, *see* Equipment Index
airborne radar, 82, 260
airborne torpedo attacks, 54
aircraft – and spark risk, 312, 365
aircraft, aspect, 241
aircraft carriers, *see* carriers
Air Council, 31
Aircraft Direction, *see* Fighter Direction
Aircraft Direction Room, 167
aircraft transporters, 392
Air Defence of Great Britain, 12
air defence of Fleet, 246
Air Defence Picket ships, 256, **257**
Air Defence Position (ADP), 312
Air Defence Research and Development Establishment (ADRDE), 80, 191, 207, 224, 226, **327**

General Index

Air Direction Room, 254
Air Ministry:
  ASVs for FAA, 28–9
  Chief of Air Staff, 31
  developments, 12
  Director of Communications
    Development (DCD), 16, 30, 192
  interest in ASE cm-radar, 208
  lead in IFF research, 277
  microwave radar, 186, 267
  Orfordness, 134
  PPI, 304
  in general fitting-out planning, **327**
  see also AMRE
Air Ministry Research Establishment, (AMRE):
  ASV, 15, 33
  Chain Home, 14–15, 278
  dependence on HMSS, 16, 107
  IFF, 278, 279
  locations, 14, 30, 278
  TRE, 189
  and Bowen, 24, 33, 111
  and SRDE Group, 14–15
  see also TRE
Airy Integral, 76
Aisenstein, A. (English Electric), 402
Ajax, HMS (cruiser), 392
Akehurst, A. G. (HMSS/ASE), **18**
Albacore aircraft, 357
Albatross, HMS (fighter catapult-ship), 392
Albion, HMS (light fleet carrier), 253
Alder, L. S. B. (HMSS/ASE), **18, 63**;
  provisional patent, *xxxiii*, 5, 11, 68, 106
Aldous, W. H. (GEC), 402
Alexander, Mr (MetVic), 403
Alexandria, 208, 312, 341
Algerine-class fleet minesweepers, 400
Allan, H. R. (ASE), 183
Allen West Ltd, Brighton:
  working with HMSS, 36, 64
  capacity, 208
  Projects and Staff, 401
  Type 271 series, 47, 196, **202**, 203, 217, 372
  Type 277T, 228
Allied aircraft, 280
Allies, 70, 282, 289
Alred, R. V. (HMSS/ASE):
  A/J, **62**, 84
  logarithmic receiver, 84

'Pig-trough' antenna, 41, 87
Yagi antenna, 25–6, 34, 37, 43
Aluminium Plant & Vessel Co., 64, 401
Alynbank, HMS (auxiliary AA ship), 396
Ambuscade, HMS (destroyer), 34, 41, 170
Ameer, HMS (escort carrier), 390
American police radio, 77
amphibious operations, 228, 266
  see also NEPTUNE
Amphion, HMS (submarine), 264
amplifiers: straight, 21
Anacostia, Washington, D.C., 290
Anderson, W. P. (HMSS/ASE), 12, 14, **18, 20**, 21
angular alignment, 316
anomalous propagation, *see* propagation
Anson, HMS (battleship):
  after-action set, 314
  alignment of radar and optics, 316
  ambient temperature in Tropics, 322
  first-fitting Type 277P, 269
  handbooks, 312
  Heightfinding Working-up, 315
  modified main-mast, 309, **309**
  radar staffing, 306, 307, 313
  remote displays, 310
  ship-fitting, 308, 309, 310
  space, 308
  and 'Bubbly', 245, 322
  sets fitted, 308, 309, 314, 322, 324, 364, 365, 380, 388
Antennae:
  and RF stage, 47
  and safety of staff, 318
  aperture, 25, 41, 88, 205, 261
  back-lobes, 26
  beam splitting, 207
  beam swinging, 32
  beam switching, 51, **59**, 369, 406; and accuracy, 152, 154, 158; and maintenance, 320; and performance, 154; and polar diagram, 154, **158**
  beamwidth, 360; gunnery sets, 39; S-band data, 271; and antenna stabilisation, 204–5; and frequency, 44, 111, 135, 188; and galactic noise, 80, 81; and target discrimination, 25
  'bedstead', 334

General Index

Antennae (*cont.*)
  'cheese': 'fan' beam, 194, 264; for wide vertical beam, 46; FD, 247; gain, 198, 220, 261; gain v beamwidth, 205; Lovell, 46, 194; side-lobe reduction, 198; stacked, 88; Swanage trials, 46, 194, 195; theory, 198, 199 TI, 57, 232; tilted, 247, **251**
  classification by code: ANA, 198; ANB, 198; ANS, 239; ANU, 377; AQR, 238; AQS, 252, 254, 256; AQT, 249, 252, 254; ATD, 362; ATE, 364, 365; ATQ, 368; AUJ, **237**, 237; AUK, **240**, 377; AUR, 232, **233**, 235, 238, 377
  common transmit/receive working (T/R), 406; 8: chronology, *xxxiv*, 269; competition for masthead, 147–8, 161; diode switch, 148; gas switch, 224; and bearing accuracy, 51; and improved gain, 225; and T/R switch, 87, 147, 161
  compatability with others, 335
  controls, 320
  cylindrical parabolic, 39, 41, 194
  design, 12, 13, 46, 136, 142, 187, 198, 204–5
  developments, 17, 25, 70, 87–9, 169
  dimensions, 13, 185
  dipole, 407; 8: basis of 'antenna gain' data, 359; initial IFF antenna, 278; multi-dipole arrays, 25, **42**, 51, 136; 'parasitic', 368; and reflectors, 25, 87
  directional, 12, 25
  director elements, 25
  distance from Radar Office, 71
  effect of height, 22, 204, 208
  experiments, 14, 17
  'fan' beam, 194, 264
  feeders: air insulated, 32; coaxial: advance on open-wire, 89; fitting, 338; losses, 26, 47, 89, 197; melted by power, 46, 90, 217; Pirelli-General, 37; polythene, 26, 37, 319, 330, 338; replaced by waveguides, 48; and rats, 310, 338; in 50-cm gunnery sets, 34; in early S-band sets, 336; *see also* pyrotenax
  losses, **59**, *see also* Appendix 2
  moisture ingress; 32, 37
  open-wire, 136, 153, 154, 302, 318
  pyrotenax: 90, 148; first RF application, 170; losses, 363; new to fitters, 330; steel sheathing, 171, 368; suppliers, 404; twin-core version, 170; unbalanced signal, 163; unusual damage, 310; and moisture, 148, 301, 310, 336, 340; in specific sets, 148, 162, 170, 171, 363, 368
  impedance matching, 38, 87
  sliprings, 365
  transmission lines, 89–90
  waveguides: brass v copper, 232; circular cross-section, 260; fitting: general, 228, 230, 336; cleanliness, 336; standard kits, 91, 230; introduction, 86, 90, 215; length of run, 224, 230; losses, 230, 232–3, 260, 270; phase-shifting, 88, 251; rotatable joints, 47, 248, 260; standing waves, 336; technical development, 185, 187, 217, 224, 258, 330; theory, 90; and ADRDE, 207; and antenna gain, 220; and magnetrons, 187, 226, 230; and moisture, 237
fitting, 335
for submarines, 174
Franklin Beam array, 25
gain, 193, 271, 359: 'cheese', 194, 198, 226, 256, 261; gunnery radar sets, **59**; 'pig-trough', 39, 41; technical data, 41, 182, 270–1; 'Yagi', 25–6, 43; and aperture, 25, 207; and beamwidth, 205; and bearing accuracy, 151; and paraboloidal reflector, 207; and radiation resistance, 87; and range, 214; and waveguide feed, 220
height, 190, 193, 194, 211, 214, 222, 348, 362, 364, 365; *see also* antennae – effect of height
lantern: integral with Radar Office, 48, 198; with teak supports, 198, 210, **211**, 336, 373, 375; side-lobes from supports, 210, **211**, 336, 373, 375; frequency 'pulling', 373
'lens', 60, 258
lightweight, 34, 237

lobes, *see* polar diagrams (below)
mountings, 60, 170, 330, 403
'nodding', 249
nutating beam, 60, 380
omnidirectional (for IFF), 278, 280, 281
on roof of radar office, 47
Outfits, 360, 474; *see also* classification by code (p. 424)
parabolic, 55
parabolic cylindrical ('pig-trough'), 39, 40, 319
paraboloid, 60, 194, 205
'pig-trough', *see* parabolic cylindrical
polar diagrams: back-lobes, 26, 170; calculation v. testing, 33; horizontal structure, **40**, **52**, 154, **158**; plotting kit, 313; side-lobes, 25, 87, 170, 198, 285, 336; vertical coverage, **236**, **238**, **244**, **251**, 256; vertical structure – factors: cause, 54; frequency, 22, 54, 154, 159, 190; antenna height, 190; plate separation, 198; local land, 315; receiver 'noise', 165; different sets, **40**, **55**, **150**, 167, 190; and Heightfinding, 74, 149; and performance of set, 149–50
polarisation, 136, 246; *see also* polarisation in general index
radar 'lens', 60, 258
radome, **212**, 219, 268
radiation patterns, *see* antennae, polar diagrams
radiation resistance/impedance, 87, 89
reflectors, 25, 39, 46, 87
rotation: chronology, 33, 124, 163; continuous, 162, 163, 182, 224, 225; early work, 21; light-weight for small ships, 34; non-continuous, 136, 158, 162, 163; rotation rate studies, 80; Selsyn drive, 158; speed, 80; technical data, 182; waveguide rotating joints, 224; and magnetron frequency changes, 91; and PPI, 163, 225
scanning: conical, 380, 382
side-lobes, 25, 37, 41
stabilisation: need, 46; chronology, 55, 205, 219, 268, 374; details (Type 295), 248; Outfit AUK, **240**; report after *Scharnhorst* Action, 223; requirements for Type 277, 228; rotary phase-changer, 88
stabilisation in: azimuth, 228; compass bearing, 210; elevation, 228, 374; pitch and roll, 57; single axis, 205 vertical, 88; full stabilisation, 247, 248
sweep rate, *see* specific Types
system development, 17
tapered array, **39**
T/R, *see* antennae – common transmit/receive working (T/R) trials, 33
two-mast working, 21, 33, 147
variants, 43
waveguide-fed horn, 48, 60, 248, **265**, 374, 384
Yagi ('fishbone'): early work, 25, 135; experimental 50-cm array, **35**; multiple arrays, 39, 43, **52**, 369; multiple dipoles per array, 136, 370; polar diagrams, **35**, **52**; unwanted lobes, 25; and Alred's design, 34, 87; and gunnery radar sets, 369, 370
Anti-Aircraft Battery, Hoy, 298
Anti-Aircraft Command, 58, 60
Anti-Aircraft cruisers, 28, 329, 335, 396
anti-aircraft – general:
 practice, Queen Bee aircraft, 7
Anti-Aircraft Range, Eastney, *xxxiii*, 26, 34, **35**, 38, 39
anti-aircraft rangefinder, 32
anti-DF facility, 146
anti-jamming:
 change of frequency, 158, 288, 317
 detail, 84, 176
 filters, 148, 165, 176, 245
 IF, 176, 245
 policy, 83
 XRC9 formed, 70
anti-minelaying, 208
anti-radar balloons, 86
anti-submarine measures:
 air escorts, 190
 Asdic, 190
 radar, 190, 192
 Swanage demonstration, 45, 191
 use of speed, 348
 5kW cm-radar, 211–15
 see also U-boats

Anti-Submarine Warfare Division, Admiralty, 45, 191
Antonio, HMS (later *Wayland*), 398
Antrim, HMS (armoured cruiser), 7
Antwerp, HMS (FD ship), 398
Apollo, HMS (minelayer), 396
Appleton, E.V. (later Sir Edward, FRS), 6, 34, 78
Application Officers, *see* HMSS, ASE and Naval Application Officers
Arbiter, HMS (escort carrier), 390
Arbutus, HMNZS (radar/radio maintenance ship), 342
Archer, HMS (escort carrier), 390
Archer-Thomson, H. (GEC), 402
'Area Radar Officer', 342
Arethusa, HMS (cruiser), 392
Argonaut, HMS (cruiser), 392
Argus, HMS (fleet carrier), 388
Ariadne, HMS (minelayer), 268, 396
Ariguani, HMS (fighter catapult ship), 392
Ark Royal, HMS (fleet carrier), 388
Ark Royal, HMS (post-WW2 carrier), 253
Armed Merchant Cruisers (AMC), 355
Armstrong, Lieutenant G.K. RNVR, 306, 312, 361
Armstrong, R.F., 403
Armstrong, Mr (Advanced Components Ltd), 401
Army radar:
  pre-WW2, 14–15
  with Air Ministry, 14, 30, 134
  liaison with HMSS/ASE, 326, **327**; Army radar used by RN, 28; Army components used by RN, 145; Naval radar used by Army, 206–7, 214–15, 228; use of Army test sites, 78, 226
  propagation, 77–8, 213
  circular-lines in transmitter tuning, 319
  high towers for CD radar, 215
  see also CD No.1 Mk.4, NT271X
Artifex, HMS (repair ship), 398
'artificial discharge line', 216
artificial load, 289, 313, 318
A-scan, *see* displays
Asdic, 45, 190, 344, 368, 374, 406
Ashford School, air raid warning and subsequent bombing, 228
'A Simple Guide to Naval Radar', 360

ASV, *see* Equipment Index
Atheling, HMS (escort carrier), 390
Athene, HMS (aircraft transporter), 392
Atlantic, Battle of, 45, 340
atmospheric 'noise', 9
atmospheric refraction, 77, 222
atmospheric water vapour distribution, 77
Attrill, G.S. (HMSS/ASE), **19**
Attacker, HMS (escort carrier), 390
Attwood, Mr (Elliot Bros), 402
Audacity, HMS (escort carrier), 390
Aurania, HMS (later *Artifex*), 398
Aurora, HMS (light cruiser), *xxxiv*, 173, 303, 392
Ausonia, HMS (repair ship), 398
Austin Smith, W. (HMSS/ASE), **18**
Australia, HMAS (heavy cruiser), 392
Australian Radar Officers, 345
Auto Barrage Unit (ABU), 51, 371
auto-following, *xxxiv*, 60
automatic gain control (AGC), 380
automatic frequency control (AFC), 92, 263
Auto-Radar Plot, *see* plots and plotting
Auxiliary Patrol, Scapa Flow, 297
Avalon, HMS (trawler), 203
Avenger, HMS (escort carrier), 390
Avenger aircraft, 357
Avometers, 179, 312–13, 318
Awards, *xxv*, 343, 346
'Azicator', *see* displays

B

Back, Captain T.H., 306, 310, 311
'back-biasing', 84
Bainbridge-Bell, L.H. (HMSS/ASE), 12, **50**, **62**
Baines, T.H. (HMSS/ASE), **18**
Baker, P.T.W. (HMSS/ASE), **18**
Baker, R. (author cited), 339
ball-and-disc integrator, 323
Ballard, A.E.R. (HMSS/ASE), **18**
balloons: barrage, 86; meteorological, 316, 381
bandwidth, *see* receivers
Bangor, HMS (Fleet minesweeper), 400
Banham, E.G. (HMSS/ASE), **19**
Bank-class (auxiliary AA ships), 355
Bareford, Dr C.F. (HMSS/ASE):
  pre-WW2, 13, 14, 15, 17, **18**, **20**
  receivers, 24, 25, **63**; auto-dyne, 26, 38
Barents Sea, 306, 321

Barfleur, HMS (destroyer), 380
Barford, N. C. (EMI), 402
Barham, HMS (battleship), 388
Barnsdale, E. A. (HMSS/ASE), **18**
Barr & Stroud, Ltd., 401
Barracuda aircraft, 259, 357
barrage fire-control radar, *see* gunnery radar
Barratt, J. V. G. (Plessey Ltd), 403
Barrel of Butter, Scapa Flow, **297**, 299, 316, 379, 381
Barrett, J. G. (HMSS/ASE), **18**
Barton, Lieutenant G., **19**
Base Radio Officer, Sydney, 348
Bassett, Warrant Telegraphist G. E., **19**
Bates, Lieutenant (Sp) H. R. K., RNVR (later Captain (L), RN), 346
Bath, Admiralty technical departments, 53, 335
Battle-class destroyers, 400
Battler, HMS (escort carrier), 390
battleships:
    numbers of sets fitted, 57
    types fitted, 41, 369, 388
    antenna height, 214
    stable platform, 204–5
    and dive-bombers, 34
    in Action, 44
Baverstock, C. O. (HMSS/ASE), **18**
Bawdsey Group (of researchers), 22
Bawdsey Manor, RAF radar research station: *xxvi*, 11, 68, 107, 134
Bay-class frigates, 400
Bayldon, Lieutenant Commander E. C., 45, 191–2
beacons, *see* radar beacons
beam-riding, 60, 406
beam-scanning, 251, **251**, 252
beam shapes, 376
beam splitting, *see* antennae
beam sweeping, 258
beam swinging, *see* antenna
beam switching, *see* antennae
beamwidth, *see* antennae
Bear Island, 321
Beaufighter aircraft, 227
Beckett, H. (HMSS/ASE), **19**
Bedford, Duchess of (aviatrix), 301
Bedford, L. H. (A.C. Cossor Ltd), 37, 58, 402
Bedruthan Steps, Cornwall, 79
Beech, T. H. (ASE), **63**
Beghian, L. E. (ASE), **63**, 78

beginnings of Naval radar, 106
Begum, HMS (escort carrier), 390
Belfast, HMS (cruiser), 84, 219, 315, 365, 374, 392
Belfast, 340, 342
Bell, J. (GEC), 110, 402
Bell, Norman (Ferranti), 402
Bell Telephone Laboratories (BTL), (USA): 81, 90
Bellona, HMS (cruiser), 392
Benham, C. M. (Painton Instruments Ltd), 403
Benjamin, R. (ASE), 56
Bennett, Lieutenant G. M., **19**
Bentley & Co., 401
Bermuda, HMS (heavy cruiser):
    maintenance, 320, 323
    spares, 314
    and Laws, 306, 310, 311, 320
    sets fitted, 311, 314, 320, 324, 392
Berwick, HMS (cruiser), 369, 392
Best, Mr (Ferranti Ltd), 402
Bethe, Hans (American physicist), 90
binomial distribution, 41, 87
Bird-class sloops, 400
Birkett, G. W. A. (HMSS/ASE), **18**
Birmingham, HMS (cruiser), 392
Birmingham University:
    chronology, 267
    CVD contract, 44, 186, 189
    Nuffield Laboratory, 44
    Randall and Boot, 44, 69, 111, 186
    resonant-cavity magnetron, 44, 111, 113, 186, 189, 215
    Sayers, 215
    'strapping' of magnetron, 215
Bishop, G. (HMSS/ASE), **18**
Bismarck (German battleship), 303, 345, 346
Biter, HMS (escort carrier), 390
Black Prince, HMS (cruiser), 392
Blake, K. W. (HMSS/ASE), **18**
Blakey, R. E. (HMSS/ASE), **19**
blast from own guns, 204, 223, 311
Blenheim, HMS (depot ship), 398
Bligh, N. (GEC), 402
blind-fire radar, *see* gunnery radar
Blumlein, A. D. (EMI), 17, 402
Blyth, 343
Board of Admiralty, *see* Admiralty
Bofors gun, 53, 58, 371, 382
Bogle, A. G., (later Professor), (HMSS/ASE), 30, 33, **62**, 183, 290

General Index

Böhm, O. M. (HMSS/ASE), 36, 70, 87–8, 198–9
Boles, Lieutenant Commander G. C., 7
Bolton, Warrant Telegraphist H., **18**
bombers, medium, 227, 231, 240
Bonaventure, HMS (light cruiser), 392
Bond, T. E. (HMSS/ASE), **19**
Bone, North Africa, 208
Bondi, H, (later Sir Hermann), (ASE), **62**, 70, 78, 81–2, 90
Booker H. G. (later Professor) (USA), 78
boom defence nets, 297
Boot, Dr H. A. H. (Birmingham University), 44, 69, 111, 186, 267
Borrow, R. L. A. (HMSS/ASE), **18**
Boston aircraft, 235, **236**
Boulogne, 207
Boulton & Paul Construction Ltd., 401
Bourne, H (BTH), 401
Bowden cable drive, 170, 171, 203, 368, 373
Bowen, E. G. (Air Ministry scientist), 12, 15, 24, 33, 111, 190
Bowen, K. C. (HMSS/ASE), **63**
Bowman, Mr (EMI), 402
Boxer, HMS (FD ship), 357, 398
Boys' Medal (of Physical Society), 118
'Brandy', Project (upgrading of warning radar in BPF), 341, 342
Bratby, Mr (English Electric), 402
breakdowns, 320–2
Brett, Mr, 403
Briggs, F. (HMSS/ASE), **18**
Briggs, J. (Defence Research Establishment), 92, 183, 272, 290
Brinkworth, A. J. (ASE), **62**
Bristol, (HMSS/ASE Extension), 79, 87, 259
Bristol University, 31, 87, 128
Bristow, H. M. (HMSS/ASE):
  Eastney, 13, 21, 177
  pre-WW2, **19**, **20**
  RC5/XRC5, **50**, **62**
British Broadcasting Corporation, 136
British Insulated Callander Cables, 401
British Pacific Fleet, 75, 150–1, 322, 341–2, 380
British Thomson-Houston Co. Ltd (BTH):
  equipment, 401
  generators, 37, 38
  magnetron development, 112–13
  modulators, 37, 38, 248

silicon-crystal diodes, 122
staff seconded to HMSS, 36, 401
thyratrons, 27–8, 114
working with HMSS, 64
and Type 295, 248
and Watson, D. S., 177
British Universities, 344
  see also Cambridge, Bristol, Birmingham, Oxford
Broadbent, R. (Sperry Ltd), 404
Broadway, L. F. (EMI), 402
Brooker, Mr (Cossor Co.), 402
Brooks, H. A. (HMSS/ASE), **18**
Brown, C. O. (EMI), 17
Brown, G. H. (USA), 87
Brown, (John) & Co. Ltd, 306, 310
Brown, R. H. J. (HMSS), **19**
Bruce, Captain the Hon J. B., 13, **18**, 306, 345
Bruce, Dr (Bruce, Peebles Ltd), 401
Bruce, Peebles Ltd., 401
Brundrett, F. W. (later Sir Fredrick) (HMSS), 6, 13, 15, 30, 31, 36
Brush Electric Co. Ltd, 401
'Bubbly', Project (upgrading of warning radar in BPF), 245, 322, 341–2
Buchanan, T. J. (HMSS/ASE), 128
Buckie, steam drifter, 298
Bull Motors, Ltd., Ipswich, 401
Bulolo, HMS (Headquarters Ship): 396
Bulwark, HMS (light fleet carrier), 58, 253
Bunyan, J. (Allen West Ltd), 401
buoys, **262**, 265, 316
Burcham, W. E., (TRE), 44
Burchill, Sub-Lieutenant H. G., RCNVR, 347
Burgess, P. L. H. (HMSS/ASE); **19**
Burns' Night Dinner, Scapa Flow, 304
Burtt, G. (HMSS/ASE), 17, **63**
Butement, W. A. S. (ADEE scientist), 15
Butt, W. S. (HMSS/ASE), **19**

C

cable testing, 313, 319
Cairns, J. F. (EMI), 402
Cairo, HMS (AA cruiser), 396
Calais, 207
Calcutta, HMS (AA cruiser), 396
Caledon, HMS (AA cruiser), 396
Calf of Flotta, 316

calibration of equipment, 314–17
  HMSS Test Department, 64
  portable oscillator, 315
  of displays, 147, 315–16, 371
  and FC system, 371
  and gunnery index error, 360, 379, 381
Callick, E. B., 131
Calpine, H. C. (HMSS/ASE), 24, 25–7, 49, **50**, **62**
Calypso, HMS (cruiser), 392
Cambridge Instrument Co., 313
Cambridge University, 30
camouflage, 85–6
Campania, HMS (escort carrier):
  Trials of Type 277, xxxv, 55, 240–2, **241**, **243**, 249–50, 269
  sets fitted, 390
Campbell, HMS (destroyer), 222
Camperdown, HMS (destroyer), 380
Canadian development and production of Naval radar sets, 264, 384
Canadian Government, 384
Canadian physics graduates, 299, 345
Canadian Marconi Company, 315
Canadian National Research Council, 61, 264
Canadian Radar Officers, 305, 345
Canberra, HMAS (heavy cruiser), 392
Cantelo, R. (Ferranti Ltd), 402
capacitors, 179, 314
Capel, Dover, 228
Capetown, HMS (light cruiser), 28, 392
capital ships, 39, 175, 181, 246
Captain-class frigates (Lend-lease), 357, 400
Captain Cook Dock,Sydney, 310
Captain, H. M. Signal School, 10, 15, 273
Caradoc, HMS (cruiser), 392
Cardiff, HMS (cruiser), 392
Cardigan Bay trials, 78
Carey, Joyce (novelist), 346
Carey, Lieutenant (Sp) Michael, RNVR (later Sir Michael), 346
Caribbean Naval Bases, 302
Carlisle, HMS (AA cruiser), 28, 33, 355, 396
Carmichael, Lieutenant (Sp), RNVR, 300
carpenters, 337
carriers:
  cm-FD radar, 246–55, 256, 269–70
  dual installation Types 79B/281BQ, 165–9

homing beacon, 106
need of warning time for interception: 54, 134
priority, 28
'time sharing' of antennae, 335
and heightfinding practice, 159, 161
and Far East, 54
Castle-class corvettes, 244, 400
Catalina aircraft, 54
Catania (VHF link), 317
catapult-ships, 392
Cathedral-class destroyers, 256
Catherine, HMS (fleet minesweeper) (Lend-lease), 400
cathode-follower, 141, 142, 159, 287
cathode-ray tube, integration effect, 73
  see also valves and displays
Cava, Scapa Flow, **297**, 315, 316
CB 3090 – Antenna outfits, 360, 361
CB 4497: 'A Simple Guide to Naval Radar', 360
C-class AA cruisers, 335
Cecil, Commander the Hon. Henry M. A., 9, 15, **18**
Centaur, HMS (light fleet carrier), 253
centimetric radar, *see* Wavelength
Centurian, HMS (battleship), 7
Ceres, HMS (cruiser), 392
Ceylon, HMS (cruiser), 392
Ceylon, 341
Chain Home (CH) (RAF radar):
  initiated, 15
  polarisation, 136
  Scapa air-cover, 142
  Stations with RN manning, 295, 300
  wavelength, 136, 278
  and first Type 281, 161
  and HMSS silica valves, 95, 97, 110
  and IFF, 278, 279
Challans, D. W. S. (HMSS/ASE), **18**
Chance, A. R. (E.K. Cole Ltd), 401
Charts for radar, 266–7
Charybdis, HMS (cruiser), 369, 392
Chaser, HMS (escort carrier), 390
Chatham Dockyard, 296, 340, 342
Cheeseman, Lieutenant W., RCNVR, 315
Chelmsford, Essex, 61, 176
Chesapeake aircraft, **175**
Chief of Defence Staff, 303
Chilcot, A. (Ferranti Ltd), 402
Childs, A. H. (HMSS/ASE), **18**
Chivers, E. W. (Army scientist), 15

Christchurch, Dorset; ADRDE, 207
Christie, H. E. (HMSS/ASE), **18**, 64
chronology of development of radar
  sets, *xxxiii–xxxv*, Appendix 2; *see
  also* specific sets
Churcher, B. A. G. (MetVic.), 403
Churchill, Sir Winston, Prime Minister,
  1940–45, 30, 31, 60, 190
Clarendon Laboratory, Oxford
  University, 87, 259
Clarke, H. A. M. (EMI), 402
Clarke, J. W. (HMSS/ASE), **18**
Clarke Chapman Ltd, 401
Clayton, R. J. (GEC), 402
Cleaver, R. F. (STC), 404
Cleopatra, HMS (cruiser), 392
Clifford, Mr (A.C.I. Ltd), 401
Close-Range Blind-Fire Director
  (CRBFD), 354, 382, 383, 404
close-range, blind-fire, radar, *xxxv*
Close-Range Director, 34, **35**
Close-Range Predictor, 371
Clough, F. (BTH), 401
clutter, *see* sea-clutter
Clyde:
  sea trials in Firth, 196, 244
  shipbuilding, 310, 331
  Maintenance Base, 340
  Port Radar Organisation, 342–3
  Sherbrooke House, 343–4
Clydebank, 306, 310
Clyde Division RNVR Gunnery
  School, 343
Coales, J. F., (later Professor), (HMSS/
  ASE):
  Author (Mono. 1), *xxxvii*
  antennae, 12
  crystal-diode mixer, 38, 44
  decimetric development, 17, **20**, 21,
    44–5
  gunnery radar, 23, 45
  guided-missiles, 58
  HF DF, 9
  long transmission-lines, 89
  low-loss cables, 26
  NPL Committees, 9
  selected 50-cm wavelength, 21
  Service, 9, 17, **18**, **20**, 50, **62**, 70
  television tests, 17
  and E773, 26–7
  and Horton, *xxxi*
    and installation in Prince of
      Wales, 49
  and Megaw, 17, 45
  and Naval Officers, 65, 306
  and secrecy, 26–7
Coast Defence radar (Army), 372
Coastal (Defence) Chain, 207, 228, 306,
  372, 374
Coastal Command, *see* Royal Air Force
coastal craft and coastal forces:
  technical data, Appendix 1, 367, 384
  see Types 268, 286U/PU, 291U, SQ;
  see also MGBs, MLs and MTBs
coastal defence, *xxxiv*, 207, 215, 268
coastal defence radar, 48, 55, 206–8,
  215
Coastal Forces radar, 264, 337
  see also Types 268, 286U, 291U, SO
coastguard cutters (ex-USCGC), 400
coaxial cables, *see* antennae,feeders
coaxial-line (in mixer), 200
Cochrane, C. A. (HMSS/ASE):
  Author (Mono.5)
  contributor, *xxxvii*, 183, 361
  Eastney, 187
  echo strength, 73
  pulse-to-pulse variation, 80
  Swanage team, 46, 193
  XRE1, **62**
  and Type 271 cable, 90
Cole, E. K. Ltd., 401
Collingwood, HMS (radar and electrical
  school), (Fareham, Hampshire):
  contributor, 131, 183, 290
  documents, 361
  radar training, 64, 345
  and Patrick, 347
Colombo, HMS (AA cruiser), 28, 396
Colombo, 341
Colony-class frigates (Lease-lend), 357,
  400
Colossus, HMS (light fleet carrier), 312,
  315, 365, 366, 388
Colossus class light fleet carriers, 312
Combined Operations,navigation
  radar, 356
Combined Research Group
  (IFF), 289–90
Committee for the Scientific Survey of
  Air Defence, (CSSAD), *see* Tizard
  Committee
common T/R working, 319, 335, 406
Communication, Theory of, 82
communications system, 17
Compass Department, Admiralty, 200

components, availability:
  development of new components/technology: 215
  electrical, 179, 360
  frequency limitations, 135, 152
  need to use available items, 152, 195
  under stress with new developments, 224
  and performance, 253
components, tolerance, 330
concentric-line circuits, 110, 116
Concord, HMS (HMSS cruiser, 1928–32), 9, 10
condensers, *see* capacitors
Condliffe, G. E. (EMI), 402
Confidential Admiralty Fleet Orders (CAFOs): 307
conical scanning, 380, 382
Connell, W. H. (EMI), 17, 402
construction programme, 329
contactors, and blast, 311
Continuous Prediction Unit, 371
Contracts Department, Admiralty, 10, 64
Controller of Navy (and Third Sea Lord):
  control of HMSS, 10, 16
  radar, 12, 14, 29, 134
  resisted centralised research, 16
  and guided missiles, 60
  and Churchill, 31
  and Somerville, 30
  post war appointment, 303
conversion kits, 245
convoy escorts:
  general, 53
  antenna height, 190
  chronology, 45, 47–8, 185, 186, 203, 244, 268
  corvettes, 190
  first operational S-band radar, 185
  maintenance, 340
  need for radar, 186
  numbers, 48, 203
  prefabrication of radar, 210, 268
  S-band for A/S work, 45, 46, 47
  sea conditions, 46
  sightings, 223
  and Type 277, 244
  in Action, 53
  sets fitted, 400
convoys – radar for escorts, 47, 191–2, 193

  – station keeping, 210
  – PQ13, 321
convoys – classified geographically:
  Atlantic, 53, 304
  East Coast, 340
  Mediterranean and Gibraltar, 44, 213
  Russia, 53, 321
convoys – German, 53
Conyers, Mr (Telegraph Condenser Co.), 404
Cook, A. H. (later Professor) (ASE), **62**, 90
Cook, Sir William, *xxxi*
Cooke & Ferguson, Ltd., 401
cooling of equipment, 308, 322, 401
Coop, G. N. (HMSS/ASE), 36, 403
Coordination of Valve Development (CVD) Committee:
  Admiralty sponsored, 31, 186, 189, 259, 267
  formation, 27
  klystrons, 44, 189
  magnetrons, 44, 188–9
  staff on loan from industry, 403
  and Birmingham University, 44, 186, 189
  and grounded-grid triode, 27
Coppercliff, HMCS (corvette), 244, 265
Cork, Mr (EMI), 17, 402
Corner, W. D. (ASE), 183
corner reflectors, 86
Cornwall, HMS (cruiser), 392
corvettes:
  antenna height, 214, 262
  as target for Type 291, **175**
  Flower-class corvette, 190, 196, 400
  integral Office/Lantern, 48, 225
  magnetic screening, 200
  numbers with radar, 195, 203, 400
  rolling in heavy seas, 220
  single radar fitted, 328
  working-up, 339
  and Type 271 A/S radar, 268
  and Type 271Q, 268
  and Type 242 IFF, 288
cosmic and galactic 'noise', 80–1, 149, 317, 362
Cossor, A.C. Ltd, 37, 64, 179, 313, 402
cost effectiveness, 143
Cottesmore, HMS (destroyer), 336
Cotton, Mr (A.C. Cossor), 402
Cottrell, Mr (Pirelli), 403
countering countermeasures, 29, 60

countermeasures, **63**, 73, 403, 408
County-class cruisers, 336
Courageous, HMS (fleet carrier), 388
Coursey, P. (Dubilier Condenser Co. Ltd), 402
Coventry, HMS (AA cruiser), 396
Coventry Gauge & Tool Co., 402
Coward, Signaller Boatswain H., **19**
Cowes, Isle of Wight, **266**
Cox, W. (HMSS/ASE), 296
Crabtree Ltd (Leeds), 402
Cragg, J. D. (MetVic.), 403
Craig, Lieutenant R. W., **18**
Crampton, C. (HMSS/ASE), 9, **18**, **20**, **63**
Crayford, Vickers-Armstrong, 404
Crofton, W. (HMSS/ASE), **18**
Crompton Parkinson Ltd, 402
Croney, Joseph (HMSS/ASE):
  recruited from industry, 30
  Receiver Division/Section, **62**, 187
  receiver P16, 26, 38, 43
  Swanage team, 46, 193
Crowe, Sir Alwyn, 60
Croydon Engineering Ltd, 402
CRS 1, 401
cruisers:
  blast from own guns, 204, 210, 311
  dive bombers, 34
  fitting policy, 329
  navigational radar, 264
  sea trials, 15
  sets fitted: gunnery, 39, 41, 49, 51, 57, 369; low air-cover, 269; TI, 232, 238–9, 269; WC, 376; WS, 269, 376, 384
  stability and antennae, 204–5
  radar in Naval Actions, 53–4
  radar staffing, 307
crystal-controlled calibration markers, 371
crystal-controlled standards, 371
Cumberland, HMS (heavy cruiser), 222, 336, 392
Curacoa, HMS (AA cruiser):
  early radar fitting, 28
  radar damaged in Action, 296
  sets fitted, 145, 396
Curlew, HMS (AA cruiser):
  air-cover at Scapa, 142
  anomalous propagation, 77
  first fitting, Type 79Z, xxxiii, 23, 28, 133, 140, 142, 181

  sunk, 300
  trials on modified Type 79, 32
  sets fitted, 396
curvature of earth, 222
CV numbers (of valves): p 475
Cyclops, HMS (repair ship), 398

D

Dahl, Major H., (ASE) (Norwegian scientist), 82, 90, 263, 377
Dalgleish, Mr (Pye Ltd), 404
Daly, E. F. (HMSS/ASE), 36
Danae, HMS (cruiser), 392
Dasher, HMS (escort carrier), 390
data transmission, 323
Dauntless, HMS (cruiser), 392
Davies J. (MetVic), 403
Davies, L. J. (BTH), 27, 401
Davies, W. T. (HMSS/ASE), **50**, 51, 92, 187
Davies, Mr (Pyrotenax), 404
Davis J. (HMSS/ASE), **19**
Davis, N. E. (HMSS/ASE) (on loan from Marconi), 36, **63**, 403
Davis, R. (Mullard Co.), 131
Davis, Mr (MetVic), 403
DCD, see Air Ministry, Director of Communications Development
Dean, J. N. (Telegraph Construction and Maintenance Co. Ltd), 404
Decca Radio Ltd., 402
decimetric radar, see wavelength
Dee, P. I. (later Professor) (TRE), 186, 192, 267
Defence Committee, 190
defence of Fleet, 134
Defence of Realm passes, 298
Defence Research Establishment, 92, 183, 272
Defiance, HMS (Torpedo School), 6
de-gaussing, 344
delay-lines, 25, 38, 85, 172–4, 367, 382
Delhi HMS (AA cruiser), 396
demarcation practices, 337
Denmark Strait, 208, 228
Department of Scientific and Industrial Research (DSIR), 34
Derby House, Liverpool, 340
design, see radar set design
Despatch, HMS (cruiser), 392
destroyers:
  antennae, 51, 169
  low air-cover, 269

maintenance, 340
need for radar, 33, 169
pre-WW2 new construction, 329
target in radar trials, **175**, 235–7
unsuitable for some sets, 203
working-up, 339
WS, 269
in Action, 53
sets fitted, 174, 226, 232, 238, 262, 400
detection of aircraft – early work, 106
detection of very low aircraft, 167
detection of ships in coastal waters, 206
Devonport, 33, 304, 308, 309, 340
Devonshire, HMS (heavy cruiser), 394
Dewar, G. D. (HMSS/ASE), **18**
DH Rapide aircraft, 300
Diadem, HMS (cruiser), 394
Dido, HMS (light cruiser):
 Type 281 trials, 33, 47, 151, 302
 sets fitted, *xxxiii*, 181, 364, 394
Dido-class light cruisers, 31
dielectric properties, 76, 98
diffraction of EM waves, 76
Diomede, HMNZS (cruiser), 394
directed projectiles, *see* guided missiles
direction finding, *xxv*, 9, 21, 29
Director of Communications Development (DCD) (Air Ministry), 30, 192
discharge lines, 216, **218**, 226, 268, 374, 376
discrimination, 407:
 range discrimination, 151–2, 167
 target discrimination, 25, 163, 185, 265–6
displays:
 general; anti-clutter device, 84; calibration marks, 147; different scan speeds, 146; evolving technology, 56; saturation, 360; suppliers, 402; *see also* RDR
 A-scan, 407; accurate ranging modification, 187; afterglow screen in anti-DF use, 146; assumptions, 72; commercial TV, 178; comparison with PPI sensitivity, 231, 234; description, 123–4; different practice in RN and RAF, 360; focusing, 139; inadequate for WA, 225; linearity, 37; tube diameter v accuracy, 37; used with fixed antenna, 229, 231; variants:

precision-bearing, 159; precision-ranging, 159, 315; warning, 159; and beam-switching, 154; and beam-width, 142; and IFF, 280; and ranging, 51; and Sector Display, 176; and 50-cm radar, 25; and 50-cm gunnery radar, 34, 36, **40**, 315; Type 79, 21, 139; Type 261, 262; Type 271X, 202; Type 277T, 226; Type 277X, 229, 231; Type 281, 159; Type 290, 174; Type 291, 174
Azicator, 242, 406
Height Position Indicator (HPI), 229, 242, 249, 377, 407
Meters, 407
Plan Position Indicator, (PPI), 124, 407; introduction, *xxxv*, 185, 282–3, 295; 'JE', 225, 229, 402; major technological advance, 124, 259; Master and Slave/ remote, 56, 225, 264, 310, 403; miscellaneous: and FD, 163; and IFF, 282–3, 285; and Tactical Plot, 57; and TI system, 57–8; and Type 279, 163; and UDU, 176; and warning radar, 55, 163; operational: in aircraft detection, 224; in 'blind' navigation, 265–6, **266**; in *Scharnhorst* action, 223; rotation, 162, 163, 225, 230, 231, 234; sea-trials, 174; shipborne version, 56, 162–3, 210; signal v range, **235**; suppliers, 162, 229, 402, 403; technical points: azimuth stabilising coils, 231; echo visibility, 73; phosphors, 124; RF slip-rings, 162, 163; sensitivity (v A-display), 228, 231; side-lobes, 285; swept-brightening modification, 322; swept-gain, 165, 232, 285
Sector Displays, 164, 165, 285, 407
Skiatron, 407; dark-trace propagation tube, 56, 125; tube supplier, 401; and earth's magnetic field, 366; and FD, 56, 163, 240, 241, 246–7, 252; and IFF, 285; and RDR, 167–8
Tactical, *see* PPI, Skiatron
Target Indication, 402

displays (cont.)
  TI Plan Display, 226, 232
  Universal Display Unit (UDU), 176, 412; see also RDR
dive-bombers, 34, 47, 54, 57, 369
Dockyards, see HM Dockyards
Dodds, J. M. (MetVic), 403
Domb, Cyril (later Professor) (HMSS/ASE):
  recruited, 70
  echoes, 73, 81
  propagation, 76, 77
  waveguides/magnetron interaction, 92
  and Böhm, 88
Dooley, A. (HMSS/ASE), **18**
Doppler effect, 83, 84
Dorsetshire, HMS (heavy cruiser), 394
Dover, *xxxiv*, 207, 222, 228, 268, 269
Dow, Commander USN (technical liaison officer), 304
Drabble, J. C. W. (HMSS/ASE), **18**
Dragon, HMS (Polish light cruiser), 394
Drury, Warrant Telegraphist H. V., **19**
Drury, W. F. (HMSS/ASE), 25, 303, 304
Dryad, HMS (Navigation School) (later Navigation Directing School), 326
Dryborough, D. A. (GEC), 403
Dubilier Condenser Co, Ltd., 402
Duck-class corvette, 400
Duckworth, Miss M. (HMSS/ASE), **18**
Dudley, Janet (DRE, Malvern), 183, 290
Duke of York, HMS (battleship):
  Actions, 53, 223, 345
  antenna height, 365
  first ABU, 51
  and Height-finding Working-up team, 315
  and VHF interference, 317
  sets fitted, 388
dummy load (for testing), 289, 313, 319
Dundee, 30, 343
Dunedin, HMS (cruiser) (later HMNZS), 22, 394
Dunham, C. R. (GEC), 402
Dunkirk, 33, 169
Dunluce Castle, SS (accommodation ship), 296, 298
Dunnet Head, 300
du Parcq, J. R. (HMSS/ASE), 30, 36, **62**
duplexer valves, 121, **122**
Durban, HMS (cruiser), 394
Durban, 341

Durnford, Captain R., 306
Durrant, R. (HMSS/ASE), 187
Dutch Radar Officers, 346
Dutch scientists at HMSS/ASE, 70
DVD generator, 323

E

Eagle, HMS (aircraft carrier), 388
Eagle, HMS (postwar fleet carrier), 89, 118, 249, **254**
early warning, see warning radar
Earp, C. W. (STC), 404
earth's magnetic field, 366
East Anglia, 215
Eastern (East Indes) Fleet, 345
Eastney, see HMSS, ASE
Eastwood, W. S. (HMSS), 13, **19**, **20**
Eckersley, T. L. (Marconi), 76, 403
E-boats, 207, 213
echoes:
  amplitude, 72, 159, 164, 222, 231
  back, 246, 406
  enhancement, 85–6
  fluctuations, 73, 79, 233, 247
  from land, 33, 246
  from rain clouds, 81
  permanent, 339
  prediction, 222
  second-trace, 77
  side, 210, 219–20, 373
  visibility, 73
'Echoes of War', 191
Eddon, Captain K., 34
Eddystone Light, 213
Edinburgh, HMS (cruiser), 394
Edison-Swan Co., 96, 402
Ediswan, 116
Edwards, Lieutenant F. W. B, **18**
Edwards, G. C. (GEC), 403
Edwards, A. T. (ASE), **63**
Edwards, Dr (Pye Ltd), 404
EDSAC 1 computer, 85
Eeles, F. (HMSS/ASE), **19**
Effingham, HMS (cruiser), 394
Ehrenstein, F. (ASE), **62**
Eide (Faroes), 208
E. K. Cole Ltd, 401
Electrical and Musical Industries Ltd (EMI), see outside contractors
Electrical Branch of RN ('L' Branch), 347
Electrical Engineering Department, Admiralty, 335
Electrical Officer, 308, 313

General Index   435

electrolytic capacitors, 314
electronic components, 360
Electronic Countermeasures (ECM),
    63, 408
Electronic Warfare Group, *see* ASE
'E' list, of spares, 313
Elliott Brothers, 402
Ellis, C. G. (HMSS/ASE), 18
Ellison, M. (HMSS/ASE), 296
Elswick, on Tyne, 39, 404
Emerald, HMS (cruiser), 320, 323, 394
Emperor, HMS (escort carrier), 390
Empress, HMS (escort carrier), 390
Emuss, Commander F. J. RNVR, 314,
    340–1
Engadine, HMS (aircraft transport), 392
English Channel, 207, 213, 214
English Electric Co. Ltd, 402
Enterprise, HMS (cruiser), 394
environmental conditions, shore
    sites, 208
environment for radar at sea:
    general, 47, 64, 110, 138, 311
    corrosion, 89, 179
    enemy action, 64
    funnel gases, 89, 318
    gun blast, 64, 302
    heat, 147, 360
    icing, 319
    lack of ventilation, 147, 319
    mechanical shocks, 180, 302
    moisture – and antenna feeders: air
        insulated, 32; aperture cover,
        237, 319; coaxial cable, 26;
        junction boxes, 41; polythene
        insulant, 26, 37, 319; Pyrotenax,
        148, 301, 310, 336
    noise, 366
    Radar Office, 308, 322
    sea motion, 46, 47, 193, 194, 200, 220,
        223
    vibration, 98, 115, 312
    weather, 64, 312, 323
    windage, 135, 136, 147
environment for radar research, 69–70
equipment:
    acceptance trials, 10
    dimensional constraints, 178, 180
Erebus, HMS (monitor), 388
errors: general, 359
    fall of shot, 379
    gunnery, 360, 371
    index, 316, 360

Espley, D. G. (GEC), 402
escort carriers, 357
escort vessels, *see* convoys
Euryalus, HMS (light cruiser), 206, 394
Euston Station, London, 296
Evans, Lieutenant (Sp), RNVR, 300
Evershed, C. R. (HMSS/ASE), 7, **8**, 15,
    18
Excellent, HMS (Gunnery School), 39,
    324, 326
Exeter, HMS (cruiser), 394
Exeter, G. A. (HMSS/ASE), **20**, **50**, 62,
    177
Experimental Commander, HMSS, *see*
    HMSS
Experimental Department, HMSS, *see*
    HMSS

F

fadeless air cover, 165, 167, 177, 247
fading, *see* propagation
Fair Isle, 300–1
fall-of-shot, *see* shell-splash spotting
Falloon, S. H. (HMSS/ASE):
    contributor, 183
    antennae, 30, 34, 90–1
    on loan from Marconi, 36, 403
    recruited, 30
    waveguide kit, 91
    XRC1, **62**
Far East, 53, 54, 57, 84, 245
Fara, Scapa Flow, **297**, 316
Fareham, Hampshire- HMS
    Collingwood, 64
Faroe Islands, 208
Farrow, J. G. (HMSS/ASE), **18**
Fawcett, Commander H. W., 45, 191,
    192
Feachem, Lieutenant (Sp) RNVR, 300
feedback, from Fleet to HMSS/ASE,
    326
feeders, *see* antenna feeders
Fencer, HMS (escort carrier), 390
Fenwick, C. E. (HMSS/ASE), 30, 183
Ferguson Radio Ltd., 402
Ferranti, Ltd, 402
Fertel, G. E. F. (HMSS/ASE), 30, 33, 87
field strength, 76, 151, 190, 213, 214
Fielding, W. H. (HMSS/ASE), **19**
fighter cover, 54
fighter direction/interception:
    feasibility, 74
    Carlisle in Norwegian campaign, 33

General Index

fighter direction/interception (*cont.*)
  cover, 165, 167, 177, **241**, 241–2, 247
  IFF, 282–3, 285
  and Type 281, 169
  FD Displays: PPI, 56, 147, 282, 283, 285; Skiatron, 240, 282
  FD in Operations, 33
  Fighter Direction Officers, 305
  fighter direction radar, *xxxv*, 163, 183, 246–59, 256, 258, 269–70, 356–7
  Fighter Direction Room (FDR), 282, 337
  Fighter Direction Ships, 167, 246–7, 342, 354
  Fighter Direction Tenders (FDT), 398
Fiji, HMS (cruiser), 394
Fiji-class heavy cruisers, 31
filters, video, 83, 84
fire-control, 37, 49, 51, 323
  radar,*see* gunnery radar
Firefly aircraft, 357
Forth: Bridge, 348; trials in Firth, 263, 265
Fisher, L. (HMSS), **19**
Fitting-Out Officer, 338, 341
Fitzgerald, L. J. (HMSS/ASE), 183, 187
Fitzrandolph, Mrs M. (ASE), **50**
Flag Officer-in-Charge, Glasgow, 344
Fleet Air Arm, 28–9, 106, 259, 279, 280
Fleet destroyers, 225, 376, 400
Fleet minesweepers, 400
Fleet Train, 342
Fleet Wireless Officer:
  and Radar Officers, 295, 296, 302–3
  at Scapa Flow, 296, 298, 299–300, 301, 302
Fleetwood, HMS (sloop), 267
Flinn, A. V. (Joseph Adamson & Co), 401
Flotta, Scapa Flow, **297**, 315, 316
Flower-class corvettes, 190, 196, 400
Flying Fortress aircraft, 227
Focke-Wulf aircraft, (German), 228
Foley, F. M. (HMSS/ASE), Author (Mono. 3), *xxxvii*, 29, 97, 128, 183
Foort, A. L. (HMSS/ASE), **18**
Foot, N. H. (HMSS/ASE), **19**
Ford, P. S. (HMSS/ASE), **19**
Formidable, HMS (fleet carrier), 161, 388
Forster, E. W. (BTH), 401
Foster Engineering Co., 402

Fort Wallington, Portsdown Hill, Naval and RAF radar site, 47, 306, 324
Fortesque, Professor C.L., 6, 7
Forth, HMS (submarine depot ship), 398
Fossey, F. A. (BTH), 36, 401
'foster-father of Naval radar', 30
Foylebank, HMS (auxiliary AA ship), 396
framecoils, *xxv*
Franklin-beam array, 25
Fraser Battery, Eastney, 51, 306, 324
Fraser, Admiral Sir Bruce, 29, 30, 341
Fraserburgh, 343
Free French Radar Officers, 345–6
Free French scientists, 70, 90
Free-Space range, **241**, 241–2, 247–50, 252–3, 256, 408
Fremlin, J. H. (STC), 404
French, Dr (Barr & Stroud Ltd.), 401
French coast, 207
frequency, *see* wavelength, 412
  see also 182, 271, Appendices 1, 2;
    frequency 'pulling', 89, 91, 373
Friend, R. G. (Army scientist), 15
frigates, 86, 369, 400
Frobisher, HMS (cruiser), 394
Fulmar aircraft, 74, 240–2, **241**, 246–7, 249–50, 357
Furious, HMS (fleet carrier), 388
fused quartz, 96
fuzes, 371, 412
Fuze-keeping clock (FKC), 370
FW 190 (German aircraft), 228
FX 1400, 356

G

G82 test set, **209**
gain, 408, *see also* – antenna; receiver
galactic noise, *see* noise, cosmic
Galatea, HMS (cruiser), 394
galvanometers, 313
Gambia, HMNZS (heavy cruiser), 394
GA radar, 353, 355–6
Garner, R. H. (HMSS/ASE), **18**
Garnons-Williams, Commander N.G., 12
gas switch, 220, **221**, 224, 374
Gates, M. J. (HMSS/ASE), **18**
Gavin, M. (GEC), 402
GB radar, 353–5

GC radar, 353–6
GCI, see RAF Ground Control of Interception Radar, 282
General Electric Company (UK): see outside contractors
generators, 37, 330
'Geneva-cross' device, 51, 320
George, Prince (later King George VI), 343
German aircraft: Focke-Wulf, 228; Ju88, 297
German Air Force (Luftwaffe), 321
German battlefleet, 53
German convoys, 53
German HF DF, *xxv*
German invasion, fear of, 169
German radar:
 comparison with British, *xxvi*, 21, 22, 85–6, 206, 306, 325
 earliest known cm-radar, 188
German shore-based aircraft, 53
German valve manufacturers, 188
Gestetner Ltd, 403
Gestetner, David, 403
Gestetner, Sigmund, 403
Ghalib, Mr (MetVic), 403
Gibbs, D. F. (HMSS/ASE), **62**, 187
Gibraltar, 208, 314, 340–1
Gibson, W. J. (STC), 404
Gilmore, C. R. (HMSS/ASE), **19**
GL, 279, 408
Gladstone, P. (Research Associates Ltd.), 404
Glasgow, HMS (cruiser), 315, 317, 365, 394
Glasgow, fitting-out staff, 320
Gleaves, USS (destroyer), 295, 304
Glen-Bott, C. L. (HMSS/ASE), **8**, **18**
Glorious, HMS (fleet carrier), 388
Glory, HMS (light fleet carrier), 315, 363, 388
Glossop, Naval Stores, 313, 330
Gloucester, HMS (cruiser), 394
G-mode (IFF), 282–3, 285
Godfrey, R. H. (STC), 404
Gold, B. S. (HMSS/ASE), 128
Gold, T. (later Professor) (ASE), **62**, 70
 antennae rotation rates, 80
 height finding, low aircraft, 88
 sea-clutter, 82–3
 target reflection characteristics, 79
 'Window', 79–80, 84–5

Goldup, T. E. (HMSS):
 HMSS, 7, **8**, **105**
 Mullard, 97, 128, 403
Gollin, E. M. (HMSS):
 pre-WW2 in HMSS, 12, 15, **18**, **20**, 25
 23-cm work, 14, 17
 split-anode magnetron in FM CW, 24, 68–9
Goodhart, Mr (Research Associates Ltd.), 404
Goodwin, E. T. (HMSS/ASE), 70, 78, 85
Gordon, W. E. (USA), 78
Gossling, B. S. (HMSS):
 recruited, 6
 valve research, 7, **8**, 96
 and GEC, 97, 128, 402
Goudie, Sub-Lieutenant R. (RNVR) (later Lieutenant Commander), 343
Govan, ship fitting, 324
Grainger, E. J. (HMSS/ASE), 6, **18**, **63**
Granton, Scotland, 343
greases, 323
Great Orme Head (N. Wales), 226, 269
Grensted, J. (HMSS/ASE), **63**
Griffith, R. (HMSS/ASE), **63**
ground-wave, 360
ground-wave suppression, 159
Grubb, Parsons & Co. Ltd, 248, 403
Grundy, E. (Ferranti Ltd), 402
GS radar, 353–6
guided-missiles, 58, 60
Guillemot, HMS (sloop), 210
Gulf of Siam, 345
gun control, see gun direction
gun direction:
 general, 33, 34–6, 134
 High-angle Control System (HACS), 309, 324, 363, 366, 371
 target allocation, 57, 58, 269
 target indication: antenna factors, 237–8, 242; concept of TI, 57–8; sea trials, *xxxv*, 269; Staff Requirements, 225–6, 232; TI Display, 232; Type 293M replaces Types 276, 293, *xxxv*
gunlaying radar, *xxxiv*, 57, 191; see also TI
Gunnery Officer, 305, 308
gunnery radar:
 general: requirements, 34, 36; characteristics, **59**; development, 36–43, 51–65;

438 General Index

gunnery radar (cont.)
    progress, 47; concept of TI
      radar, 57
    aircraft, HA or combined HA/LA,
      33, 54, 57
    barrage-fire, 43–4, 51, 53, 371
    blind-fire, 51, 53, 381, 406
    by wavelength: 50 to 70cm
      wavelength, 21, 25, 110, 163,
      186; decimetric, 23–4; 10-cm and
      below, 45, 53, 57
    close-range AA, 260
    laying-on of other radars, *xxxiv*, 57,
      163
    main armament, 32, 34
    surface, 57
    TI concept, 57
    tracking, 53
    weapon allocation, 56
Gunnery School, see *Excellent*
Gutton, H. (French scientist), 111
gyro, 210, **240**, 312, 365

H

H44, HMS (submarine), **262**
HA Fire-Control radar, see gunnery
    radar
Haine, M. E. (MetVic), 403
Halcyon, HMS (fleet minesweeper), 400
Hall (J & E) & Co. Ltd, 403
Hall, W. (HMSS/ASE), **63**, 403
Hampden bomber aircraft, 79
Hancock, J. (Defence Research
    Establishment), 183, 290
handbooks, 39, 298, 311, 312, 326
Hannover (German liner), see *Audacity*
Hansford, R. F. (HMSS/ASE), 49
Harben, Mr (Elliott Bros.), 402
Harbour Defence radar, 206–8
Hardy, A. M. (HMSS/ASE), **18**
Hargaback Mark, Scapa Flow, 316
Harley, A. (Coventry Gauge & Tool
    Co.), 402
Harris, G. W. (HMSS), **18**, 37
Harris, N. L. (GEC), 402
Hart, Mr (Pyrotenax Cables Ltd), 404
Hartlepool, 343
Harwich, 340
Haslemere, 61, 330
Hastings-class sloops, 400
Hatston Naval Air Station, 297, **297**,
    300
Hawke, Mr (Pye Ltd), 404

Hawkins, HMS (cruiser), 394
Hayes, EMI works, 58
H-class submarines, 223
Heath, C. (Permali Ltd), 403
Heaton-Armstrong, L.S. (STC), 404
Hecla, HMS (destroyer depot ship),
    341, 398
Hecla-class depot ships, 31
Heemskerck, (Dutch cruiser), 356, 396
height determination/estimation/etc.
    'Azication', 242
    calibration, 74–5, 161, 164, 299, 312,
      314–15
    early methods, 74–5
    HPI, 229, 242, 249, 377
    low-flying aircraft, 88
    Performance Meters, 150–1, 161, 164
    Polar Diagrams, use of, 33, 54, 74,
      159, 299–300
    post-WW2, 246, 258
    practice, 161
    Report (1940) on techniques, 299
    Sector Display, 164–5, 285
    separate Heightfinding set, 75
    specifications, 269
    trials, 242, 295
    Types 79/281 together, 54, 57, 335–6
    Type 277 sweeping, 55, 229, 231, 240
    Vertical Coverage Diagrams, **150**,
      241, **244**, **251**, 256
    and Type 295, 248, 249, 269
    and Type 980/981, 270
Heightfinding Working-up Team, 315
Height Position Indicator (HPI), see
    displays
helicopters, 316, 381
Hellcat aircraft, 357
Hemingway, A. V. (RRDE), 58
Henderson, S. T. (EMI), 402
Henderson, Mr (Vickers
    Armstrong), 404
Henley (W.T.) Cables Ltd, 403
Herbert, L. (BTH), 401
Herbert, P. G. M. (HMSS), **19**
Hermes, HMS (fleet carrier), 388
Hermione, HMS (cruiser), 268, 394
Hesperus, HMS (destroyer), 211
Hey, J. S. (TRE), 81
HF DF, see high-frequency direction-
    finding
H.H. Wills Laboratory, Bristol
    University, 31
Hickin, C. E. M. (GEC), 402

Hicky, Mr (GEC), 402
high-angle director, 33, 39, 43
high-frequency direction-finding (HF DF): *xxv*, 9, **233**, 335, 403, 404
Hilary, HMS (Headquarters Ship), 396
Hill, E. G. (HMSS/ASE), **18**
Hill, W. S. (HMSS/ASE), 49
Hitler, Adolph, 53
HM Dockyards:
  and fitting of radar, 10, 31, 325, **328**, 332, 334, 335
HM Mining School, 69
HM Nautical Almanac Office, 76
HM Signal School (HMSS) (later ASE):
  anomalous propagation, 77
  A/S applications, 45
  Application Officers, 9, 10, 13, 15
  authorised to work on radar, *xxxiii*, 34, 68, 134, 181
  Bristol Laboratory, 114
  Captain, Signal School, 13, 15, 16, 28, 31, 134, 196
  Departments/Divisions/Groups/ Sections: 50-cm Group, 49; centimetric Development Group, 186, 191, 303; Coales' Group, 23, 36, 47, 49, 70; Design Department, 29–30; Display Division, 187; DF Section, 17; Infra-red Section, 9, 14, **20**, **63**; Instructional Department, 10; Landale's Groups, 46–7, 53, 186, 187, 191, 193–4, 303; modulator/ transmitter Group/Divisions, 187; Production Department, 30; R1 Division, 17, 70; R2 Division, 17, 23, 36, 47, 49, 70, 72; R3 Division, 17, 70; Radar Countermeasures Group, 36; Radar Department, 69–70, 187; Rawlinson's Group, 32, 33, 70; 'R' Department, 16, **20**; RE2 (Gunnery) Section, **50**, 58; Receiver Division, 187; Receiver Section, 30; Ross's Group, 21, 33, 36, 70; Sutton's Group, 36, 44, 189; Symond's Group, 17, 21, 53, 70; Transmitter Development Group, 267; Transmitter Section/Group, 12, 13, 15–16, 17, **18**, **20**, 187; Valve Section/ Group, 29, 31, 32, 87, 95, 96, 140, 189; Valve Laboratory, Bristol, 259; X2, *see* Transmitter Section; Yeo's Group, 12–13, 14
dive bombers, 34
early days, 6–7
Eastney: 177–80; general, 30; AA range, 26, 34, **35**, 38, 39; 'Eastney' Group, 17, 33; equipment built, 14, 33, 195–6, 203, 217; Fraser Battery, 51; gunnery antenna, 369; metric radar, 13–14, 33, 161, 172, 173, 186; modulation of 10-cm magnetrons, 174; O-i-C, 23, **50**; radar "in action", 47, 161; RAF 'Admiralty' trailer, 45–7, 194, 195; RM Barracks, 12, 134; shipborne PPI, 210; testing post-WW2 Marine Radar, 267; Transmitter Section and new valves, 152
Equipment Design and Production: Design, 10, 29–30, 64, 175, 325, 326; Drawing Office, 10, 64, 143, 161, 180; Workshops: expansion, 30; experience in seaborne equipment, 64, 139, 178–9, 180; production method, 180; manufacture of prototypes, 47, 139, 180
Experimental Commander: appointments, 12, 15, 16, **18**, 30, 345; and project control, 10, 21; and RDF security, 12
Experimental Department: administration, 15; formation, 7; location, 9; organisation, **8**, 9, 16, 17, **18–19**; programme selection, 10; resources, 9–10; scope, 9; staffing, 7, 9–10; Superintending Scientist, 13, 31; tenders (auxiliary craft), 9, 229; valve research laboratory, 189
gunnery radar, 45, 49, 369
Handbooks/Manuals, 10, 39, 326
Haslemere, 61, 330
height-finding, 299
IFF, 29
industry, 110, *see also* outside contractors
inter-War work, 7, 12, 14, 15
liaison: Armed Services, 9, 295, 326; other Countries, **50**, 295; other researchers, 12, 13, 24, **50**

HM Signal School (cont.)
  missed opportunities, 68
  Naval Applications Group, 191
  Nutbourne (Antenna and Propagation Group): Antenna Group, 30; gunnery antennae, 25–6; propagation, 76; single-mast working, 148, 162; staff, 17, 30, 36, 88, 90; Type 79 antenna, 17, 21, 136; Type 79 receiver, 23, 141; Type 281, 33; Type 286, 34, 48, 169, 170, 171; waveguide kits, 91
  Onslow Road, 36, 47, 186, 306
  organisation charts, **8**, **18–19**, **20**
  original concepts, 68
  patents, *xxxiii*, 96
  priorities: radar not highest, *xxvi*, 28, 135; upgraded, *xxvi*, 14, 15, 28, 135, 203; 'urgent requirement', 169
  remit on radar, 326
  reorganisation, 9, 16–21, **18–19**, **20**
  secrecy and security, 12, 17, 26–7, 134
  ship-fitting, 22, 47–8, **327**, **328**
  staffing and recruitment: pre-1935, 7, 9–10, 97; post-1935 increases, 12, 13, 16, 21; shortage of engineers, 28; shortage of development staff, 107; wartime increases, 29–30, 48, 178; centralised recruiting, 31; 'leave' re-introduced, 46
  Standards Laboratory, 179
  Swanage, 10-cm trials, 45, 192–3
  Tantallon, *see* ASE
  training Radar Officers, 345
  valve research, in HMSS: Valve Section, 7, 29, 31, 32, 95, 96, 140, 177, 189; staff, 7, 29, 31, 36, 68–9, 87, 97, 99, 107, 125, 128; resisted Watson-Watt take-over, 30–1; klystrons, 44, 113–14, 118, 189, 191, 259; magnetrons, 69; new seals, 32; patents, 96; silica technology, 97, 98–104, *see also* valves, silica; skiatron tube, 125; thoriated-tungsten filaments, 99–100, 140; valves for pre-WW2 radar, 107; valves for Type 79X, 138; valves for Type 281, 152–3; valve T22, 104
  Witley, *see* ASE

Hobart, HMAS (cruiser), 394
Hodge, Lieutenant John. R. RNVR, (later Sir John Hodge,Bart), 306
Hodgson, B. (HMSS/ASE):
  Receiver Section (1919–22), 7, **8**
  joined Mullard, 97, 128
  to HMSS from retirement, 30, 178
  liaison, **50**, **62**
Hogben, H. E. (HMSS/ASE), **18**, 30, **50**, **62**, 177
'Hohlraum', 27
Holleyoak, W. J. B. (HMSS/ASE), **18**
Hollingsworth, Dr. D. T. (B.I.C.C.), 401
Home Fleet:
  Base and anchorage, 296, **297**, 316
  located by ASV in poor weather, 15
  and height calibration, 314–15
  and Russian convoys, 53
  and *Scharnhorst*, 222–3
Hong Kong, 306, 313
Hood, HMS (battlecruiser):
  first Type 79B, *xxxiv*, 147, 148, 181, 303
  short wave radio, 106
  lost, 346
  sets fitted, **153**, 388
Hoolahan, F. H. P. (HMSS/ASE), **18**
Hopkinson, R. G. (GEC), 402
Horton. C. E. (HMSS/ASE):
  frontispiece, Tribute, *xxv-xxxi*
  Churchill's support, 31
  DF between Wars, *xxv*, **8**, 9, **18**
  head-hunted, 31
  management style, *xxvii*, *xxviii-xxix*, 17
  post-WW2, *xxix-xxxi*
  propagation, *xxv*, 9
  priorities, *xxvi*
  radar for WW2, *xxv*, *xxvi*, *xxvii*
  recruited, 6, 9
  R Department, 16, **20**, 69
  re-organisations, *xxviii*, **20**, **50**, 51, 61, 69, 70, 135
  research decisions, 21, 24, 32
  Superintending Scientist, 31
  Swanage demonstrations, 45, 192
  to DSR, 61
  and radar secret, 26–7
  and Type 79X, *xxvii*
  and Type 281, 151
  and Type 286, 169, 170
Houton Signal Station, 316
Howard, R. F. (Plessey), 403

General Index

Howard, Grubb, Parsons Ltd, 248, 403
Howe, HMS (battleship), 388
Howell, J. (Plessey), 403
Howlett, B. A. (HMSS/ASE), **19**
Howse, H. D.: Author (Appendices 1, 3), *xxxviii*, 361
Hoxa, Scapa Flow, 297, **297**
Hoy, Scapa Flow, **297**, 297–8, 315
Hoyle, F. (later Sir Fred) (HMSS/ASE):
  analysis of WA radars, 72
  height estimation, 74–5
  'noise', 81
  propagation, 77
  recruited, 70
  RC8/XRC8, **50, 62**
  target reflection, 79, 85
  and Böhm, 88
  and mercury delay-lines, 85
  and stabilisation, 88
  and Staff Requirements for TI systems, 72
Hs.293, 356
'huff-duff', *see* HF DF
Hughes, H. G. (HMSS/ASE): **105**
  Head of Valve Section, **8**, 97, 128
  administration, **8, 19, 20, 63**
Hughes, J. V. (HMSS/ASE), 187
Humber, Mr (Ferranti), 402
Humby, A. M. (HMSS/ASE), **63**
Hunt-class destroyers, 264, 336, 384, 400
Hunter, HMS (escort carrier), 390
Hurricane fighter aircraft, 240
hydraulic control, 401

I

Iceland, 208, 228, 306
Identification Friend or Foe (IFF), radar:
  general, Mono.6, 334, 409
  Air Ministry, 29
  AMRE, 278
  antennae, 87, 165, 288
  common Allied IFF, 281–2
  development, 29, 60
  G-mode, 282, 285
  HMSS, 29
  maintenance, 317
  Mark I, 278–9
  Mark II, 277, 279–80, 362, 364
  Mark IIN, 277, 280–1, 367
  Mark III, 277, 278, 281–3, 362, 364, 367
  Mark IIIN, 283–5
  Mark IV, 289

  Mark V, 277, 289–90
  on naval aircraft, 279
  personal codes, 165, 176
  polarisation, 281, 288
  power supply, 281
  RAF, 277
  Sector Display, 165, 285
  sliprings for antenna, 163
  technical data, 354, 362, 364, 367
  see also IFF sets listed in Appendix 1
Illingworth, Captain (E) J.H., 341
Illustrious, HMS (fleet carrier):
  first carrier with radar, 296
  trials of radar sets, 161, 246, 269, 296
  and FD, 74
  and Carey, 346
  sets fitted, 390
Illustrious-class fleet carriers, 28, 31
impedance, 25, 38, 87, 89
Implacable, HMS (fleet carrier), 390
Indefatigable, HMS (fleet carrier), 390
Indomitable, HMS (fleet carrier), 357, 390
industry, *see* outside contractors
inert gas, 53
Information Centre, *see* AIC
information handling, 56–7, 58
information theory, 82
infra-red, 9, 14, **63**, 125
Institution of Electrical Engineers (IEE): 61, 78, 188, 189
instruments:
  impedance meter, 87
  performance meter, 150–1
  waveform monitor, 154
  wavemeter, 154
  misc., 179, 188, 312–13, 319, 340
integrated circuits, 360
integrator, ball-and-disc, 323
interception of aircraft, *see* fighter direction/interception
interference:
  electrical jammimg, 83–4
  mutual interference, own sets, 43, 49, 142, 158, 167, 303, 369
  making jamming by enemy harder, 142, 158, 369
  Radio Interference Suppression equipment: 38, 167
  unwanted land-echoes, 33
  VHF communications, 317
  'Window', 73, 79–80, 84–5, 412
intermediate frequency (IF):
  general, **221**

intermediate frequency (*cont.*)
  specific, 5MHz, 362; 6MHz, **59**, 370; 12MHz, 159, 365; 13.5MHz, 377; 30MHz, 61, 367, 382; 31MHz, 384; 45MHZ, 191, 228, 367; 60MHz, **59**, 373, 377, 378, 380
interrogators, 277, 280, 285, 287, 354, 356, 409
Inter-Service:
  CVD valve research, 186
  guided-missiles, 58–60
  propagation, 78
  recruitment to Research Establishments, 31
  10-cm research, 44–5
invasion of Normandy, *see* OVERLORD and NEPTUNE
Invergordon, 343
ionosphere, *see* propagation
Ireland (r), 306
Irish Sea, 304
Iron Duke, HMS (battleship), 17, **297**, 297–303
Isaac Sweers (Dutch destroyer), 356
Isle of Man, 231
Isle of May, 348
Isle of Wight, 83, **266**
isotropic radiator, 359
Israeli Navy, 37
Italian aircraft, 44
Italian Fleet: submarines, 213; two-man torpodoes, 208
Itchen, HMS (frigate), 220

J

J. and E. Hall & Co., 403
Jack, H. (BTH), 49, 401
Jackson, Admiral of the Fleet Sir Henry, FRS: 6, 56
Jackson, Professor Willis, 37
Jacob, Commander C. M., 296, 305, 345
Jamaica, HMS (cruiser), 315, 394
James, E. G. (GEC), 402
James, Mr (Allen West Ltd), 401
jammimg, own, *see* interference
Janus, HMS (destroyer), *xxxv*, 232–3, **233**, 235, 269
Japan, Allied attack on mainland, 342
Japanese aircraft, 54
Japanese radar, 84
jar, (unit of capacitance), 179
'JE', *see* Displays – PPI
'Jellicoe' Express, 296

Jenkins, G. V. (HMSS/ASE), **18**
Jenkinson, Lieutenant R. A. RNVR, 315, 317, 323, 361
Jesty, L. C. (GEC), 402
Jodrell Bank, 81
Jofeh, L. (Cossor), 402
John Brown & Co. Ltd, 306, 310
Johnson & Phillips Ltd, 403
Jolliffe, H. G. (HMSS/ASE), **19**
Jolliffe, Dr S. A. W. (Marconi Osram Valve Co.), 403
Jollin, Miss (HMSS), **19**
Jones, G. W. (HMSS/ASE), 36, 51
Jones, H. (HMSS/ASE), **63**
Jones, M. J. (HMSS/ASE), 30
Jones, T. J. (HMSS/ASE), 29, 99, 107, 128
Joseph, Mr (AGI Ltd), 36, 401
Joseph Adamson & Co. Ltd, 401
junction box, 337, 340
Junkers aircraft (German), 297

K

K Stations, of Coastal Chain, 207
Kardex, 331
Kay, R. H. (ASE), 92
K-band, 61
Keall, O. E. (HMSS/ASE), **63**, 403
Kempenfelt, HMS (destroyer), 237
Kent, HMS (cruiser), 394
Kent, R. R. (HMSS/ASE), **18**
Kentsbeer, E. J. (HMSS/ASE), **20**
Kenya, HMS (cruiser), 203, 204, 394
Kersey, Eda (musician), *xxxi*
Kete, South Wales, (RN Aircraft Direction Centre), **250**
Kettlewell, E. (GEC), 402
Khedive, HMS (escort carrier), 390
Kiel harbour, Germany, 188
King, A. (BTH), 401
King, G. (HMSS/ASE), 187
King, P. G. R. (HMSS/ASE), 125, 128
King George V, HMS (battleship), 303
  experimental PPI, 210, 295, 304
  first prototype Type 284/5, 41
  sea trials, *xxxiv*, 43, 47, 219, 222, 223, 225, 238, 268
  sets fitted, 203, 204, 369, 388
King George V-class battleships, 31, 329, 337
King George VI, 343
Kingfisher-class corvette, 400

General Index

Kingsley, F. A. (HMSS/ASE): Editor, *xxxviii*
Kinman, T. H. (BTH), 122, 401
Kirkham, A. L. (HMSS/ASE), **19**, 128
Kirkpatrick, Lieutenant Commander, USN, 304
Kirkwall, 297, **297**, 299, 300
Kitchingman, Lieutenant Commander J.V., 329
klystron, *see* valves
Knight, H. de B. (BTH), 27, 38, 114, 401
'Knobbly'. Project, (fitting out of long range LSFs for Pacific), 342
Korean War, *xxix*
Kola Inlet (Murmansk) Russian base, 322
Kuhn, Lieutenant S. (Polish Forces) (HMSS/ASE), 90, 187
Kühnhold, Dr R. (German scientist), 188

L

L22 (HM submarine), 9, 10
Lamont, Dr H. R. L. (GEC), 402
lamp-load, 319
Landale, Dr S. E. A. (HMSS/ASE):
  career, 30, **50**, 61, **62**, 178
  cm-wave Development Group, 186, 303
  Modulator and Transmitter Group, 187
  sea trials, 302, 303
  Swanage cm-party, 46, 47, 191–5
  technology transfer, 195
  Transmitter Development Group, 267
  Tribute, 273
  and specific sets, 32, 53, 178, 302, 303
Landing Craft, Headquarters (LCH), 266–7
Lane, G. C. W. (HMSS/ASE), **18**
Largs, HMS (Headquarters Ship), 396
Latham, R. (BTH), 401
Latona, HMS (minelayer), 396
Laurence Scott Ltd (engineers), 39, 403
Laurence, Lieutenant (Sp) R.F. RNVR, 343, 346
Law, Lieutenant (Sp) M. RNVR, 346
Laws, C. A. (HMSS/ASE), 56, 187, 296
Laws, Lieutenant Commander R. A. RNVR: Author (Mono. 8, Appendix 2), *xxxviii*, 306, 311, 316, 346

laying-on of one radar by another, 163, 382
L-Branch, *see* Electrical Branch of Royal Navy
LCR Bridge, 179
Lea, N. (Marconi), 403
Leander, HMNZS (light-cruiser), 394
Lee, E. N. (HMSS/ASE), **19**
Leeson House, 191, 194
Lecher-line/wire:
  with split-anode magnetron, 106
  in Type 79/X/Y/Z, 138–9, 141, 362
  in Type 281/B/BQ, 152, 364
  in Type 290, 172
  in IFF Mk.3 interrogator, 287
Le Fanu, Lieutenant M. (later Admiral Sir Michael), 303
Legion, HMS (destroyer), 171
Le Havre, deception in NEPTUNE, 86
Leith, 343
le Rossignol, R. (GEC Laboratories), 26, 402
lethal voltages, 319
Levin, Dr N. (HMSS/ASE), **63**, 403
Lewin, L. (HMSS/ASE), 88
Lewis, D. S. G. (Mullard Radio Valve Co.), 403
Lewis, F. G. H. (HMSS/ASE), **18**
Lewis, Mr (Pyrotenax Cables Ltd), 404
Leydene, Petersfield (HMS *Mercury I*), **50**, 64
liaison:
  Allies, **50**, 281–2, 289–90, 304
  UK Armed Forces, 9–10, 191, 295, 326
  other Research Establishments, 12–13, 24, **50**, 193, 195
Liberator aircraft, 54
light cruisers, 237
limiting factors in Naval radar, *see* environment for radar at sea, maintenance
Lively, HMS (destroyer), 206
Liverpool, HMS (cruiser), 315, 363, 394
Liverpool, 208, 340, 342
Liverpool University, 76
Llandaff, HMS (Air Direction frigate), **257**
lobes, *see* Antennae, polar diagrams
local oscillator, 91–2, 268
Loch-class frigates, 400
Logie, K. R. (RRDE), 58

444 *General Index*

London, HMS (cruiser), 394
London air-raid, 301
London Underground trains, 87
Londonderry, 340, 342
Long Range Navigation System (Loran) (USA): 317, 324, 409
Longhope, Scapa Flow, 297, **297**, 303
Loran, *see* Long Range Navigation System
Lorenz GmBH, (German firm), 24
Lothian, HMS (Headquarters Ship), 396
Lough Foyle, 231
Loveband, Commander A. W., **18**
Lovell, A. C. B. (later Sir Bernard), (TRE):
  contributor, 24, 191
  first 10-cm aircraft echoes, 44
  Swanage demonstrations, 45–6, 192
  'cheese' antenna for Naval use, 48, 194
  radio-astronomy, 81
low air-cover, 241–2, 246, 248–9, 250
Lunt, Dr. (Parkinson Cowan), 403
Luxford, G. (HMSS/ASE), **63**
Lyddon Spout, Dover, 207
Lynch,Sub-Lieutenant 'Mac', RCNVR (later
  Rear Admiral, RCN), 347
Lyness, Orkney, 296–304, **297**
Lythall, B. W. (HMSS/ASE): Author (Preface, Tribute, Significant Milestones, Mono. 2), *xxxix*, 92, 187
Lythe Hill, **50**

M

Macleod gauge, 104
Macrorie, Captain A. K., (HMSS), 128
Madden, Captain A. C. G., 306, 322
Madge, H. A. (HMSS), 6, 7, **8**
magnesium oxide, 310
magnetic compass, 200
magnetic field strength:
  S-band magnetrons, 200, 215, 216, 217, 372, 374, 376
  X-band magnetron, 260–1, 263
magnetron, *see* valves
magslips, 323, 365, 383
Maidstone HMS (submarine depot-ship), 341, 398
Mainguy Report, 347
Mainland, Orkney, **297**, 315

maintenance of radar at sea:
  general, 61, 64, 196–7, Monos.7, 8, 9
  design factors, 175, 245, 318
  'Fleet Train' in BPF, 342
  preventative maintenance, 317–19
  pyrotenax, 148
  see also handbooks and environment for radar at sea
maintenance bases, 340
Malaya, HMS (battleship), 388
Malayan waters, 54
malfunctions, 319–20
Mallinson, W. D. (HMSS/ASE), 30, 36
Maloney, P. (HMSS/ASE), **19**
Malta, 33, 208, 317, 341
Malvern: TRE, 60
Manchester, HMS (cruiser), 394
Manchester Institute of Science and Technology, 37
Manifold, M. B. (EMI), 402
Manuals, *see* Handbooks
Manxman, HMS (minelayer), 396
Manxman-class minelayers, 342
Maplin, HMS (fighter catapult-ship), 392
Marconi Ltd, *see* outside contractors
Marconi-Osram Valve Co., 403
Marigold, HMS (corvette), *xxxiv*, 219, 223, 268
marine navigation, 266–7, 384
Marsh, Miss M. G. (HMSS), **18**
Marshall, J. H. (HMSS/ASE), **63**
Martello Tower, Rinnigill, 316
Mason, J. J. A. (HMSS/ASE), **18**
Massachussets Institute of Technology, (MIT), 82, 83
Matapan, Battle of Cape, 346
Matchless, HMS (destroyer), 313
materiel procurement: Mono.9
Matthews, C. (HMSS/ASE), **18**, **63**
Mauritius, HMS (cruiser), 394
Maxwell's Field Equations, 90
Mayo, B. J. (EMI), 402
McCarthy, J. C. (HMSS/ASE), **63**
McClelland, Lieutenant J. W., **18**
McCormick, Sub-Lieutenant (Sp) H.D., RCNVR, (later Captain RCN), 347
McGee, Mr (EMI), 402
McGill University, Canada, 345
McLaren, J. C. (HMSS/ASE), **19**
McMillan, R. (HMSS/ASE), **18**
McPherson, A. (HMSS), **8**, **19**
M Division, ASE, 330

Mediterranean, 53, 208, 246
Medway, HMS (submarine depot ship), 398
Meek, J. M. (MetVic), 403
Megaw, Dr Eric C. S., (GEC):
　GEC, 17, 402, 403
　forward scattering in troposphere, 78
　IEE, 78
　magnetrons, 17, 24, 44, 106, 111, 186
　propagation studies, 77, 78
　and Coales, 17
Megger, 313, 340
Mellor Bromley Ltd, 403
Merchant Navy, 266–7, 384
merchant ships, 29, **175**, 266
Mercury I, HMS (Signal School, Leydene), 64
Mercury II, HMS (ASE), 64, 313
Merren, W. J. R. (HMSS/ASE), **18**
Metadyne servo system, 248
Metcalf, C. (EMI), 402
Meteor aircraft, 252, **253**
meteorological balloons, 316, 381
meteorological conditions, *see* environment for radar at sea
Meteorological Office, 78
Methil, Fife, 340
'Metres to Microwaves' (book reference), 131
metric radar, *see* wavelength
Metropolitan-Vickers Electrical Co. (MetVic), 64, 208, 228, 248, 403
'micropup', *see* valves
Middleton Report, 347
Miles, Colonel (Admiralty), 192
Miller, J. C. P. (Liverpool University), 76
Mills, Lieutenant Commander (Sp) G.W.B. RNVR: 343
Minches, The, 304
minesweepers, 33, 86, 264, 328, 337, 384
Ministry of Aircraft Production, 192, 334
Ministry of Supply, 78
Ministry of Supply RDF Applications Committee, Ultra-Short-Wave Propagation Panel, 78
Ministry of Transport, 267
Minotaur HMS (cruiser) (later HMCS *Ontario*), 394
Minns, H (HMSS), **19**
missed opportunities, 67, 68
Mitchell, J. A. (HMSS/ASE), **19**

M-type motors, 323, 371
Mobile Naval Radar Stations: 'Monrads', 355; Mobile Radar Trailer, *xxxiv*, 207
modifications, 322, 326
modulators and modulation:
　hard-valve modulators, 114, 216, 372
　spark-gap modulators, 115–16, 187
　pulse modulators, 114–16
　thyratron modulators, 25, 36, 38, 53, 114–15, **115**
　series-triode modulator, 107
　anode modulation, 367, 370
　cathode modulation, 362, 364
　discharge-line modulation, 38, 224, 226, 268, 374, 376
　modulation of magnetrons, 174
　fast warm-up modulators, 187
　the '1MW modulator', 216, 224, 226, 269
　components, 404
molybdenum, 99
Mombasa, 341
monitoring instruments, 403
monitorscope, 141, 154
'Monrad', 355
Montcalm, HMS, 398
Montclare, HMS (submarine depot ship), 398
Morey, C. H. (HMSS/ASE), **19**
Morey, P. S. B. (HMSS/ASE), 296, 339
Morley, Mr (Nash & Thompson), 403
Morris-Airey, H. (HMSS), 6, 7, **8**, 128
morse, *see* wireless telegraphy
Mosquito aircraft, 252
Moss, H. (A.C. Cossar), 402
Moss, S. J. (HMSS/ASE), **18**, **63**
Moss, Lieutenant W. E., **18**
motor alternators, 323
Motor Gun Boats (MGB), 258, 262, 264
Motor Launches (ML), **262**, 264, 337
Motor Torpedo Boats (MTB):
　need for radar, 169
　radio-control, 7
　ship-fitting, 337
　sets fitted, 174, 258, 262, 264, 288
　and propagation trials, 76
Moule, Mr (ACI Ltd), 401
Moullin, E. B. (HMSS/ASE), 30, 70, 87
Mountbatten, Lord Louis, 65
Mount Snowdon, 78
Moxon, L.A. (HMSS/ASE), 30, 32, **50**, **62**, 80–1, 317

Mullard Radio Co. Ltd:
  see outside contractors
Mullard S. R. (HMSS), 96, 128
mu-metal, 200
Murmansk, 314, 321–2
Murray, Captain A.J.L. : Captain of HMSS 1937, 15
Mussett, R. D. G. (HMSS/ASE), **63**
Mynall, D.J. (BTH), 401

N

Nab Tower as reference echo source, 28, 216
Nabob, HMS (escort carrier), 390
NachrichtensVersuchAnstalt, (NVA), 188, 410
Naiad, HMS (cruiser), 394
Nairana, HMS (escort carrier), 390
Nash, Lieutenant C. C., **19**
Nash & Thomson, 403
National Physical Laboratory (NPL), Slough, 9, 11, 78
National Research Council of Canada, 61, 264
Naval airborne radar, 357
Naval Aircraft, *see* Fleet Air Arm
  see also Albacore, Avenger, Barracuda, Firefly, Fulmar, Hellcat, Swordfish
Naval Air Station, Hatston, 297, **297**
Naval Application Officers, 9, 10, 15
Naval Artificers, 30
Naval Construction Department, Admiralty (DNC), 10, 86, 248
Naval Contracts Department,Admiralty, 329, 330
Naval Electrical Artificers, 178
Naval Ordnance Department, Admiralty, 7, 12, 34, 39, 53, 58
Naval Meteorological Service, 78
Naval Radar RDF Panel, 28, 29, 31
Naval Radar Trust, *xix-xxi*
Naval reaction to radar, 326
Naval Research Laboratory (NRL) (USN), 290
Naval Review, Spithead, 1937, *xxv*
Naval Staff, 6, 28, 29, 31, 140, 144
Naval Staff Requirements, 411
  assessment of attainability, 72
  cm FD radar, 246–7, 258, 269
  gunnery and FC radar, 34, 369
  Performance Meter, 150–1
  priorities, 28
  ranging on dive-bombers, 369
  target allocation, 58
  target indication, 72
  WA/WS radars, 13, 15, 134
  X-band against aircraft, 259
  and Type 267W, 263
  and Type 281, 151
  and Type 282, 34
Naval Stores, 313–14, **328**, 330, 331, 332
Naval Stores Officer, 313
navigation, *xxxv*, 264, 265–7, 266, 384
Navigation (Direction) School: *see Dryad*
Navigator, 305, 308
Neill, Carol, 344
Neill, Commander O. S. RNVR, 343–4, **343**
Nelson, HMS (battleship), *xxxiii*, 41, 302, 369, 388
Nelson, Admiral Lord H. (1758–1805), 320
neon, 87
neon tube (for testing), 188
Neptune, HMS (cruiser), 394
NEPTUNE, Operation (assault phase of OVERLORD, invasion of N. Europe), 84, 86
Netherbutton, Orkney (RAF CH station), **297**, 300, 316
Netherlands Forces, 346
Newcastle, HMS (cruiser), *xxv*, 394
Newcastle, 342
Newfoundland, HMS (cruiser), 394
Newfoundland, 302
New Zealand Radar Officers, 345
New Zealand scientists at HMSS/ ASE, 30
Nickels, L. J. I. (STC), 404
Nigeria, HMS (cruiser), 205, 268, 394
night-fighter radar, 44
Nind, E. A. (EMI), 402
Nissen huts, 344
Noble, H. (HMSS/ASE), 7, **19**, 64
noise:
  in receivers, 9, 71, 80–2
  grounded-grid valve operation, 149, 165
  specific Noise-Factors, 362, 364, 365
  see also cosmic noise
Norfolk, HMS (cruiser), 43, 53, 315, 365, 378, 394
Normandy Invasion, *see* NEPTUNE, OVERLORD

North Atlantic, 65
North Cape (Norway), Battle of, 53, 322
North Channel, 244
North, Dr. J. D. (Boulton & Paul), 401
Northover, F. H. (ASE), **63**, 78
Norway and Norwegian coast, 53, 264, 296
Norwegian:
  Campaign, 33, 300
  Forces, 346
  Radar Officers, 346
  scientists, 70, 82
  ships, 223
Nuffield Laboratories, Birmingham University, 44
nutating beam, 60, 380
Nutbourne, see HMSS, ASE
Nutter, Mr (MetVic), 403

O

Obedient, HMS (destroyer), 235
Ocean, HMS (light fleet carrier), 58, 357, 390
O'Dell, D. T. (HMSS/ASE), **19**, 128
Oersteds, 200, 215, 216, 217, 260, 263, 372, 374, 376
Officer of the Day, 318
Officer of the Watch, 318
Offord, W. S. (HMSS/ASE), **18**
Ogilvy, Commander F. C. A., 6
O'Hagan, B. (ASE), **62**
Oil, Castor, AP109, 321
O'Kane, B. J. (GEC), 403
Oldham, Naval Store, 330
Oliphant, Professor M. L. E. (later Sir Mark) (Birmingham University), 44, 113
Oliver, H. (HMSS/ASE), **18**
'one-shot' multivibrator, 146
Onslow Road, Southsea, see HMSS
Ontario, HMCS (heavy cruiser), 315, 365, 394
Operating and Maintenance Instructions, 10
operational requirements, see Naval Staff Requirements
Operational Research, 214
Operators, radar:
  antennae, manual rotation, 174, 320
  assistance, 32, 203
  dedicated TI, 57
  general, 147, 180, 305, 326, 340, 347
  heightfinding – skill dependence, 75
  honoured, 346
  numbers, 307, 336
  performance, 72, 174
  Sector Display, 164
  training, 315, 344
  and PPI, 234
  and UDU, 176
Ora, Mr (EMI), 402
Orchis, HMS (corvette):
  sea trials of Type 271X, *xxxiv*, 47, 196, 198, 200, 202, 203, 211, 213, 214, 216, 268
  and 'under the beam' hypothesis, 213, 215
Orfordness, 12, 13–14, 106, 134
Orion, HMS (cruiser), 394
Orkney Islands, 297, 301
Osborne, HMS (Training School), 343
oscilloscopes, see cathode ray oscilloscopes
Outfits, 474
outside contractors – general, 10, 27, 36, 64, 330, Appendix 4
outside contractors – specific firms:
  ACI Ltd, 401
  Adamson, (Joseph) & Co Ltd, 401
  Advanced Components Ltd, 64, 401
  Aeronautical & General Instruments Ltd, 401
  Allen West, 47, 64, 196, 208–9, 401
  Aluminium Plant and Vessels, 64, 401
  Barr & Stroud Ltd, 401
  Bentley & Co., 401
  Boulton & Paul Construction Ltd, 401
  British Insulated Callender Cables, 401
  British Thomson-Houston Co. Ltd (BTH): generators, 37, 49; mass-produced silicon crystals, 122; modulators, 37, 38; products, 401; staff and secondment, 36, 401; working with HMSS/ASE, 64; and magnetrons, 113; and thyratron CV22, 114
  Bruce, Peebles Ltd, 401
  Brush Electric Co. Ltd, 401
  Bull Motors Ltd, 401
  Clarke Chapman Ltd, 401
  Cole, (E.K.) Ltd., 401
  Cooke & Ferguson Ltd, 401
  Cossor, (A.C.) Ltd, 37, 64, 402

outside contractors (*cont.*)
  Coventry Gauge & Tool Co., 402
  Crabtree Ltd. (Leeds), 402
  Crompton Parkinson Ltd, 402
  Croydon Engineering Ltd, 402
  Decca Radio Ltd, 402
  Dubilier Condenser Co. Ltd, 402
  Edison Swan Company, 402
  Ediswan, 116
  Electrical & Musical Industries Ltd (EMI): Air Ministry contract, 267; equipment, 402; Naval PPI, 162–3, 210, 225, 229; staff, 402; TV shipboard trials, 17; and CVD research, 186; and klystron, 259; in radar production, 64; specific projects, *xxxiv*, 58, 260
  Elliott Bros. (London) Ltd, 402
  English Electric Co. Ltd, 402
  Ferguson Radio Ltd, 402
  Ferranti Ltd, 402
  Foster Engineering Co., 402
  General Electric Company (UK): Air Ministry contract, 186, 267; Communications equipment, 14, 17; CVD Valve Research Programme, 27, 186; duplexer valves, 121, 161–2; early radar equipment, 64; graded-glass seals, 103; gunnery radar, 36, 38; long-range propagation, 77, 78; magnetron development and production: split-anode, 24, 106; resonant cavity, 111, 113, 186; E1198, 191; E1263, 263; meeting of Industrialists at HMSS, 34; metal/glass valve development, 109, 110; 'micropup', 110; modulator valves, 114; oxide cathodes, 110; planar-diode mixer, 122; power valves, 116; projects, 402–3; radar secret, 26, 109, 110; staff/secondment, 17, 64, 402–3; thyratrons, 114; and BTH, 113; and E773, 26, 27, 37
  Gestetner Ltd, 403
  Grubb Parsons Ltd, 403
  Hall. (J & E), & Co., 403
  Henley (W.T.) Cables Ltd, 403
  Johnson & Phillips Ltd., 403
  Laurence Scott Ltd, 39, 403
  Marconi Ltd: enlisted in radar work, 34, 36, 64; high-power diode, 176; products for RN, 403; receivers for gunnery sets, 37; staff and secondment, 36, 70, 403; and Böhm, 36, 70; and G. M. Wright, 36, 61; and specific sets, 176, 217, 228
  Marconi-Osram Valve Co., 403
  Mellor Bromley, 403
  Metropolitan Vickers Ltd, 64, 208, 403
  Mullard Radio Value Co.: EF 50, 116, 370; products, 96, 403; RF heating, 109; silicon technology, 98; staff, 96, 97, 128, 403; valves, 32, 96, 97, 106, 116, 370
  Nash & Thompson, 403
  Painton Instruments Ltd, 403
  Parkinson, Cowan, 403
  Parmeko Ltd., 403
  Permali Ltd., 37, 348, 363, 365, 403
  Philips, 103, 123, 145
  Pirelli General Cable Co., 64, 403
  Plessey Radio Ltd, 403
  Precision Engineering (Haslemere), 403
  Pye Ltd., 404
  Pyrotenax Cables Ltd., 89, 404 *see also* antennae – Feeders
  Rediffusion Ltd., 404
  Research Associates Ltd, 404
  Reyrolle Electric Ltd, 404
  Rose Brothers Ltd, 39, 404
  Sperry Ltd, 404
  Standard Telephones & Cables Ltd (STC), 24, 27, 34–8, 140–1, 404
  Telegraph Condenser Co. Ltd, 404
  Telcon, 64
  Telegraph Construction & Maintenance Co. Ltd, 404
  Teneplas (Pangbourne), 404
  Thermal Syndicate Co., 96–104, 404
  Vickers-Armstrong Ltd, 39, 404
  W.H. Smith (Engineers) Ltd, 48, 210, 404
outside contractors – specific sets:
  Type 73X, 14
  Type 262, *xxxiv*, 58
  Type 281, 161
  Type 284, 36–7, 38
  Type 292, 58
Overall Efficiency, 151

General Index 449

OVERLORD, Operation, (invasion of Northern Europe), 228, 266, 323 *see also* NEPTUNE
Owen, C. (HMSS), 46, 193
Owen, A. R. G. (HMSS/ASE), 70, 77, 88
Oxford Joint Recruiting Board, 295
Oxford University, 30, 114, 259

P

P & O Shipping Co., 341
Pacific, 58, 342
Pacific Fleet, *see* British Pacific Fleet
Paddick, Mr (Allen West Ltd), 36, 47, 401
Paddon, Sub-Lieutenant Stuart E. RCNVR, (later Rear-Admiral, RCN), 345, 347
Painton Instruments Ltd, 403
Palomares, HMS (FD ship), 357, 398
Parkinson, Cowan, 403
Parmeko Ltd., 403
Partridge, Mr (EMI), 402
Patia, HMS (fighter catapult-ship), 392
Patents:
    Alder – radar: *xxxiii*, 11, 106
    HMSS – valve construction, 96
    Mullard – silica valves, 96
    Thermal Syndicate – valve seals, 96, 102
Patrick, A. M.. Author (Mono. 9), *xxxix*, 343, 348
Patroller, HMS (escort carrier), 390
pattern recognition, 255
Patterson, C. C. (GEC), 402
Paul, Commander R. T., (later Captain), **19**
Peake, W. S. (HMSS/ASE), **8**, **18**, 64
Pearce, A. F. (EMI), 402
Pearl Harbor, 281, 304
Pegasus, HMS (fighter catapult-ship), 392
Pegler, Major G. D. REME (ASE), 315
Pemberton, N. (HMSS/ASE), **18**, 64
Pemberton, Mr (EMI), 402
Pembroke W/T Station, 105
Penelope, HMS (cruiser), 394
Penny, E. W. (HMSS/ASE), **18**
Penstemmon, HMS (corvette), 213, 214
Pentland Firth, 232
performance meter, 74–5, 150–1, 164
performance prediction, 256
Periwinkle, HMS (corvette), **198**

'Permali', 37, 348, 363, 365, 403
persistence of vision, 255
Perth, HMAS (cruiser), 394
Peter Peregrinus Ltd, 131
Peveril Point, 194, 195
PF, *see* prefabrication
phase-changers, 88–9, 251
Philips Ltd. (Holland), 103, 123, 145
Phillips, A. S. C. (HMSS/ASE), **18**
Phillips, G. J. (HMSS/ASE), **63**
Philoctetes, HMS (depot ship), 398
Phoebe, HMS (cruiser), 43, 394
phosgene, 312
phosphors, 124–5, 139
physical properties, valve materials, 98
Physical Research Department, Admiralty, *xxxi*
Physical Society, 118
Picken, W. H. (Marconi), 403
Pinewood, ASE extension, **62**
Pirelli, 26, 37, 38, 64, 403
Pitcairn, Commander R. F., **18**
Pitts, W. J. (Plessey Co.), 403
'plan-packed' radar sets, 331–2, **332**, **333**
Plan Position Indicator (PPI), *see* displays
Platt, J. (Laurence Scott Ltd), 403
Plessey Radio Ltd., 403
plots and plotting:
    Auto Radar Plot, 403
    congestion, 56
    digital system (post-WW2), 56
    manual plot, 167
    Plotting Tables, 56, 125
    Tactical Plot, 57
    see also Displays – PPI, Skiatron
Plymouth, 342
polar diagram, *see* Antenna
polarisation:
    reasons for choice, 22, 76, 83, 136
    studies, 22, 75–6
    Airy Integral, 76
    effect of 'Window', 85
    first 10-cm Naval set using vertical, 246
    horizontal, 22, 76, 82–3, 133, 136, 229, 248, 281; and sea-reflections, 248
    vertical, 76, 82–3, 246, 281, 288
    horizontal v. vertical, 22, 192
    and IFF, 281, 288
Polish Navy, 304

Polish scientists at HMSS/ASE, 70, 90
Pollard, P. E. (Army scientist), 15
Pollockshields, Glasgow, 344
Pollux, HMS (ex-French minelayer), 264–6
polystyrene, 37
polythene:
  insulation in cables, 26, 37, 330
  manufacturers, 26, 37, 403
  and rats, 310, 338
pom-poms, 34, 58, 369, 370, 371
Port Radar Officers, general, 326, 332, 338–9, 341–3, 348
Port Radar Officers, specific:
  Glasgow, 344; Newcastle, 343; Rosyth, 343, 348
Port Radar Organisation, 340, 341, 342–3
Port Technical Officer, 341
Port Technical Officer, Gibraltar, 314
Port W/T Officer, 348
portable 'after-action' radar set, 314, 324, 357
Portsmouth:
  air raids, 31, 33, 47
  dentist's equipment, and Type 281, 170
  HMSS, 12–13, 97, 177, 306
  Maintenance Base, 340
  misc. locations, 306
  Naval Barracks, 7, 9, 177, 178
  Port Radar Officer Organisation, 342
  Portsmouth Command, 29
  Royal Marine Barracks, 12, 134, 186
  and dispersal, 31
post-detector integration, 152
potentiometers, 145, 403
Pout, H. W. (HMSS/ASE), 36, 49, 361
power output, *see* transmitters and 182, 271; *see also* Appendices 1, 2
power supplies, 323, 360
  EHT, 177
  IFF, 281, 287
  voltage control, 49
  12-phase, 49
  220V DC, 138, 323, 360
  50 Hz, 138, 323
  500 Hz, 49, 287, 323
  2000 Hz, 287
  and missing motor-alternators, 308
Pozarica, HMS (auxiliary AA ship), 396
PPI, *see* displays
pre-amplifier ineffective, 362

Precision Engineering, Haslemers, 403
precision panel, *see* RBL10,ll on p. 474
prefabrication, 48, 210, 268, 372
Prendergast, Commander J. F., **8**
Premier, HMS (escort carrier), 390
Preston, Captain L. G., 128
Pretoria Castle, HMS (escort carrier), 390
pre-tuned valve circuit units, 110
Prime Minister, *see* Churchill
Prince of Wales, HMS (battleship):
  in Action, 213, 303
  detection of submarines, 213–14, 223
  lost, 54, 304, 345
  multi-set installation: *xxxiv*, 49, 303
  prototype Type 281, 48, 151
  radar sea trials, 222, 268, 303
  sets fitted, 161, 388
Principal Medical Officer (PMO), 303
Pringle, C. O. (HMSS/ASE), **19**, 128
printed circuit boards, 360
priorities, *xxvi-xxvii*, 21, 196, 203, 313, 329
Pritchard, W. H. (HMSS/ASE):
  metric development, 13, 21, 177
  Deputy O-i-C, Witley, **50**
  administration, **19, 20, 50, 62**
private firms, *see* outside contractors
probability of signal detection, 241–2, **243**, 252–3, **253**, 255, 256
procurement of equipment: Mono.9, 329–30
production of equipment, 22–3, 64–5
Project Coordination Party (post-WW2): *xxx*
propagation of EM waves:
  absence of cm data, 260
  anomalous, 41, 72, 77, 142, 214, 312
  Cardigan Bay trials, 78
  early studies, 106
  fading, 58, 78
  Free-Space radar equation, 212–13
  GEC, 403
  hypothesis on S-Band, 213–4
  meteorological conditions, 72, 77, 78
  propagation field, 190
  radio propagation, 75–9
  transmission path: atmospheric, 71; ionospheric, *xxvi*, 9, 106; tropospheric, 78
Prosperine, HMS (shore base, Lyness), 298, 302
Pryce, M. H. L. (later Professor) (HMSS/ASE), **50**, 70, 72, 76–7, 90–1

General Index 451

Public Records Office, 208
Puckle, O. S. (Cossor Ltd), 324, 402
Puleston, R. (EMI), 402
pulse details:
  length: Appendix 2; gunnery radar, 59; metric warning radar, 182; S-band warning radar, 271; Type 79/279 series, 21, 32, 139, 145, 163; Type 281/BQ, 32, 151, 163-4;
  power, 21;
  repetition frequency, 21, **59**, 72-3, 83, 138, 164
  shape: pulse forming networks, 153, 402; shape preservation, 89; rectangular, 378, 380; square, **59**, 382; triangular, **59**, 369;
  pulse-to-pulse variation, 79-80; *see also* technical data; 182, 271, App. 2
pulse generators, 153
pulse transformers:
  air-cored, 173
  iron-cored, 173, 174
  heavy duty, 187, 216, 224
  oil-immersed, 226
  and Gibbs, 187
  and IFF, 289
  and specific types, 173, 174, 226, 367
Pumphrey, Dr R. J. (HMSS/ASE), **50**
Puncher, HMCS (escort carrier), 390
Pursuer, HMS (escort carrier), 390
Puttick, T. (Precision Engineering), 403
Pye Ltd, 404
pyrene, 311
pyrotenax, *see* Antenna, feeders
Pyrotenax Ltd., 89, 404

Q

'Q', (impedance), 89
Q-meter, 179
quarter-wave line, 51
quasi-optical techniques, 190
Quebec, HMCS (heavy cruiser), 396
Queen, HMS (escort carrier), 390
'Queen Bee' radio-controlled aircraft, 7
Queen Elizabeth, HMS (battleship), 388
'*Queen Mary*', RMS, 310

R

'R' Department, *see* ASE
R1 Division, *see* HMSS

R2 Division, *see* HMSS
R3 Division, *see* HMSS
radar – applications: *see also* beginnings of Naval radar, coast defence, dive-bombers, gunnery, height determination, merchant ships, navigation, station keeping, submarines, warning radar
radar – general, 410
  invention, demonstration, 6, 11-12, 68, 106
  patent, 11, 68, 106
  'radiolocation', 11, 410
  RDF, 17, 21, 26, 410
  advantage of higher frequency, 185-6, 188
  birth of Naval 10-cm radar, 44-6
  basic scientific research, naval radar: Monograph 2
  the environment for research, 69-70
  preparing for war, 21-3
  no longer a novelty, 225
radar, anti-submarine, 45, 190, 192, 204, 211-15, 223
radar beacons, 13, 100, 289, 402, 410
Radar 'Branch' of Navy, 307, 347
radar breakdowns, 320-2
radar calibration, *see* calibration of equipment
radar camouflage, 85-6
radar countermeasures, **63**, 73, 403
'Radar Days', 190
Radar Display Room (RDR), 410
  air-conditioning, 322
  assembly of remote displays, 167
  early example, 168, **168**
  and remote control, 176
  as an Information Centre, 254
radar, effect of own gunfire, 223, 302
radar and the environment, *see* environment for radar at sea
'Radar Equation', 72, 152, 216, 222, 255
radar equipment: designation standardised, 204
radar for Armed Merchant Cruisers, 355
radar for coastal forces, 354, 355, 356, 357
radar for laying other radar on target, 163, 382
radar for small ships, 33-4, 169, 193, 287, 288

radar for submarines, 61, 354, 355, 356
radar for surveying, 356
radar, German and UK parallels, *xxvi*, 21, 22, 85–6, 206
radar hut, *see* Radar Office
radar installation, 338–42
   see also ship fitting
radar interference, *see* interference
radar lantern, *see* antenna
Radar Maintenance School, *see Collingwood*
radar mechanics:
   introduced, 334, 340
   Awards, 346
   duties, 315
   'hostilities only', 347
   numbers in ships, 307, 336
   training, 344
   tribute, 346
   WRNS, 340
Radar Officers:
   general, 180, 305, 306–8, 346
   'Branch', 347
   Commonwealth and other Nationals, 299, 305, 345–6
   'Hostilities Only', 347
   post-WW2, 347
   staffing, 336
Radar Officer Organisation, 344–7
Radar Offices/huts and their environment: 268, 287, 308, 336 *see also* environment for radar at sea
Radar operators, *see* Operators
radar performance, 70, 71–2, 74
radar performance prediction, 71–3, 81
radar priorities, 21
radar rangefinder, *xxxiii*, 32, 34, 39, 49
radar reflectors, 86, 379, 381
Radar Research and Development Establishment, (RRDE) (Army), 58
radar equipment, resistance to enemy action, 296
radar secrecy, *see* secrecy and security
radar set design, 64, 175–6, 180, 245
radar spares, 313–14;
   shortages, 298, 301, 311, 332
radar shipboard support staff, 307, 336, 339
radar set details, *see* Equipment Index
radiation resistance, 87, 89
radio astronomy, 81
Radio Corporation of America (RCA), 87

radio direction finding, *see* direction finding and HF DF
Radio Equipment Department, Admiralty, **327**, **328**, 329–30
radio fingerprints (RFP), 63
radio guidance:
   see guided-missiles and FX 1400
Radio Interference Suppression, 38, 167
Radio Location, *xxxiii*; *see also* radar, general
Radiolocation Convention, 1946, 61, 188, 189
Radio Mechanics, *see* radar mechanics
radio remote control, 7
Radio Research Station, 106
radio 'silence', 146
radio telegraphy (R/T), 278, 285
radiology, 170
radome, 211, 219, 268, 410
Raffery, Warrant Telegraphist G., **18**
Railway Transport Officers, 302
railway warrants, 301
Rajah, HMS (escort carrier), 390
Ramillies, HMS (battleship), 388
Ramsay, H. T. (GEC), 114, 402
Randall, Dr J. T. (later Professor) (Birmingham University), 44, 69, 111, 112, 186, 267
Randall, R. L. (HMSS/ASE), **18**, **20**
Randall, Dr (Telegraph Construction & Maintenance Co. Ltd), 404
Ranee, HMS (escort carrier), 390
range of detection:
   general, 54, 152, 154
   aircraft: philosophy, 134; provisional Staff Requirements, 13, 134; attainment, 23; early AMRE results, 12, 15; early HMSS results, 14, 15; specific sets, *see* Equipment Index; *see also* low-flying aircraft, below
   E-boats, 207, 213
   landmarks, 28
   low-flying aircraft: wavelength selection, 15, 22, 54; 50-cm results, 31–2, 34; specific sets, 43, **175**, 246
   schnorkels, 244–5
   ships: Staff Requirements, 13, 134; early results, 15, 31; Swanage trials, 46, 194; antenna height, 204, 208; use of shorter wavelengths, 32; *Scharnhorst*

Action reports, 223; specific set
    results, *see* Equipment Index
shell splashes, 43
shells in flight, 43
submarines, surfaced: Swanage
    trials, 45, 192, 268; state of sea,
    196; 5kW cm radars, 211–15;
    Type 271 series, 47, 196, 211–15,
    223–4; Type 273 series, 211–15,
    223–4; Type 277 series, 244–5
submarines, periscope or Schnorkel
    only, 47, 244–5
range-finding equipment:
    radar, 51; visual, 196, 206
range-gating, 83; *see also* sea clutter
range measurement, 15
range-scales, calibration, 315–16
range-tracking panels, 323
Range Transmission Unit (RTU), 371,
    373
range-rate, *see* gunnery
ranging panels:
    Army GL1 panel, 145, 335
    call for accurate panel for
        S-band, 206
    conversion of Type 972, 266
    manufacturers, 402, 403, 404
Ranpura, HMS (depot and repair
    ship), 398
Rapide (de Haviland aircraft), 300
Ras al Tin, 208
rate-aiding, 382, 410
rate-aided handwheel control, 51, 371
rats, 310, 338
Ratsey, O. L. (HMSS/ASE):
    pre-WW2, 14, **19**, **20**, 36
    delay-line development, 38
    sea trials, modified GL1, 33
    Symonds' Group, 17, 21, 177
    RC2/XRC2, **50**, **62**, 187
Ravager, HMS (escort carrier), 390
Rawlings, Commander H. H., **19**
Rawlinson. J. D. S. (HMSS/ASE):
    O-i-C, Eastney, 23, 46, 178
    O-i-C, Witley, **50**
    pre-WW2, 7, **18**, 70
    'Rawlinson's Group', 32, 33
    remote radio-control of ships, 7
    and Far East, 23
Rawlinson, W. F. (HMSS/ASE), 7, **8**,
    18, **20**
Rayleigh, Lord J. W. S. (1842–1919),
    (British physicist), 90

Raytheon (USA), 113
RE2, *see* ASE, HMSS
Reaper, HMS (escort carrier), 390
receivers:
    AFC, 176, 378, 380
    AGC, 83, 378, 380
    'autodyne', 26, 38
    'back-biassing', 84
    bandwidth/bandpass: general, 87,
        163, 182, 245, 271, 288; specific:
        140kHz, 141, 145, 182, 362;
        150kHz, 159, 182, 364; 0.5MHz,
        271, 376; 1MHz, 145, 182, 271,
        362; 1.25MHz, 271, 374, 376;
        1.5MHz, 159, 182, 364; 2MHz,
        182, 368, 372; 2.5MHz, 271, 372;
        4MHz, 271, 282, 288, 369, 374,
        378, 380, 382; 8MHz, 384
    ground-wave, 159, 360
    IF, 38, 182, 245, 271, 288, 365, 380
    IFF, 288
    instability, 140, 141
    'logarithmic', 84
    low-power RF valves, 116–21
    maintenance at sea, 317
    noise-factor: crystal-mixer, 38, 245,
        269; grounded-grid triode, 27,
        118, 149, 165; gunnery radar
        sets, **59**; low-noise RF stage, 80;
        theoretical study, 81; and
        parametric amplifier, 58; values
        cited, **59**, 80, 118, 149, 165, 202,
        204, 220, 235, 269, 271, App.2
    parametric amplifier, 58
    receiver panel in use, **40**
    saturation, 360
    sensitivity test, 151
    suppliers, 402, 403
    swept gain, 165, 220, 232, 285
    type of receiver: 'straight', 21, 139,
        140; superheterodyne, 23, 25, 26,
        32, 38, 76, 185, 189, 365
    video-frequency filters, 84
    wavelength: S-band, 47, 189, 226,
        245, 268, 271; decimetric, 24, 25,
        26, 38, 43; metric, 14, 21, 23, 32,
        142, 177, 182, 365
    for specific receivers P11, P12, P14,
        P16, P19, P29, *see* Equipment
        Index
recruitment, *see* staffing
Rediffusion Ltd, 404
Redgment, P. G. (HMSS/ASE), **63**

454 *General Index*

Reeves, H. J. W. (HMSS/ASE), 121, 183
reflections from sea, 54, 76, 193, 233–4, 242, 248
reflections from ground, 76
reflection properties of 'targets', *see* target reflection characteristics
Reith, A. M. (HMSS/ASE), 128
Renown, HMS (battlecruiser), 388
Renwick, R. (Rediffusion Ltd), 404
Renwick, W. (ASE), 83
repair of silica valves, 104
reporting information, 203
Repulse, HMS (battlecruiser):
  access of Radar Officer to Captain, 306
  repairs, 303
  short-wave radio, 106
  sunk, 54, 304
  target in Type 273 trials, 206, 222
  sets fitted, 388
re-radiation, 11
Research Associates Ltd, 404
resistors, 179
Resolution, HMS (battleship), 304, 38
resonant cylinder, 370
resonant lines, 319
Resource, HMS (repair ship), 398
responsers, 277, 410
Revenge, HMS (battleship), 388
Reynolds, F. (Thermal Syndicate), 97, 102
Reynolds, L. G. (ASE), 87
Reyrolle Electric Ltd, 404
Rho-metal, 174
Rhys-Jones, Mr (Plessey Co.), 403
Rice, S. O. (BTL), 81–2
Ridge, E. J. (HMSS/ASE), **18**
Ridgeway, D. V. (HMSS/ASE), 290
ring main, 323
Rinnigill, Martello tower, 316
River-class frigates, 400
Robb, Mr (Allen West Ltd), 401
Robb, Mr (Marconi), 403
Roberts, HMS (monitor), 86, 388
Roberts, T. C. (HMSS/ASE), **18**
Robinson, Commander C.V. (HMSS Ship Fitting Section), 330, **343**
Robinson, E. Y. (Edison Swan Co.), 402
Robinson, Lieutenant Commander J. D. M., **18**
Roche, Lieutenant P. E., 306
Rocke, A. F. L. (HMSS/ASE), **63**
Rockwood, HMS (destroyer), **262**

Rodney, HMS (battleship):
  first pre-WW2 fitting of Type 79Y, *xxxiii*, 5, 22, 28, 32, 133, 135, 139, 334
  radar performance report, 29
  air-cover at Scapa, 300
  sets fitted, 388
Rogers, D. C. (STC), 404
Rose Brothers, 39, 404
Rose, Mr (Rose Bros.), 404
Rosinski, W. (HMSS/ASE), **63**
Roskill, Commander S.W. (later Captain), 34, 190
Ross, Alfred W. (HMSS/ASE):
  antennae development, 17, **20**, 21, 32, 33, 70, 87–9, 138
  A/J, **50**, 70, 83–4
  contributor, 183
  DF, **18**
  echo-enhancement in NEPTUNE, 86
  general research, **50**, 70
  long transmission-lines, 89
  polarisation, 22, 76, 82, 83, 136
  prediction of performance, 71
  propagation, 75–6
  Ross's Group, 21, 33
  sea-clutter, 22, 82–3
  video-frequency filter, 84
  and height estimation, 74
  and Horton, *xxxi*
  and RX Department, 61, **62**
  and Type 79/X, 17, 22
Rosta Drydock, Russia, 322
Rosyth, 49, 208, 331–2, 342, 343
Rosyth Escort Force, 340
rotary capacity-switch, 51, **52**, 403
Rother, HMS (frigate), 225
Rowe, A. P., (Air Ministry), 107
Rowe, E. G. (MetVic), 403
Rowley, B. G. H., (HMSS/ASE):
  Author: Monograph 7, *xxxix*
  radar career, **50**, 295–6, 302–3, 304, 305, 306
  Heightfinding lecture, 299–300
  liaison, **50**
  split responsibility, 301
  and FWO, 295, 296, 305
  and North Atlantic duty, 304
Royal Air Force:
  general, *xxvi*, 228
  Coastal Command, 54, 191
  collaboration with RN and ASE, 23, 109, 195, 326, 334

General Index 455

Fighter Command, 161, 306, 324
   radar, 15, 30, 97, 300
   Stations, 79
Royal Aircraft Establishment (RAE), 7
Royal Artillery, 295
Royal Australian Navy, 345
Royal Canadian Navy, 347
Royal Canadian Navy Volunteer Reserve, 299, 345
Royal Dockyards, *see* HM Dockyards
Royal Indian Navy, 304
Royal Marines, 300, 308
Royal Marine Barracks, Eastney, 12, 134, 186
Royal Military College of Science, 315
Royal Naval Air Station, Kete, **250**
Royal Naval Barracks, Portsmouth, 17
Royal Naval Fleet Repair Base, Sydney, 341
Royal Naval Radar School, Eastney, 306, 324
Royal Naval Scientific Service (RNSS), *xxxi*
Royal Naval Volunteer Reserve (RNVR), 47, 295, 301, 343, 344–5, 347
Royal Navy, 53, 300
   Branches, *see* 'Radar' and Torpedo Branches
Royal New Zealand Navy, 345
Royal Oak, HMS (battleship), 297, 316, 388
Royal Society, Fellows of (FRS), *xxxviii*, 56, 70
Royal Sovereign, HMS (battleship), 388
Royalist, HMS (light cruiser), 394
Ruler, HMS (escort carrier), 390
Rundle, W. (HMSS/ASE), **8**, **18**
Rushforth, L. (BTH), 401
Russell, E. S. (HMSS/ASE), **18**
Russian warship, 86
RX Department, *see* ASE
Ryanson, Lieutenant Commander C. J. W., **19**
Ryde, J. W. (GEC), 403
Ryde, Isle of Wight, **266**
Ryle, M. (later Professor Sir Martin, Astronomer Royal) (TRE), 81

S

SA, 31, 411
Saebol, Iceland, 208, 228
S-band, 411; *see also* wavelength (10cm)

St. Eval, RAF Station, 79
St Ninian, SS , 296, 302
Salmond, Captain J.S.C. RN, 11
Saltburn, HMS (trials vessel, sloop):
   sea trials: Type 79X, *xxxiii*, 14, 21–2, 135, 136, 139, 177, 181, 362; Type 277X, *xxxv*, 55, 229, 230, 269; Type 261, 260, **261**; Type 291, 174, 175
   and polarisation, 22, 82, 136
   and sea clutter, 82
   and low-flying aircraft, 22
Sand, Mr (Silica Syndicate), 102
Sandhurst, HMS (repair ship), 398
Sardonyx, HMS (destroyer), 9, 10, 28, 31, 369
Sargent, Dr (English Electric), 402
saturation of receivers and displays, 360
Saunders, Captain L. S., 306
Savory, R. (HMSS), **18**
Sayer, Mr (Ferguson Radio Ltd), 402
Sayers, Dr H. J. (later Professor) (Birmingham University), 112, 215
Sayers, Commander Tel. W. A., **19**
Scaife, F.N. (HMSS/ASE), 50, 62, 88, 187
scanning loss, 72–3
Scapa Flow, 297–9, **297**
   maintenance base, 340
   radar calibration, 316
   working-up, 49, 339
   surveillance radar, 23, 142, 300
   anomalous propagation, 77
   trials, 206, 219
   and IFF, 279
   and Radar Officers, 296, 305
Scharnhorst, (German battle cruiser), 53, 222–3, 346
Scheldt, River estuary, 267
schnorkel, 85, 244, 265
Schoenberg, Dr (EMI), 402
Schofield, Dr (Telegraph Construction & Maintenance Co. Ltd), 404
Scholes, Dr (MetVic), 403
Schwarz, Mr (Decca Radio Ltd), 402
'Science at War', 191
Scientific Research and Experiment Department, Admiralty (DSR):
   allocation of new staff, 31
   first news of radar, 12
   graduate recruitment, 30
   highest priority requested, 29
   industry and radar secret, 27
   Naval radar programme, 12

(DSR) (cont.)
  rejected Alder's proposal, 68
  resisted centralised research, 16, 31
  review of HMSS progress, 15
  Tizard Committee, 12
  and AMRE dependence on HMSS, 107
  and Horton, 61
  and HMSS staffing levels, 16
  and Randall and Boot, 111
  see also Sir Charles Wright
Scott, D. C. (HMSS/ASE), **62**
Scott, Lieutenant (Sp) RNVR, 300
Scott-Taggart, J. (HMSS), 128
Scrabster, 296
Scylla, HMS (cruiser), 394
Sczaniecki, A. (Polish Army) (ASE), 128
sea clutter, 71, 82–3
  polarisation, 82, 83, 136
  swept gain, 165
  Type 277 on Mount Snowdon, 78, 82
Seaford, 83
Sea Lion, HMS (submarine), 213
Searby, N. (Ferranti Ltd), 402
'Search' radar, *see* warning radar
Searcher, HMS (escort carrier), 390
sea-reflection coefficient, 233–4, 242, 248
sea-sickness, 312
Seaslug, surface-to-air missile, 60
sea trials:
  Acceptance Trials, new construction, 339
  Organisation, 9, 10, 326
  polarisation, 22
  propagation, 213
  Radar Sea Trials Party, 315, 339
  vessels in radar trials: *see Campania, Concord, Dido, Dunedin, Janus, King George V, L22, Legion, Marigold, Nigeria, Orchis, Penstemmon, Saltburn, Sardonyx, Southdown, Tuna, Tuscan*
  working-up after installation, 312, 339
  for specific Types *see* Equipment Index
Sea-Trials Party, *see* ASE
secrecy and security:
  the radar secret, 12, 13, 27, 107, 348
  within HMSS/ASE, 13, 14, 84
  within Radar Department, 73
  Eastney, a secure location, 134

HMSS, a secure source of silica valves: 97, 107
manufacture of radar sets, 330
some contractors privy to secret, 17, 27, 109, 110, 179
within the Navy, 298, 299, 341
in Dockyards and shipyards, 329, 335
in IFF coding, 290
Sector displays, *see* displays
Sedgefield, Major P. (AA Command), 58, 60
SeeKay wax, 310
Selsey Bill, 208, 226
selsyn torque transmission:
  manual antenna positioning, 138, 147, 158, 363
  continuous antenna rotation, 229, 365
  reversible antenna motion, 320
Service(s) Valve Number, 145
Services Electronics Research Laboratory (SERL), 128
servo system, 411
  antenna control, 229, 248, 249
  suppliers, 401, 403, 404
setting up radar on ship, *see* ship-fitting
SFR, Paris, 111
SG, 357, 474
Shah, HMS (escort carrier), 390
Shannon, C. E. (Bell Laboratories), 82
Sharpe, J. (GEC), 402
Shayler, J. S. (HMSS/ASE):
  Author (Mono.4, Mono.6), *xl*
  contributor, 272, 361
  first radar on a carrier, 296
  Type 281 transmitter, 32, 296
  valve development, 128, 177
  XRC4, **62**
Shearing, Dr G. (HMSS),
  Superintending/Chief Scientist, 8, 12–13, **20**, 61, 128
  pre-radar days, 7, **8**, 10
  start of naval radar, 12–13
  re-organisation, 15, 16, **18**, **20**
  retiral, 61
  in valve research, 128
Shearston, C. R. (HMSS/ASE), 20, 187
Sheffield, HMS (cruiser):
  first Type 79Y fitted pre-WW2, *xxxiii*, 5, 22–3, 28–9, 32, 133, 135, 139, 177, 181, 334
  sets fitted, 394

General Index

Sheldrick, J. E. (HMSS/ASE), **18**
shell fuzes, 34
shell-flight observation, 43, 302
shell-splash spotting and ranging, 43, 61, 219, 302, 379
Shepherd, G. F. (HMSS/ASE), **19**
Sherbrooke House, Glasgow, 340, 343–4, 348
Sherrin, Mr (Allen West), 401
Shetland, 301
shipboard environment, *see* environment for radar at sea
shipbuilders, 10, **328**; *see also* shipyards
ship design, 86
ship-fitting – general, 307–10, 334–8
  at sea, 65
  calibration, 314–17
  first RDF on a carrier, 296
  fitting-out bases, 48
  'Fitting-Out' Data, 10
  new problems, 196–7
  priorities, 10, 28, 327, 329
  Project 'Bubbly', 245
  radar installation and testing, **327–8**, 338–9
  supervision, 329
  The Task, 334–8
  waveguides, 91, 228, 230
  specific ships, 22, 49, 210
shipping losses, 189
shipwrights, 337
shipyards, commercial, 325, 329, 334, 335
  see also shipbuilders
shop steward, 339
Short, A. W. (HMSS), **19**
'shot effect', 82
Shropshire HMS (heavy cruiser), 394
Shuttleworth, Norman (HMSS/ASE), 30, 49
Signal Department/Division, Admiralty, 6–7
  responsibility for HMSS, *xxvi*
  equipment acceptance authority, 10
  Administration of HMSS, 10, 12, 14–16, 28, 134
  and Swanage, 192
  Trials Vessel, 260
  and Type 79Y ship reports, 29
signal detection probability, 241–2, 253, **253**, 255–6
signal generators, 313, 315, 404

signal-to-noise ratio, 227, 255
Signal School, *see* HM Signal School
signal transmission circuits, 308
Signals Research and Development Establishment, (SRDE) (Army), 14, 78
silica, properties, 98
Silica Syndicate, 102
silica valves, Mono.3, 136, 318
silicon, aluminium-doped (mixer diode), 123
Sirius, HMS (cruiser), 396
Sisson, G. M. (Grubb Parsons Ltd), 403
Skiatron, *see* displays
Skinner, H. W. B. (TRE):
  Air Ministry cm-wave Group, 44–5, 186
  crystal-mixer valve, 38, 44, 46, 122, 189, 200
  Swanage trials, 44–5, 192, 194, 195
  and Coales, 45
  and S-band for Naval use, 45
Slinger, HMS (escort carrier), 390
slip-rings, 162, 163, **164**, 170, 283–5, 289
sloops, 400
Small, D. F. (HMSS), **19**
Smiter, HMS (escort carrier), 390
Smith, Dr H. (HMSS/ASE), **8**, **18**, **20**, **63**
Smith J. B. (Plessey), 403
Smith, S. B. (Marconi), 403
Smith, W. Austin (HMSS/ASE), **18**
Smith, W. H., *see* W. H. Smith Ltd
Smith, W. T. (Telegraph Construction & Maintenance Co. Ltd), 404
Smyth, C. N. (STC), 116, 404
Snowdon Mountain Railway, 78
Solent, 265, **266**
Solomon, A. K. (RRDE), 58
Somerville, Admiral Sir James F, *xxvii*, 7, **8**, 11, 30–2
'Sonar', *see* Asdic
Soper, R. E. (Pirelli General Cable Co.), 403
South African Naval Force, 346
South African Radar Officers, 345–6
Southampton, HMS (cruiser), 396
Southampton Water, **266**
Southdown, HMS (destroyer), *xxxiv*, 43, 47, 369
Southend, 340
South Ronaldsay, **297**, 297, 300
Southsea, 30, 186
Southsea Castle, 14, 17, 28, 186

South Walls W/T masts, Scapa Flow, 316
Sowan, Mr (Mullard Co.), 131
Sowter, R. J. (HMSS), **8**
Spagnoletti, P. H. (STC), 404
spares, *see* radar spares
spark-gap, 53, 115–16, 162, 319
sparks from radar, 312
spark technology, 15
Spartan, HMS (light cruiser), 396
Speaker, HMS (escort carrier), 390
Special Branch of Navy, *see* RNVR
special intelligence, 411
specular reflection, 233, 248
Sperry Ltd, Brentford, 404
Spilling, J. F. (HMSS/ASE), **19**, 128
Spithead, *xxv*, 264
splash-spotting, *see* shell splash spotting
Sporberg, H. N. (BTH), 401
Springbank, HMS (auxiliary AA ship), 396
SQ radar (US), 314, 324, 357
squegger, 366
SS radar, 31, 411
stabilisation of antennae, *see* antennae
Stabilized tachymetric anti-aircraft gun, (STAAG), 354, 382–3, 404
Staff Requirements, *see* Naval Staff Requirements
staffing of HMSS/ASE, 12–13, 15–16, 30, 61
Stainsby, A. G. (GEC), 402
Stalker, HMS (escort carrier), 390
Standard Telephones & Cables Ltd (STC):
    see outside contractors
standing-waves, 336
standing-wave detectors, 313
standing-wave ratio, 163, 230, 376, 412
Standards Laborarory, ASE, 179
Stanley, C.O. (Pye Ltd), 404
starters, Watford automatic, 323
Stanmore, (RAF), 228, 306, 324
station-keeping, 210
'stealth' ships and aircraft, 86
Stephenson, Commodore G., 339
Stephenson, J. D. (Mullard Radio Valve Co.), 403
Stevenson, Mr (Pye Ltd), 404
Story, J. G. (HMSS/ASE), **18**
Strang, Dr (Laurence Scott), 403

Striker, HMS (escort carrier), 390
Stromness, 297, **297**
Strong, C. E. (STC), 404
Strong, Mr (Allen West Ltd), 401
Struszynski, W. (HMSS/ASE), **63**
Stuart Prince, HMS (fighter-direction ship), 398
Stubbs, Lieutenant (Sp) B. RNVR, 303, 346
submarines:
    Allied – specific: L22; 9, 10; H44; **262**
submarine radar, *xxxv*, 61, 187, 262, 337, 402; *see also* Types 261W, 267/W/MW/PW, 286W, 291/W
submarine threat, *see* U-boats
Suffolk, HMS (cruiser), 28, **42**, **212**, 306–7, 324, 396
Sumburgh, 301
Sunderland, 343
Sunderland flying boats, 224
sunspot activity, *see* propagation
Superb, HMS (cruiser), 396
Superintending Naval Stores Officer (SNSO), 313
Surdin, M. (later Professor) (Free French Forces) (ASE), 90, 263
surface gunnery fire-control, *see* GS radar
surface warning radar, *see* warning radar
surveillance radar, *see* warning radar
surveying, 266, 356
Sussex, HMS (cruiser), 315, 365, 396
Sutton tube, (local oscillator), 121
Sutton, R. W. (HMSS/ASE):
    recruited from industry, 29
    administration, **20**, **63**, 189
    Bristol, 31, 44, 128, 189
    reflex klystron, 44, 46, 118–20
Swanage, Dorset, *xxxiii*, 44–5, 189, 191–5, 211, 214
swept gain, *see* receivers
Swiftsure, HMS (cruiser), 378, 396
switch-gear, 323, 402
Swordfish aircraft, 259, 357
Sydney, HMAS (light cruiser), 396
Sydney, Australia, 310, 322, 341–2
Symonds, Dr A. A. (HMSS/ASE):
    Eastney, 21, 177, 178
    pre-WW2, 17, **18**, **20**, 70
    spark-gap modulator, 187
    XRE1, **62**

General Index

Type 79, 17
Type 274, 53, 187

T

Ta'Bois, P. L. (HMSS/ASE), **18**
Tactical Plot, *see* plots and plotting
tactical radar: Mono 5, 259
Tanner, S. R. (HMSS/ASE), 187
Tantallon, ASE Extension, *see* ASE
Tapir, HMS (submarine), 264
target allocation, *see* gunnery
target discrimination, *see* discrimination
target indication, *see* gunnery
target indication radar, *see* specific sets indicated on p. 356
Target Indication Unit (TIU), 376, 379, 381
  specific TIUs, *see* Equipment Index
target reflection characteristics, 11, 13, 79–80, 212–13, 255
Task Force 37 (USN 3rd Fleet), 342
TB cell, 121, 411
T-class submarines, 264
Technical Exchange Agreement, UK-USA, 48
technical data sheets, 182, 271, Appendices 1, 2
technological breakthroughs, 140
technology of early days, 179, 360
technology transfer, 96, 98, 104, 192, 193–4
Telcon, 64
Telecommunications Research Establishment (TRE): formerly AMRE, 38, 44, 189
  'Admiralty' Trailer, 46
  'Cm-Wave Application Group', 186, 191, 267
  crystal-mixer valve, 38, 44, 185, 189
  HMSS copy of S-band equipment, 45–6, 194
  IFF, 60
  PPI development, 56, 162, 210
  Swanage and S-band equipment, *xxxiii*, 46, 189, 190, 191–4, 204–5, 268
  trials against submarine, 45, 191–3
  T/R switch, 87
  and fitting-out programme, **327**
  and HPI, 229
  and radio-astronomy, 81
  and waveguides, 86
Telefunken Company (Germany), 36, 70, 87
Telegraph Construction & Maintenance Co. Ltd, 404
Telegraphists, 305
telephone circuits, 308
teleprinter room, 317
television, use of commercial equipment, 17, 25, 38, 116, 136, 178
tenders (auxiliary ships), 9
Teneplas (Pangbourne), 404
Terror, HMS (monitor), 388
Terry, G. A. (HMSS/ASE), **18**
test apparatus, 245, 339
  see also artificial load, Avometer, cable testing, galvanometers, LCR Bridge, Megger, monitorscope, neon tube, performance meter, Q meter, signal generator, standing-wave detector, wavemeter
Thames, River, estuary survey, 266
Thane, HMS (escort carrier), 392
'The Good Shepherd' (motor drifter), 301
Thermal Syndicate Co., *see* outside contractors
Third Sea Lord, *see* Controllor of the Navy
Thompson, F. H. (English Electric Co. Ltd), 402
Thompson, G. (ASE), **63**
Thomson, Dr J. (HMSS/ASE), 128
Thurso, 296, 302
thyratron, *see* valves
tilt-test calculation, 335
time-base, 146
time-sequencing of radar set operation, 49
Tipper, E. E. (HMSS/ASE), **19**
Tiptoe, HMS (submarine), 357
Titlark (launch), 194
Tizard, Sir Henry T., *xxvii*, 12, 48, 107
Tobermory, Scotland, 219, 223, 339
Toller-Bond, D. H. (HMSS/ASE), **18, 63**
Tomlinson, Mr (Pyrotenax Cable Ltd), 404
Tomlyn, Mr (STC), 404
Toothill, J. S. (Ferranti Ltd), 402
top-weight compensation, 335
Toronto University, 305

torpedo bombers, 57
Torpedo Branch of Royal Navy, 347
Torpedo School: *see Vernon*
Torpedo and Mining Establishment, 6
Torpedoes and Mining Department, Admiralty, 6
torpedoes, manned, 208
Torpedo Officer, 307, 308
Town-class destroyers (ex-USN), 400
Toy, E. J. (HMSS/ASE), 303
Tracker, HMS (escort carrier), 392
tracking of targets, 146
tracking ships in coastal waters, 206
'Trade Protection' Committee, 190, 191, 195
training of personnel, 10, 104, 196, 305–6, 344–5
  see also Collingwood, Excellent, Mercury, Victory
Training and Staff Duties Division, Admiralty, 34
transformers, 313, 314, 403
transistors, 360
transmission lines, *see* Antennae, feeders
Transmit/Receive (T/R) switch:
  uses, 51, 147
  diode, 87, 148, **149**, 176
  gas-switch, 87, 121, 171
  spark-gap, 319
  suppliers, 401, 402
  and IFF, 282, 288
  see also common T/R working
transmitters:
  general: adaptation of conventional unit, 21; technical data, 61, 182, 271, 318–9; Appendices 1, 2; W/T, 7; and earlier valve development, 7
  oscillators, 27, 362
  Performance Meter, 150
  power: technical data, 182, 271; mean v. peak pulse power, 152; monitoring, 150–1; metric, 21, 23, 32, 121, 153; decimetric, 24, 27; S-band, 89–90
  pulse-length/prf, 21, 25, 271
  valves, 21, 25, 26
Transmitting Station, (TS), 371, 379, 412
transponders, 277, 278–80, 282, 317, 411
trawlers, 208, 337, 367
TR cell, 121, 412

Treasury, H.M., 12
Trevaskis, E. (HMSS/ASE), **18**
trials of radar sets, *see* sea trials and specific Types
Tribal-class destroyers, 400
Tricker, Dr R. A. R. (HMSS/ASE), 79, 125, 128
Trier, P. (ASE), **63**
trigatron, 403
Trigle, S. E. (HMSS/ASE), **18**, 37
Trincomalee, 341
Trinidad, HMS (cruiser):
  equipment problems, 308, 319, 320, 321–2
  hit by own torpedo, 322
  sunk, 306, 313
  sets fitted, 324, 396
Trinidad, 208
Trippe, S. H. (HMSS/ASE), **18**
tropospheric scattering, *see* propagation
Trouncer, HMS (escort carrier), 392
Troup, Admiral J. A. G., 344
Trump, HMS (submarine), 357
Trumpeter, HMS (escort carrier), 392
Tuna, HMS (submarine), *xxxv*, 263–4
Turnbull, Mr (EMI), 402
Turner, L. B. (HMSS/ASE), 30, 70
Turpin, HMS (submarine), 264
Tuscan, HMS (destroyer), *xxxv*, 237, 269
Tustin, A. E. (MetVic), 403
Tyne HMS (destroyer depot ship), 341, 398
Tyneside, 331
Tynwald, HMS (auxiliary AA ship), 396
Type Number variants, notation, 359
Type 23 frigates, 86

U

U-boats – general:
  threat, 53–4, 190
  A/S radar called for, 185, 190
  A/S radar applications, 45, 46–8, 186
  surface detections at night, 185, 186, 210
  'first-sighting' ranges, 213–14, **214**, 223
  Royal Oak sunk, 297
U-class submarines, 192
Uganda, HMS (heavy cruiser), 396
Ulrich, E. H. (STC), 404
Ulster Queen, HMS (fighter direction ship), 398

Ultra-Short-Wave Propagation Panel, 78
'under the beam' hypothesis, 213, 215, 216, 225
Unicorn, HMS (aircraft maintenance ship), 390
'Unipivot' galvanometer, 313
United Kingdom Conference on Radio for Marine Transport, 267
United States Navy:
   coastguard cutters, 400
   coordination of Allied IFF, 289–90
   radar camouflage, 86
   Technical Liaison Officer, 304
   3rd Fleet, 342
   and heightfinding, 75
   and Loran, 317
United States of America:
   American radar in British ships, 357
   Exchange, destroyers/bases, 302
   IFF, 60, 277, 281–2, 289
   prototype Type 271, 48
   Technical Exchange Agreement, 48
   valves, 123, 260, 264
US-UK technical exchange agreement, 48
Universal Display Unit (UDU), *see* displays
Universities, 344–5; *see also* Birmingham, Bristol, Cambridge, Oxford and Toronto
Ure, W. (HMSS/ASE):
   Transmitter Section pre-WW2, 7, **8**, 12, 13, 16
   radar review, 15
   reorganisation, 16, **18**, **20**
Uredd (Norwegiam ship), 223
'user-friendly' radar sets, 175–6, 180
Usher, Mr (Nash & Thompson), 403
Usk, HMS (submarine), 45, 192, 211, 213

V

VACNA, 341
Valiant, HMS (battleship), 28, 388
Valve Section, HMSS, (Bristol University), 87
valves – general, 110, 123, 367
valves – specific:
   for specific numbers, 476–6
   'Acorn', 117, 405; IFF, 278; mixer, 122; pentode, 122, 141; triode,
116; and Type 79, 139; and Type 279, 116, 145
CRT: afterglow screen, 146, 163, 164; technical, 123–5, **125**, **126**, **127**; manufacturers, 402
crystal, 215, 374, 401, 403
crystal-mixer; spares on *Bermuda*, 314; and Coales, 38; and Dahl, 90, 377; and Skinner, 38, 44, 46; and TRE, 185; on back of antenna, 373
detector(s), 38, 159, 368
diode: mixer, 26, 38, 122, 368; point-contact, 122, **122**; manufacturers, 401, 403; as 2nd detector, 368; in T/R switching, 87, 121, 162, 176, 363, 365
'Doorknob', 24
duplexer, 121, **122**
gas-switch, 215
klystron, 409: cooperative effort, 259; CVD sponsorship, 44, 189; HMSS/ASE development, 44, 118, 189, 259; local oscillator, 46, **119**, 185, 191; manufacturers, 259–60, 402; power klystron, 113; 'Sutton' tube, 44, 46, 118–21, 197; technical aspects of development, 189; USA work, 264; wideband frequency agility, 114; X-band, 259; and T/R, 87; in specific sets, 197, 203–4
magnetron, 409: split-anode, 24, 69, 102, 106, 111, 188; resonant-cavity, 5, 69, 111, 185, 186, 189, 191; development, 44–5, 113–14, 188–9, 267; chronology, 68, 109, 185, 188; power: up to 10kW, 46, 200; 10–100kW; 112, 185, 216, 260, 263; 100kW–1MW, 90, 185, 187, 216, 224, 267; search for 2MW, 58, 109, 247, 249, 258; wavelength, 17, 113, 191, 263, 264; delay-lines, 174; 'fat' oxide cathode, 111; frequency instability, 91; German research, 188; HMSS early idea, 69; magnetic field strengths, 186, 200, 215, 216, 217, 260, 263; mass production, 113; modulation, 174; outside contractors, 24, 44–5, 106, 111, 113, 186, 401, 402; pulse transformers, 174, 216;

valves – specific (*cont*)
  secondary emission cathodes, 90; 'strapping', 90, 112, **113**, 215, 259–60, 374; *see also* specific valves, 475–6
'micropup', 110, **111**, 172–3, 173–4
microwave, 111–14
'millimicropup', 267
mixer, 122, 123, 189, 200–201, **201**
modulator: cathode-modulation, 140, 153; fast warm-up required, 187; pulse modulators, 114–16; series modulation, 107, 109, 200; suppliers, 401
oscillator, 114, 118, **119**, 203
output power, 109–14
pentode, 116, 141, 159
radio frequency, 116–21
rectifier, 105
silica: advantages, 104, 138, 318; disadvantages, 14, 21, 98, 135, 301; early days, 7, 96–7; for radar, 106–9; for radio, 104–6; HMSS, the UK specialists, 6, 12, 16, 69, 97; superseded, 110; technology, 97, 98–104; thoriated-tungsten filaments, 23, 32, 140; and Type 79, 136, 138, 140; and Type 281, 32, 153, 177
skiatron, dark-trace tube, 125, **127**
spark-gap, 162, 187
tetrode, 107, 109, 216
thyratron, 114–15, 411: long warm-up time problem, 53; manufacturer, 401; as output stage, 114; in modulators, 38, 114, 172–3, 370, 374
transmitter: CVD work, 44; manufacturers, 402; silica, 96; Type 79/279, 138, 301; Type 281, 152–3; E773, 37
triode: pulsed triode at 50-cm, *xxxiii*; need for development, 24, 26; grounded-grid, 27, 165, 369; first valve at Bawdsey, 107; modulation of Naval radar, 114, 140–1, 153, 226, 376; disc-seal, 116, **118**; silica, 129–30; gas triode, 226, 376; in IFF, 287, 289; *see also* 'micropup', 'minimicropup'
Van der Kasteele, Lieutenant J., 306
Vanguard, HMS (battleship), *xxxv*

'Variac', 321
Varian, R. H. (Stanford University USA), 114
Varian, S. F. (Stanford University USA), 114
Venerable, HMS (light fleet carrier), 390
Vengeance, HMS (light fleet carrier), 390
ventilation, 308
Ventnor, Isle of Wight, 226
Vernon, HMS (Torpedo School), 5, 6, 7
Verry, H. R. (HMSS/ASE), **19**
Vertical Coverage Diagram: *see* antennae, polar diagrams
Very High Frequency (VHF) radio, 163, 317
Vetch, HMS (corvette), 213–14, 223
Veteran, HMS (WW1 destroyer), 210
Vickers, Mr (Marconi Company), 403
Vickers Armstrong Ltd, 39, 404
Victorious, HMS (fleet carrier), 258, 357, 390
Victory, HMS (Portsmouth Naval Barracks), 9
video-frequency filters, 83, 84
Vidler, N. V. (HMSS/ASE), **62**
Villiers, Lieutenant J. M., **19**
Vindex, HMS (escort carrier), 392
Vindictive, HMS (repair ship), 398
visual signalling, 56
voice pipe, 203
voltage, lethal, 319
voltage control equipment, 49
voltage regulation, 38, 49, 401

W

WA radar, 186; *see also* sets indicated in Appendix 1
Walden, S de (HMSS/ASE), **63**
Walker, F. M. (A.C. Cossor), 402
Walker, J. (BTH), 401
Wallace, HMS (AA destroyer), 268
Walrus, amphibious aircraft, 224, 357
'War at Sea', 190
War Office, 68, 208
Warlow, A. J. P. (HMSS/ASE), **18**
warming up time for radar sets, 53
warning radar:
  general: Monographs 4, 5; chronology, 181, 267–70; technical data, 154, 271
  aircraft: chronology, 21, 134, 176–7, 267–70; cm-radar in "all height" WA role, 233; development of

Type 281, 32; discriminating echoes from 'Window', 79; estimation, 10-cm performance, 226; metric wavelength suitability, 23; priority, 21, 28–9; pre-WW2, 21; Staff Requirements, 134; Type 79Y at Scapa, 300; and Type 960, 177; *see also* heightfinding
  low-flying aircraft: need, 31, 224, 225; chronology, 31, 54–5, 82, 185, 225, 228; antenna rotation and PPI, 224, 225; detectability in clutter, 82; range v power, 54; wavelength suitability, 31, 54; specific sets, 54–5, 224, 228
  submarines, 54, 185, 211–15, 223–4
  surface craft: chronology, 267–70; priority, 21; requirement, 134; the need, 31–3; S-band, 185, 193, 206, 215–31; X-band, 259–67; *see also* WS, WC sets in Appendix 1
warning radar in searchlight control, 29
Warrant Officers' galley, 308
Warren, G. W. (GEC), 402
Warspite, HMS (battleship), 345, 388
Watford automatic starters, 323
Watson, D. Stewart (HMSS/ASE):
  anomalous propagation, 77
  contributor, *xxxi*, 183
  from industry, 177
  modulator for Type 79Y, 23, 177
  Type 281 sea trials, 302
  RC4/XRC4, **50, 62**
Watson, Professor G. N. (Birmingham University), 76
Watson, Mr (Vickers Armstrong Ltd), 404
Watson-Watt, R. A. (later Sir Robert):
  demonstrated radar potential, *xxxiii*, 6, 11, 68, 95, 134
  Admiralty support, pre-WW2, *xxv-xxvi*
  advice on ship antennae, *xxvii*
  pressed for centralised research, 16, 30
  propagation study cruise, 9
  and Horton, 9, 31
  and HMSS valves, 97, 106
  and NPL, 9
  and Orfordness, 12
  and Somerville, 30
waveguides, *see* antennae feeders

wavelength:
  (frequency shown in brackets after wavelength, *see* 412)
  K-band: around 1.25cm (24,000 MHz), 61, 409
  L-band: around 50cm (600MHz), 369
  S-band: around 10cm (3,000MHz), 411
  X-band: around 3cm (10,000MHz), 412
  centimetric radar: (before WW2 the term 'centimetric' included all wavelengths below 1m); accurate elevation on high-flying aircraft, 51; birth of Naval 10-cm radar, 44–6; cm-development Groups, 46, 186, 191, 193; cm-FD proposals, 246–9; dramatic change, 189; earliest experiments, 188; first operational cm-radar, 195–7; power increase, 89; unexpectedly long ranges, 76; and low-flying aircraft, 22
  1.0 cm (30,000MHz), 113
  1.25cm (24,000MHz) (K-band), 61
  3cm (10,000MHz) (X-band): chronology, 38, 53, 58, 61, 113, 259–60; auto-follower gunnery radar, 58; blind-fire, close-range AA, 53; Canadian development, Type 268, 264; crystal-detector, 38; propagation trials, 78; rejected as WS in larger ships, 262; WS in small ships, 262; X-Band WS radar, 259–66; and magnetrons, 113; and sea-clutter, 83; for airborne use, 259–60; for submarines, 174–5, 187
  3.17 – 3.23cm (9, 460 – 9, 290MHz), 384
  8.1 – 8.7cm (3, 700 – 3, 448MHz), 380
  9cm (3, 333MHz), 38
  decimetric (1m to 10cm) radar: air warning, 22, 23 chronology, 69–70; generation, 24; German decision, 188; HMSS work, 17, 21, 70; valves, 110; *see also* sets noted in Appendix 1
  8.88 – 9.28cm (3, 380 – 3, 230MHz), 378
  9.80 – 10.2cm (3,060 – 2, 940MHz), 372, 376

wavelength (cont.)
  10cm (3,000MHz) (S-band):
    applications, U-Boat war, 46–8,
    192; birth of Naval radar, 44–6;
    characteristics of Naval radar,
    271; chronology, 267–70; cm-
    Group in HMSS, 186, 191;
    concentric-cable losses, 47; data
    on specific sets, 271; Appendix
    1, Appendix 2; feasibility, 185,
    188; first Naval radar, 185,
    189–95, 195–7; gas in T/R
    switches, 87; propagation, 78,
    213; pulse-to-pulse variation, 79,
    247; sea clutter, 83; Swanage
    trials, xxxiii, 44–6, 191–5 TI
    radar, 57; valves, 44, 111–14,
    118, 186; waveguide directional
    couplers, 90; and blind-fire, 53;
    and low-flying aircraft, 167; and
    Schnorkel, 244–5; see also
    waveguides and sets noted in
    Appendix 1
  20 – 30cm, 267
  20cm (1, 500MHz), 14
  23cm (1, 304MHz): pre-WW2, 14;
    Gollin and Bareford, 14, 17, 24;
    performance, 15, 17, 24, 28;
    staffing, 15; work abandoned,
    21, 24
  25cm (1, 200MHz), xxxiii, 267
  26 – 32cm (1150 – 950MHz), 290
  48cm (625MHz), 188
  50cm (600MHz) (L-band): airborne
    radar possibility, 24; antennae,
    87; choice, xxxiii, 21, 186;
    chronology, 369; early
    developments, 23–9; gunnery
    and FC radar, 25, 34–6, 36–49,
    186; modulation, 36; priority, 21;
    rangefinding, xxxiii, 186; short-
    range capability, 25; sea trials,
    31–2; valves, 26, 110, 114, 116,
    122; 'world-first', 27; see also sets
    noted in Appendix 1
  49 – 52cm (610 – 580MHz), 369
  60cm (500MHz), 14, 15, 17–18, 24
  75cm (400MHz), 289
  metric radar: RN warning radar;
    Mono. 4; characteristics, 182;
    chronology, 181; HMSS
    organisation, 70; valves, 109–10,
    122; and aircraft detection, 23,
    54, 163; and FD, 163; and IFF,
    280; in RN pre-WW2, 69; see also
    sets noted in Appendix 1
  1.25m (240MHz), 109
  1.4m (214MHz): AMRE ASV
    adaptation to Type 286, 24, 33,
    169, 186, 262; CHL, 300; IFF
    Mk.II, 279–80, 364; and
    pyrotenax, 89
  1.4 – 1.5m (220 – 195MHz), 280
  1.5m (200MHz), 13, 14, 15, 44; see also
    sets noted in Appendix 1
  1.6 – 1.7m (179 – 173MHz), 280
  1.6 – 1.9m (187 – 157MHz), 281, 282,
    288
  1.7m (176 – 179MHz), 280, 282, 364,
    367
  1.8m (171 – 167MHz), 282
  2.4m (125MHz), 188
  3m (100MHz), 24, 57, 107–8
  3.2m (94MHz), 317
  3.2 – 3.5m (86 – 94MHz), 32, 364
  3.3m (90MHz), xxxiii, 152
  3.5m (86MHz), 32, 33, 186, 364; see
    also Type 281 series
  3.7 – 5.5m (82 – 54MHz), 279
  4m (75MHz), 13–15
  5m (60MHz): IF, 373, 377, 378, 380
  5.8 – 7.7m (52 – 39MHz), 279, 280
  6.7m (45MHz): IF, 367
  6.7 – 8.6m (45 – 35MHz), 279
  7m (43MHz), xxxiii, 177
  7.5m (40MHz), 17, 21, 135, 136; see
    also Types 79/279
  7.1 – 7.7m (42 – 39MHz), 362
  7.9 – 13.6m (38 – 22MHz), 279
  9.7m (31MHz): IF, 384
  10m (30MHz): IF, 367, 382
  10 – 13.6m (30 – 22MHz), 278–9
  22.2m (13.5MHz): IF, 377
  25m (12MHz), 13–15; IF, 365
  27m (11MHz): IF, 288
  50m (6MHz), 106; IF, 370
  60m (5MHz): IF, 362
waveguides, see Antennae, feeders
wavemeter, 141, 154, 313
Wayland, HMS (repair ship), 398
Waymouth, Commander G. R., **18**
WC radar, 412
  see also sets noted in Appendix 1
Weapon Allocation, see gunnery
Weapon Control Systems, see gunnery
Webb, C. H. (HMSS/ASE), **18**

General Index

Webber, J. W. (HMSS/ASE), **19**, 128
Welbourn, Dr (BICC), 401
Wellington bomber, 175
Welshman, HMS (minelayer), 396
Wembley, GEC premises, 45
Western Approaches Command, 340
Western Electric Company, USA, 24–5
Whalley, Mr (MetVic), 403
Wheeler, Commander A., **19**
Whitaker, Captain G. C. P., 345
White, E. L. C. (EMI), 402
Whiteley, A.L. (BTH), 38, 49, 401
W.H. Smith Ltd., Old Trafford, 48, 210, 404
Wild, E. (HMSS/ASE), **62**
Wilkins, Arnold F., (Group Leader at Bawdsey and Orfordness), 6, 11–12
Wilkinson, A. (HMSS), 8, 19, 20
Wilkinson, K. J. R. (BTH), 38, 401
Willett, Captain Basil R., **343**:
 Experimental Commander, 15, 16
 Experimental Captain, 31, 296
 reorganisation, 16
 Swanage demonstrations, 45, 192–3
 SS performance of Type 79, 32
 and Allen West Ltd, 36
'William', 356
Wills (H. H.) Laboratory, 31
Willshaw, W. E. (GEC), 402
Winder, Mr (Vickers Armstrong), 404
'Window', *see* interference
Winterburn, J. A. (Thermal Syndicate Co.), 131
Wireless Section (Torpedo and Mining Establishment), 6
wireless 'silence', 146
wireless telegraphy (W/T – morse), 6, 9, 178, 230
Witley, ASE radar laboratories/ workshops *see* Admiralty Signal Establishment
Witt, B. J. (HMSS/ASE), **62**
Wolfe, HMS (submarine depot ship), 398
Women's Royal Naval Service (WRNS), 324, 331, 340
Wood, A. B. (HMSS), 12, 14, 17, **19**, 68–9
Wooloomooloo, Australia, 341
Woolwich, HMS (destroyer depot ship); 398
Wootton, Mr (EMI), 402
work with/by private firms, *see* outside contractors

'working-up', 49, 412
workshops, *see* HMSS, equipment design and production
Worship, K. (Laurence Scott), 403
Wright, Charles S., (later Sir Charles): Director of Scientific Research at Admiralty (1934–46), 12, 16, 27, 31, 44, 345
Wright, G. M. (HMSS/ASE), 36, 61, 192
Wright, G. Paul (HMSS/ASE), 36, 128
Wright, S. T. (HMSS/ASE), **257**:
 contributor, 183, 272, 361
 Eastney, 187
WS radar, *see* warning radar – against surface craft; *see also* sets listed in Appendix 1
W. T. Henley Cables Ltd, 403
Wykeham, W. A. P. (HMSS/ASE), 30
Wylie, Commander F. J. (Experimental Commander, HMSS, 1935–7), 12–13, **18**

X

X2, *see* HMSS, Transmitter Section
XRC4, *see* ASE
XRC8, *see* ASE
XRC9, *see* ASE
X-band, *see* wavelength – 3cm
X-ray machine, 170

Y

Yagi, H., 25
Yagi, *see* antennae
Yarmouth, HMS (Second Class cruiser), 9
Yates, Captain Andrew V.S., 345
Yeo, R. A. (HMSS/ASE):
 first HMSS radar worker, 12
 development programme, 12, 15
 Orfordness, 12
 RDF team in RM Barracks, 12, 134
 technical, 14
 reorganisation, 17, **18**, **63**
Y-fed dipole, 278
York, HMS (cruiser), 396
Yorke, I. (HMSS/ASE), 36, 401
Young, A. J. (English Electric Co. Ltd), 402
Young, Warrant Officer H.W. RNV (W)R, 301
Younger's brewery, 30, 178

Equipment Index

Brief descriptive notes are given in Appendix 1. Technical details are given on pp. 129–30, 182, 271 and in Appendix 2.

Valve supplement on pp. 475–6

7DX, communications transmitter, 14
7H, 10m Naval radio, 106
7K, shortwave aircraft radio, 106
19D, Antenna Control Unit, 320, 363
20D, Antenna Control Unit, 320, 321, 365–6
36, W/T transmitter , 7
73X, communications set, 60cm, 14
79, WA radar:
  air warning, 168–9
  development, xxxiii, 17, 401
  equipment, **143**, **144**
  technical data, 138, 354, 362–3
  expected performance , 71–2
  range of detection , 74
  anomalous propagation , 77
  and IFF, 279–80, 282
  antenna: common transmit/receive, 335–6; compromises, 87; control gear, 320; equipment, **137**, 362–3; fitting, 335; long serving, 138; open-wire feeders, 89, 136, 318; polarisation, 22, 82, 136; retractable, 335; vertical lobe pattern, 167, 336, 364
  magslips, 323
  oil-filled components, 314
  receivers, 80, 141, **144**, 168, **168**
  valves, 116, 121, 162, 301
  production methods, 180
  ship-fitting, numbers, 31, 331, 334
  ships fitted – general, 301
  ships fitted – specific, 334, 388–98
  modified to Type 279, 32
  replaced, *xxxv*
79B, WA radar, 147–51, 354
  chronology, *xxxiv*, 181, 303, 335, 362
  common transmit/receive, 335–6
  dual installation with Type 281BQ, 165–9

technical data, 354, 362–3
ships fitted – specific, *xxxiv*, 181, 303, 388–98
79BX, WA radar, 22
79M, WA radar, 354
79X, WA radar:
  chronology, *xxxiii*, 14–16, 135, 139, 177, 181
  details, 135–9
  antenna, 17, 136, 138
  manufacture, 139
  polarisation, 22, 136
  priority, 15
  range of detection, aircraft, 14, 139
  requirements, 15
  sea trials, 21, 135
  supervision, 16
  improved version (Type 79Y), 22, 135
  and air-raids on Portsmouth, 47
79Y, WA radar, 135–40, 181
  first operational radar, *xxxiii*
  numbers, 23, 28
  priority in fitting, 28
  range of detection, 22–3, 139
  receiver , 76
  Report on performance, 29
  ships fitted – specific, 135, 139, 300
79Z, WA radar, 140–3
  chronology, *xxxiii*, 140, 141, 142, 181
  characteristics, 182
  countering jamming, 142
  modified from Type 79Y, 23
  modulator, 141
  numbers, *xxxiii*, 36, 143
  range of detection – aircraft, 23, 142
  valves, 107
  in *Curlew*, 23, 142
  in *Suffolk*, 306
  and AGI, 36
  ships fitted – specific, *xxxiii*, 324

467

91, radar jammer, 354, 401, 403
241, interrogator, 280, 354, 364
242, interrogator, 278, 287–8, 354, 367, 401
242M, interrogator, 288–9, 354, 376
242P, interrogator, 376
243, interrogator:
  antenna, **137**, 282, **283**, **284**
  details, 282
  technical data, 354
  triggering, 285
  working with specific sets, 362, 364, 378, 380
  and G-Mode, 283, 285
  and RDR, 168, **168**
  and Type 281BQ, 165, **166**,
  and Type 941, 283, 285
244, interrogator, 354
245, interrogator, 354
251, shipborne beacon, 280, 354
251M, shipborne beacon, 324, 354, 402
251P, shipborne beacon, 354
252, IFF transponder, 280, 354
253, IFF, 289, 354
253P, IFF, 354
253S, IFF, 354
255, beacon, 354
256, shore beacon, 354
257, carrier-controlled approach radar, 354
258, shore radar beacon, 354
259, shipborne beacon, 354
261, WS radar, 260–2, **261**, 354
261W, WS radar, 263, 354
262, GC radar (search-and-lock-on), *xxxiv*, 57–8, 113, 354, 382–3, 401–4
263, GB radar, 354
267, submarine WS/WC radar, 113, 175, 259
267W, submarine WS/WC radar, *xxxv*, 263–4, 354
267MW, submarine WS/WC radar, 264, 354
267PW, submarine WS/WC radar, 113, 264, 354
268, WS radar, 258, 264, **265**, 354, 384–5
269, coastal forces GS radar, 354
271, WS radar:
  antenna, **197**, **199**, 215
  chronology, *xxxiv*, 195, 268, 303, 372
  comparison with Type 261, 262
  crystals hand-made in ASE, 122
  design, 401
  designation, 195, 204, 268
  display, 401
  escort vessels, *xxxiv*
  first operational 10–cm radar, 195–7
  IF panels used in Type 261, 260
  improvements, 54
  manufacturers, 372, 401
  Mark 4, *xxxiv*
  modifications, 54
  numbers, *xxxiv*, 208–9
  obsolescence, 262, 266
  Office/lantern installation, **198**
  Operational Reports, 213–4
  panels used in CD No.1 Mk.5, Mk.6, 207
  power, 90
  performance, and antenna height, 214–15
  range of detection: surface vessels, 222; submarines, 213, 223
  reconditioned for Merchant Navy, 266
  replaced, 217–19, 225
  side echoes/lobes, 210, **211**, 336
  space, 204
  successors, 54
  technical data, 204–5, 220, 271, 355, 372–3
  valves, 114, 121, 215, 216
  ships fitted – general, *xxxiv*, 303, 400
  ships fitted – specific, 204, 213, 223, 389–99
  and battery FC, 207
  and CD, 268
  and IFF design, 287
  and *Scharnhorst*, 222
  and station-keeping, 210
  and U-boats, 210
  in larger ships, 204–5
271M, WS radar, 355
271P, WS radar:
  chronology, *xxxiv*, 209, 219, 268
  numbers, *xxxiv*, 209
  panels, **209**
  prefabrication, 48, 210, 372, 404
  prototype, *xxxiv*
  side echoes, 210
  technical data, 271, 355, 372–3
  ships fitted – specific, 210
271Q, WS radar, 217–24, **218**, **219**
  chronology, *xxxiv*, 219–20, 262, 268
  detection range, 223
  developed from Type 271 Mk4, *xxxiv*

Equipment Index

power, 222, 271
replaced Type 271, 262, 266
technical tables, 271, 355, 374–5
trials, *xxxiv*
waveguide feed, 90
and Type 276, 231
271S, WS radar (shore use), 208, 372
271X, WS radar:
  antennae, 198–200
  CD operations ashore, 206–7
  general, 196, **202**
  installation, 48
  manufacturers, 47, 196, 203
  mixer unit, 200–2
  mu-metal box, 200
  numbers, 195–6, 203
  performance, 47, 196, 211–13
  prototypes, 47, 195
  sea trials, *xxxiv*, 47, 196, 268
  technical data, 197–203, 271
  terminology, 195
  training, 196
  USA-UK Technical Exchange, 48
  and waveguides, 48
  ships fitted – general, 48
  ships fitted – specific, 203, 204
272, WS radar:
  started, *xxxiv*
  technical data, 268, 271, 355, 372–3
  antenna separate from Office, 203, 220
  advantage of higher antenna, 214–15
  local oscillator, 204
  Mark Numbers, *xxxiv*, 204, 220, 225
  X-band replacement, 260
  ships fitted – general, 400
  ships fitted – specific, 268, 389–99
272M, WS radar, 355
272P, WS radar, *xxxiv*, 209, 268, 271, 355
273, WS radar:
  antenna, **205**, **212**
  antenna gain, 208
  blast from own guns, 311
  blast-tested radome, 211
  chronology, *xxxiv*, 268
  earth curvature limits range, 222
  integral Office/lantern, 225
  lantern, 336
  Mark IV, *see* Type 273Q
  Mark V becomes Type 277, *xxxiv*, 225
  Operational Report, 213

polar diagram, **55**
precision as rangefinder, 206
sea trials, 205, 206, 213, 222, 303
shore defence, 208
spares, 314
superseded, 57
surface warning, 54, 206, 213, 222, 223
technical data, 271, 355, 372–3
up-graded, *xxxiv*, 219
warning of low-flying aircraft, 54
and gun control, 206
and shipborne PPI development, 210, 304
ships fitted, 205, 206, 213, 222, 223, 303, 304, 311, 314, 324, 336, 389–99
273M, WS radar, 208, 355
273P, WS radar, *xxxiv*, 209, 219, 268, 271, 355
273Q, WS radar:
  Action Report, 223
  antenna, **221**
  experimental development, *xxxiv*, 217–24
  improvement on Type 271, 222–3
  manufacturer, 403
  need for PPI, 225
  performance, 219, 223, 230, 235–8, **236**
  sea trials, *xxxiv*, 219, 222, 268
  shell-splash spotting, 219
  technical data, 271, 355, 374–5
273S, WS radar (shore use), 208, 226, 228, 268, 271, 372
274, GS radar:
  technical data, 355, 378–9
  development, *xxxiv*, 53, 57
  installation, *xxxv*
  auxiliary set used for fall of shot, 61
  fast warm-up modulator needed, 53, 187
  mounting, manufacturer, 404
  parallel development, radar & mounting, 53
  spark-gap modulator, 187
  thyratron, 53
  waveguides and frequency stability, 91
  ships fitted – specific, 324, 389–99
275, GA radar, **257**
  accurate elevation, 57
  alignment, 316
  awaited, 51
  chronology, *xxxiv*, *xxxv*, 380

275, GA radar (cont.)
  designed along with mounting, 53
  magnetron instability, 91
  manufacturers, 403, 404
  technical data, 355, 380–1
  ships fitted – specific, 309, 316, 324, 380, 389–99
276, WS radar, (formerly Type 272 Mark 5)
  description, 231, **237**
  development, *xxxiv*
  technical data, 271, 355, 376–7
  performance: expected, 72, 231–2, 237, attained, 237–8, **238**
  sea-trials, *xxxv*, 73, 237, **238**, 269
  antenna, **237**
  obsolescent, 238
  replaced, *xxxv*
  vertical coverage diagram, **238**
  waveguide kit, 91
  and IFF, 288
  and TI, *xxxv*, 226, 232, 238, 269
  ships fitted – general, 400
  ships fitted – specific, 237, 389–99
277, WS radar:
  description, 226, 269
  development, xxxiv, 225
  into service, xxxv, 239, 244
  technical data, 271, 355, 376–7
  performance, 72, 231, **241**, 241–2, 247, 269
  antenna, 80, 239, **240**, 312
  comparison with Type 281, **244**
  frequency stability, 92
  heightfinding, 57, 242
  low air-cover, 241–2, 247
  overheating, 322
  sea-clutter trials, 78, 82
  sea-trials, 73, 240, 244
  target acquisition probability, 241–2, **241**, 243
  valves, 114, 121
  waveguides, 91, 92
  and 'Bubbly', 341
  and IFF, 288
  and Schnorkel, 244, 265
  up-graded, 245–6
  superseded, *xxxv*
  ships fitted – general, *xxxv*, 400
  ships fitted – specific, 230, 312, 322, 324, 389–99
277A, WS radar (RAF low cover variation), 228

277P, WS radar:
  chronology, *xxxv*, 254, 269
  modification of Type 277, 245–6, 341
  technical data, 269, 271, 355, 376–7
  and 'Bubbly', 341
  ships fitted – specific, 269
277Q, WS radar, 246, **257**, 269, 271, 355
277S, WS/low air cover radar, 228, 355, 401
277T, WS/low air cover radar: 226–8, **227**
  chronology, *xxxiv*, 55, 269
  magnetron tuning, 230
  manufacturers, 228, 401, 403
  operational locations, 228, 269
  range of detection – aircraft, 227–8
  technical data, 226, 271, 355, 376
  trials, 226–8
  and Ashford School, 228
  and FD, 246–7
  in amphibious landings, 228
277X, WS radar, *xxxv*, 55, 228–31, **229**, 239, 269
279, WA radar:
  general, 144–7, 181
  technical data, 138, 182, 355, 362–3
  development, 32, 145
  lobe-structure limitations, 54
  full cover with Type 281, 54, **55**
  transmitter, 147, 301
  modulator, 141, 147
  ranging unit, 145–6
  receiver, 76, 116, 141–2, 145
  antenna, 89, 121, **137**, 138, 147, 163, 302
  valves, 107, 114, 116, 121, 301
  display, 146, 147
  anti-DF facility, 146
  and FC, 363
  and IFF, 280, 282, 287
  method of production, 180
  manufacturers, 401
  replaced, 57
  ship-fitting numbers, 143
  ships fitted – general, 301, 302
  ships fitted -specific, 302, 388–98
279B, WA radar, 147–51, **149**, **150**, 181, 182
280, WA/GA radar, (adapted from GL1), 33, 355
281, WA radar:
  description, 32, 54
  requirements, 32

Equipment Index 471

technical data, 138, 151–2, 182, 355, 364–6
design, 32
expected performance, 71–2
range of detection, **150**, 154, 158, 161, 223
air-cover gap, 242, **244**
development, *xxxiii*, 151, 169, 334
prototype, 48
production, *xxxiii*, 161, 180
numbers, *xxxiii*, 33, 47, 48, 151, 161
sea-trials, 33, 47, 302
antenna: general, 138, 154, **156**, **157**, **284**; common transmit/receive, 87; control, 320, 365–6; drive, 158, 162, 163; feeders, 302, 318; gearbox, frozen, 321–2; telescopic, 348
beam switching, 51, 154, **158**, 158
coaxial cable, 89
display, 159, **160**
magslips, 323
modulator, 153, 178
oil-filled components, 314
Performance Meter, 74–5
power supplies, 138, 323
receiver, 80, 159, **160**, 177
sparking to deck, 312, 365
transmitter, 152–3, **155**
valves, 80, 107–8, 114, 116, 121
vertical polar diagram, **55**, **158**
and 'Bubbly', 341
and FD, 169, 246–7, 269
and gunnery, 366
and heightfinding, 57, 74–5, 159–61, 312, 335–6
and IFF, 280, 282, 285, 289, 354
and RAF Chain Warning System, 161
and *Scharnhorst* action, 223
superseded, *xxxv*
ships fitted – specific, *xxxiii*, *xxxiv*, 48, 139, 151, 161, 302, 324, 337, 364, 365, 388–98
281B, WA radar, *xxxiv*, 161–2, 181–2, 355, 364–6, 388–98
281BM, WA radar, 355
281BP, WA radar, 355, 356
281BQ, WA radar:
general, *xxxv*, 162–5, **164**, **166**, 168, **168**, 181, **286**
antenna gearbox fitting, 308
dual installation with Type 79B, 165–9

modifications for FD, 163
range of detection, aircraft, 176
space requirements, 308
Tantallon trials, 165
technical data, 182, 355, 364–6
with Type 277 in FD role, 246
and 'Bubbly', 341
and VHF interference, 317
and Type 940/1 interrogator, 356
ships fitted – specific, 308, 317, 324, 364
282, GC radar:
description, 369
start, *xxxiii*
common panels with Types 283–5, 318
trials, 39, 41, 49
technical data, 355, 369–71
antenna, 39, 41
Office equipment, **40**
transmitter tuning, 319
display, 315
index-error calibration, 316
malfunctions, 319
overheated alternators, 323
range-tracking panel, 323
rotary capacity-switch, 403
transformer problem, 314
and FC, 371
numbers, *xxxiii*, 34, 334
ship-fitting, 336
ships fitted – general, 43, 49, 400
ships fitted – specific, 49, 308, 336, 389–99
282M, GC radar, 318, 355, 369
282P, GC radar, 369
282Q, GC radar, 355
283, GB radar, 369
A-scan, 315
Barrage Director, 402
chronology, 43–4, 334, 369
index error, 316
installation, 336
interchangeable panels, 318
power supply, 308, 323
range-tracking panel, 323
technical data, 355, 369–71
transmitter tuning, 319
ships fitted – specific, 324, 369, 389–99
283M, GB radar, 355
284, GS radar:
development, 36–43

284, GS radar (*cont.*)
  technical data, 355, 369–71
  common panels with Types 282–5, 318
  sea-trials, *xxxiv*, 43, 47
  to sea, 303
  antenna, 39–41, **39**, **40**, **42**, 404
  modulator, 38
  receiver, 38
  display, 315
  Office layout, **40**
  range of detection, 43, 49, 51
  rotary capacity-switch, 51, 52, 403
  index-error calibration, 316
  vertical polar diagram, 371
  breakdowns and malfunctions, 319–20, 321
  and FC, 371
  numbers, 41, 47, 334
  ship-fitting, 336
  ships fitted – general, 41, 43, 369
  ships fitted – specific, 49, 303, 308, 324, 369, 389–99
284M, GS radar, 319, 355
284P, GS radar, *xxxiv*, 53, 355, 369
284X, GS radar, 369
285, GA radar:
  description, 369
  beam switching, 51
  common panels with Types 282–4, 318
  technical data, 355, 369–71
  detection range, 43
  antenna, **42** mounting, 43, 404
  sea-trials, *xxxiv*, 43, 47
  transmitter, 319
  display, 315
  Office equipment, **40**
  rotary capacity-switch, 51, 52, 403
  shells in flight, 43
  shell splash, 43
  beam switching, 51, 320
  index-error calculation, 316
  malfunctions, 319–20
  modifications, 369
  overheated alternators, 323
  and FC, 323, 371
  and high-flying aircraft, 51
  numbers, 41
  ships fitted – general, 41, 43, 49, 400
  ships fitted – specific, 41, 49, 308, 320, 324, 336, 389–99
285M, GA radar, 319, 355

285P, GA radar, *xxxiv*, 51, 355, 369
285Q, GA radar, 355, 371
286, WC radar:
  general, 169–72
  antenna pedestal, 170–1
  antenna, retractable, 48
  crash programme, 279, 334
  coaxial-cable antenna feed, 89
  into service, 171–2, 181
  power supply, 287
  pyrotenax, 170, 171, 336, 340
  RAF ASV modified, *xxxiii*, 33–4, 169
  ship-fitting, 336
  small-ship set, 33–4
  technical data, 182, 367–8
  variants, 172, 262
  X-ray analysis of cable moulding, 170
  and Mark IIN IFF, 280
  in large ships as IFF, 280
  numbers, 48, 334
  ships fitted – general, *xxxiii*, 336, 400
  ships fitted – specific, 388–98
286M, WC radar, 172, 182, 287, 301, 367, 355
286P, WC radar, 34, 172, 181, 182, 287, 355, 367
286PQ, WC radar, 172, 182, 262, 355, 367
286PU, WC coastal forces radar, 367
286PW, WC submarine radar, 367
286U, WC coastal forces radar, 355
286W, WC submarine radar, 355
287, minewatch radar, 355
288(1), GC radar, 355
288(2), GC radar, 356
289, GA radar, 356
290, WC radar, 172–3
  antenna, **171**, 173, 336, 340
  first installation, *xxxiv*, 181
  numbers, 48
  replaced Type 286;*xxxiv*, 279
  ships fitted – general, 400
  ships fitted – specific, *xxxiv*, 303, 388–98
  technical data, 182, 356, 367–8
291, WC radar, 173–5
  antennae, 138, 171, **171**
  associated beacons, 354
  chronology, 181, 367
  developed from RAF ASV, 262
  fittings, *xxxiv*, 173
  gas T/R switch, 87
  general, 173–5

Equipment Index

pyrotenax, 89, 336, 340
range of detection, **175**
replacement for Type 286/P, *xxxiv*, 34
sea trials, 174, **175**
submarine version, 173, 174–5
superseding Type 290, 334
technical data, 182, 356, 367–8
variants, 262
and early PPI, 174
and IFF, 279, 287
ships fitted – general, 174
ships fitted – specific, 389–99
291M, WC radar, 356
291U, WC coastal forces radar, 356
291W, WC submarine radar, 356
292, TI radar, 58
293, WC/TI radar:
 technical data, 271, 356, 376–7
 expected performance, 72
 Naval Staff Requirements, 225–6, 232
 WS performance, 235–7
 development, 225–6
 sea-trials, *xxxv*, 73, 235
 detection range, 235–7
 antenna, 57, 80, 377
 high transmitter temperature, 322
 display, 377
 swept-brightening modification, 322
 waveguide kit, 91
 and IFF, 288
 and TI, 57, 226, 232, 235, 377
 and 'Bubbly', 341
 ships fitted – general, 226, 232, 238, 377, 400
 ships fitted – specific, 322, 324, 389–99
293M, WC/TI radar:
 antenna, **239**, 239
 chronology, *xxxv*
 technical data, 271, 356, 376–7
 and 'Bubbly', 245
 and Outfit AQR, 238
 ships fitted – specific, 245
293P, WC/TI radar, *xxxv*, 245, 341, 376–7
293Q, WC/TI radar, 239, **257**
293X, WC/TI radar, 232–7, **233–4**, **236**, 269
294, WC/FD radar, *xxxiv*, 73, 247, 356
295, WC/FD radar, 247–8, 269
 antenna, 75, 251
 chronology, *xxxiv*, 269
 height-measuring system, 248, 249

R & D under contract, 248
 range, 248
 replaced, 88
 technical data, 356
 and waveguide rotating-joint, 248
 and predicted performance, 73
650, jammer, 356, 403
651, jammer, 356, 403
901, WA radar, 60
930, GS/splash radar, 356
931, Canadian GS/splash radar, 356
940, interrogator, 356
941, interrogator:
 antenna, **286**
 dual working with Type 243, 285
 G-mode, 283
 panels, 165, **166**, 168, **168**
 side-lobes and swept-gain, 285
 slip-rings, 163
 technical data, 356
 triggering, 285
951, beacon, 356
952, beacon, 356
960, WA radar: 175–7
 antenna, 177, **250**, **257**
 chronology, *xxxv*, 133, 175, 177, 181
 first installation, *xxxv*
 range of detection, 176
 replaced Types 279/281, 57, 175
 technical data, 175–6, 356
 'user-friendly' design, 175–6
 ships fitted – specific, *xxxv*
 and UDU, 176
 and Type 980/981, 254
960P, WA radar, 252, 254
961, CCA radar, 356
970, WS radar, 266, 356
971, WS radar, 267, 356
971M, WS radar, 356
972, WS radar, 265–7, 356
972M, WS radar, 266
974, navigational radar, **257**
980, WC/FD radar:
 chronology, *xxxv*, 252–3, 270
 antenna, 75, 88–9, 250–1
 electronic beam-scanning, 251
 performance, 58, 73, 254, 255–8
 renamed, 253–4
 technical data, 271, 356
 trials, 252, **253**
 and spectrometer, 92
981, heightfinder radar:
 technical data, 271, 356

981, heightfinder radar (*cont.*)
  plans, *xxxv*
  performance estimates, 73, 249–50, 270
  development, 75, 253–4
  sea-trials, 252
  antenna, 249
  put on target by Type 960P, 252
  and FD, *xxxv*, 249–55
  and heightfinding, 270
  ships fitted – specific, 253
982, FD radar (postwar designation of Type 980), 89, **250**, 253–8, 270, 271
982M, FD radar, 256, **257**
983, FD radar, (postwar designation of Type 981), **250**, 253–4, 255–8, 270, 271
984, FD radar, 258
990, WC radar, 73, 248–9, 356
992, TI radar, 109, 356
ACH, 279
AI, RAF air interception (airborne) radar: 15, 44, 189, 354, 357
AIA radar, 357
AN/APS-4, (US airborne radar), 357
AN/APS-6, (US airborne radar), 357
ASB, (US airborne radar), 357
ASH, (US airborne radar), 357
ASV, (RAF and Naval versions):
  AMRE development, 15
  AMRE sets in ships, 334
  basis of small-ship radar, 33, 169–70, 262, 279
  detects Home Fleet in poor weather, 15
  first X-band Staff Requirement, 259
  fitted in FAA, 28–9
  Mark I, 171–2, 279
  Mark II, 171–2, 279–80, 356, 364
  Mark IIN, 357
  Mark XI, 357
  technical data, 357
  and CCA, 356
  and cm-waveband, 189
  in Types 241, 243, 364
CA No.1(Army radar set), 356
CD No.1, Mk.IV, (Army radar set), *xxxiv*, 207, 268
CD No.1, Mk.V, (Army radar set), 207
CD No.1, Mk.VI, (Army radar set), 207
FX.1400, German radio guided bomb, 356
G82 tuning test set, **209**
GCI, RAF ground control interception radar, 282
GL1, Army gunlaying set, 28, 32, 33, 145, 159
GL1(T), 402
GMY4, guided-missile control system, 60
H2S, RAF airborne radar, 266, 356
H2X, RAF airborne radar, 267
L12, ranging panel, 371
L17, precision ranging panel, 373, 374
L22, ranging panel, 371
L24, precision ranging panel, 371
L31(B), bearing display panel, 379
L31(RS), range display panel, 379
L34, precision ranging panel, 371
Loran, (US navigational equipment), 317, 324, 409
M68, receiver, 403
Mark 4 (Types 271Q, 273Q), 217–24
Mark 5, 224–6
  shore trials (Type 277T), 226–8
  shipborne trials (Type 277X), 228–31
NT271X, (Army designation of Type 271X), *xxxiv*, 207, 222, 268
Outfit ANS, (antenna), 239
Outfit ANU, (antenna), 377
Outfit AQR, (antenna), 238
Outfit AQS, (antenna), 252, 254, 256
Outfit AQT, (antenna), 249, 252, 254
Outfit ATD, (antenna), 362
Outfit ATE, (antenna), 364, 365
Outfit ATQ, (antenna), 368
Outfit AUJ, (antenna), 237, **237**
Outfit AUK, (antenna), **240**, 377
Outfit AUR, (antenna), 232, **233**, 235, 238, 377
Outfit JE, (PPI), 210, 225, 229, 402
P11, receiver, 141, 145, 362
P12, receiver, 145, 177, 362
P14, receiver, 26, 38, 43
P16, receiver, 26, 38, 43
P19, monitor receiver, 401
P29, monitor receiver, 401
RBL 10, precision ranging panel, 363
RBL 11, precision ranging panel, 366
S25, (DF frame coil), 401
SA, WA radar, (US), 357, 400
SG, WS radar, (US), 357, 390–2
SJ, WS radar, (US), 357
SK, WA radar, (US) 357, 390–2
SL, WS radar, (US), 354, 357, 400
SM-1, FD radar, (US), 78, 357

Equipment Index

SO, WS radar, (US), 357
SQ, WS portable radar, (US), 314, 324, 352

Target Indication Unit (TIU) Mark 2, 377
Valves, specific, see Valve Supplement, below
'William' (Army radar set): 356

VALVE SUPPLEMENT

4279A, (modulator triode, NT87), 141
BM313, (cavity magnetron), 382
BM325, (cavity magnetron), 260
CV12, (modulator thyratron), 114, 226
CV22 (thyratron modulator), 114–5, **115**, 374
CV35, (LO klystron), **119**, 121, 203–4
CV43, (LO klystron), 121
CV53, (grounded-grid triode), 116–18, **118**
CV56, (cavity magnetron), 216–17, 224, 268, 270, 374
CV57, (modulator thyratron) **120**
CV63, (triode), 287
CV76, (cavity magnetron)
 strapped magnetron, 217
 500 kW output, 215, 224
 long-waveguide operation, 230, 269
 technical data, 270, 376
 and Type 277, 114
 and Type 277T, 226
CV91, (triode), 116
CV94, (diode), 121, 162
CV101, (crystal mixer), **122**
CV199, (triode), 289
CV208, (cavity magnetron), 260
CV209 (cavity magnetron), 260
CV214, (cavity magnetron), 113, 260
CV313, (silica tetrode modulator), 109, 130
CV1091, (HF pentode, EF50, VR91), 116, 123
CW10, (split-anode magnetron), 24
DET 12, (triode), 27
E316, ('Doorknob' triode), 24, 25
E773, ('micropup' triode), 26–7, 37
E1189, (cavity magnetron), 111, **112**
E1198, (cavity magnetron), 191, 200
E1263, (cavity magnetron), 263
EF50, (HF pentode, CV1091, VR91), 116, 123, 145–7, 159, 370, 403
NC7, (CRT), 371

NC17, (Skiatron CRT), 125, **127**
NR54, (acorn pentode), 122
NR89, (LO klystron, CV35), **119**, 121, 189, 197, 203, 373
NT22B, (silica triode), **100**, 129
NT22C, (silica triode), 129
NT23B, (silica triode), 129
NT24, (silica triode), 129
NT30, (silica triode), 129
NT31, (silica triode), 129
NT32B, (silica triode), 129
NT33, (silica triode), 129
NT34, (silica triode), 107
NT35, (silica triode), 129
NT41, (silica triode), 107
NT41A, (silica triode), 107, 129
NT43, (silica triode), 129
NT45A, (silica triode), 129
NT46R, (silica triode), 130
NT52, (silica split-anode magnetron), 106, 130
NT54, (silica triode), 130
NT57, (silica triode), 21, 107, 130, 138, 162
NT57D, (silica triode), 130
NT57T, (silica triode), 107, **108**, 130, 140, 141, 362
NT59A, (silica pentode), 130
NT69, (silica pentode), 130
NT77, (silica tetrode), 130
NT77A, (silica tetrode), 107, 109
NT78, (silica triode), 108, 114, 153
NT78A, (silica triode), 108, 114, 130
NT84, (silica triode), 130
NT86, (silica triode), 32, 108, 114, 130, 152–3, 364
NT87, (modulator triode, 4279A), 114
NT88, (silica triode), 109
NT90, (silica triode), 130
NT98, (cavity magnetron): 111–12, **112**
 chronology, 189, 215, 216
 efficiency, 215
 technical data, 200
 unstrapped, 259, 372
 and Type 271X, 200
NT99, ('micropup' triode), 110, **111**
NT100, (modulator tetrode), 114, 200, 216, 372
NU22, (silica rectifier), 130
NU23, (silica rectifier), 130
NU24, (silica rectifier), 130
NU25, (silica rectifier), 130
NU26, (silica rectifier), 130

NU28, (silica rectifier), 130
NU29, (silica rectifier), 130
NU30, (silica rectifier), 130
SP41, (HF pentode), 116
T1, (power triode), 96
T22, (power triode), 96, 97, 104, 105
T23, (power triode), 105

U21, (rectifier), 105
VCR97, (CRT), 124, **125**
VCR516, (CRT), 124, **125**
VR65, (pentode), 116
VR78, (diode), 122
VR91, (HF pentode, EF50, CV1091), 145, 370